T0348650

Methods in Cell Biology

Microtubules, *in Vitro* 2nd Edition

Volume 115

Series Editors

Methods in Cell Biology

Microtubules, *in Vitro*
2nd Edition

Volume 115

Edited by

John J. Correia
Department of Biochemistry
University of Mississippi Medical Center
Jackson, Mississippi, USA

Leslie Wilson
Department of Molecular, Cellular and Developmental Biology
University of California
Santa Barbara, California, USA

AMSTERDAM • BOSTON • HEIDELBERG • LONDON
NEW YORK • OXFORD • PARIS • SAN DIEGO
SAN FRANCISCO • SINGAPORE • SYDNEY • TOKYO
Academic Press is an imprint of Elsevier

Academic Press is an imprint of Elsevier
525 B Street, Suite 1800, San Diego, CA 92101-4495, USA
225 Wyman Street, Waltham, MA 02451, USA
The Boulevard, Langford Lane, Kidlington, Oxford, OX5 1GB, UK
32 Jamestown Road, London NW1 7BY, UK
Radarweg 29, PO Box 211, 1000 AE Amsterdam, The Netherlands

First edition 2009
Second edition 2013

ISBN: 978-0-12-407757-7
ISSN: 0091-679X

For information on all Academic Press publications visit our website at store.elsevier.com

Contents

Contributors xiii
Preface xix

CHAPTER 1 Site-Specific Fluorescent Labeling of Tubulin **1**
Kamalika Mukherjee, Susan L. Bane
Introduction 2
1.1 Material 4
1.2 Method 5
Acknowledgments 11
References 11

CHAPTER 2 Measuring Microtubule Persistence Length Using a Microtubule Gliding Assay **13**
Douglas S. Martin
Introduction 14
2.1 Theory 15
2.2 Methods 17
2.3 Discussion 22
Summary 23
Acknowledgments 23
References 23

CHAPTER 3 Structural Studies of the Doublecortin Family of MAPs **27**
Franck Fourniol, Mylène Perderiset, Anne Houdusse, Carolyn Moores
Introduction 28
3.1 Rationale 30
3.2 Methods 32
3.3 Discussion 43
Acknowledgments 44
References 45

CHAPTER 4 Detection and Quantification of Microtubule Detachment from Centrosomes and Spindle Poles **49**
Anutosh Ganguly, Hailing Yang, Fernando Cabral
Introduction 50
4.1 Choice of Cell Line 51

4.2 Microtubule Detachment in Spindles 52
4.3 Labeling Microtubules 53
4.4 Measuring Detachment 54
4.5 Microtubule Fragments as a Surrogate for Detachment 57
4.6 Mechanism of Detachment 59
Summary 60
Acknowledgments 60
References 60

CHAPTER 5 Regulation of Tubulin Expression by Micro-RNAs: Implications for Drug Resistance 63

Sharon Lobert, Mary E. Graichen

Introduction and Rationale 64
5.1 Methods and Materials 65
5.2 Results and Discussion 68
Summary 70
References 73

CHAPTER 6 Determining the Structure–Mechanics Relationships of Dense Microtubule Networks with Confocal Microscopy and Magnetic Tweezers-Based Microrheology 75

Yali Yang, Megan T. Valentine

Introduction 76
6.1 Protocols 78
Summary 93
Acknowledgments 93
References 94

CHAPTER 7 Studying Mitochondria and Microtubule Localization and Dynamics in Standardized Cell Shapes 97

Andrea Pelikan, James Sillibourne, Stephanie Miserey-Lenkei, Frederique Carlier-Grynkorn, Bruno Goud, Phong T. Tran

Introduction 98
7.1 Design of the Photomask 99
7.2 Manufacturing of Micropatterned Coverslips 100
7.3 Seeding and Spreading of Cells 101
7.4 Mitochondria and Microtubule Visualization 102
7.5 Materials 104
Conclusion 106
Acknowledgments 107
References 107

CHAPTER 8 **Going Solo: Measuring the Motions of Microtubules with an *In Vitro* Assay for TIRF Microscopy** 109
Kris Leslie, Niels Galjart
Introduction 110
8.1 Protocols 111
Concluding Remarks 123
Acknowledgments 123
References 123

CHAPTER 9 **Analysis of Microtubules in Isolated Axoplasm from the Squid Giant Axon** 125
Yuyu Song, Scott T. Brady
9.1 Preparation of Axoplasm 126
9.2 Analysis of Axoplasmic Microtubule Dynamics 131
9.3 Biochemistry of Axoplasmic Microtubules 133
9.4 Immunohistochemistry of Axoplasmic Microtubules 135
References 136

CHAPTER 10 **Imaging GTP-Bound Tubulin: From Cellular to *In Vitro* Assembled Microtubules** 139
Hélène de Forges, Antoine Pilon, Christian Poüs, Franck Perez
Introduction 141
10.1 Imaging GTP Islands in Permeabilized Cells 142
10.2 Imaging GTP Caps and GTP Islands Using Centrosome-Based Microtubule Assembly or Endogenous Microtubule Elongation in Permeabilized Cells 145
10.3 Imaging GTP Islands in Microtubules Assembled *In Vitro* 147
10.4 Discussion and Future Prospects 150
Acknowledgments 151
References 152

CHAPTER 11 **Tubulin-Specific Chaperones: Components of a Molecular Machine That Assembles the α/β Heterodimer** 155
Guoling Tian, Nicholas J. Cowan
Introduction 156
11.1 Methods 159
11.2 Discussion 167
References 169

CHAPTER 12 Heterotrimeric G Proteins and Microtubules **173**
Witchuda Saengsawang, Mark M. Rasenick
Introduction 174
12.1 General Protocols 174
12.2 Buffer Compositions 187
Concluding Comments 188
Acknowledgments 188
References 188

CHAPTER 13 Purification and Biophysical Analysis of Microtubule-Severing Enzymes *In Vitro* **191**
Juan Daniel Diaz-Valencia, Megan Bailey, Jennifer L. Ross
Introduction 192
13.1 Methods 194
Discussion and Summary 211
Acknowledgments 211
References 211

CHAPTER 14 Measurement of *In Vitro* Microtubule Polymerization by Turbidity and Fluorescence **215**
Matthew Mirigian, Kamalika Mukherjee, Susan L. Bane, Dan L. Sackett
14.1 Background and Theory 216
14.2 Materials and Equipment 223
14.3 Methods 225
References 228

CHAPTER 15 Live-Cell Imaging of Microtubules and Microtubule-Associated Proteins in *Arabidopsis thaliana* **231**
Jessica Lucas
Introduction 232
15.1 Protocols 234
15.2 Materials 242
Concluding Comments 243
Acknowledgment 243
References 243

CHAPTER 16 Investigating Tubulin Posttranslational Modifications with Specific Antibodies **247**
Maria M. Magiera, Carsten Janke
Introduction 248
16.1 Observations 249

16.2 Materials and Methods....255
16.3 Buffer Composition....261
Concluding Remarks....263
Acknowledgments....263
References....264

CHAPTER 17 Purification and Assembly of Bacterial Tubulin BtubA/B and Constructs Bearing Eukaryotic Tubulin Sequences....269
José M. Andreu, María A. Oliva
Introduction and Rationale....270
17.1 Materials: Genes, Constructs, and Expressed Proteins....272
17.2 Methods....273
Conclusions....280
Acknowledgments....280
References....280

CHAPTER 18 Microtubule-Associated Proteins and Tubulin Interaction by Isothermal Titration Calorimetry....283
P. O. Tsvetkov, P. Barbier, G. Breuzard, V. Peyrot, F. Devred
Introduction....284
18.1 Isothermal Titration Calorimetry....286
18.2 Tubulin and MAPs Sample Preparation....290
18.3 Results: Tubulin/MAPs by ITC....294
Conclusion....298
Acknowledgment....299
References....299

CHAPTER 19 Methods for Studying Microtubule Binding Site Interactions: Zampanolide as a Covalent Binding Agent....303
Jessica J. Field, Enrique Calvo, Peter T. Northcote, John H. Miller, Karl-Heinz Altmann, José Fernando Díaz
Introduction....304
19.1 Materials....305
19.2 Methods and Results....307
Summary....324
Acknowledgments....324
References....324

CHAPTER 20 Studying Kinetochore-Fiber Ultrastructure Using Correlative Light-Electron Microscopy 327
Daniel G. Booth, Liam P. Cheeseman, Ian A. Prior, Stephen J. Royle
Introduction 328
20.1 Materials 329
20.2 Methods 331
20.3 Discussion 338
Acknowledgments 341
References 341

CHAPTER 21 Fluorescence-Based Assays for Microtubule Architecture 343
Susanne Bechstedt, Gary J. Brouhard
Introduction 344
21.1 Materials 347
21.2 Methods 348
21.3 Discussion 352
Acknowledgments 352
References 352

CHAPTER 22 Structure–Function Analysis of Yeast Tubulin 355
Anna Luchniak, Yusuke Fukuda, Mohan L. Gupta
Introduction 356
22.1 Reagents and Equipment 358
22.2 Introducing Tubulin Mutations into Yeast 360
22.3 Analysis of Tetrad Viability 363
22.4 Assessing Microtubule Stability by Drug Sensitivity 366
22.5 Direct Analysis of Microtubule Dynamics *In Vivo* 367
22.6 Localization of Microtubule-Associated Proteins *In Vivo* 370
Acknowledgments 372
References 373

CHAPTER 23 Using MTBindingSim as a Tool for Experimental Planning and Interpretation 375
Julia T. Philip, Aranda R. Duan, Emily O. Alberico, Holly V. Goodson
Introduction 376
23.1 MTBindingSim 376
23.2 Experimental Designs in MTBindingSim 376
23.3 Binding Models in MTBindingSim 377
23.4 Using MTBindingSim Example 1: Tau–MT Binding 379

23.5 Using MTBindingSim Example 2: Does Human EB1 Bind to the MT Seam or Lattice? 381
Conclusion....383
References....383

CHAPTER 24 Imaging Individual Spindle Microtubule Dynamics in Fission Yeast....**385**
Judite Costa, Chuanhai Fu, Viktoriya Syrovatkina, Phong T. Tran
Introduction....386
24.1 Methods....387
Conclusion....393
Acknowledgments....393
References....394

Index....**395**

Contributors

Emily O. Alberico
Department of Chemistry and Biochemistry, Interdisciplinary Center for the Study of Biocomplexity, University of Notre Dame, Notre Dame, Indiana, USA

Karl-Heinz Altmann
Department of Chemistry and Applied Biosciences, Institute of Pharmaceutical Sciences, Swiss Federal Institute of Technology, Zürich, Switzerland

José M. Andreu
Centro de Investigaciones Biológicas, CSIC, Madrid, Spain

Megan Bailey
Molecular and Cellular Graduate Program, University of Massachusetts Amherst, Amherst, Massachusetts, USA

Susan L. Bane
Department of Chemistry, Binghamton University, State University of New York, Binghamton, New York, USA

P. Barbier
Aix - Marseille University, Inserm, CR02 UMR_S 911, Faculté de Pharmacie, 13385 Marseille, France

Susanne Bechstedt
Department of Biology, McGill University, Montreal, Quebec, Canada

Daniel G. Booth
Department of Cellular & Molecular Physiology, Institute of Translational Medicine, University of Liverpool, Liverpool, and Wellcome Trust Centre for Cell Biology, University of Edinburgh, Edinburgh, United Kingdom

Scott T. Brady
Marine Biological Laboratory, Woods Hole, Massachusetts, and Department of Anatomy and Cell Biology, University of Illinois at Chicago, Chicago, Illinois, USA

G. Breuzard
Aix - Marseille University, Inserm, CR02 UMR_S 911, Faculté de Pharmacie, 13385 Marseille, France

Gary J. Brouhard
Department of Biology, McGill University, Montreal, Quebec, Canada

Fernando Cabral
Department of Integrative Biology and Pharmacology, University of Texas Medical School, Houston, Texas, USA

Enrique Calvo
Unidad de Proteómica, Centro Nacional de Investigaciones Cardiovasculares, Madrid, Spain

Frederique Carlier-Grynkorn
Cell Biology, Institut Curie, UMR 144 CNRS, Paris, France

Liam P. Cheeseman
Department of Cellular & Molecular Physiology, Institute of Translational Medicine, University of Liverpool, Liverpool, United Kingdom

Judite Costa
Cell and Developmental Biology, University of Pennsylvania, Philadelphia, Pennsylvania, USA, and Cell Biology, Institut Curie, UMR 144 CNRS, Paris, France

Nicholas J. Cowan
Department of Biochemistry and Molecular Pharmacology, New York University Langone Medical Center, New York, New York, USA

José Fernando Díaz
Centro de Investigaciones Biológicas, CSIC, Madrid, Spain

Hélène de Forges
Institut Curie, and CNRS UMR144, Paris Cedex 05, France

F. Devred
Aix - Marseille University, Inserm, CR02 UMR_S 911, Faculté de Pharmacie, 13385 Marseille, France

Juan Daniel Diaz-Valencia
Department of Physics, University of Massachusetts Amherst, Amherst, Massachusetts, USA

Aranda R. Duan
Department of Chemistry and Biochemistry, Interdisciplinary Center for the Study of Biocomplexity, University of Notre Dame, Notre Dame, Indiana, USA

Jessica J. Field
School of Biological Sciences, Victoria University of Wellington, Wellington, New Zealand

Franck Fourniol
Institute of Structural and Molecular Biology, Birkbeck College, London, United Kingdom

Chuanhai Fu
Biochemistry, University of Hong Kong, Pokfulam, Hong Kong

Yusuke Fukuda
Department of Molecular Genetics and Cell Biology, University of Chicago, Chicago, Illinois, USA

Niels Galjart
Department of Cell Biology, Erasmus Medical Center, P.O. Box 2040, Rotterdam, The Netherlands

Anutosh Ganguly
Department of Physiology and Pharmacology, Snyder Institute for Chronic Disease, University of Calgary, Calgary, Alberta, Canada

Holly V. Goodson
Department of Chemistry and Biochemistry, Interdisciplinary Center for the Study of Biocomplexity, University of Notre Dame, Notre Dame, Indiana, USA

Bruno Goud
Cell Biology, Institut Curie, UMR 144 CNRS, Paris, France

Mary E. Graichen
School of Nursing, University of Mississippi Medical Center, Jackson, Mississippi, USA

Mohan L. Gupta, Jr.
Department of Molecular Genetics and Cell Biology, University of Chicago, Chicago, Illinois, USA

Anne Houdusse
Structural Motility, Institut Curie, Centre National de la Recherche Scientifique, Unité Mixte de Recherche 144, Paris Cedex 05, France

Carsten Janke
Institut Curie, CNRS UMR3306, INSERM U1005, Orsay, France

Kris Leslie
Department of Cell Biology, Erasmus Medical Center, P.O. Box 2040, Rotterdam, The Netherlands

Sharon Lobert
School of Nursing, University of Mississippi Medical Center, Jackson, Mississippi, USA

Jessica Lucas
Department of Biology, Santa Clara University, Santa Clara, California, USA

Anna Luchniak
Department of Biochemistry and Molecular Biology, University of Chicago, Chicago, Illinois, USA

Maria M. Magiera
Institut Curie, CNRS UMR3306, INSERM U1005, Orsay, France

Douglas S. Martin
Lawrence University, Appleton, Wisconsin, USA

John H. Miller
School of Biological Sciences, Victoria University of Wellington, Wellington, New Zealand

Matthew Mirigian
Program in Physical Biology, Eunice Kennedy Shriver National Institute of Child Health and Human Development, NIH, Bethesda, Maryland, USA

Stephanie Miserey-Lenkei
Cell Biology, Institut Curie, UMR 144 CNRS, Paris, France

Carolyn Moores
Institute of Structural and Molecular Biology, Birkbeck College, London, United Kingdom

Kamalika Mukherjee
Department of Chemistry, Binghamton University, State University of New York, Binghamton, New York, USA

Peter T. Northcote
School of Chemical and Physical Sciences, Victoria University of Wellington, Wellington, New Zealand

María A. Oliva
Centro de Investigaciones Biológicas, CSIC, Madrid, Spain

Andrea Pelikan
Cell Biology, Institut Curie, UMR 144 CNRS, Paris, France

Mylène Perderiset
Structural Motility, Institut Curie, Centre National de la Recherche Scientifique, Unité Mixte de Recherche 144, Paris Cedex 05, France

Franck Perez
Institut Curie, and CNRS UMR144, Paris Cedex 05, France

V. Peyrot
Aix - Marseille University, Inserm, CR02 UMR_S 911, Faculté de Pharmacie, 13385 Marseille, France

Julia T. Philip
Department of Chemistry and Biochemistry, Interdisciplinary Center for the Study of Biocomplexity, University of Notre Dame, Notre Dame, Indiana, USA

Antoine Pilon
EA4530, Dynamique des microtubules en physiopathologie, Faculté de Pharmacie, Université Paris-Sud, Châtenay-Malabry, and Unité d'Hormonologie et Immunoanalyse, Pôle de Biologie Médicale et Pathologie, Hôpitaux Universitaires Est Parisien, APHP, Paris, France

Christian Poüs
EA4530, Dynamique des microtubules en physiopathologie, Faculté de Pharmacie, Université Paris-Sud, Châtenay-Malabry, and Laboratoire de Biochimie-Hormonologie, Hôpitaux Universitaires Paris-Sud, APHP, Clamart, France

Ian A. Prior
Department of Cellular & Molecular Physiology, Institute of Translational Medicine, University of Liverpool, Liverpool, United Kingdom

Mark M. Rasenick
Departments of Physiology & Biophysics and Psychiatry, University of Illinois at Chicago, and The Jesse Brown VA Medical Center, Chicago, Illinois, USA

Jennifer L. Ross
Department of Physics, University of Massachusetts Amherst, Amherst, Massachusetts, USA

Stephen J. Royle
Department of Cellular & Molecular Physiology, Institute of Translational Medicine, University of Liverpool, Liverpool, and Centre for Mechanochemical Cell Biology, Division of Biomedical Cell Biology, Warwick Medical School, University of Warwick, Coventry, United Kingdom

Dan L. Sackett
Program in Physical Biology, Eunice Kennedy Shriver National Institute of Child Health and Human Development, NIH, Bethesda, Maryland, USA

Witchuda Saengsawang
Department of Physiology, Faculty of Science, Mahidol University, Bangkok, Thailand

James Sillibourne
Cell Biology, Institut Curie, UMR 144 CNRS, Paris, France

Yuyu Song
Department of Genetics and Howard Hughes Medical Institute, Yale School of Medicine, New Haven, Connecticut, and Marine Biological Laboratory, Woods Hole, Massachusetts, USA

Viktoriya Syrovatkina
Cell and Developmental Biology, University of Pennsylvania, Philadelphia, Pennsylvania, USA

Guoling Tian
Department of Biochemistry and Molecular Pharmacology, New York University Langone Medical Center, New York, New York, USA

Phong T. Tran
Cell Biology, Institut Curie, UMR 144 CNRS, Paris, France, and Cell and Developmental Biology, University of Pennsylvania, Philadelphia, Pennsylvania, USA

P.O. Tsvetkov
Engelhardt Institute of Molecular Biology, Russian Academy of Sciences, Moscow, Russia

Megan T. Valentine
Department of Mechanical Engineering, and The Neuroscience Research Institute, University of California, Santa Barbara, California, USA

Hailing Yang
MD Anderson Cancer Center, Houston, Texas, USA

Yali Yang
Department of Mechanical Engineering, University of California, Santa Barbara, California, USA

Preface

This is the fifth *Methods in Cell Biology* volume that focuses on microtubules and microtubule-associated proteins. Two volumes in the 1980s that were published when knowledge of the cytoskeleton was in its infancy covered the entire cytoskeleton, *The Cytoskeleton* (Volume 24 Part A and Volume 25 Part B, ed. L. Wilson). These two volumes covered a wide range of microtubule, actin, myosin, intermediate filament, flagella, mitotic spindle, and sea urchin topics. Our knowledge of the cytoskeleton has expanded greatly and now cytoskeletal methods books have become focused on the individual filaments and their associated regulatory proteins. With respect to microtubules, Volume 95 *Microtubules, in vitro* (ed. J.J. Correia and L. Wilson) appeared in 2010 and covered five broad microtubule topics: (1) isolation of microtubules and associated proteins and the biochemistry and characterization of antibodies and tubulin isotypes; (2) microtubule structure and dynamics; (3) drugs; (4) interactions with motors and maps; and (5) functional extracts and force measurements. Volume 97 *Microtubules: in vivo*, edited by Lynn Cassimeris and Phong Tran, appeared that same year and covered a wide range of cellular and species topics including microtubule dynamics, fission and budding yeast, *Drosophila*, *Giardia*, *Dictyostelium*, *C. elegans*, plants, melanophores, zebrafish, and cryo-electron tomography. While these volumes provided an extensive compilation of the methods used for studying microtubules, microtubule-associated proteins, and antimitotic drugs in numerous contexts, they were by no means exhaustive and the field has continued progressing rapidly with the introduction of many new methodologies and approaches. This current and second volume on *Microtubules, in vitro* (ed. J.J. Correia and L. Wilson) contains 24 chapters that continue our attempts to inform and educate microtubule researchers and new investigators now entering the field on the rapidly expanding progress in the field.

The 24 chapters in the current volume (*Microtubules, in vitro*, Part 2, ed. J.J. Correia and L. Wilson) cover a diverse range of topics. Susan Bane's group describes biochemical methods for specific fluorescent labeling of tubulin and tubulin/microtubule-targeting drugs (Chapter 1), while Gary Brouhard's group describes a fluorescence-based assay for tubulin structure analysis (Chapter 21). Dan Sackett's group compares various modern methods for measuring tubulin polymerization *in vitro* (Chapter 14). Three chapters describe methods to evaluate microtubule motion. Douglas Martin's group looks at gliding assays for persistence length analysis (Chapter 2), while Kris Leslie and Niels Galjart describe how to examine microtubule dynamics by TIRF microscopy (Chapter 8). Megan Valentine's group describes how to use magnetic tweezers to visualize 3D deformations in microtubule networks (Chapter 6). Liam Cheeseman and collaborators present kinetochore ultrastructure by correlation light-electron microscopy (Chapter 20). Jessica Lucas' lab presents imaging methods for plant microtubules (Chapter 15). Two groups describe methods for production and analysis of microtubule antibodies. Franck Perez's lab looks at recombinant antibodies for assaying microtubule structure, conformational states of tubulin, and dynamics (Chapter 10 and cover art). Carsten Janke and Maria Magiera describe how to use antibodies to obtain quantitative information about tubulin posttranslational modifications with an emphasis on carboxy-terminal modifying enzyme effects (Chapter 16). Jennifer Ross's group describes the expression, purification, and biophysical characterization of katanin, a microtubule-severing enzyme (Chapter 13). Carolyn Moores' group presents biochemical and structural methods for studying the binding of doublecortin to the microtubule lattice (Chapter 3). Holly Goodson's group presents computational methods for simulating various protein or drug binding modes to the microtubule lattice (Chapter 23). Francois Devred and collaborators present ITC methods for studying tubulin–MAP interactions (Chapter 18). Jessica Field and her many international collaborators present structural methods for studying the interaction of a novel covalent drug Zampanolide (isolated from a marine sponge) to a tubulin–stathmin complex (Chapter 19). Scott Brady describes methods for the analysis of microtubule motor-driven transport in giant squid axoplasm (Chapter 9). Sharon Lobert's group describes methods for studying the complex role of micro-RNA in regulating tubulin isotype expression and the implications for antimitotic resistance (Chapter 5). Nicholas Cowan describes the role of a tubulin chaperone machine that assembles α/β heterodimers (Chapter 11). Mark Rasenick' group describes the evidence that tubulin interacts with and regulates G proteins (Chapter 12). Fernando Cabral's group presents methods for studying the drug-induced microtubule detachment from centrosomes and spindle poles (Chapter 4). Phong Tran presents two chapters: one on the dynamics of mitochondria localization to microtubules (Chapter 7) and a second on imaging individual spindle microtubule dynamic in fission yeast (Chapter 24). Finally, José Andreu's lab describes methods for the preparation of bacterial tubulin BtubA/B bearing eukaryotic tubulin sequences (Chapter 17).

This rather eclectic group of methods chapters cover biochemistry, physical chemistry, binding, antimitotic drugs, enzymology, fluorescence, structure, microscopy, imaging, assembly, dynamics, accessory proteins, molecular motors, chaperones, antibodies, signal transduction, centrosomes, kinetochores, spindles, mitochondria, and bacterial tubulin. As any inquiring microtubule student or investigator will attest, this appears to be the appropriate reality when describing the methods and fields of investigation required to know and understand the general and specific aspects of microtubule, their regulation, and their varied functions. Attempting to enter this field both broadens your base of cell biology knowledge and stimulates your scientific imagination. We thank the authors for the hard work that went into preparing a volume like this collectively successful. (Once again some authors could not meet the deadline and we invite them to participate in future volumes, especially if they start now.) We thank Shaun Gamble and Zoe Kruze at Elsevier for their excellent technical support and constant reminders during this process. As usual we also thank the patience and understanding of our collaborators, colleagues, and family during our single-minded pursuit of this task—we promise to not do it again, too soon.

John (Jack) Correia
UMMC, Jackson, MS
Leslie Wilson
UCSB, Santa Barbara, CA

CHAPTER

Site-Specific Fluorescent Labeling of Tubulin

1

Kamalika Mukherjee and Susan L. Bane

Department of Chemistry, Binghamton University, State University of New York, Binghamton, New York, USA

CHAPTER OUTLINE

Introduction **2**
1.1 Material **4**
1.1.1 TLL Expression and Purification 4
1.1.2 Labeling Tubulin 4
1.1.3 Detection of Labeled Tubulin 5
1.1.4 Detection of Polymerization Activity 5
1.2 Method **5**
1.2.1 Protein Isolation and Purification 5
1.2.1.1 Tubulin 5
1.2.1.2 Tubulin Tyrosine Ligase 6
1.2.2 Tubulin Detyrosination 6
1.2.3 Coupling Tyrosine Derivative to the C-terminus of α-tubulin 6
1.2.3.1 Ligand Preparation 6
1.2.3.2 Coupling Reaction 7
1.2.4 Fluorescent Labeling of Tubulin 7
1.2.4.1 Preparation of Catalyst Stock 7
1.2.4.2 Preparation of Fluorophore Stock 7
1.2.4.3 Preparation of fluorophore-labeled Tubulin 8
1.2.5 Coupling Biotin to the C-terminus of α-tubulin 8
1.2.5.1 BH Stock Preparation 8
1.2.5.2 Biotinylation of Tubulin 8
1.2.6 Detection of Modification in the C-terminus of Tubulin 8
1.2.6.1 Detection of Tyrosine Or 3-fomyltyrosine in the C-terminus of α-tubulin 8
1.2.6.2 Detection of fluorophore-labeled Tubulin 9
1.2.6.3 Detection of Biotinylated Tubulin 9
1.2.7 Evaluation of Tubulin Assembly Activity 9
Acknowledgments **11**
References **11**

Methods in Cell Biology, Volume 115

ISSN 0091-679X
http://dx.doi.org/10.1016/B978-0-12-407757-7.00001-3

Abstract

Fluorescent tubulin can be prepared in which a fluorophore is covalently bound to the protein at only the carboxy terminus of the α-subunit of the αβ-tubulin dimer. This two-step procedure consists of an enzymatic reaction followed by a bioorthogonal chemical reaction. In the first step of the process, the enzyme tubulin tyrosine ligase is used to attach a reactive tyrosine derivative, 3-formyltyrosine, to the protein. In the second step of the procedure, a fluorophore possessing a complementary reactive functional group, such as a hydrazine, hydrazide, or hydroxylamine, is allowed to react with the protein under conditions that are compatible with native tubulin. Polymerization-competent, fluorescently labeled tubulin can be prepared in just a few hours using this protocol. The method described here should be useful for attaching virtually any probe or material to tubulin at this site.

INTRODUCTION

Attaching an exogenous reporter molecule to tubulin is normally accomplished with a reactive probe that targets a naturally occurring amino acid such as lysine. Many lysine-reactive probes are commercially available, and protocols for performing the labeling reactions can be found in the literature (Hyman, et al., 1991; Peloquin, Komarova, & Borisy, 2005; Wadsworth & Salmon, 1986). The conditions necessary for efficient reaction between the probe and the protein (e.g., high pH) sharply decrease the ability of tubulin to assemble into microtubules. It is necessary to perform a cycle of assembly–disassembly to obtain polymerization-competent protein. Although active, fluorescent tubulin is accessible by this method, the overall yield of labeled protein is low, and the position of the labels and the degree of labeling cannot be controlled.

We have developed a method by which a single probe can be attached to just the C-terminus of α-tubulin through a two-part procedure (Banerjee, et al., 2010). In part 1, a tyrosine derivative possessing a reactive functional group is attached to the carboxy terminus of α-tubulin using the enzyme tubulin tyrosine ligase (TTL). The synthetic amino acid 3-formyltyrosine (3fY) is used as the substrate, resulting in the retyrosinated protein that now possesses a reactive aromatic aldehyde. In part 2, a probe containing a complementary reactive functional group such as a hydrazine, hydrazide, or hydroxylamine is allowed to react with the 3-formyltyrosinated protein. Since the two functional groups are chosen to have orthogonal reactivity with respect to endogenous amino acid residues, the probe is attached to tubulin at a single location.

An important factor to consider in part 1 is that the TLL uses both the tyrosine derivative and the carboxy terminal peptide of α-tubulin as substrates; therefore, the structure of each component is significant to the ligation reaction. Consider tubulin: purified tubulin from most sources contains a mixture of isotypes and isoforms. Isotypes of α-tubulin that terminate in E-E-Y are subject to the tyrosination/

detyrosination cycle by the endogenous enzymes, and these are the major isotypes found in most tissues (Erck, Frank, & Wehland, 2000). Tyrosinated tubulin (Tyr-tubulin) in this population can be effectively detyrosinated by carboxypeptidase A (CPA) to yield "Glu-tubulin." *In vivo*, Glu-tubulin may be further processed by an endogenous glutamylase to form Δ-2 tubulin, which is not a substrate for TTL (Janke & Bulinski, 2011). Fortunately, Glu-tubulin is a much poorer substrate for CPA than Tyr-tubulin and prolonged treatment of αβ-tubulin with CPA does not produce Δ-2 tubulin.

A number of studies have demonstrated that TTL exhibits some flexibility in the structural requirement for the amino acid substrate. In general, tyrosine derivatives with small substituents ortho- to the —OH group of tyrosine are tolerated by the enzyme (Coudijzer & Joniau, 1990; Joniau, Coudijzer, & Decuyper, 1990; Kalisz, Erck, Plessmann & Wehland, 2000; Monasterio, Nova, Lopezbrauet & Lagos, 1995). At least two of these tyrosine derivatives possess reactive functional groups suitable for bioorthogonal chemical labeling. The tyrosine analog 3-azidotyrosine has been used as a photoinhibitor of TTL. Conceivably, tubulin containing 3-azidotyrosine can be labeled using "click" chemistry with alkyne-containing reagents, many of which are commercially available. This possibility has not been demonstrated in the literature, and we have not attempted it in our lab. Instead, we have synthesized and studied the tyrosine derivative 3fY. We selected a formyl group rather than the more commonly used acyl group for the 3-position because of its smaller size and also because of the higher reactivity of aldehydes with nucleophiles versus ketones with nucleophiles. The tyrosine derivative is a poorer substrate than tyrosine for the ligation reaction, but complete retyrosination can be achieved with 3fY, given sufficient time and substrate concentration.

The final component of the first part of the procedure is TTL. We use a recombinant human TTL with a GST tag, prepared from a commercial vector. We have found that it is not necessary to remove the GST from the enzyme; the procedures described here are for the TTL–GST fusion protein. This enzyme can be stored at −50 °C for many months without observable loss of activity.

The second part of the protocol is the bioorthogonal ligation reaction. In the procedures described here, the probe of interest possesses a hydrazide or an aromatic hydrazine as the reactive functional group. Few aromatic hydrazine-containing probes are commercially available at this time, but many different hydrazide probes can be purchased. In recent years, it has become common to use a hydroxylamine-containing probe as the nucleophile in these coupling reactions (Dirksen, Hackeng, & Dawson, 2006). Owing to the similar reactivity of the nucleophiles, it is likely that a hydroxylamine-containing probe can be successfully substituted for a hydrazide probe in these procedures.

Coupling reactions between hydrazides, hydrazines, or hydroxylamines and aldehydes are typically fastest at acidic pH: pH 4.5 and lower (Cordes & Jencks, 1962). The rate of the reaction drops sharply with increasing pH. Optimally, the coupling reaction should be performed under conditions that retain the activity of the protein, and purified tubulin is most stable near neutral pH and at cold temperatures.

Fortunately, it is possible to perform the coupling reaction at higher pH by using an aromatic amine as a catalyst. Aniline at a final concentration of 100 mM is the most common catalyst (Dirksen, et al., 2006). We do not use aniline for catalyzing reactions with tubulin, however, as it also serves as a denaturant at concentrations necessary for efficient catalysis. We have found that the commercially available 4-aminophenylalanine satisfactorily catalyzes the coupling reaction at neutral pH (Blanden, Mukherjee, Dilek, Loew, & Bane, 2011). Furthermore, the catalyzed reaction can be performed efficiently at low temperature (4 °C), which further protects tubulin from denaturation during the labeling process.

The protocol presented here describes tubulin labeling with fluorophores. The method likely to be amenable to other applications (such as attaching tubulin to a bead, a surface, a nanoparticle, etc.) provided that the reacting partner with an appropriate functional group can be obtained. One such example, biotinylation of tubulin, is described.

1.1 MATERIAL

1.1.1 TLL expression and purification

1. EX-T8583-B05 (Genecopoeia, MD)
2. Luria Bertani (LB) broth and LB agar (Sigma-Aldrich, MO)
3. Ampicillin (Sigma-Aldrich)
4. Glutathione Sepharose 4B resin (GE Healthcare, Sweden)
5. 1 M Isopropyl β-D-1-thiogalactopyranoside (IPTG) (Gold Biotechnology, MO) solution in water
6. TE buffer (10 mM Tris, 1 mM EDTA, pH 8) containing 0.3 mg of PMSF (phenylmethanesulfonyl fluoride) per mL
7. PBS (140 mM NaCl, 2.7 mM KCl, 10 mM Na_2HPO_4, 1.8 mM KH_2PO_4, pH 7.4)
8. Elution buffer (50 mM Tris, 20 mM reduced glutathione, pH 8.0)

1.1.2 Labeling tubulin

1. PME buffer (0.1 M PIPES, 1 mM $MgSO_4$, 2 mM EGTA, pH 6.90)
2. Sephadex G50 and G75 resin (GE Healthcare)
3. CPA from bovine pancreas (Type II-PMSF treated), $\geq$ 50 units mg^{-1} protein, an aqueous suspension (Sigma-Aldrich, Catalog # C9268)
4. TTL buffer (25 mM MES, 150 mM KCl, 27 μM $MgCl_2$, 2.5 mM ATP, 1 mM DTT, 1.5% glycerol, pH 6.8)
5. 3fY

   Note: This unnatural amino acid is presently not available commercially. It can be prepared by a two-step synthesis process from commercially available starting material (BOC-L-tyrosine) and common laboratory chemicals (Banerjee, Panosian, et al., 2010)
6. EZ-linked biotin hydrazide (BH) (Thermo Scientific, IL)

7. Texas Red hydrazide (TxRed, 90% single isomer) (Life Technologies, CA)
8. 7-Hydrazinyl-4-methylcoumarin (coumarin hydrazine, CH)

Note: This fluorophore is not commercially available, but it is not complicated to synthesize (Banerjee, Panosian, et al., 2010). A commercially available hydrazide such as 7-(diethylamino)coumarin-3-carbohydrazide, CZ, can be substituted for CH.

9. 4-Amino-DL-phenylalanine hydrate; 97% (4aF) (Acros Organic, NJ)

1.1.3 Detection of labeled tubulin

1. Sample buffer, 3 × stock (3.75 mL glycerol, 3.0 mL 10% (w/v) SDS (Catalog # L5750, Sigma-Aldrich), 0.3 mL 0.5% (w/v) bromophenol blue, 0.325 mL water, 1.875 mL, 0.5 M Tris, pH 6.8). Add 7.5% β-mercaptoethanol to the sample buffer prior to running the gel
2. SDS-PAGE gel (8% acrylamide resolving layer and 6% acrylamide stacking layer)
3. Resolving buffer (1.5 M Tris, pH 9.8)
4. Stacking buffer (0.5 M Tris, pH 6.8)
5. PBST (10 mM phosphate buffer at pH 7 with 0.1% Tween-20)
6. Blocking buffer (PBST with 5% bovine serum albumin)
7. Monoclonal anti-tubulin, tyrosine antibody TUB-1A2 (Sigma-Aldrich)
8. Streptavidin–HRP (Life Technologies)

1.1.4 Detection of polymerization activity

1. Taxol
2. DMSO (Sigma)

1.2 METHOD

1.2.1 Protein isolation and purification

1.2.1.1 Tubulin

Isolate bovine brain tubulin by two cycles of assembly/disassembly followed by phosphocellulose chromatography (Williams & Lee, 1982). Add 1–2 mM $MgSO_4$ and drop-freeze the purified tubulin into liquid nitrogen and store until use. Gently thaw the frozen pellets and equilibrate the protein in PME buffer. We use Sephadex G-50 in 1 mL syringe spin columns following the method of Penefsky (1979). Determine tubulin concentration spectrophotometrically by using an extinction coefficient of 1.23 $(mg/mL)^{-1}$ cm^{-1} at 278 nm in PME.

Owing to the conserved nature of both tubulin and the ligase, tubulin from other sources can be substituted for bovine brain tubulin. We have successfully performed labeling procedure on chicken erythrocyte tubulin, purified as described previously (Sharma, et al., 2010).

1.2.1.2 Tubulin tyrosine ligase

Transform competent *Escherichia coli* BL21 with expression vector EX-T8583-B05 (pReceiver 05 × containing GST–TTL gene and ampicillin-resistant gene) and grow them on 100 μg mL^{-1} Ampicillin containing LB (LB-Amp) agar. Inoculate transformed cells in LB-Amp broth and allow them to grow overnight at 37 °C. Mix the overnight culture with fresh medium (1:10) and incubate in a shaker incubator at 37 °C. At OD_{600} = 0.4–0.5, induce protein expression with 1 mM IPTG. Return to the shaker incubator for 20 h and then centrifuge the broth at 3000 rpm for 20 min at 4 °C. At this point, the pellet may be stored at −50 °C for many months.

Resuspend the cell pellet by sonication in TE buffer containing 0.3 mg of PMSF per mL. (We use 20–30 mL buffer for cell pellet obtained from a 400 mL culture.) Perform six cycles of sonication. Each cycle should include 10 s of sonication followed by 15 s incubation on ice. Centrifuge the resultant lysate at 3000 rpm for 20 min to remove the debris. TTL–GST can then be isolated by affinity chromatography.

Wash glutathione Sepharose 4B with PBS following the manufacturer's instructions. Mix the cleared lysate with glutathione Sepharose 4B (~300 μL for the 400 mL culture volume) and incubate on a rocker at 4 °C for 45 min. Remove unbound protein by batch-washing in the following manner. Pellet the resin by centrifuging at 1300 rpm or 300 × *g* for 2 min. Add approximately 3 mL of PBS per mL of resin volume, mix by gentle inversion three times, and centrifuge the suspension at 1300 rpm or 300 × *g* for 2 min. Repeat this step four times. Resuspend the resin in PBS (approximately 2 mL PBS per mL of resin) and pour the mixture into a 5-mL column. Elute the bound protein with five resin volumes of elution buffer. Collect elution fractions equivalent to the resin volume and check each fraction for protein. Pool the fractions containing purified protein and determine the protein concentration. Store the protein as aliquots at −50 °C until use.

The GST portion of the expressed protein can be removed by TEV protease (Sigma). We use the entire fusion protein (GST–TTL) for further experiments.

1.2.2 Tubulin detyrosination

Detyrosinate tubulin by treating purified tubulin in PME buffer with CPA (final concentrations: 0.1 mg per mL CPA, 3 mg per mL tubulin) for 10 min at 37 °C. Submit the mixture to rapid gel filtration with Sephadex G-75 in TTL buffer. We use 1 mL syringe spin columns following the method of Penefsky (1979). This process also removes the cleaved tyrosine and CPA.

1.2.3 Coupling tyrosine derivative to the C-terminus of α-tubulin

1.2.3.1 Ligand preparation

1.2.3.1.1 L-Tyrosine

Make 100 mM tyrosine stock solution by dissolving appropriate mass of tyrosine in 250 mM NaOH solution.

1.2.3.1.2 3-Formyltyrosine

Prepare a 30 mM solution of 3fY in PME buffer. This concentration is near the limit of the molecule's solubility, and dissolution of the solid requires repeated vortexing and sonication. Centrifuge the solution at 5600 × *g* for 10 min to remove any undissolved solute. Determine the actual concentration of the clarified 3fY stock spectrophotometrically using an extinction coefficient of 10,715 M^{-1} cm^{-1} at 257 nm. Store the solution at 4 °C until use. (This solution can be stored at 4 °C for months.) Equilibrate the solution to room temperature and vortex well before use.

1.2.3.2 Coupling reaction

Allow the detyrosinated tubulin from section 1.2.2 in TTL buffer to react with excess ligand (Y or 3fY) in the presence of GST–TTL (final concentrations: 0.2 mg per mL GST-TTL, 1 mM Y or 2 mM 3fY) for 20–30 min at 37 °C. Next, remove the excess ligand by rapid gel filtration into PME buffer using G-50 resin. Note that the elutant will contain both 3-formyltyrosinated tubulin (3fY-Tb) and GST–TTL. This material can be drop-frozen in liquid nitrogen and stored in liquid nitrogen until use.

The presence of GST–TTL in the tubulin solution at this ratio does not affect taxol-induced tubulin assembly, and no further processing of the protein mixture is performed for experiments in which the presence of GST–TTL is unimportant. To obtain pure tubulin, remove GST–TTL using glutathione Sepharose 4B in PME. We find it convenient to add the appropriate volume of resin to the reaction mixture before removing the excess ligand through rapid gel filtration; thus, this step removes both enzyme and ligand together.

1.2.4 Fluorescent labeling of tubulin

1.2.4.1 Preparation of catalyst stock

Prepare 50 mM stock solution of 4aF in PME, pH 6.9. Repeated sonication and vortexing is required to dissolve all 4aF at this concentration. Store the stock at 4 °C until use. Warm the solution to room temperature and vortex well before use.

1.2.4.2 Preparation of fluorophore stock

Texas Red hydrazide: Dissolve Texas Red hydrazide in DMSO to a final concentration of 5 mM or higher. Determine the concentration spectrophotometrically using an extinction coefficient of 109,000 M^{-1} cm^{-1} at 588 nm. The stock solution can be stored at −20 °C. Vortex well before use. Keep in the dark when not in use.

CH: Just before use, prepare CH in PME to a final concentration of 500 μM or higher. Dissolve CH in PME by repeated vortexing and sonication. Centrifuge at 5600 × *g* for 10 min to remove any undissolved solute. Check the concentration of the solution after centrifugation by UV–Vis spectroscopy using extinction coefficient of 19,000 M^{-1} cm^{-1} at 346 nm. Keep in the dark when not in use.

7-(Diethylamino) coumarin-3-carbohydrazide: Dissolve the commercially available CZ in DMSO to a final concentration of 5 mM or higher. Determine the concentration spectrophotometrically using an extinction coefficient of 46,000 M^{-1} cm^{-1} at 420 nm. The stock solution can be stored at −20 °C. Vortex well before use. Keep in the dark when not in use.

1.2.4.3 Preparation of fluorophore-labeled tubulin

Prepare a solution of the 3fY ligated tubulin (3fY-Tb) in PME containing 10 mM 4aF. Allow this solution to react with a fluorophore of choice in 1:10 molar ratio (protein:fluorophore) in the dark, at 4 °C for 60 min or at room temperature for 2 h (Banerjee, Panosian, et al., 2010; Blanden, et al., 2011). It would be prudent to empirically determine the incubation times for the selected fluorophore the first time the procedure is performed. Desalt fluorescent protein into PME or a chosen buffer to remove excess fluorophore and 4aF. This material can be drop-frozen and stored in liquid nitrogen until use.

1.2.5 Coupling biotin to the C-terminus of α-tubulin

1.2.5.1 BH stock preparation

Make a 50 mM BH stock in DMSO. Store the stock at 4 °C until use. Vortex well before use.

1.2.5.2 Biotinylation of tubulin

Prepare 3fY-Tb in PME containing 10 mM 4aF. Allow this solution to react with BH in 1:10 molar ratio (protein: BH) at 4 °C for 2 h. Desalt biotinylated protein into PME or a chosen buffer to remove excess BH and 4aF. This material can be drop-frozen and stored in liquid nitrogen until use.

1.2.6 Detection of modification in the C-terminus of tubulin

1.2.6.1 Detection of tyrosine or 3-fomyltyrosine in the C-terminus of α-tubulin

Mix the protein sample with sample buffer and run SDS-PAGE according to the procedure described by Banerjee, Bovenzi, et al. (2010). Standard protocols often will not separate the αβ dimer. Equilibrate the gel in transfer buffer for 45 min and then transfer the protein samples for Western blot onto a polyvinylidene fluoride membrane. Incubate the membrane in blocking buffer for 2 h at room temperature on a rocker. Then probe the membrane with TUB-1A2 antibody (1:2000) for 90 min at room temperature. Wash off the excess antibody with PBST and then probe the membrane with goat antimouse secondary antibody HRP conjugate (1:5000) for 60 min at room temperature. Wash three times with PBST and develop with the blot with TMB solution (Fig. 1.1).

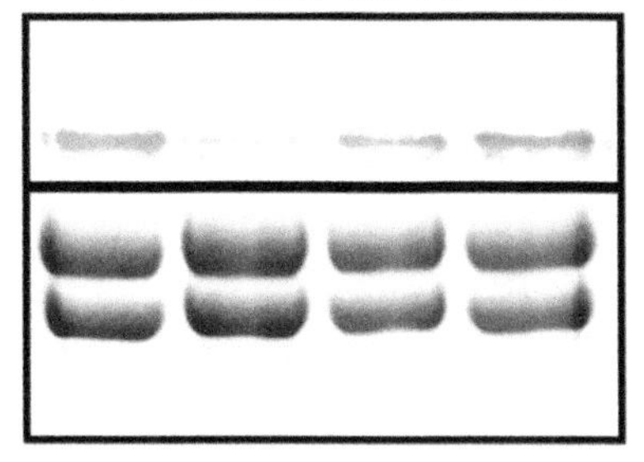

FIGURE 1.1

Detection of tyrosine or 3-fomyltyrosine in the C-terminus of α-tubulin: Detyrosinated tubulin was incubated with tubulin tyrosine ligase in the presence of either L-tyrosine or 3-formyltyrosine as described and then subjected to SDS-PAGE. Top: Western blot with tyrosinated α-tubulin antibody (Sigma-Aldrich TUB-1A2). Bottom: Coomassie stain. The first protein band in the gel is α-tubulin and the second band is β-tubulin. The mass of protein loaded into each well was 1 μg for the Western blot and 10 μg for the protein stain. (See color plate.)

1.2.6.2 Detection of fluorophore-labeled tubulin

Separate αβ dimer by SDS-PAGE and observe α-tubulin in a dark room using long-wavelength UV lamp (Fig. 1.2).

The ratio of the fluorophore to tubulin dimer may be estimated by UV spectroscopy and protein assay. From the UV spectrum of the final product, calculate the concentration of the fluorophore using extinction coefficients listed above. Determine the protein concentration by protein assay (e.g., BCA). The overall labeling efficiency is typically up to 0.3 mol of fluorophore per mol of tubulin.

1.2.6.3 Detection of biotinylated tubulin

To detect biotinylated tubulin, perform a Western blot as described previously and probe the blot with streptavidin–HRP conjugate (1:10,000) (Fig. 1.3).

1.2.7 Evaluation of tubulin assembly activity

Equilibrate protein samples (5 μM final concentration) in PME buffer at 37 °C and record the absorbance at 350 nm to obtain the baseline. Add 5 μM taxol in DMSO (final concentration 2% v/v) to initiate polymerization. Detect the extent of polymerization by monitoring apparent absorption at 350 nm over time (Fig. 1.4).

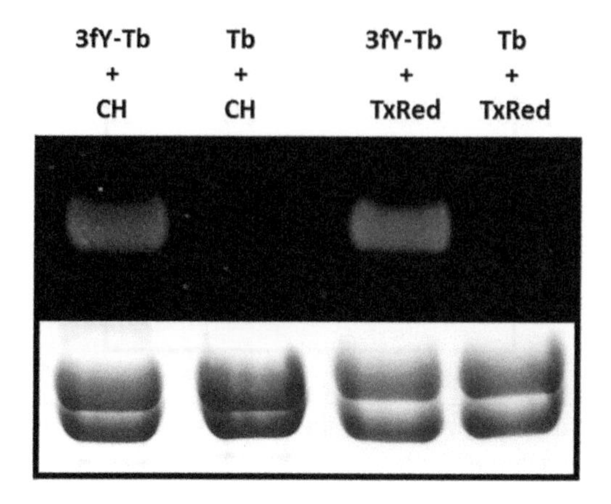

FIGURE 1.2

Detection of fluorophore-labeled tubulin: 3-Formyltyrosinated tubulin (3fY-Tb) or unmodified tubulin (Tb) in PME buffer was incubated with CH or TxRed in the presence of 10 mM 4aF as described prior to SDS-PAGE analysis. The gel was then visualized under long-wavelength UV light (top) and stained with Coomassie blue (bottom). The mass of protein loaded into each well was 25 μg. Photographic exposure time for the fluorescent gel = 10 s. (See color plate.)

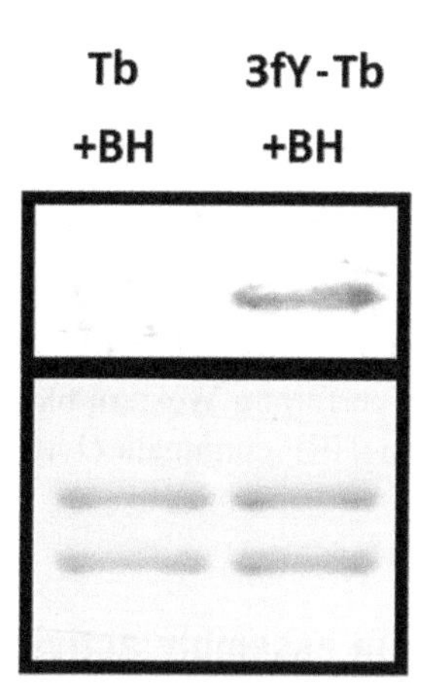

FIGURE 1.3

Detection of biotinylated tubulin: 3-Formyltyrosinated tubulin (3fY-Tb) or unmodified tubulin (Tb) was incubated with biotin hydrazide (BH) in the presence of 4aF as described prior to SDS-PAGE analysis. Top: Western blot with streptavidin–HRP conjugate (Life Technologies). Bottom: Coomassie stain. The mass of protein loaded into each well was 1 μg for the Western blot and 5 μg for the protein stain. (See color plate.)

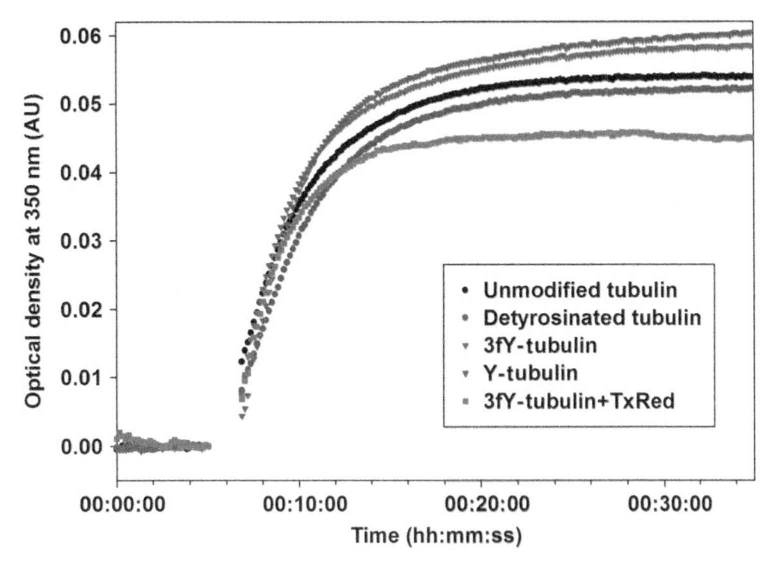

FIGURE 1.4

Polymerization activity of labeled tubulin: Tubulin from different stages of the labeling process (5 μM final concentration) in PME buffer was equilibrated at 37 °C and the absorbance at 350 nm was recorded to obtain the baseline. Taxol was added to a final concentration of 5 μM to initiate polymerization. The extent of polymerization was detected by monitoring apparent absorption at 350 nm over time. GST–TTL was not removed from the samples containing 3fY-tubulin. (See color plate.)

Acknowledgments

We would like to thank Dr. Dan L. Sackett and Dr. Rebecca M. Kissling for helpful scientific discussions and suggestions, Dr. Abhijit Banerjee for synthesizing 7-hydrazinyl-4-methylcoumarin, Mr. Daniel Wu for synthesizing 3-formyltyrosine, Mr. Robert Ingraham and Mr. Matthew D. Antalek for purification of 3-formyltyrosine, Mr. Paul G. Yoffe and Mr. Richard Law for assisting in standardizing this protocol, and Mr. David Tuttle for scientific photography. This work was supported by NIH Grants GM093941 and GM102867.

References

Banerjee, A., Bovenzi, F. A., & Bane, S. L. (2010). High-resolution separation of tubulin monomers on polyacrylamide minigels. *Analytical Biochemistry*, *402*, 194–196.

Banerjee, A., Panosian, T. D., Mukherjee, K., Ravindra, R., Gal, S., Sackett, D. L., et al. (2010). Site-specific orthogonal labeling of the carboxy terminus of alpha-tubulin. *ACS Chemical Biology*, *5*, 777–785.

Blanden, A. R., Mukherjee, K., Dilek, O., Loew, M., & Bane, S. L. (2011). 4-Aminophenylalanine as a biocompatible nucleophilic catalyst for hydrazone ligations at low temperature and neutral pH. *Bioconjugate Chemistry*, *22*, 1954–1961.

Cordes, E. H., & Jencks, W. P. (1962). The mechanism of Schiff-base formation and hydrolysis. *Journal of the American Chemical Society*, *84*, 832–837.

Coudijzer, K., & Joniau, M. (1990). 3-Azido-L-tyrosine as a photoinhibitor of tubulin:tyrosine ligase: Role of thiol-groups. *FEBS Letters*, *268*, 95–98.

Dirksen, A., Hackeng, T. M., & Dawson, P. E. (2006). Nucleophilic catalysis of oxime ligation. *Angewandte Chemie (International ed. in English)*, *45*, 7581–7584.

Erck, C., Frank, R., & Wehland, J. (2000). Tubulin-tyrosine ligase, a long-lasting enigma. *Neurochemical Research*, *25*, 5–10.

Hyman, A., Drechsel, D., Kellogg, D., Salser, S., Sawin, K., Steffen, P., et al. (1991). Preparation of modified tubulins. *Methods in Enzymology*, *196*, 478–485.

Janke, C., & Bulinski, J. C. (2011). Post-translational regulation of the microtubule cytoskeleton: mechanisms and functions. *Nature Reviews Molecular Cell Biology*, *12*, 773–786.

Joniau, M., Coudijzer, K., & Decuyper, M. (1990). Reaction of alpha-tubulin with iodotyrosines catalyzed by tubulin-tyrosine ligase—Carboxy-terminal labeling of tubulin with I-125 monoiodotyrosine. *Analytical Biochemistry*, *184*, 325–329.

Kalisz, H. M., Erck, C., Plessmann, U., & Wehland, J. (2000). Incorporation of nitrotyrosine into alpha-tubulin by recombinant mammalian tubulin-tyrosine ligase. *Biochimica et Biophysica Acta—Protein Structure and Molecular Enzymology*, *1481*, 131–138.

Monasterio, O., Nova, E., Lopezbrauet, A., & Lagos, R. (1995). Tubulin-tyrosine ligase catalyzes covalent binding of 3-fluoro-tyrosine to tubulin—Kinetic and F-19 NMR studies. *FEBS Letters*, *374*, 165–168.

Peloquin, J., Komarova, Y., & Borisy, G. (2005). Conjugation of fluorophores to tubulin. *Nature Methods*, *2*, 299–303.

Penefsky, H. S. (1979). Preparation of nucleotide-depleted F1 and binding of adenine nucleotides and analogs to the depleted enzyme. *Methods in Enzymology*, *55*, 377–380.

Sharma, S., Poliks, B., Chiauzzi, C., Ravindra, R., Blanden, A. R., & Bane, S. (2010). Characterization of the colchicine binding site on avian tubulin isotype beta VI. *Biochemistry*, *49*, 2932–2942.

Wadsworth, P., & Salmon, E. D. (1986). Preparation and characterization of fluorescent analogs of tubulin. *Methods in Enzymology*, *134*, 519–528.

Williams, R. C., & Lee, J. C. (1982). Purification of tubulin from brain. *Methods in Enzymology*, *85*, 376–385.

CHAPTER

Measuring Microtubule Persistence Length Using a Microtubule Gliding Assay

2

Douglas S. Martin
Lawrence University, Appleton, Wisconsin, USA

CHAPTER OUTLINE

Introduction **14**
2.1 Theory **15**
2.2 Methods **17**
2.2.1 Polymerization of Fluorescently Labeled Microtubules 18
2.2.2 Microtubule Gliding Assay 19
2.2.3 Image Analysis and Persistence Length Calculation 20
2.3 Discussion **22**
Summary **23**
Acknowledgments **23**
References **23**

Abstract

The mechanical properties of microtubules have been an area of active research for the past two decades, in part because understanding the mechanics of individual microtubules contributes to modeling whole-cell rigidity and structure and hence to better understanding the processes underlying motility and transport. Moreover, the role of microtubule structure and microtubule-associated proteins (MAPs) in microtubule stiffness remains unclear. In this chapter, we present a kinesin-driven microtubule gliding assay analysis of persistence length that is amenable to simultaneous variation of microtubule parameters such as length, structure, or MAP coverage and determination of persistence length. By combining sparse fluorescent labeling of individual microtubules with single particle tracking of individual fluorophores, microtubule gliding trajectories are tracked with nanometer-level precision. The fluctuations in these trajectories, due to thermal fluctuations in the microtubules themselves, are analyzed to extract the microtubule persistence length. In the following, we describe this gliding assay and analysis and discuss two example microtubule variables, length and diameter, in anticipation that the method may be of wide use for *in vitro* study of microtubule mechanical properties.

Methods in Cell Biology, Volume 115

ISSN 0091-679X
http://dx.doi.org/10.1016/B978-0-12-407757-7.00002-5

INTRODUCTION

Microtubules, a component of the cytoskeleton, are micrometer long, 25-nm wide polymers of tubulin (Lodish et al., 2007). Microtubules are key players at various stages in cellular life: cell division, cell movement, and intracellular transport. Each of these functions requires microtubules which are stiff on cellular length scales—that is, microtubules that bend relatively little over micrometer lengths. As a result, the mechanical properties of microtubules have been an area of active research for the past 2 decades (for a recent review, see Hawkins, Mirigian, Selcuk Yasar, & Ross, 2010), because understanding the mechanics of individual microtubules contributes to modeling whole-cell rigidity and structure and hence to better understanding the physical processes underlying motility and transport. Other work has explicitly focused on the details of microtubule stiffness in the context of microtubule function: the length-dependence of the mechanical properties of microtubules may govern their role in whole-cell mechanics (Mehrbod & Mofrad, 2011); microtubule-associated proteins (MAPs) can increase or decrease the flexibility of microtubules (Cassimeris, Gard, Tran, & Erickson, 2001; Felgner et al., 1997; Portran et al., 2013); and microtubule stiffness variation may affect whole-cell morphology (Topalidou et al., 2012), to name a few examples.

In addition, the intrinsic flexibility of microtubules shows substantial heterogeneity. Experimentally, *in vitro* measurements of mechanical rigidity (characterized by the persistence length or, alternatively, the Young's modulus) of microtubules have varied by an order of magnitude, with persistence lengths varying from 1 to 10 mm (Brangwynne et al., 2007; Cassimeris et al., 2001; Gittes, Mickey, Nettleton, & Howard, 1993; Janson & Dogterom, 2004; Valdman, Atzberger, Yu, Kuei, & Valentine, 2012; van den Heuvel, Bolhuis, & Dekker, 2007; van Mameren, Vermeulen, Gittes, & Schmidt, 2009). Surprisingly, a number of experiments have found that very short microtubules have substantially shorter persistence lengths, on the scale of 0.02–0.1 mm (Pampaloni et al., 2006; Taute, Pampaloni, Frey, & Florin, 2008; Van den Heuvel, de Graaff, & Dekker, 2008). And, experiments on identical microtubules show mechanical rigidity variations, which are substantially broader than those expected from experimental imprecision alone (Valdman et al., 2012). Theoretical models have been developed to describe these effects (Bathe, Heussinger, Claessens, Bausch, & Frey, 2008; Heussinger, Bathe, & Frey, 2007; Pampaloni et al., 2006; Tounsi, Heireche, Benhassaini, & Missouri, 2010), but these models differ in key predictions, predictions which still require experimental examination.

As a result of this interest in microtubule mechanical properties, a number of experimental techniques have been developed to measure microtubule flexibility; in particular, active techniques which involve bending the polymer using optical traps or electric fields (Kikumoto, Kurachi, Tosa, & Tashiro, 2006; Van den Heuvel et al., 2008) and passive techniques which measure the fluctuations of free polymers in solution (Brangwynne et al., 2007; Gittes et al., 1993). The active measurements, however, require specialized setups to implement known forces on the micrometer scale, while the free-fluctuation measurements can be challenging due to diffusion out of the plane of focus of the microscope used.

In this chapter, we describe a complementary, passive technique to measure microtubule persistence length. The technique involves kinesin-driven gliding assays, which ensure that the microtubule always remains in the microscope focal plane. Moreover, it involves tracking single fluorophores attached permanently to the polymer of interest so that specific locations along the polymer are well characterized. By varying the density of kinesin used in the gliding assay, the effective length of the microtubule can be changed. The primary restrictions on the addition of other substances which may modify microtubule persistence length, such as paclitaxel or MAPs, are that they must support kinesin motility.

2.1 THEORY

One way to characterize flexibility (or stiffness) of microtubules, and polymers in general, is the persistence length, the length of polymer which bends by approximately 1 rad under thermal fluctuations at ambient temperature: a rigid or stiff polymer has a long persistence length; a flexible polymer has a short persistence length. Persistence length l_p is heuristically defined as the length over which polymers bend significantly (Fig. 2.1); mathematically, it is defined as

$$\langle \cos \theta_s \rangle = e^{-s/l_p} \tag{2.1}$$

where s is the distance between two points on the polymer, θ_s is the angle between tangent vectors to the polymer, and $\langle \rangle$ indicates an average over all positions a distance s apart (Phillips, Kondev, & Theriot, 2008). For very stiff polymers, $s/l_p \ll 1$ and $\theta_s = d/s$ are small, so a first-order Taylor expansion gives

$$\langle \theta_s^2 \rangle \approx \left\langle (d/s)^2 \right\rangle = s/l_p \tag{2.2}$$

with d the lateral displacement of the polymer with respect to a straight line.

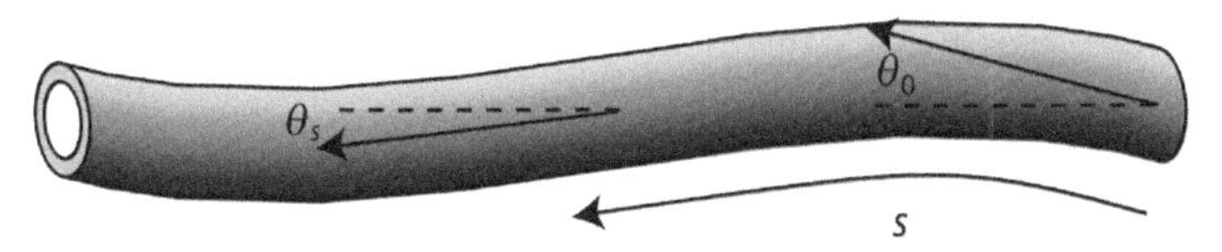

FIGURE 2.1

Cartoon of curved microtubule, showing tangent vectors and angles. The gray tube is a cartoon microtubule (25 nm diameter). Straight arrows represent tangent vectors at two locations along the microtubule axis; s is the path length along the microtubule between the two tangent vectors. Each tangent vector makes an angle θ with respect to the horizontal; the difference between θ_0 and θ_s is used to calculate the persistence length following Eq. (2.1).

The persistence length calculated following Eq. (2.1) is related to the Young's modulus, E, of the microtubule by

$$l_p = \frac{EI}{2kT} \tag{2.3}$$

where I is the moment of inertia of the microtubule and kT is the thermal energy of the solution. The goal, then, is to measure θ_s along an individual microtubule and use Eq. (2.1) to calculate the persistence length.

However, in solution, microtubules fluctuate rapidly, making high-precision measurements of θ_s difficult. Our gliding assay uses the motor protein kinesin-1 to propel a microtubule over a glass surface (Fig. 2.2). If the density of kinesin is high enough, the microtubule will be bound, on average, to many kinesins. However, the microtubule tip will fluctuate before attaching to the next kinesin the tip comes in contact with. The tip fluctuations are well described by Eq. (2.1), and the rest of the microtubule simply follows the same path as the tip since kinesin is processive (remains bound to the microtubule for many ATP turnovers). Thus, the trajectory of the microtubule is a frozen-in fluctuation of the tip—the persistence length of the trajectory reports the persistence length of the tip (Duke, Holy, & Leibler, 1995). Gliding assays have been used previously to calculate microtubule persistence lengths (van den Heuvel et al., 2007); this method uses a different (and, arguably, simpler) method of reconstructing the microtubule trajectory.

In order to reconstruct the microtubule trajectory, we first attach single fluorophores sparsely to the microtubule surface. Using a microscope capable of imaging single fluorophores (a total internal reflection fluorescence, TIRF, microscope in our case), we track all individual fluorophores on a microtubule (Crocker & Grier, 1996)

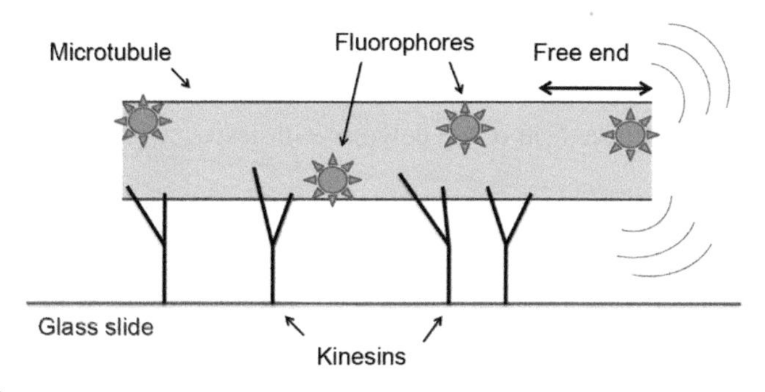

FIGURE 2.2

Cartoon of kinesin gliding assay for microtubules. Kinesins are specifically attached to a glass slide by the coiled-coil, leaving the motor domains free to contact microtubules. Microtubules are propelled by kinesins; the free microtubule end fluctuates due to thermal fluctuations. Single fluorophores are sparsely attached to the microtubule; individual fluorophores are imaged using a TIRF microscope. (See color plate.)

and then combine the trajectories of fluorophores from a single microtubule into one, whole-microtubule, trajectory (Lee, 2000). It is this trajectory we use to calculate θ_s and l_p.

2.2 METHODS

Most materials used in this method are available commercially; for the in-house materials, we offer references for preparation and possible commercial substitutes.

Table of materials used:

Name	Source
Biotinylated kinesin	Berliner, Young, Anderson, Mahtani, and Gelles (1995); Alternative: Cytoskeleton KR01
Bovine brain tubulin	Anderson and Martin (2011); Alternative: Cytoskeleton TL238
Fluorescent GTP	Jena Bioscienes NU-820-TAM; Alternative: fluorescent tubulin: Cytoskeleton TL620M
Bovine Serum Albumin (BSA)	Calbiochem 126615
BSA, biotinylated	Thermo Fisher Pierce 29130
α-Casein	Sigma-Aldrich C6780
Streptavidin (SA)	Thermo Fisher Pierce 21125
Dithiothreitol (DTT)	Sigma-Aldrich D0632
Paclitaxel (Taxol)	LC-Labs P-9600
Glucose oxidase	Sigma-Aldrich G2133
Catalase	Sigma-Aldrich C100
Glucose	Sigma-Aldrich G8270
Sucrose	Sigma-Aldrich S0389
ATP	Sigma-Aldrich A2383
GTP	Sigma-Aldrich G8877
2-Mercaptoethanol	Sigma-Aldrich M3148
Imidazole	Sigma-Aldrich I2399
Potassium chloride	Sigma-Aldrich P9541
Magnesium chloride	Sigma-Aldrich M8266
EGTA	Sigma-Aldrich E3889
PIPES	Sigma-Aldrich P6757
24 × 60 mm No. 1 1/2 cover glass	VWR 48393-252
22 × 22 mm No. 1 cover glass	VWR 48366-067

Biotinylated kinesin and bovine brain tubulin were purified in-house following standard protocols. The biotinylation of kinesin is helpful, but not essential, for gliding assays; a potential substitute is commercially purified kinesin.

The following buffer stock solutions should be prepared and stored (storage temperature is suggested) prior to beginning an experiment. Flash freeze all protein stock solutions in liquid nitrogen prior to storing at −80 °C.

Tubulin:	50 μl, 0.5 mg, in 80 mM PIPES with ~0.2 mM GTP (−80 °C)
Glycerol–MT buffer:	80 mM PIPES, 10 mM $MgCl_2$, 1 mM EGTA, pH 6.7, 30% Glycerol (4 °C)
GTP:	5 μl, 150 mM in water, pH 7 (−80 °C)
Fluorescent GTP:	5 μl, 1 mM in water (−80 °C)
Paclitaxel:	10 μl, 4 mM in DMSO (−80 °C)
Assay buffer (AB):	50 mM imidiazole, 50 mM KCl, 4 mM $MgCl_2$, 2 mM EGTA, pH 6.7 (4 °C)
AB+30% sucrose:	AB with 30% w/v sucrose (4 °C)
BSA:	117 mg/ml (4 °C)
Biotin-LC-BSA:	100 μl, 2 mg/ml in AB (−80 °C long term, 4 °C short term)
α-Casein:	250 μl, 5 mg/ml in AB (−80 °C long term, 4 °C short term)
SA:	20 μl, 10 mg/ml in AB (80 °C long term, 4 °C short term)
DTT:	100 μl, 200 mM in water (−20 °C, use within a day of thawing)
Oxygen scavenger:	10 μl, (Yildiz et al., 2003) (−80 °C long term, 4 °C short term)
Glucose:	100 μl, 120 mg/ml in water (−20 °C long term)
ATP:	10 μl, 150 mM in water (−80 °C)
Kinesin:	5 μl, ~1 μM in AB (−80 °C, use within a day of thawing)

The oxygen scavenging solution is prepared as described in Yildiz et al. using glucose oxidase and catalase.

2.2.1 Polymerization of fluorescently labeled microtubules

The purpose of this protocol is to polymerize microtubules with a fluorescent GTP analog bound permanently between the β-tubulin of one dimer and α-tubulin of the next dimer. The target fluorophore density is approximately one fluorophore per micrometer of microtubule length. The steps of this part of the procedure follow:

1. Remove an aliquot of tubulin from the −80° C freezer, thaw rapidly in hand (37 °C) until just thawed, and then place on ice for 5 min.
2. Spin at 14k RPM for 20 min to remove insolubles.
3. Remove supernatant, add to 50 μl glycerol–MT buffer to make 100 μl, and mix gently with pipette.
4. Divide 100 μl into two 50 μl aliquots in 1.5-ml microcentrifuge tubes.
5. Dilute GTP to 10 mM in MilliQ water or AB (e.g., 14 μl buffer to 1 μl GTP).
6. To one tube, add 0.5 μl fluorescent GTP and 1.0 μl 10 mM GTP to make 1:20 dye-GTP MTs.

7. To the second tube, add 0.75 μl GTP for control, if needed.
8. Wrap dye-GTP tube in aluminum foil to keep light out.
9. Incubate both tubes at 37 °C for 30 min.
10. Add 0.5 μl paclitaxel to each, mix well with pipette (40 μM paclitaxel final).
11. Incubate both tubes at 37 °C for 20 min.
12. During last incubation, add 100 μl AB + sucrose to two 1.5-ml microcentrifuge tubes, warm to room temperature, and add 1 μl paclitaxel, mixing well.
13. Very carefully layer MTs on top of sucrose cushion with pipette.
14. Spin at room temperature at 14k RPM in microcentrifuge for 20 min.
15. Carefully aspirate supernatant, changing tips often to reduce residual dye. Once dye is gone, before cushion is gone, rinse cushion once with AB and then aspirate cushion as well.
16. Gently rinse pellet with 50 μl room temperature AB + paclitaxel.
17. Resuspend pellet in 50 μl room temperature, ~5 mg/ml final concentration.
18. Store, light protected, at room temperature for up to 2 weeks.

If fluorophore density on the surface is too low (one or two fluorophores per microtubule) or too high (individual fluorophores cannot be resolved), adjust fluorescent GTP concentration in step 6.

2.2.2 Microtubule gliding assay

The microtubule gliding assay requires stock solutions prepared on the day of the experiment, as described below. These stock solutions will generally keep on ice for that day but should be discarded at the end of the day. Of particular concern are solutions containing DTT and ATP. Finally, the solutions should all be warmed to room temperature before using in the gliding assay since microtubules depolymerize more rapidly at cold temperatures.

AB + 2 mM DTT:	1 ml AB, 10 μl DTT. *Use as AB*
Biotin-LC-BSA:	70 μl, 1 mg/ml in AB (35 μl each of bio-BSA stock and AB)
BSA:	700 μl, 1 mg/ml in AB (694 μl AB + 6.0 μl BSA stock)
SA:	60 μl, 0.5 mg/ml in AB (57 μl AB + 3 μl SA stock)
CBAB:	375 μl, α-casein 1 mg/ml, BSA 1 mg/ml in AB (300 μl BSA and 75 μl α-casein stock)
Fluorescence anti-bleach (FAB):	100 μl CBAB + 1/3 μl oxygen scavenger, 1 μl β-ME, 1 μl paclitaxel, 2.5 μl glucose (add last)

In preparing for the gliding assay experiment, we prepare slides with a sequential coating from the glass slide up as follows. We begin with precleaned 24 × 60 mm microscope glass slides and create flow-lanes on the slides using either vacuum grease or double-sided tape. We cover these flow-lanes with 22 × 22 mm microscope cover slips. The ends of the lanes are left open so that new solutions can be washed through, using a Kim-wipe type tissue to wick out fluid from the opposite side of the flow lane. A particularly important side note: be sure no bubbles flow through the lanes; the surface-bound proteins must always remain in solution. We build the gliding assay by first coating the surface with biotinylated BSA, blocking the surface with BSA, adding SA to create a biotin–SA layer, removing excess SA, adding biotinylated kinesin to form a bio-SA–bio-kinesin layer, washing out free kinesin, and washing in microtubules in the appropriate solution. The steps to this protocol are as follows:

1. Construct flow cell with ~4 lanes.
2. Wash bio-BSA into each lane (10–15 μl/wash depending on size of lane).
3. Wash lane 3× with BSA (~15 μl/wash).
4. Wash SA into lane.
5. Wash 3× with BSA.
6. Wash 1× with CBAB.
7. Dilute kinesin 1:100 in CBAB, wash ~15 μl/lane, wait no less than 5 min (10 min seems optimal, but up to 2 hours works), and make FAB during this wait time.
8. Wash 1× with CBAB + 1/100 paclitaxel (40 μM paclitaxel).
9. Dilute microtubules 1:100–1:1000 in FAB + ATP (~1 mM works well), wash in, and observe.

We observe the fluorescently labeled microtubules using a home-made TIRF microscope (Friedman, Chung, & Gelles, 2006); however, any microscope capable of imaging single fluorophores with approximately 20 nm precision at 200 ms will work. At 1 mM ATP, kinesin propels microtubules at approximately 0.5 μm/s, depending on the temperature of the experiment. We observe microtubules in a 50 × 50-μm field of view, so an individual microtubule is observable for approximately 100 s. Using an EM-CCD camera (Andor iXon DV-897), we image the microtubules at 5 Hz, choosing an illumination intensity that does not photobleach the fluorophores over the 100 s of observation. The typical fluorophore intensity we observe is approximately 1000 photons per fluorophore per 200-ms image. See Fig. 2.3A for a typical image of microtubules in the gliding assay.

2.2.3 Image analysis and persistence length calculation

The first step in the analysis is extracting the trajectories of individual fluorophores from the sequence of 500 images (100 s, 5 Hz) collected in the Section 2.2. We use a modified version of software developed by Crocker and Grier (1996) for this analysis, which results in trajectories for each fluorophore (i.e., an ordered set of positions with time for each position). As an aside, an excellent web tutorial on particle tracking, useful for exactly this analysis, is maintained by Weeks (2013). If this resource is accessible, it provides an ideal gateway into single fluorophore tracking.

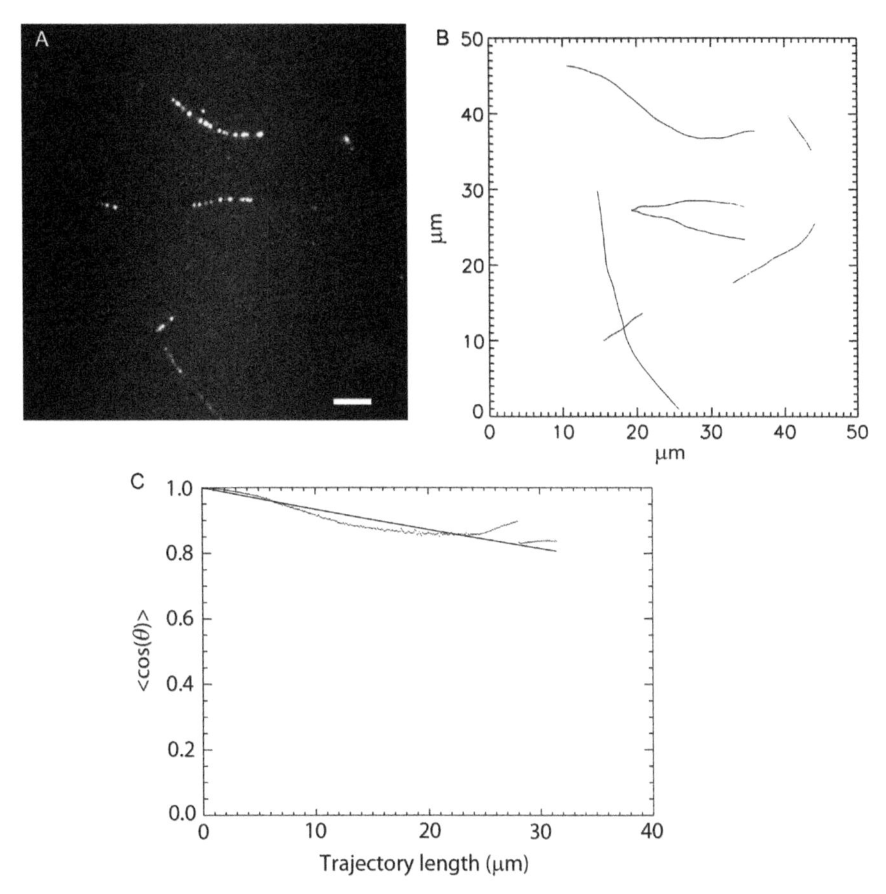

FIGURE 2.3

Gliding assay calculation of microtubule trajectories and persistence length. (A) TIRF image of microtubules sparsely decorated with fluorescent GTP analogs, taken following Section 2.2. Scale bar is 5 μm. (B) Microtubule trajectories from (A) reconstructed as described in Section 2.3. Where these trajectories appear to cross, timing information is associated with each trajectory, permitting unique trajectory identification. (C) Average of cos θ_s as a function of path length, *s*, (dots) and fit to Eq. (2.1) (solid line). The data shown are an average across all trajectories in (B). Fewer independent data for long trajectory lengths lead to statistical fluctuations. The persistence length for these microtubules is 150 μm, consistent with very short tip-length microtubules.

After individual fluorophore trajectories are calculated, the trajectories of all fluorophores on one microtubule are automatically combined into a single microtubule trajectory, with ordered positions and times. The software we have developed for this task simply takes one fluorophore trajectory and, for each point in that trajectory, searches for all other trajectories which come within a micron of that point—any trajectories meeting this criteria are candidate trajectories. Crossing trajectories, that is, those at angles greater than 5° to the original trajectory, are rejected. After running through all trajectories and combining those which meet the above criteria, unique microtubule trajectories are numbered, albeit with many overlapping fluorophores. These are thinned into one, averaged, ordered microtubule trajectory following Lee (2000). Example trajectories are shown in Fig. 2.3B.

We then calculate the tangent angle θ as a function of path length, s, along the trajectory. We compute the angle difference between all points separated by a given distance s and then compute the average of the cosines of these angle differences, again using custom software written in IDL. A fit of Eq. (2.1) to these cosine data returns the persistence length, as shown in Fig. 2.3C.

2.3 DISCUSSION

The protocol described in this chapter provides a fast and efficient method of measuring microtubule persistence lengths. We have found that we can measure the persistence lengths of hundreds of microtubules per experiment and finish all analysis within a few hours using a desktop computer. In our lab, the protocol has been used to measure the persistence length of microtubules as a function of length, by varying the density of kinesin on the surface. Simply changing the average concentration of kinesin added in step 7 of the microtubule gliding assay permits length-dependent experiments. Calibration of the density of kinesin is straightforward with low-density kinesins or very short microtubules (Duke et al., 1995): for very dilute surface coatings, the average trajectory length scales with microtubule length until the microtubule length is longer than the average spacing between kinesins.

In addition, our lab has also extended this technique to measure microtubule persistence length as a function of microtubule diameter. Microtubules with different diameters have different numbers of protofilaments (13 is typical for *in vitro* assembled microtubules, and in many *in vivo* contexts). Thirteen protofilament microtubules have protofilaments parallel to the long axis of the microtubule (Wade, Meurer-Grob, Metoz, & Arnal, 1998), while microtubules with different numbers of protofilaments have protofilaments skewed relative to the microtubule axis into a super-helical pitch, with a pitch length on the scale of micrometers. Because kinesin follows single protofilaments, microtubules with skewed protofilaments will rotate in a gliding assay (Marchuk, Guo, Sun, Vela, & Fang, 2012; Nitzsche, Ruhnow, & Diez, 2008). We use this analysis to correlate the rotation rate of microtubules (and, hence, the number of protofilaments and diameter) with the persistence length of individual microtubules.

These two examples indicate one advantage of the technique described above: it permits the connection of microtubule mechanical and structural properties. We anticipate, as well, that the use of a motility (gliding) assay at the same time as a mechanical (persistence length) assay will permit other experiments examining connections between motility and mechanical properties.

On the other hand, there are limitations to this technique. Microtubules must be stable on a time-scale of hundreds of seconds, meaning that they must be either stabilized by small molecules such as paclitaxel (as in this experiment) or slowly hydrolyzing GTP analogs (GMP-CPP or GTP-γS), or stabilized by high temperatures on a microscope designed to operate at 37 °C. In addition, this experiment requires that the microtubules support kinesin motility; hence, the uses of conditions which disrupt motility (such as certain MAPs or temperatures) are not appropriate for this experiment. Finally, while equipment required for single-fluorophore imaging is in wide use (in the form of TIRF and confocal microscopes), the up-front costs of such equipment compared to a standard fluorescence microscope can be substantial.

Ultimately, however, we hope that this combination of rapid experiments and analysis, which are reproducible from day to day, will make this technique useful to a range of workers studying biophysical properties of microtubules. Indeed, the analysis technique described here is not limited to microtubules driven by kinesin; actin polymers in myosin-driven gliding assays are amenable to exactly the same technique.

SUMMARY

We have described a technique to measure microtubule mechanical properties using kinesin-driven gliding assays. This technique is complementary to existing techniques relying on active forces or passive, nondriven, microtubule fluctuations. The key advantages it offers are a simple, well-characterized, biochemical system (kinesin motility); high precision through the tracking of single fluorophores; and the ability to combine mechanical and structural measurements on microtubules. The software necessary to perform this analysis is relatively straightforward to use and available from the author upon request.

Acknowledgments

We would like to thank Melissa Klocke for help preparing Fig. 2.2, and Anna Ratliff for substantially simplifying the gliding assay described in Section 2.2. This work was supported by the Research Corporation for Science Advancement.

References

Anderson, E. K., & Martin, D. S. (2011). A fluorescent GTP analog as a specific, high-precision label of microtubules. *BioTechniques*, *51*, 43–48.

Bathe, M., Heussinger, C., Claessens, M. M., Bausch, A. R., & Frey, E. (2008). Cytoskeletal bundle mechanics. *Biophysical Journal*, *94*, 2955–2964.

Berliner, E., Young, E. C., Anderson, K., Mahtani, H. K., & Gelles, J. (1995). Failure of a single-headed kinesin to track parallel to microtubule protofilaments. *Nature*, *373*, 718–721.

Brangwynne, C. P., Koenderink, G. H., Barry, E., Dogic, Z., MacKintosh, F. C., & Weitz, D. A. (2007). Bending dynamics of fluctuating biopolymers probed by automated high-resolution filament tracking. *Biophysical Journal*, *93*, 346–359.

Cassimeris, L., Gard, D., Tran, P. T., & Erickson, H. P. (2001). XMAP215 is a long thin molecule that does not increase microtubule stiffness. *Journal of Cell Science, 114*, 3025–3033.

Crocker, J. C., & Grier, D. G. (1996). Methods of digital video microscopy for colloidal studies. *Journal of Colloid and Interface Science, 179*, 298–310.

Duke, T., Holy, T. E., & Leibler, S. (1995). "Gliding assays" for motor proteins: A theoretical analysis. *Physical Review Letters, 74*, 330–333.

Felgner, H., Frank, R., Biernat, J., Mandelkow, E. M., Mandelkow, E., Ludin, B., et al. (1997). Domains of neuronal microtubule-associated proteins and flexural rigidity of microtubules. *The Journal of Cell Biology, 138*, 1067–1075.

Friedman, L. J., Chung, J., & Gelles, J. (2006). Viewing dynamic assembly of molecular complexes by multi-wavelength single-molecule fluorescence. *Biophysical Journal, 91*, 1023–1031.

Gittes, F., Mickey, B., Nettleton, J., & Howard, J. (1993). Flexural rigidity of microtubules and actin filaments measured from thermal fluctuations in shape. *The Journal of Cell Biology, 120*, 923–934.

Hawkins, T., Mirigian, M., Selcuk Yasar, M., & Ross, J. L. (2010). Mechanics of microtubules. *Journal of Biomechanics, 43*, 23–30.

Heussinger, C., Bathe, M., & Frey, E. (2007). Statistical mechanics of semiflexible bundles of wormlike polymer chains. *Physical Review Letters, 99*, 048101.

Janson, M. E., & Dogterom, M. (2004). A bending mode analysis for growing microtubules: Evidence for a velocity-dependent rigidity. *Biophysical Journal, 87*, 2723–2736.

Kikumoto, M., Kurachi, M., Tosa, V., & Tashiro, H. (2006). Flexural rigidity of individual microtubules measured by a buckling force with optical traps. *Biophysical Journal, 90*, 1687–1696.

Lee, I. (2000). Curve reconstruction from unorganized points. *Computer Aided Geometric Design, 17*, 161–177.

Lodish, H., Berk, A., Kaiser, C. A., Krieger, M., Scott, M. P., Bretscher, A., et al. (2007). *Molecular cell biology* (6th ed.). New York: W. H. Freeman and Company.

Marchuk, K., Guo, Y., Sun, W., Vela, J., & Fang, N. (2012). High-precision tracking with non-blinking quantum dots resolves nanoscale vertical displacement. *Journal of the American Chemical Society, 134*, 6108–6111.

Mehrbod, M., & Mofrad, M. R. (2011). On the significance of microtubule flexural behavior in cytoskeletal mechanics. *PloS One, 6*, e25627.

Nitzsche, B., Ruhnow, F., & Diez, S. (2008). Quantum-dot-assisted characterization of microtubule rotations during cargo transport. *Nature Nanotechnology, 3*, 552–556.

Pampaloni, F., Lattanzi, G., Jonas, A., Surrey, T., Frey, E., & Florin, E. L. (2006). Thermal fluctuations of grafted microtubules provide evidence of a length-dependent persistence length. *Proceedings of the National Academy of Science of United States of America, 103*, 10248–10253.

Phillips, R., Kondev, J., & Theriot, J. (2008). *Physical biology of the cell*. New York: Garland Science.

Portran, D., Zoccoler, M., Gaillard, J., Stoppin-Mellet, V., Neumann, E., Arnal, I., et al. (2013). MAP65/Ase1 promote microtubule flexibility. *Molecular Biology of the Cell, 24*, 1964–1973.

Taute, K. M., Pampaloni, F., Frey, E., & Florin, E. L. (2008). Microtubule dynamics depart from the wormlike chain model. *Physical Review Letters, 100*, 028102.

Topalidou, I., Keller, C., Kalebic, N., Nguyen, K. C., Somhegyi, H., Politi, K. A., et al. (2012). Genetically separable functions of the MEC-17 tubulin acetyltransferase affect microtubule organization. *Current Biology*, *22*, 1057–1065.

Tounsi, A., Heireche, H., Benhassaini, H., & Missouri, M. (2010). Vibration and length-dependent flexural rigidity of protein microtubules using higher order shear deformation theory. *Journal of Theoretical Biology*, *266*, 250–255.

Valdman, D., Atzberger, P. J., Yu, D., Kuei, S., & Valentine, M. T. (2012). Spectral analysis methods for the robust measurement of the flexural rigidity of biopolymers. *Biophysical Journal*, *102*, 1144–1153.

van den Heuvel, M. G., Bolhuis, S., & Dekker, C. (2007). Persistence length measurements from stochastic single-microtubule trajectories. *Nano Letters*, *7*, 3138–3144.

Van den Heuvel, M. G., de Graaff, M. P., & Dekker, C. (2008). Microtubule curvatures under perpendicular electric forces reveal a low persistence length. *Proceedings of the National Academy of Science of United States of America*, *105*, 7941–7946.

van Mameren, J., Vermeulen, K. C., Gittes, F., & Schmidt, C. F. (2009). Leveraging single protein polymers to measure flexural rigidity. *The Journal of Physical Chemistry. B*, *113*, 3837–3844.

Wade, R. H., Meurer-Grob, P., Metoz, F., & Arnal, I. (1998). Organisation and structure of microtubules and microtubule-motor protein complexes. *European Biophysics Journal*, *27*, 446–454.

Weeks, E. (2013). Particle tracking using IDL. http://www.physics.emory.edu/~weeks/idl/ Accessed 15.05.13.

Yildiz, A., Forkey, J. N., McKinney, S. A., Ha, T., Goldman, Y. E., & Selvin, P. R. (2003). Myosin V walks hand-over-hand: Single fluorophore imaging with 1.5-nm localization. *Science*, *300*, 2061–2065.

CHAPTER

3 Structural Studies of the Doublecortin Family of MAPs

Franck Fourniol*,[1], Mylène Perderiset†, Anne Houdusse† and Carolyn Moores*

**Institute of Structural and Molecular Biology, Birkbeck College, London, United Kingdom*

†Structural Motility, Institut Curie, Centre National de la Recherche Scientifique, Unité Mixte de Recherche 144, Paris Cedex 05, France

CHAPTER OUTLINE

Introduction **28**

3.1 Rationale **30**

3.2 Methods **32**

3.2.1 Expression and Purification of Recombinant Human DCX 32

3.2.1.1 Introduction 32

3.2.1.2 Expression of DCX 33

3.2.1.3 Purification of DCX 33

3.2.2 Cryo-ET and Evaluation of DCX–MT Architecture 37

3.2.2.1 Introduction 37

3.2.2.2 Sample Preparation 37

3.2.2.3 Data Collection 38

3.2.2.4 Structure Determination 38

3.2.3 Cryo-EM and Subnanometer Resolution 3D Structure Determination 39

3.2.3.1 Introduction 39

3.2.3.2 Sample Preparation and Data Collection 39

3.2.3.3 Setup and Automated Processing 39

3.2.3.4 Initial 3D Model and Reference Projections 39

3.2.3.5 Finding the Seam and Image Selection 39

3.2.3.6 Reconstructions of DCX–K–MTs 40

3.2.3.7 Pseudo-atomic Model Building 40

3.2.3.8 Insight into MT Structure in the Absence of Stabilizing Drugs 42

[1]Current address: Cancer Research UK, London Research Institute, Lincoln's Inn Fields Laboratories, 44 Lincoln's Inn Fields, London WC2A 3LY, United Kingdom.

Methods in Cell Biology, Volume 115

ISSN 0091-679X

http://dx.doi.org/10.1016/B978-0-12-407757-7.00003-7

3.3 Discussion 43
Acknowledgments 44
References 45

Abstract

Doublecortin (DCX) is a microtubule (MT)-stabilizing protein essential for neuronal migration during human brain development. Missense mutations in DCX cause severe brain defects. This implies that the many other MT-stabilizing proteins in neurons cannot compensate for DCX function. To understand the unusual properties of DCX, we expressed the recombinant human DCX in Sf9 cells and undertook structural characterization of its interaction with MTs using cryo-electron microscopy. DCX specifically nucleates 13-protofilament (13-pf) MTs, the architecture of human MTs *in vivo*. Cryo-electron tomography (cryo-ET) of DCX-nucleated MTs showed that they are primarily built from B-lattice contacts interrupted by a single discontinuity, the seam. Because of this asymmetry, we used single-particle reconstruction and determined the 8 Å structure of DCX-stabilized 13-pf MTs in the absence of a stabilizing drug. The DCX-binding site, at the corner of four tubulin dimers, is ideally suited to stabilize both lateral and longitudinal tubulin lattice contacts. Its precise geometry suggests that DCX is sensitive to the angle between pfs, and thereby provides insight into the specificity of DCX for 13-pf MT architecture. DCX's precise interaction at the corner of four tubulin dimers also means that DCX does not bind the MT seam. Our work has provided mechanistic insight into the evolutionarily conserved DCX family of MT-stabilizing proteins and also into more general regulatory mechanisms of the MT cytoskeleton.

INTRODUCTION

The neuronal microtubule (MT)-associated protein (MAP) doublecortin (DCX) was first characterized in 1998, when it was discovered that mutations in the X-linked doublecortin (*DCX*) gene cause a syndrome called double cortex (or subcortical band heterotopia, SBH) in females and lissencephaly in males (des Portes et al., 1998; Gleeson et al., 1998). Lissencephaly and SBH patients exhibit a range of symptoms—including epilepsy, intellectual impairment, and infant death—that result from abnormal development of the cerebral cortex. Mutation of the DCX protein causes defective neuronal migration, such that the precisely structured layers of the cortex are poorly organized. Because the *DCX* gene is on the X chromosome, females with a *DCX*± genotype exhibit random inactivation of one of the two X chromosomes; this ensures that half of the cells have a functional copy of the gene and migrate correctly into a layered cortex. In contrast, the other half lack a functional copy and subsequently

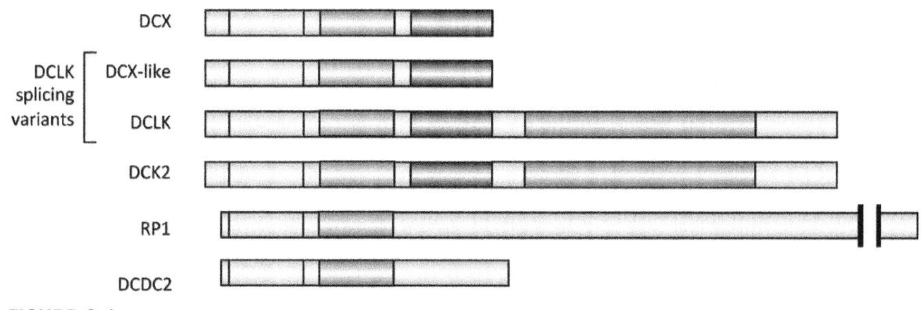

FIGURE 3.1

Major human paralogs of doublecortin. The paralogs illustrated here all have two DC domains (N-DC depicted in yellow, C-DC in orange). Two of the four isoforms of DCLK are represented: DCX-like and full-length DCLK, which has a kinase domain (green). DCLK full-length has a very close homolog called DCK2. The serine/proline rich domain is colored blue. (See color plate.)

stop half-way through their journey through the developing cortex, creating a heterotopic band of gray matter lying between the cortex and the ventricle, the so-called double cortex. *DCX* y/- patients possess no functional copy of the DCX protein, resulting in lissencephaly, in which their cortex is abnormally thick and composed of four poorly ordered layers, and these patients exhibit more severe symptoms (Gleeson, 2000). Thus, DCX is essential for migration and differentiation in human neurons. It is enriched at the distal ends of neuronal processes and may regulate MTs in response to extracellular signals in these distal zones to facilitate path finding during development (Francis et al., 1999; Tint, Jean, Baas, & Black, 2009). DCX is also essential for neurogenesis in the adult brain (Jin, Wang, Xie, Mao, & Greenberg, 2010).

DCX is now known to be a member of a much larger family of MAPs involved in MT regulation during cell division, migration, and differentiation (Fig. 3.1). DCX–MAPs are widely distributed evolutionarily and are often essential (Reiner et al., 2006): in worms, the DCX–MAP ZYG-8 is important for MT stabilization during asymmetric cell division in embryos (Gönczy et al., 2001), and in flies, a DCX–MAP is essential for the development of mechanotransduction machinery (Bechstedt et al., 2010), demonstrating the conservation and importance of these MAPs. Doublecortin-like kinase (DCLK), initially known as doublecortin and CaM kinase-like 1 (DCAMKL1) (Burgess, Martinez, & Reiner, 1999), is the closest homologue to human DCX. It has several splicing isoforms, with a DCX-like isoform that is 72% identical to DCX. Full-length (FL) DCLK is a 729-amino-acid protein, with a C-terminal serine/threonine–protein kinase domain, similar to CaM kinase II. The similarity of DCX and DCLK is highlighted by their functional compensation in mice, where the activity of both proteins must be perturbed to recapitulate the severity of the human lissencephaly phenotype (Deuel et al., 2006; Kerjan et al., 2009; Koizumi, Tanaka, & Gleeson, 2006). In humans, DCX paralogs also include RP1, a protein that is mutated in retinitis pigmentosa (a common form of inherited blindness) and whose MT-stabilizing activity is essential for photoreceptor cell development (Liu, Zuo, & Pierce, 2004). In addition, *Dcdc2*, another DCX–MAP encoding gene, is linked with developmental dyslexia (Meng et al., 2005), and its protein product is involved with the control of primary cilia size and activity in neurons (Massinen et al., 2011).

The DCX 366-amino-acid sequence (40 kDa) shows no homology to other classical neuronal MAPs. In fact, sequence analysis reveals that DCX is built from an N-terminal tandem of homologous 90-amino-acid (11 kDa) domains which were accordingly named DC domains; these domains are separated by a well-conserved but presumed unstructured linker and are followed by a presumed intrinsically unstructured C-terminal serine/proline-rich (S/P-rich) domain (Fig. 3.1). Point mutations causing lissencephaly cluster within the two DC domains and modify the DCX–MT interaction in transfected cells (Bahi-Buisson et al., 2013; Sapir et al., 2000; Taylor, Holzer, Bazan, Walsh, & Gleeson, 2000). These data established the DC domains of DCX as MT-binding domains and reinforced the importance of DCX–MT binding during neuronal cortical migration.

An NMR study revealed the solution, β-grasp-like structure of the recombinant N-terminal DC domain (N-DC), and initiated the exploration of structure–function relationships of disease-causing point mutations in this domain (Kim et al., 2003). This analysis allowed discrimination between mutations of buried residues—affecting folding and stability of the protein—and of surface residues affecting putative interactions with DCX's binding partners, including tubulin and MTs. However, an equivalent study of C-DC and of longer constructs proved technically challenging, leaving many aspects of DCX's unique MT-binding mechanism unresolved.

3.1 RATIONALE

We were fascinated by the idea that although neurons are full of MT-stabilizing proteins, DCX has unique properties that cannot be functionally compensated for by other neuronal MAPs. To address this mystery, we used established structural approaches for studying MAPs (Amos & Hirose, 2007; Hoenger & Gross, 2008).

Our early work revealed several distinctive properties of DCX and its interaction with MTs (Moores et al., 2004, 2006). An emerging area of interest is the sensitivity of MAPs to tubulin-bound nucleotide (Maurer, Bieling, Cope, Hoenger, & Surrey, 2011; Zanic, Stear, Hyman, & Howard, 2009); however, we found that DCX stabilizes MTs independent of the bound nucleotide, it does not affect the intrinsic GTPase of tubulin polymer, and the lattice parameters of the DCX-stabilized MTs reflect the nucleotide that is bound in the lattice (Fourniol et al., 2010). In our hands, DCX does not enhance MT growth rates but blocks depolymerization (Moores et al., 2006). All of these effects occur at substoichiometric or close to stoichiometric ratios of DCX:tubulin and appear to operate independent of MT bundling, which is the primary and apparently nonspecific outcome of DCX–MT interactions at superstoichiometric ratios of DCX both *in vivo* and *in vitro* (Sapir et al., 2000).

One of the most striking properties of DCX is its ability to specifically nucleate and bind MTs with a 13-protofilament (13-pf) architecture (Bechstedt & Brouhard, 2012; Moores et al., 2004). This is particularly important because although the number of pfs within MTs polymerized *in vitro* varies between 8 and 19, the almost exclusive architecture found *in vivo* is the 13-pf MT (McIntosh, Morphew, Grissom, Gilbert, & Hoenger, 2009; Tilney et al., 1973; Wade & Chrétien, 1993). Cryo-electron microscopy (cryo-EM) continues to be an invaluable method for elucidating mechanism and function in the MT cytoskeleton and its binding partners (Amos & Hirose, 2007; Hoenger & Gross, 2008), and the stabilization mechanism and architecture specificity of DCX were in large part explained by cryo-EM reconstructions.

Our early, low-resolution (~30 Å) reconstruction was calculated using helical analysis of a few, rare 14-pf paclitaxel-stabilized MTs that could be decorated with a truncated construct of DCX (t-DCX, DCX 1–275, lacking the S/P-rich domain; Moores et al., 2004). Because of its architectural preference, FL DCX could not be analyzed using this method. This first reconstruction showed a globular density wedged between pfs and in contact with four tubulin monomers. The DCX density corresponds to only one DC domain while additional weak density that may be attributable to the rest of the protein was observed at higher radius. It was immediately obvious that DCX's binding between pfs represents an excellent way to "staple" pfs together and thus increases MT stability. By binding at the junction between four tubulin monomers, DCX has the potential to strengthen both lateral contacts between pfs and also longitudinal contacts along the MT. In addition, this unique binding mode suggests the key to DCX's specificity for 13-pf MTs: the width of the inter-pf valley varies with pf number and DC domain may have evolved to fit this binding site in the 13-pf MT wall (Fig. 3.2A).

This binding site does not overlap with those of the MT-based motors kinesin and dynein and, thus, would be predicted not to impede movement of these motors (Fig. 3.2B). Indeed, in an ensemble gliding assay, DCX–MTs were found to support kinesin motility and with only slightly decreased speed (Moores et al., 2006). This suggests that, *in vivo*, MTs stabilized by DCX (and its relatives) may act as tracks for intracellular transport; the ramifications of this for cargo delivery control are an active area of ongoing research (Deuel et al., 2006; Liu et al., 2012).

Yet, after this initial work, several key questions remained and have been the focus of our more recent research efforts. For example, the biochemical stoichiometry of binding to tubulin for both t-DCX and FL DCX is estimated to be 1:1, suggesting that each density in our reconstruction corresponds to a single DC domain, with the rest of the molecule disordered and therefore invisible in the reconstruction. However, it is possible that adjacent binding sites are occupied by N-DC and C-DC from the same molecule—the inter-DC linker is long enough that either a longitudinal or lateral configuration is possible. Our hypothesis was that in working with t-DCX—in the absence of the S/P-rich domain that is critical for conferring MT architecture specificity—on non-13-pf MTs at low resolution, we were compromising our structural experiment. In addition, the low resolution of the reconstruction prevented us from determining whether DCX binds between two dimers or at the corner of four

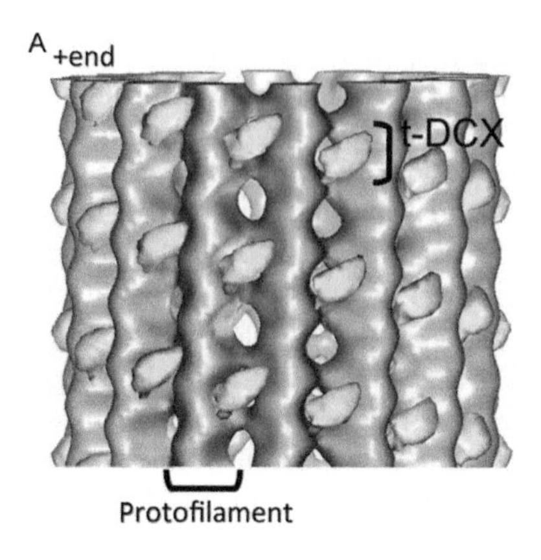

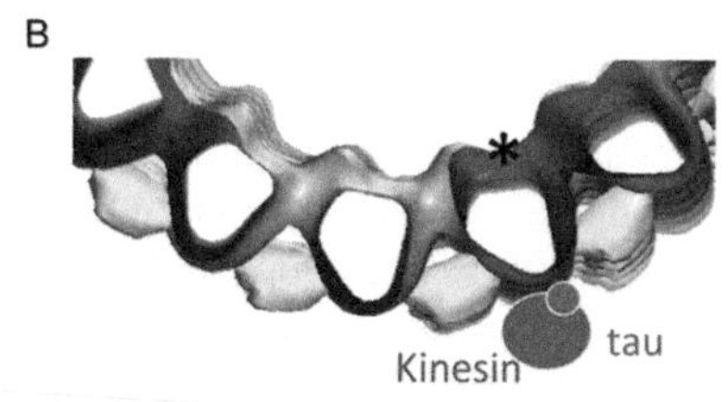

FIGURE 3.2

Low-resolution cryo-EM reconstruction reveals doublecortin's unique MT-binding site between protofilaments. (A) Front view of the 3D difference map showing density attributable to t-DCX (yellow), wedged between the pfs and over one of the two distinct fenestrations of the undecorated MT 3D map (blue). (B) View from the MT plus-end, illustrating how DCX fits into the inter-pf valley. The asterisk marks the paclitaxel-binding site. The green and pink densities show, respectively, kinesin- and MAP2/tau-binding sites on one pf. Scale bars = 20 Å (Moores et al., 2004). (See color plate.)

dimers. We also wanted to understand how DCX would stabilize MTs in the absence of the stabilizing drug paclitaxel. Therefore, we have focused our efforts on subnanometer resolution structural studies of 13-pf MTs nucleated and stabilized by FL DCX.

3.2 METHODS

3.2.1 Expression and purification of recombinant human DCX

3.2.1.1 *Introduction*

Early biochemical studies used recombinant DCX expressed in *E. coli* (Sapir et al., 2000; Taylor et al., 2000). However, when we started our work on DCX, our bacterially expressed protein produced primarily large 3D bundles of MTs that were impossible to study structurally, even at low-protein concentrations. On the other hand, DCX protein expressed in *Spodoptera frugiperda* insect cell cultures was stable and could be readily studied in the variety of ways described below. Others have had success subsequently with molecular studies of bacterially expressed DCX including recapitulation of DCX's selectivity for 13-pf MTs (Bechstedt & Brouhard, 2012), but the proteins' concentrations used were typically lower than required for our structural studies.

3.2.1.2 Expression of DCX

The 366-amino-acid isoform of human DCX was cloned in the Invitrogen Bac-to-Bac system by Dr Fiona Francis, Institut du Fer à Moulin, Paris. This isoform is a splicing variant from the 360-amino-acid form, containing five residues (GNDQD) inserted after residue 310 and one Val inserted after residue 342 (nomenclature of the 360-amino-acid form; NCBI Reference Sequence database); the significance of these differences is not understood. The engineered construct includes in its N-terminus a 6-Histidine affinity tag (His-tag), a TEV (tobacco etch virus) protease cleavage site (ENLYFQG, cleavage occurs between Q and G), and a Flag-tag (DYKDDDDK) used originally for immunofluorescence in transfected cells (Sapir et al., 2000).

The baculovirus genome containing the tagged DCX gene was transfected into Sf9 cells and amplified using standard methods (Bac-to-Bac Baculovirus Expression System, Life Technologies). Test expressions of DCX constructs were performed in small flasks; fewer cells with larger average diameters are indicative of baculovirus infection. Sf9 cells are usually round but the expression of DCX constructs causes their shape to change: they become lemon shaped or even grow a long cytoplasmic extension, which is presumably due to extensive MT bundling by DCX (up to 200 μm) (Fig. 3.3A).

For large-scale protein expression, cells were passaged to 450-mL cultures in 2.5-L spinners containing fresh medium, to a final concentration of 1×10^6 cells/mL. They were grown for 24 h, so that they reached $\sim 2 \times 10^6$ cells/mL at the infection time point 24 h later. In our hands, 50 mL P2 was added to a 450-mL culture to achieve a multiplicity of infection (MOI = number of virions/number of cells) of 3–10, which stops cell multiplication and induces protein expression. Cultures are incubated for 48 h at 27 °C. Cells are harvested by centrifuging in 1-L tubes at $700 \times g$ for 20 min at 4 °C in a Beckman Avanti J-20 I centrifuge. Pelleted cells were washed with 40 mL cold PBS per 1 L of culture and centrifuged in 50-mL tubes at $700 \times g$ another 20 min in an Eppendorf 5810R centrifuge at 4 °C. Pellets were frozen in liquid nitrogen and stored at −80 °C until protein extraction.

3.2.1.3 Purification of DCX

The protocol originally described in Moores et al. (2004) was modified slightly to improve the solubility and minimize protein degradation, together leading to better yields. All protein purification steps and centrifugations were performed at 4 °C to limit protease activity. Buffers used are listed in Table 3.1.

Extraction of cytoplasmic proteins is performed by incubating frozen cell pellets (typically 20 mL for 2-L cultures) with an equal volume of hypotonic lysis buffer

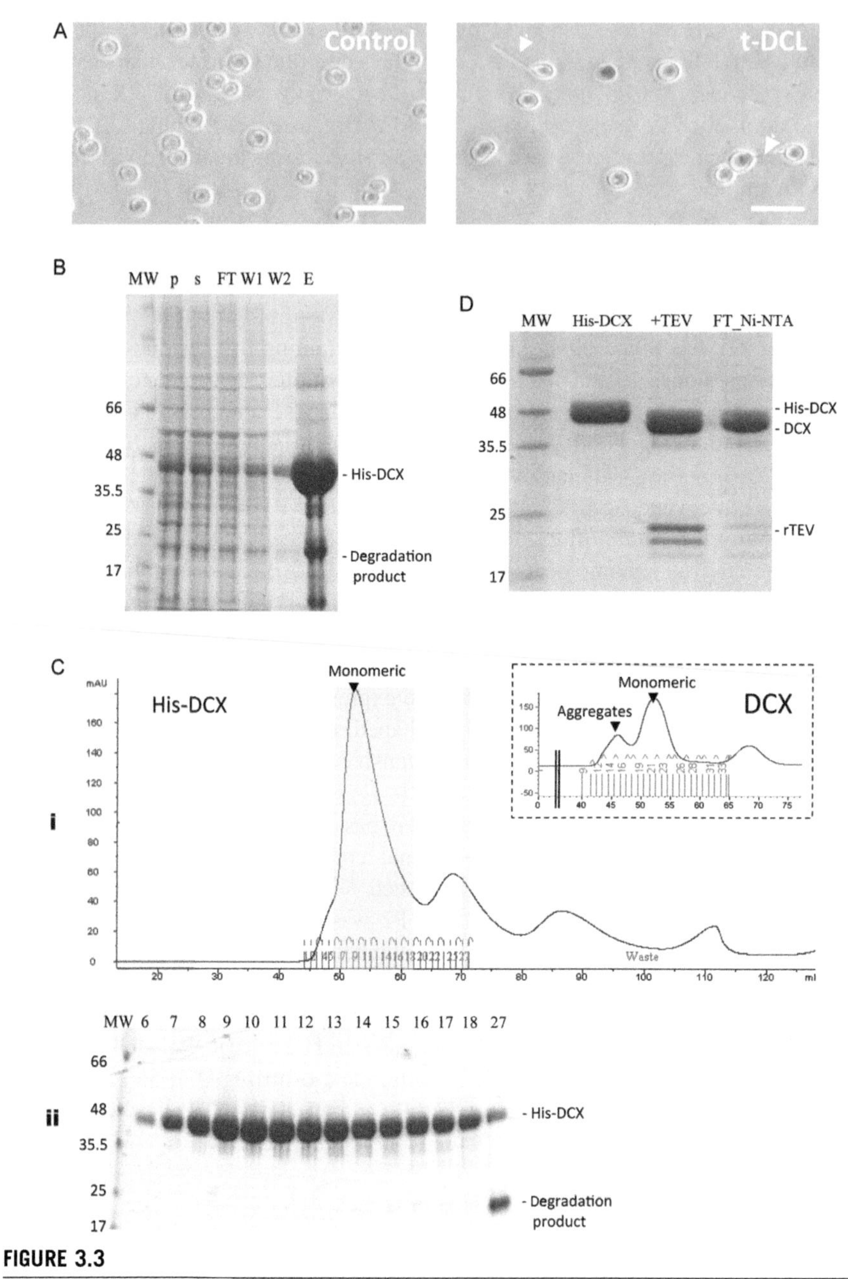

FIGURE 3.3

Expression and purification of DCX. (A) Sf9 cells observed by phase contrast light microscopy. Infected cells stop dividing and have a larger average diameter than control cells. Recombinant expression of constructs of human DCX and DCLK (t-DCL, truncated DCX-like) produces cytoplasmic elongations (arrows) in Sf9 cells, probably due

(lysis—low salt in Table 3.1), for 30 min. The combined action of defrosting, osmolysis, and detergent (Triton) lyses the plasma membrane and releases cytoplasmic content. Freeze–thawing cycles cannot be employed as DCX degrades. A number of inhibitors are used to limit the action of a broad spectrum of proteases, and β-mercaptoethanol is indispensible to increase the solubility of DCX, which contains nine cysteines.

To extract DCX from large MT bundles, the salt concentration is brought to ~350 mM NaCl by addition of high-salt lysis buffer and incubation for 15 min. To maximize extraction, the lysate is homogenized using a Dounce homogenizer (Wheaton) and the resulting lysate is centrifuged for 45 min at 45,000 × g to pellet cell debris and inclusion bodies (Fig. 3.3B). A lipid fraction is present at the top of the supernatant. The nonlipid supernatant contains soluble proteins and is removed for further purification.

His-tagged DCX is purified by affinity chromatography: 1 mL Ni-NTA resin (Qiagen) per 1 L of culture is equilibrated with wash buffer, then added to the soluble lysate, and incubated for 1 h on a rotating wheel at 4 °C. The relatively high imidazole concentration (50 mM) reduces nonspecific binding. The resin is then deposited on a small gravity column and the flow-through discarded (Fig. 3.3B). Two successive 5-column volume washes help minimize the contaminants bound to the resin. The His-DCX is eluted with 250 mM imidazole. This step both purifies and concentrates the protein of interest, yielding typically 20–30 mg in 3-mL buffer, starting from a 2-L culture. However, His-tagged degradation products cannot be separated, and other contaminants may remain.

The purified 42 kDa His-DCX is loaded on a HiLoad Superdex 75 16/60 column (GE Healthcare) and elutes with a peak at 53 mL, monitored by AKTA FPLC (GE Healthcare) (Fig. 3.3C). This step purifies the protein and allows buffer exchange but dilutes it ~10-fold, typically yielding 8–12 mg protein in 10-mL buffer.

---

to extensive MT bundling. Scale bars = 100 μm. (B) Extraction of DCX and initial purification of His-DCX by affinity chromatography. SDS-PAGE and Coomassie staining. MW, protein molecular weight marker (NEB); p and s, pellet and supernatant of cell lysate centrifuged 45 min at 45,000 × g; FT, flow-through of first Ni affinity column; W1, wash 50 mM imidazole; W2, second wash; E, elution 250 mM imidazole. (C) Gel filtration profile of His-DCX monitored by UV absorbance and SDS-PAGE. (i) UV absorbance read at the exit of an S75 16/60 size-exclusion column injected with 3 mL His-DCX Ni-NTA elution, run at 0.5 mL/min. The 1 mL fractions analyzed by SDS-PAGE are highlighted in yellow. The inset shows a gel filtration profile of cleaved DCX, with a bigger aggregates peak (first peak). (ii) SDS-PAGE and Coomassie staining of fractions 6–18 and 27. Fractions 7–17 were pooled. (D) TEV cleavage of the His-tag. SDS-PAGE and Coomassie staining. MW, protein marker; His-DCX, gel filtration pooled fraction; +TEV, same with addition of rTEV and overnight dialysis—observe the band shift; FT_Ni-NTA, flow-through of the second Ni affinity column which retains uncleaved protein and His-tagged rTEV. (See color plate.)

Table 3.1 List and composition of the buffers used for DCX purification

	Buffer (pH 7.2)	Composition
1	Lysis—low salt	40 mM HEPES 50 mM NaCl 50 mM Imidazole 10% Glycerol 1% Triton X-100 10 mM β-Mercaptoethanol 0.5 mM EDTA 1 mM PMSF One protease inhibitor tablet 0.015 mg/mL Aprotinin, leupeptin, pepstatin 0.15 mg/mL DNase I
2	Lysis—high salt	25 mM HEPES 1 M NaCl 50 mM Imidazole 10 mM β-Mercaptoethanol 0.5 mM EDTA 1 mM PMSF
3	50 mM Imidazole Ni-NTA wash	25 mM HEPES 300 mM NaCl 50 mM Imidazole 10 mM β-Mercaptoethanol 0.5 mM EDTA 1 mM PMSF
4	250 mM Imidazole Ni-NTA elution	25 mM HEPES 300 mM NaCl 250 mM Imidazole 0.5 mM EDTA 10 mM β-Mercaptoethanol
5	Gel filtration	25 mM HEPES 300 mM NaCl 0.5 mM EDTA 2 mM DTT

Finally, the His-tagged DCX is cleaved using recombinant tobacco etch virus protease (rTEV; gift from EMBL Heidelberg). Complete cleavage is achieved by adding 1% w/w rTEV (protease/fusion protein) and incubating at 4 °C overnight (Fig. 3.3D). A second Ni-NTA column (300 μL resin) retains uncleaved His-DCX as well as the rTEV; addition of 20 mM imidazole in the column buffer reduces unspecific interaction of the cleaved protein with the resin. In the original protocol, gel filtration was performed last but the modified order improved yields by reducing degradation and aggregation (see Fig. 3.3Cii). The His-tag may make the protein more soluble, while performing the gel filtration at an early stage helps eliminate contaminants including proteases that might cause both instability and degradation.

Finally, the ~10 mL dilute flow-through is concentrated in a 10,000-molecular-weight cut-off concentrator (Vivaspin 15R; Sartorius). Centrifugation at 4000 × *g*, always at 4 °C, for 1 h concentrates DCX approximately 5 × (~2 mL at 5 mg/mL). The concentrated protein is aliquoted and snap-frozen in liquid nitrogen and keeps for several years at −80 °C.

3.2.2 Cryo-ET and evaluation of DCX–MT architecture

3.2.2.1 Introduction

In the different MT architectures polymerized *in vitro*, lateral contacts between pfs are usually homotypic—α–α and β–β—forming a so-called B-lattice. However, heterotypic contacts—α–β and β–α, the so-called A-lattice—also occur (Wade & Chrétien, 1993). Although the precise *in vivo* significance for MT architecture is far from clear, crucially for our studies, 13-pf MTs show a line of discontinuity called the seam, where the dominant B-lattice contacts are disrupted by a line of A-lattice contacts (Kikkawa, Ishikawa, Nakata, Wakabayashi, & Hirokawa, 1994). However, 13-pf MTs built entirely of A-lattice contacts have also been reported (des Georges et al., 2008).

Subnanometer resolution structure determination by single-particle processing requires an initial 3D reference model, which implies some prior knowledge of the structure of interest. In our case, it was well established that copolymerized DCX–MTs are 13-pf, but it was unclear whether they were made of mainly A- or B-lattice contacts or a mixture. Although previous reports showed that truncated DCX bound to paclitaxel-stabilized MTs that are built of a helical B-lattice (Moores et al., 2004), we considered the hypothesis that DCX–MTs might contain a proportion of A-lattice contacts. Cryo-ET permits 3D reconstruction from a single macromolecular complex without averaging or model bias and is thus ideally suited to determine the lattice parameters of a population of MTs. McIntosh et al. (2009) successfully applied cryo-ET to MTs extracted from cells and decorated with a kinesin motor domain, and observed primarily B-lattice MTs. We took a similar approach, using $K340_{T93N}$, a 340-residue Thr93Asn mutant motor domain of rat conventional kinesin to emphasize the underlying lattice of our DCX–MTs (Crevel et al., 2004). This mutant has a very strong affinity for MTs facilitating full decoration but did not alter the stoichiometry of DCX binding (validated by cosedimentation assay, data not shown).

3.2.2.2 Sample preparation

DCX–MTs were polymerized by incubating bovine tubulin (Cytoskeleton, Inc.) and human recombinant DCX in equimolar amounts (10 μM) at 37 °C for 1 h. To maximize decoration of MTs with DCX and kinesin, DCX–MTs were diluted 1:1 in the low-ionic-strength buffer BRB20 (20 mM PIPES, pH 6.8, 1 mM EGTA, 1 mM $MgCl_2$) with 2 mM TCEP and adsorbed to glow-discharged lacey carbon grids (Agar). A second solution containing 30 μM DCX and 5 μM kinesin ($K340_{T93N}$) in BRB20, 2 mM TCEP was mixed 3:1 with colloidal gold (Sigma) and applied to the grids. Grids were then transferred into a Vitrobot (FEI) set to 37 °C and 100% humidity, to prevent evaporation and consequent changes in ionic strength. They were blotted for 2 s and instantaneously vitrified by rapid plunging into liquid ethane.

3.2.2.3 Data collection

For each tomogram, 61 images of the sample tilted from −60° to +60° were recorded on a 2k × 2k CCD (Gatan) on a Polara microscope (FEI Company) operating at 300 kV, at 6–8 μm defocus. Seven tomograms were reconstructed using the FEI software Inspect3D. Data collection and processing were performed with the generous help of Dr. Dan Clare (Birkbeck College).

3.2.2.4 Structure determination

A total of 5 μm of DCX–kinesin–MT tomographic reconstructions were visually inspected and all were found to be B-lattice 13-pf MTs (Fig. 3.4). This validated the use of B-lattice parameters in subsequent single-particle structure determination of DCX–MTs.

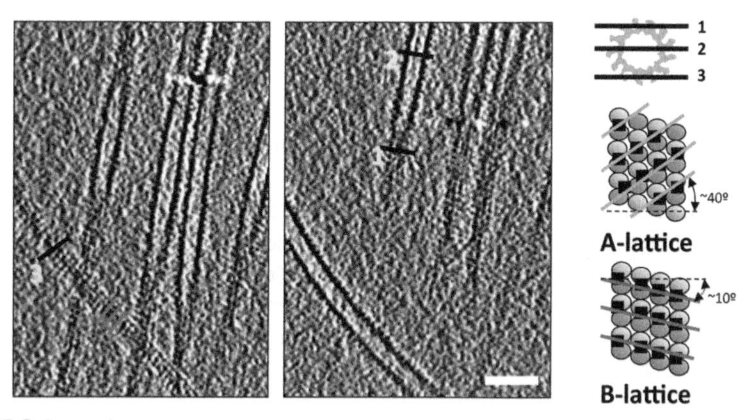

FIGURE 3.4

Cryo-electron tomography of DCX–K–MTs. Two slices of one tomogram containing longitudinal sections of MTs at different heights (1, 2, and 3), emphasized by a black line in the pictogram of an MT cross-section on the right. Tangential sections of decorated MTs (1 and 3) show a distinctive pattern of shallow stripes (red lines) due to the shallow helical path of kinesin motor domains in B-lattice MTs (see diagrams of A- and B-lattice on the right of the tomograms; McIntosh et al., 2009). Scale bar = 50 nm (Fourniol et al., 2010). (See color plate.)

3.2.3 Cryo-EM and subnanometer resolution 3D structure determination

3.2.3.1 Introduction

Using cryo-ET, we demonstrated that DCX promoted the assembly of B-lattice 13-pf MTs. These MTs contain a discontinuity—the seam—that means that they are not amenable to helical reconstruction. Instead 13-pf MTs can be reconstructed using so-called single-particle image processing. The method developed by Charles Sindelar (Yale, CT; Sindelar & Downing, 2007, 2010), referred to as "Chuff," yielded an ~8 Å resolution reconstruction of kinesin-decorated, DCX-stabilized 13-pf MTs, the same sample that we had validated by cryo-ET (Section 2.2) (referred to below as DCX–K–MTs; Fourniol et al., 2010).

3.2.3.2 Sample preparation and data collection

Sample preparation was as described in Section 2.2.2. Low-dose images were collected on a Tecnai F20 FEG electron microscope (FEI Company), operating at 200 kV, 50,000 ×, and 0.8–2.9 μm defocus. Micrographs were recorded on film (SO-163; Kodak) and were digitized (SCAI scanner; Carl Zeiss, Inc.) to a final sampling of 1.4 Å/pixel.

3.2.3.3 Setup and automated processing

Chuff is a fully automated set of scripts that simply require the user to set up the data and parameter files. To speed up image processing, and because finer sampling did not significantly improve resolution, micrographs digitized to 1.4 Å/pixel were binned by a factor of 2 to a final sampling of 2.8 Å/pixel. For our study, DCX–K–MTs were boxed with the EMAN Boxer program (option helix/normal; Ludtke, Baldwin, & Chiu, 1999) into particles/segments of 432 × 432 pixels, containing 7–8 tubulin dimer axial repeats. The input for the DCX–K–MT reconstruction was 4772 segments coming from 172 MTs and 63 cryo-electron micrographs.

3.2.3.4 Initial 3D model and reference projections

We used the default initial 3D model of kinesin-decorated 13-pf B-lattice MT generated by Chuff for alignment. In particular, the kinesin head density bound every dimer greatly facilitates identification of the position of the MT seam. The absence of DCX in the initial model gave us confidence that additional density observed in the output 3D reconstruction is indeed a faithful reflection of the MT-bound DCX in our sample. To further minimize the risk of model bias, high-resolution information was deleted using a low-pass filter with a 15 Å cut-off. Two-dimensional projections of the model with φ varying between 0° and 359°, θ between 75° and 105°, both with a 1° increment, resulted in a set of 11,160 references.

3.2.3.5 Finding the seam and image selection

In Chuff, a first round of reference-based alignment is performed in SPIDER (Frank et al., 1996), where MT segments are cross-correlated to the references. Each segment cut along one MT image is assigned a seam orientation (φ angle). Because the seam runs parallel to the axis of the MT, one would expect that the φ values assigned to segments from the same 13-pf MT would be very similar. Typically, however, they vary by multiples of ~28° (360/13) because cross-correlation scores computed between references whose seam is rotated by an integer numbers of pfs and relatively noisy cryo-EM images are very similar; hence, the best score does not always correspond to the actual orientation of the seam in the image (Sindelar & Downing, 2007). Alternatively, variations of the φ angle can arise from MTs with different architectures, and Chuff contains AWK scripts that analyze the output φ angles from reference-based alignment and decide whether each MT has consistent enough seam orientations (φ values) to be kept for the reconstruction. This decision depends on a number of parameters that can be defined by the user, in particular, the size of the angular window where φ values can be considered to be consistent between one another (default: 20°), and the minimum fraction of the boxes that must be in this window (default: 20%). The default selection parameters were successfully used to process DCX–K–MTs and approximately 90% of the dataset passed this selection process. If the MT is accepted, the angles assigned are edited so that all segments in the MT have φ values within the previously determined angular window. Subsequent rounds of reference-based alignment are restricted to this φ window and performed to refine alignment parameters and Euler angles before the final 3D reconstruction.

3.2.3.6 Reconstructions of DCX–K–MTs

Reconstruction of DCX–K–MTs with no symmetry imposed yielded a 13.5 Å resolution 3D map (FREALIGN option helical_subunits = 0 (Grigorieff, 2007); Fig. 3.5A; EMDB ID 1787). This asymmetric reconstruction confirmed that the MT (cyan) is occupied by both kinesin motor domain (red) and DCX (yellow), which binds at the corner of four tubulin dimers, stabilizing both lateral and longitudinal tubulin–tubulin contacts. Remarkably, DCX does not bind at the A-lattice seam of the 13-pf MT (right panel, arrow), but only at the 12 B-lattice inter-pf grooves (Fourniol et al., 2010).

To gain further insight, the 12 inter-pf valleys bound by DCX were averaged together. The averaging of approximately 168,000 decorated tubulin dimers generated an 8.2-Å resolution map (FREALIGN option helical_subunits = 12; Fig. 3.5B; EMDB ID 1788). At this resolution, secondary structures are resolved: alpha helices and beta sheets appear, respectively, as rods and sheets of density.

3.2.3.7 Pseudo-atomic model building

The detail in our 8.2 Å structure of DCX–K–MTs allowed us to dock atomic structures of each constituent subunit into our reconstruction in order to generate a pseudo-atomic model of DCX-stabilized MTs. Such modeling provides invaluable data about the binding interfaces between the components and the conformational changes they undergo in the context of the physiologically relevant macromolecular complex. We focus here on the interactions between DCX and tubulin; the implications of kinesin binding on DCX have recently been discussed elsewhere (Liu et al., 2012).

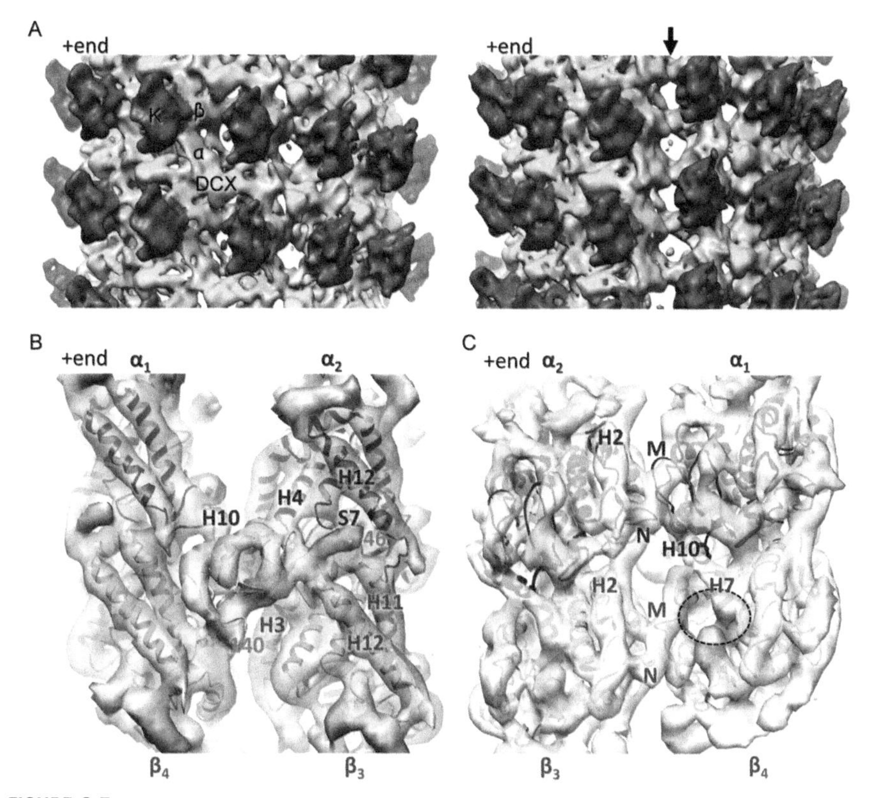

FIGURE 3.5

Structure of DCX–K–MTs determined using Chuff. (A) Asymmetric reconstruction of DCX–K–MTs (Fourniol et al., 2010; EMDB ID 1787). Side views, 180° apart, of a cryo-EM reconstruction of DCX–MTs decorated with kinesin motor domain, low-pass filtered with a 13 Å cut-off, density thresholded at 3σ. Each kinesin motor domain (red) binds one αβ-tubulin heterodimer (blue). DCX (yellow) binds at the interface between four tubulin dimers. The density for DCX is visible in the interprotofilament valleys all around the MT up to a threshold of 5σ, except at the seam (no density above 3σ; arrow, right panel). (B) Averaged view of DCX-binding site with the pseudo-atomic coordinates docked within the cryo-EM envelope. Tubulin secondary structural elements are labeled in blue, and the N- and C-terminal residues of the docked DCX coordinates are labeled in orange (EMDB ID 1788, PDB ID 2XRP; N-DC in gold, alpha-tubulin in blue, and beta-tubulin in cyan). (C) Averaged view of the inside surface of the DCX–MT reconstruction, highlighting the inter-pf lateral contacts that form in the MT wall. The empty paclitaxel-binding pocket is indicated (dotted circle). The chimeric lateral loops used to generate the α-tubulin pseudo-atomic model are shown in pink and tubulin secondary structural elements and loops are labeled in blue. (See color plate.)

UCSF Chimera (Pettersen et al., 2004) was used for visualization of 3D models and rigid-body fitting of atomic structures in the cryo-EM volumes. Several crystal structures of α- and β-tubulin (1JFF.pdb (Löwe, Li, Downing, & Nogales, 2001), 3HKE.pdb (Dorléans et al., 2009)) were fitted independently in their respective densities in the DCX–K–MT reconstruction because we wanted to evaluate the conformation of each monomer separately in our DCX-stabilized MTs formed in the absence of paclitaxel. The quality of the fits was assessed by calculating the cross-correlation between our structure and a simulated 8 Å model for each atomic structure: β-tubulin from 1JFF gave the best score of 0.651 in our β-tubulin density (3HKE β-tubulin chain B gave 0.599), α-tubulin from 1JFF and 3HKE chain A gave similar scores, respectively, 0.605 and 0.607, whereas a chimeric structure increased the cross-correlation slightly to 0.614 (3HKE chain A residues 31–61, 69–92, and 275–298, corresponding to the N, H2S3, and M loop regions, respectively, involved in lateral contact formation) were substituted into 1JFF chain A. To build the pseudo-atomic model of the DCX–MT complex, four tubulin subunits and the N-DC domain of DCX (1MJD.pdb, model 11, residues 46–140) were placed independently in the EM density using Chimera, and Flex-EM was used for the refinement of the multiple subunit fitting (Topf et al., 2008). The best multicomponent fit had a cross-correlation value of 0.819 (PDB ID 2XRP). This pseudo-atomic model allowed us to visualize the precision with which the DC domain contacts four tubulin dimers at its binding site and revealed several disease-causing mutations in both DCX and tubulin at their binding interface (Bahi-Buisson et al., 2013; Fourniol et al., 2010).

3.2.3.8 *Insight into MT structure in the absence of stabilizing drugs*

High-resolution information about MTs and straight pf structures has mainly been derived from paclitaxel-stabilized tubulin assemblies (Li, DeRosier, Nicholson, Nogales, & Downing, 2002; Löwe et al., 2001). In fact, paclitaxel appears to be a promiscuous stabilizer, binding MTs with varying pf numbers, sheets of antiparallel and inverted pfs (Nogales, Wolf, & Downing, 1998), or short-straight pfs in solution (Elie-Caille et al., 2007). These data imply that paclitaxel stabilizes a conformation of tubulin that favors straight pfs but that is not necessarily specific to MTs. However, because there has been no high-resolution model of MTs in the absence of paclitaxel, it has been unclear whether the conformation stabilized by the drug is a native state of tubulin or is induced by drug binding. We found that the paclitaxel-binding pocket is present in our reconstructions, demonstrating that paclitaxel stabilizes a native conformation of polymerized tubulin (Fig. 3.5C). Our data thus reveal a native structure of MTs bound solely by cellular ligands, where the straight pf structure appears virtually identical to that found in zinc-induced sheets.

Our structure confirms the importance of the lateral contacts formed by the M, N, and H2S3 loops previously visualized by Li et al. (2002) and more recently by Sui and Downing (2010). While the procedure employed by Downing and colleagues in these studies averaged together α- and β-tubulin, our reconstruction discriminates between them. This enabled us to clearly verify that α–α and β–β lateral contacts—and at a lower resolution, the seam α–β and β–α lateral contacts—have similar densities. This is despite the divergence of the M and N loop sequences between α- and β-tubulin.

These native tubulin–tubulin contacts within GDP MTs can now be compared with other MT reconstructions. We have recently visualized another structural state of MTs in the absence of stabilizing drugs, but containing the nonhydrolyzable GTP analog GTPγS and bound by another MAP—the fission yeast end binding protein (EB) Mal3 (Maurer, Fourniol, Bohner, Moores, & Surrey, 2012). GTPγS acts as a static mimic of the otherwise dynamic binding site that is specifically recognized by Mal3 and other EBs at growing MT ends (Maurer et al., 2011). The comparison of this MT end-like structure, with the GDP MT lattice structure in the presence of DCX, shows that growing MT ends possess an additional layer of lateral contacts at higher radius, which likely explains their action as a stabilizing cap (see also Yajima et al., 2012). Thus, the comparison of cryo-EM maps of MTs at secondary-structure resolution is revealing the structural basis of MT dynamic instability.

3.3 DISCUSSION

Cryo-EM has been an essential tool in shedding light on MT stabilization by DCX. Nevertheless, critical aspects of our current data continue to limit our understanding of the molecular mechanism of this essential protein. One intriguing aspect of our reconstructions continues to be that we currently only visualize a single DC-shaped density in our structures, that is, ~1/4 of the FL DCX molecule. This is consistent with the original low-resolution cryo-EM reconstruction of paclitaxel-stabilized MTs bound with the t-DCX construct (Fig. 3.2; Moores et al., 2004), thereby disproving the hypothesis that our ability to visualize the entire DCX molecule was limited by the original experimental conditions. In addition, our recent analysis of the structure of DCX–MTs in the absence of bound kinesin has revealed a specific conformation for the linker regions on either side of the bound DC domain that strongly suggests the density visualized in all our reconstructions corresponds to N-DC (Cierpicki et al., 2006; Liu et al., 2012).

Recent single-molecule studies have revealed the cooperative nature of MT binding by DCX, implying that DCX molecules contact each other when present at close to stoichiometric concentrations on the MT lattice (Bechstedt & Brouhard, 2012). Unfortunately, our structures currently provide no information about these interactions: N-DC—the only domain of DCX we have visualized—is not by itself sufficient either for nucleation, stabilization, or 13-pf specification, for which C-DC and the C-terminal domain are required (Kim et al., 2003; Moores et al., 2004; Sapir et al., 2000; Taylor et al., 2000). One explanation is that the samples used for our subnanometer resolution reconstructions are end-points of the DCX-mediated nucleation and stabilization process. Therefore, some of our current efforts are directed toward a structural understanding of early species in the process of

DCX-mediated MT nucleation. A parallel possibility is that the missing ~75% of the DCX molecules are not sufficiently ordered for us to visualize using our current averaging procedures. By computationally investigating the conformational variability of DCX within our samples, we aim to gain further insight into its location and structure, albeit with a likely compromise of resolution in the resulting reconstructions.

Based on the availability of the structure C-DC of DCDC2 (PDB ID 2DNF), it seems likely that the conserved DC fold is important for C-DC function and may involve tubulin interaction and control of DCX cooperativity (Bechstedt & Brouhard, 2012; Kim et al., 2003). However, structural insight into the C-terminal S/P-rich domain of DCX—previously shown to be essential for DCX's highly specific binding of 13-pf MTs (Moores et al., 2004)—remains lacking. As it is predicted to be disordered and highly charged due to phosphorylation (Reiner et al., 2004), one hypothesis is that the C-terminal domain does not make specific contact with tubulin but rather modulates the DCX–MT interaction indirectly. This would predict that single point mutations in the S/P-rich domain would only indirectly affect the interaction and in fact hardly any disease-causing mutations in this domain have been reported (Bahi-Buisson et al., 2013). To further test this hypothesis, it would be interesting to generate a DCX construct with a scrambled C-terminal sequence, keeping the same phosphorylation sites and overall charge.

Our structural insight into the DCX–MT interface provides a framework to localize and predict the severity of the gradually accumulating examples of disease-causing point mutations in DCX. Our results suggest that a spectrum of cellular effects would be seen, with some mutations totally disrupting the DCX–MT interaction, while others might have only minor effects on the MT interface and exert their mutagenic effects elsewhere within the network of DCX binding partners, including via intracellular trafficking control (Liu et al., 2012). It will be informative to continue to test the effects of these mutations *in vitro* to investigate their consequences on the DCX structure and function (Bechstedt & Brouhard, 2012). Clinically, the rarity of these mutations presents a challenge for establishing a statistically robust diagnostic link between genotype and phenotype; *in vitro* studies could provide functional parameters to predict the exact nature and severity of the disease based on the location and nature of the mutation (Bahi-Buisson et al., 2013). Multiple molecular pathways to disease may emerge from such studies with DCX forming a MT-based regulatory hub.

Acknowledgments

We thank Charles Sindelar (Yale University) for sharing his MT reconstruction scripts and members of the Birkbeck EM group for helpful discussions and advice. We are supported by The Wellcome Trust, New Life and Fédération pour la Recherche sur le Cerveau.

References

Amos, L. A., & Hirose, K. (2007). Studying the structure of microtubules by electron microscopy. *Methods in Molecular Medicine*, *137*, 65–91.

Bahi-Buisson, N., Souville, I., Fourniol, F. J., Toussaint, A., Moores, C. A., Houdusse, A., et al. (2013). New insights into genotype-phenotype correlations for the DCX-related lissencephaly spectrum. *Brain*, *136*, 223–244.

Bechstedt, S., Albert, J. T., Kreil, D. P., Müller-Reichert, T., Göpfert, M. C., & Howard, J. (2010). A doublecortin containing microtubule-associated protein is implicated in mechanotransduction in Drosophila sensory cilia. *Nature Communications*, *1*, 11.

Bechstedt, S., & Brouhard, G. J. (2012). Doublecortin recognizes the 13-protofilament microtubule cooperatively and tracks microtubule ends. *Developmental Cell*, *23*, 181–192.

Burgess, H. A., Martinez, S., & Reiner, O. (1999). KIAA0369, doublecortin-like kinase, is expressed during brain development. *Journal of Neuroscience Research*, *58*, 567–575.

Cierpicki, T., Kim, M. H., Cooper, D. R., Derewenda, U., Bushweller, J. H., & Derewenda, Z. S. (2006). The DC-module of doublecortin: Dynamics, domain boundaries, and functional implications. *Proteins*, *64*(4), 874–882.

Crevel, I. M., Nyitrai, M., Alonso, M. C., Weiss, S., Geeves, M. A., & Cross, R. A. (2004). What kinesin does at roadblocks: The coordination mechanism for molecular walking. *EMBO Journal*, *23*(1), 23–32.

des Georges, A., Katsuki, M., Drummond, D. R., Osei, M., Cross, R. A., & Amos, L. A. (2008). Mal3, the Schizosaccharomyces pombe homolog of EB1, changes the microtubule lattice. *Nature Structural & Molecular Biology*, *15*(10), 1102–1108.

des Portes, V., Pinard, J. M., Billuart, P., Vinet, M. C., Koulakoff, A., Carrie, A., et al. (1998). A novel CNS gene required for neuronal migration and involved in X-linked subcortical laminar heterotopia and lissencephaly syndrome. *Cell*, *92*, 51–61.

Deuel, T. A., Liu, J. S., Corbo, J. C., Yoo, S. Y., Rorke-Adams, L. B., & Walsh, C. A. (2006). Genetic interactions between doublecortin and doublecortin-like kinase in neuronal migration and axon outgrowth. *Neuron*, *49*, 41–53.

Dorléans, A., Gigant, B., Ravelli, R. B., Mailliet, P., Mikol, V., & Knossow, M. (2009). Variations in the colchicine-binding domain provide insight into the structural switch of tubulin. *Proceedings of the National Academy of Sciences of the United States of America*, *106*, 13775–13779.

Elie-Caille, C., Severin, F., Helenius, J., Howard, J., Muller, D. J., & Hyman, A. A. (2007). Straight GDP-tubulin protofilaments form in the presence of taxol. *Current Biology*, *17*, 1765–1770.

Fourniol, F. J., Sindelar, C. V., Amigues, B., Clare, D. K., Thomas, G., Perderiset, M., et al. (2010). Template-free 13-protofilament microtubule-MAP assembly visualized at 8 A resolution. *The Journal of Cell Biology*, *191*(3), 463–470.

Francis, F., Koulakoff, A., Boucher, D., Chafey, P., Schaar, B., Vinet, M. C., et al. (1999). Doublecortin is a developmentally regulated, microtubule-associated protein expressed in migrating and differentiating neurons. *Neuron*, *23*, 247–256.

Frank, J., Radermacher, M., Penczek, P., Zhu, J., Li, Y., Ladjadj, M., et al. (1996). SPIDER and WEB: Processing and visualization of images in 3D electron microscopy and related fields. *Journal of Structural Biology*, *116*, 190–199.

Gleeson, J. G. (2000). Classical lissencephaly and double cortex (subcortical band heterotopia): LIS1 and doublecortin. *Current Opinion in Neurology*, *13*, 121–125.

Gleeson, J. G., Allen, K. M., Fox, J. W., Lamperti, E. D., Berkovic, S., Scheffer, I., et al. (1998). Doublecortin, a brain-specific gene mutated in human X-linked lissencephaly and double cortex syndrome, encodes a putative signalling protein. *Cell*, *92*, 63–72.

Gönczy, P., Bellanger, J. M., Kirkham, M., Pozniakowski, A., Baumer, K., Phillips, J. B., et al. (2001). zyg-8, a gene required for spindle positioning in C. elegans, encodes a doublecortin-related kinase that promotes microtubule assembly. *Developmental Cell*, *1*, 363–375.

Grigorieff, N. (2007). FREALIGN: High-resolution refinement of single particle structures. *Journal of Structural Biology*, *157*, 117–125.

Hoenger, A., & Gross, H. (2008). Structural investigations into microtubule-MAP complexes. *Methods in Cell Biology*, *84*, 425–444.

Jin, K., Wang, X., Xie, L., Mao, X. O., & Greenberg, D. A. (2010). Transgenic ablation of doublecortin-expressing cells suppresses adult neurogenesis and worsens stroke outcome in mice. *Proceedings of the National Academy of Sciences of the United States of America*, *107*, 7993–7998.

Kerjan, G., Koizumi, H., Han, E. B., Dubé, C. M., Djakovic, S. N., Patrick, G. N., et al. (2009). Mice lacking doublecortin and doublecortin-like kinase 2 display altered hippocampal neuronal maturation and spontaneous seizures. *Proceedings of the National Academy of Sciences of the United States of America*, *106*, 6766–6771.

Kikkawa, M., Ishikawa, T., Nakata, T., Wakabayashi, T., & Hirokawa, N. (1994). Direct visualization of the microtubule lattice seam both in vitro and in vivo. *The Journal of Cell Biology*, *127*, 1965–1971.

Kim, M. H., Cierpicki, T., Derewenda, U., Krowarsch, D., Feng, Y., Devedjiev, Y., et al. (2003). The DCX-domain tandems of doublecortin and doublecortin-like kinase. *Nature Structural Biology*, *10*, 324–333.

Koizumi, H., Tanaka, T., & Gleeson, J. G. (2006). Doublecortin-like kinase functions with doublecortin to mediate fiber tract decussation and neuronal migration. *Neuron*, *49*, 55–66.

Li, H., DeRosier, D. J., Nicholson, W. V., Nogales, E., & Downing, K. H. (2002). Microtubule structure at 8 A resolution. *Structure*, *10*(10), 1317–1328.

Liu, J. S., Schubert, C. R., Fu, X., Fourniol, F. J., Jaiswal, J. K., Houdusse, A., et al. (2012). Molecular basis for specific regulation of neuronal kinesin-3 motors by doublecortin family proteins. *Molecular Cell*, *47*, 707–721.

Liu, Q., Zuo, J., & Pierce, E. A. (2004). The retinitis pigmentosa 1 protein is a photoreceptor microtubule-associated protein. *Journal of Neuroscience*, *24*, 6427–6436.

Löwe, J., Li, H., Downing, K. H., & Nogales, E. (2001). Refined structure of alpha beta-tubulin at 3.5 A resolution. *Journal of Molecular Biology*, *313*(5), 1045–1057.

Ludtke, S. J., Baldwin, P. R., & Chiu, W. (1999). EMAN: Semiautomated software for high-resolution single-particle reconstructions. *Journal of Structural Biology*, *128*, 82–97.

Massinen, S., Hokkanen, M. E., Matsson, H., Tammimies, K., Tapia-Páez, I., & Dahlström-Heuser, V. (2011). Increased expression of the dyslexia candidate gene DCDC2 affects length and signaling of primary cilia in neurons. *PLoS One*, *6*, e20580.

Maurer, S. P., Bieling, P., Cope, J., Hoenger, A., & Surrey, T. (2011). GTPgammaS microtubules mimic the growing microtubule end structure recognized by end-binding proteins (EBs). *Proceedings of the National Academy of Sciences of the United States of America*, *108*(10), 3988–3993.

Maurer, S. P., Fourniol, F. J., Bohner, G., Moores, C. A., & Surrey, T. (2012). EBs recognize a nucleotide-dependent structural cap at growing microtubule ends. *Cell*, *149*, 371–382.

McIntosh, J. R., Morphew, M. K., Grissom, P. M., Gilbert, S. P., & Hoenger, A. (2009). Lattice structure of cytoplasmic microtubules in a cultured Mammalian cell. *Journal of Molecular Biology*, *394*(2), 177–182.

Meng, H., Smith, S. D., Hager, K., Held, M., Liu, J., Olson, R. K., et al. (2005). DCDC2 is associated with reading disability and modulates neuronal development in the brain. *Proceedings of the National Academy of Sciences of the United States of America*, *102*, 18763.

Moores, C. A., Perderiset, M., Francis, F., Chelly, J., Houdusse, A., & Milligan, R. A. (2004). Mechanism of microtubule stabilisation by doublecortin. *Molecular Cell*, *18*, 833–839.

Moores, C. A., Perderiset, M., Kappeler, C., Kain, S., Drummond, D., Perkins, S. J., et al. (2006). Distinct roles of doublecortin modulating the microtubule cytoskeleton. *EMBO Journal*, *25*, 4448–4457.

Nogales, E., Wolf, S. G., & Downing, K. H. (1998). Structure of the alpha beta tubulin dimer by electron crystallography. *Nature*, *391*(6663), 199–203.

Pettersen, E. F., Goddard, T. D., Huang, C. C., Couch, G. S., Greenblatt, D. M., Meng, E. C., et al. (2004). UCSF Chimera—A visualization system for exploratory research and analysis. *Journal of Computational Chemistry*, *25*(13), 1605–1612.

Reiner, O., Coquelle, F. M., Peter, B., Levy, T., Kaplan, A., Sapir, T., et al. (2006). The evolving doublecortin (DCX) superfamily. *BMC Genomics*, *7*, 188.

Reiner, O., Gdalyahu, A., Ghosh, I., Levy, T., Sapoznik, S., Nir, R., et al. (2004). DCX's phosphorylation by not just another kinase (JNK). *Cell Cycle*, *3*, 747–751.

Sapir, T., Horesh, D., Caspi, M., Atlas, R., Burgess, H. A., Wolf, S. G., et al. (2000). Doublecortin mutations cluster in evolutionarily conserved functional domains. *Human Molecular Genetics*, *9*, 703–712.

Sindelar, C. V., & Downing, K. H. (2007). The beginning of kinesin's force-generating cycle visualized at 9-Å resolution. *The Journal of Cell Biology*, *177*, 377–385.

Sindelar, C. V., & Downing, K. H. (2010). An atomic-level mechanism for activation of the kinesin molecular motors. *Proceedings of the National Academy of Sciences of the United States of America*, *107*(9), 4111–4116.

Sui, H., & Downing, K. H. (2010). Structural basis of interprotofilament interaction and lateral deformation of microtubules. *Structure*, *18*(8), 1022–1031.

Taylor, K. R., Holzer, A. K., Bazan, J. F., Walsh, C. A., & Gleeson, J. G. (2000). Patient mutations in doublecortin define a repeated tubulin-binding domain. *Journal of Biological Chemistry*, *275*, 34442–34450.

Tilney, L. G., Bryan, J., Bush, D. J., Fujiwara, K., Mooseker, M. S., Murphy, D. B., et al. (1973). Microtubules: Evidence for 13 protofilaments. *The Journal of Cell Biology*, *59*, 267–275.

Tint, I., Jean, D., Baas, P. W., & Black, M. M. (2009). Doublecortin associates with microtubules preferentially in regions of the axon displaying actin-rich protrusive structures. *Journal of Neuroscience*, *29*, 10995–11010.

Topf, M., Lasker, K., Webb, B., Wolfson, H., Chiu, W., & Sali, A. (2008). Protein structure fitting and refinement guided by cryo-EM density. *Structure*, *16*(2), 295–307.

Wade, R. H., & Chrétien, D. (1993). Cryoelectron microscopy of microtubules. *Journal of Structural Biology*, *110*, 1–27.

Yajima, H., Ogura, T., Nitta, R., Okada, Y., Sato, C., & Hirokawa, N. (2012). Conformational changes in tubulin in GMPCPP and GDP-taxol microtubules observed by cryoelectron microscopy. *The Journal of Cell Biology*, *198*, 315–322.

Zanic, M., Stear, J. H., Hyman, A. A., & Howard, J. (2009). EB1 recognizes the nucleotide state of tubulin in the microtubule lattice. *PLoS One*, *4*(10), e7585.

CHAPTER

Detection and Quantification of Microtubule Detachment from Centrosomes and Spindle Poles

4

Anutosh Ganguly*, Hailing Yang† and Fernando Cabral‡

**Department of Physiology and Pharmacology, Snyder Institute for Chronic Disease, University of Calgary, Calgary, Alberta, Canada*

†*MD Anderson Cancer Center, Houston, Texas, USA*

‡*Department of Integrative Biology and Pharmacology, University of Texas Medical School, Houston, Texas, USA*

CHAPTER OUTLINE

Introduction **50**
4.1 Choice of Cell Line **51**
4.2 Microtubule Detachment in Spindles **52**
4.3 Labeling Microtubules **53**
4.3.1 Rhodamine Tubulin Microinjection 53
4.3.2 GFP-tubulin Expression 54
4.3.3 GFP–MAP4 Expression 54
4.4 Measuring Detachment **54**
4.4.1 Equipment 54
4.4.2 Cell Preparation 55
4.4.3 Data Collection 55
4.5 Microtubule Fragments as a Surrogate for Detachment **57**
4.6 Mechanism of Detachment **59**
Summary **60**
Acknowledgments **60**
References **60**

Abstract

Microtubule detachment from microtubule organizing centers is an important cellular process required for normal cell proliferation. When cells enter mitosis, microtubule turnover increases along with a concurrent increase in microtubule

Methods in Cell Biology, Volume 115
ISSN 0091-679X
http://dx.doi.org/10.1016/B978-0-12-407757-7.00004-9

detachment. MCAK, a kinesin-related protein whose abundance is highest during the early stages of mitosis, has been shown to regulate microtubule detachment. Abnormal increases or decreases in the frequency of detachment interfere with spindle function and inhibit cell division. It has been shown that drugs able to promote microtubule assembly (e.g., paclitaxel, epothilones) prevent cell division by suppressing microtubule detachment from centrosomes. Conversely, cytotoxic concentrations of microtubule destabilizing drugs (e.g., vinblastine, nocodazole), tubulin mutations that cause paclitaxel resistance, and specific β-tubulin isotypes increase the frequency of microtubule detachment. In this chapter, we describe a method to calculate the frequency of microtubule detachment by transfecting cells with EGFP–MAP4 and directly observing detachment by live cell imaging.

INTRODUCTION

During interphase, microtubules are nucleated at the centrosome, an organelle located near the nucleus. Their plus ends grow out toward the cell periphery, while their minus ends remain embedded in the centrosome, thereby establishing a cytoplasmic microtubule network. Despite their static appearance in photographs, microtubules are actually very dynamic structures that exhibit stochastic episodes of growth and shortening of their plus ends. This behavior, referred to as dynamic instability, allows microtubules to constantly probe their environment and remodel in response to changing cellular conditions and morphology (Kirschner & Mitchison, 1986; Mitchison & Kirschner, 1984). When cells enter mitosis, there is a dramatic rearrangement of the microtubules. The pair of centrosomes formed by duplication during interphase split apart and migrate to opposite sides of the cell to become spindle poles (Hinchcliffe & Sluder, 2001). The cytoplasmic microtubule complex disassembles and is reorganized into the mitotic spindle apparatus by nucleation of new microtubules at the spindle poles and the capture of some of these new filaments by kinetochores on the condensed chromosomes. When compared to interphase, spindle microtubules have a shorter average length and are more dynamic. The more rapid dynamics were initially discovered by measuring the time required for microinjected-labeled tubulin to incorporate into spindle and interphase microtubules and by fluorescence recovery after photobleaching (Saxton et al., 1984). The results were more recently extended by directly observing the behavior of microtubule plus ends in GFP-tubulin-expressing cells (Rusan, Fagerstrom, Yvon, & Wadsworth, 2001).

Given that they are composed of the same tubulin subunits that are present during interphase, spindle microtubule structural and behavioral changes are likely to result from differences in the rate of nucleation at the spindle poles versus the interphase centrosome (Piehl, Tulu, Wadsworth, & Cassimeris, 2004) as well as from the binding and activity of microtubule-interacting proteins (Jiang & Akhmanova, 2011). In addition, it has been shown that chromosomes are involved in nucleating microtubule assembly in a ran-dependent manner. This process plays a major role in the formation of meiotic spindles and may play a significant role in forming mitotic spindles as well (O'Connell & Khodjakov, 2007). More recently, it was reported that a protein complex called "augmin" is able to nucleate new microtubules away from the centrosome and may thus contribute to the high density of these polymers in the area between the spindle poles and chromosomes (Goshima, Mayer, Zhang, Stuurman, & Vale, 2008). Additional microtubule fragments for spindle assembly may be generated through detachment. It has recently been shown that a sevenfold increase in the rate of microtubule detachment from spindle poles occurs as early as prophase (Yang, Ganguly, & Cabral, 2010) and a similar and perhaps larger increase occurs during anaphase but the mechanisms appear to differ (Rusan & Wadsworth, 2005).

Although the detachment of microtubules from interphase centrosomes occurs at a very low rate (Yang et al., 2010), the increased frequency in mitotic cells suggests that it is a cell cycle-regulated event that may be of vital importance for spindle structure and cell division. In support of this hypothesis, paclitaxel-dependent CHO cells were shown to have a highly elevated frequency of detachment. When grown without the drug, these cells have fragmented microtubules, much reduced polymer levels, defective mitotic spindle assembly, and an inability to divide into daughter cells (Cabral, 1983; Cabral, Wible, Brenner, & Brinkley, 1983; Ganguly, Yang, & Cabral, 2010; Schibler & Cabral, 1986). All of these problems are prevented by the presence of paclitaxel or other microtubule stabilizing drugs that also correct the abnormally high rate of microtubule detachment. The results with paclitaxel-dependent mutants suggested that microtubule destabilizing drugs such as colchicine, vinblastine, and nocodazole might act by increasing the frequency of microtubule detachment, and this prediction was confirmed in a number of cell lines (Yang et al., 2010). The observation that drugs modulate the frequency of microtubule detachment at the same concentrations that interfere with cell division suggests that microtubule detachment is an important process that is necessary for spindle assembly and function. The additional observation that tubulin mutations and overexpression of β-tubulin isotypes that interfere with cell division also alter the microtubule detachment frequency and that detachment is corrected at drug concentrations that allow the cells to proliferate, further supports the link between microtubule detachment and spindle function (Bhattacharya, Yang, & Cabral, 2011; Ganguly, Yang, Pedroza, Bhattacharya, & Cabral, 2011; Ganguly et al., 2010). Given the growing awareness of the importance of detachment for understanding cell division and antimitotic drug action, we will present what we have learned in the process of studying these events as a guide for further investigations into this understudied and poorly understood phenomenon.

4.1 CHOICE OF CELL LINE

The main challenge for calculating microtubule detachment is the high density of microtubules surrounding the centrosome that make it difficult to resolve single microtubules. A related factor is the high fluorescence intensity in that area that requires a wide dynamic range to effectively capture much less intense single microtubules in a single image. These limitations vary considerably among different cell lines making some of them more optimal than others for detecting microtubule detachment events. Ideal cell lines should be thin and well spread, and they should have a centrosomal area that is populated with relatively few microtubules. We have had good experience using Chinese hamster ovary (CHO) and vascular endothelial cells from human umbilical cord (HUVEC) for these measurements.

One of the first published studies describing microtubule detachment involved PtK1 cells, a very flat cell line, but one with a relatively high microtubule density (Keating, Peloquin, Rodionov, Momcilovic, & Borisy, 1997). To optimize the visualization of detachment events, the authors recommended using cells in which the centrosome is situated on the ventral surface with microtubules that run under the nucleus, a suggestion that we have adopted. Using cells with this orientation effectively reduces the three-dimensional space to two dimensions while also providing better contrast for image capture (Fig. 4.1).

4.2 MICROTUBULE DETACHMENT IN SPINDLES

Mitotic cells present an especially difficult challenge because they are frequently round, lack a nucleus, and have two spindle poles rather than a single centrosome nucleating microtubules. Mitotic cells are also thicker than interphase cells thus causing poor contrast and difficulty tracking individual microtubules in the three-dimensional space. Other cells such as PtK2 remain well attached during mitosis but are still difficult to image because of the high density of microtubules in the

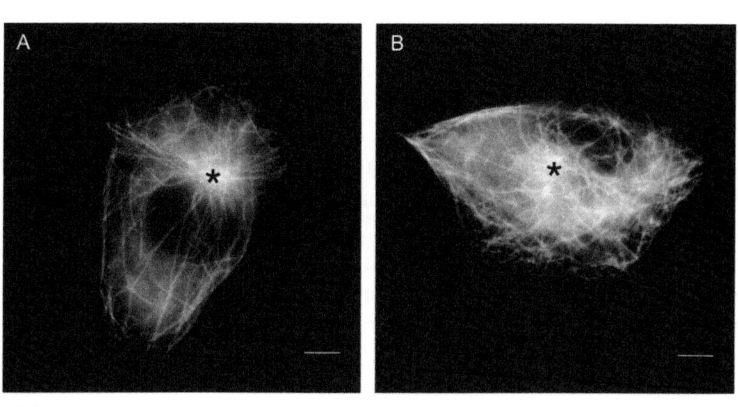

FIGURE 4.1

CHO cells transfected with EGFP-MAP4. Panel A shows an interphase cell with the centrosome (marked with an asterisk) on the bottom (ventral) surface and microtubules clearly visible under the dark nuclear area. Panel B shows a cell in the same dish with the centrosome near the upper surface. Microtubule density and overall fluorescence intensity are too high to detect individual microtubules at the centrosome in panel B. Scale bar = 5 μm.

region near the spindle poles. We have been able to partially overcome problems associated with the rounded morphology of mitotic CHO cells by examining those cells during prophase, a stage at which they remain well attached. Despite being early in mitosis, detachment events are already elevated during prophase compared to interphase (Yang et al., 2010). We have not been able to reliably measure detachment at later stages of mitosis, and so we cannot say with any confidence that detachment remains elevated following prophase. However, others have reported a high detachment frequency during anaphase in LLCPK1 cells, and they proposed that detachment is required for microtubule rearrangements that may be occurring during that stage of mitosis (Rusan & Wadsworth, 2005). It is possible that detachment plays a similar role during prophase to hasten the conversion of the cytoplasmic microtubules into mitotic spindle fibers. Alternatively, detachment may be taking place throughout mitosis to provide microtubule fragments for the construction and dynamics of the mitotic spindle apparatus (Ganguly et al., 2010; Yang et al., 2007).

Given the problems associated with the measurement of microtubule detachment from spindle poles, we have used interphase cells as a substitute to follow changes that are likely to also take place during mitosis. The rationale for this approach comes from the observation that perturbations such as drug treatment, mutant tubulin expression, overexpression of class V β-tubulin, and overexpression of MCAK that interfere with spindle function and cell division alter the detachment frequency not only during mitosis but also during interphase (Bhattacharya et al., 2011; Ganguly, Yang, & Cabral, 2011b; Ganguly, Yang, Pedroza, et al., 2011; Ganguly et al., 2010; Yang et al., 2010).

4.3 LABELING MICROTUBULES

4.3.1 Rhodamine tubulin microinjection

To detect microtubule detachment and other forms of microtubule behavior such as dynamic instability, the cellular microtubules must be selectively labeled. This can be achieved by a variety of methods. One of the earliest approaches used microinjection of fluorescently labeled tubulin. While this methodology is effective in labeling the microtubules, it suffers from a requirement for specialized microinjection equipment, mechanical injury to the cell, and a change in the cellular composition of tubulin. This latter problem can be especially significant because tubulin is commonly purified from brain tissue which has a different tubulin isotype composition from most cultured cell lines (Leandro-Garcia et al., 2010; Luduena, 1998). Class III β-tubulin found in brain is the most troublesome isotype in this regard because it has been shown to reduce microtubule assembly, alter drug sensitivity, and counteract drug effects on dynamic instability (Ganguly, Yang, & Cabral, 2011a; Hari, Yang, Zeng, Canizales, & Cabral, 2003; Kamath, Wilson, Cabral, & Jordan, 2005). Moreover, we have found that expression of this isotype increases the frequency of detachment events (Ganguly, Yang, Pedroza, et al., 2011).

4.3.2 GFP-tubulin expression

A more recent trend has used the expression of GFP-tubulin to label microtubules. Although this approach eliminates many of the disadvantages of microinjection, caution again needs to be used because incorporation of this chimera can also potentially alter the behavior of the microtubules and possibly cause their disruption leading to problems in cell proliferation. We have found, for example, that elevated expression of GFP-tagged α- or β-tubulin can interfere with cell division (unpublished studies). Thus, this approach works best when the level of expression is kept very low so that the labeled tubulin is only present in trace amounts.

4.3.3 GFP–MAP4 expression

A third method for labeling microtubules, and the one that we most commonly use, involves the expression of GFP-labeled MAP4 (available from Life Technologies, Grand Island, NY). MAP4 is a ubiquitous protein that binds to microtubules, but it is not a structural component of the filaments. Thus, unlike the other approaches, the subunit composition of the microtubules is not altered. Although this protein has been reported to stabilize microtubules (Nguyen et al., 1997), we found no effects of high overexpression on microtubule assembly, drug sensitivity, or proliferation of CHO or HeLa cells (Barlow, Gonzalez-Garay, West, Olmsted, & Cabral, 1994). Similarly, another group found no phenotype when they prevented MAP4-microtubule interactions by microinjection of antibodies into human 356 fibroblast and monkey LLCMK2 epithelial cells (Wang, Peloquin, Zhai, Bulinski, & Borisy, 1996). Consistent with a lack of MAP4 effects on microtubule assembly, we found that the residence time of MAP 4 on microtubules is very short (<5 s, unpublished studies), suggesting that it is not a structural component of the filaments; we found that measurements of dynamic instability in mammalian cells expressing GFP–MAP4 are nearly identical to those obtained using microinjection of rhodamine-labeled tubulin (Ganguly et al., 2010; Kamath et al., 2005; Yang et al., 2010). Ultimately, it is likely that all the methods work and give similar results provided that labeled tubulin is kept at sufficiently low levels. Because GFP–MAP4 does not incorporate into microtubules, maintaining a low level of expression is less critical. However, keeping its expression low has the advantage that most of the protein will be associated with the microtubules, thereby providing better contrast.

4.4 MEASURING DETACHMENT

4.4.1 Equipment

The quantification of detachment involves counting the number of such events within a defined period of time in cells that are growing under normal or experimental conditions. To accomplish this, a reliable inverted fluorescence microscope equipped with high power/high numerical aperture objectives, a sensitive digital camera, and a cell incubation system are required. In our experience, a large environmental chamber that encloses the stage and objectives works best. Although it takes longer to equilibrate temperature, humidity, and CO_2 levels compared to smaller alternatives, temperature control and stability of the optics can be maintained more precisely. Alternatively, mini-incubators such as those from Tokai-Hit can be used, but temperature is less well controlled, and the incubator must be disassembled and reequilibrated every time samples are changed.

Although any fluorescence microscope can potentially be used for these experiments, we prefer wide field systems over scanning laser confocal microscopes because we encounter less severe problems associated with photobleaching and photodamage to the cells. To minimize exposure to the UV light source, it is imperative to use an automatic shutter that only opens during image capture and to equip the microscope with a sensitive digital camera that minimizes exposure times. In addition, it is useful to have a mechanical stage for the precise placement of cells into the field of view and software that allows on-screen placement and focus of cells using short-exposure image capture. We have tested a number of systems for this work and favor the solutions provided by Deltavision (Applied Precision, Issaquah, WA). Typically, we use an Olympus 100× objective with a numerical aperture of 1.35.

4.4.2 Cell preparation

We typically plate cells into 35-mm dishes at a density that will grow to 50–60% of confluence after an overnight incubation. EGFP–MAP4 plasmid DNA is then transfected, and the cells are again grown overnight, or until the cell density is about 70–80% of confluence. This density usually gives well spread and healthy cells for imaging. If the cell density is too high, the cells should be subcultured and grown to the proper density before continuing. For the high power objectives required in these studies, glass-bottom dishes such as those from glass-bottom-dishes.com will be needed. Alternatively, we have found that primary cells such as human vascular endothelial cells attach much better to plastic. In these cases, plastic dishes of sufficient optical quality (e.g., μ-dishes from Ibidi LLC, Verona, WI) are used (Ganguly, Zhang, Sharma, Parsons, & Patel, 2012). Once the cells are ready for imaging, the medium is replaced with fresh medium containing 25 mM HEPES to help maintain the proper pH. Although this medium may contain phenol red, better contrast is usually obtained in dye-free medium. Note that the use of HEPES-containing medium is not necessary if the stage incubator is equipped with CO_2 control.

4.4.3 Data collection

Before collecting images, we scan the dish with a low power objective to find regions containing a high number of cells with GFP-labeled microtubules, low background fluorescence, and a clearly discernable centrosome on the ventral surface as shown in Fig. 4.1A. We then switch to a 100× objective and focus on the centrosome by

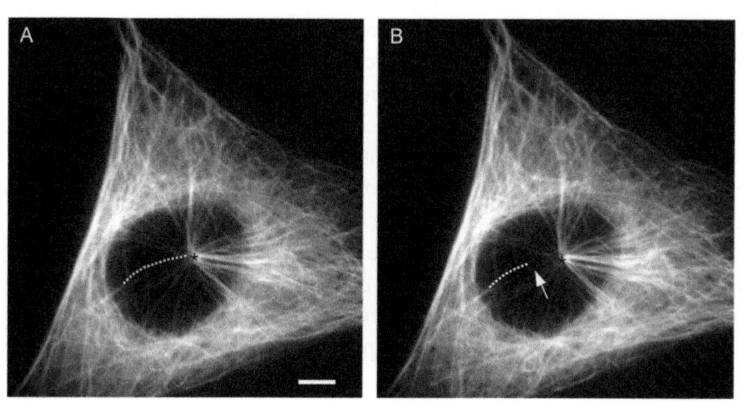

FIGURE 4.2

Microtubule detachment. An early prophase CHO cell expressing EGFP–MAP4 is shown in successive frames taken 5 s apart. A detaching microtubule traced with a dotted line (or colored red in the online version) is seen to shorten at its minus end (arrow, panel B). Only one spindle pole is in the plane of focus. Scale bar $=5\ \mu m$. (For interpretation of the references to color in this figure legend, the reader is referred to the online version of this chapter.)

adjusting the focus knob while taking images of the cell with brief illumination. Once in good focus, images are taken every 5 s for 55 frames. More frequent sampling may cause bleaching of the fluorescence as well as cell damage depending on exposure times and intensity of the light source. Image stacks may be deconvolved, if necessary, to improve contrast.

Detachments are detected by examining successive frames in each stack for a clear release of one or more microtubules from the centrosome region (e.g., see Fig. 4.2). Microtubules with free ends near the centrosome that suddenly appear in a frame and were not clearly attached in the previous frame are not counted because these may have moved into the plane of focus by chance, and it is often difficult to establish whether the free end is the plus or minus end. Any released microtubules are observed for several more successive frames to determine whether they exhibit the typical behavior of growth and shortening of the plus end (when visible), shortening (but not growth) from the minus end, and translocation away from the centrosome. To minimize drift, manual adjustments are made during image capture to keep cellular markers such as intersections between microtubules in sharp focus. We generally accumulate data for 10–20 cells in this way and quantify the results as the number of detachment events that we see divided by the total recording time examined. For wild-type CHO cells, we typically see a detachment rate of 0.17/cell/min during interphase and a rate seven times as high in prophase cells (Yang et al., 2010). Other published rates have varied from 0.02 to 1.5/cell/min during interphase (Keating et al., 1997; Waterman-Storer & Salmon, 1997). Some of these differences may arise from the use of different cell lines, but there are also likely to be differences that arise from the methodologies that are used in different laboratories. Nonetheless, the effects of various treatments on microtubule detachment can be accurately assessed provided that appropriate controls are used, and the procedure is standardized within a given laboratory.

4.5 MICROTUBULE FRAGMENTS AS A SURROGATE FOR DETACHMENT

Direct measurements of microtubule detachment can be tricky and time consuming. However, we have seen that cells with high frequencies of detachment invariably exhibit microtubule fragments in the cytoplasm. For example, paclitaxel-dependent CHO cells with mutations in tubulin not only have been found to have high frequencies of microtubule detachment when they are grown without the drug, but they also have fewer than normal microtubules as well as abundant microtubule fragments (Ganguly et al., 2010). Similar results have been seen in cells that overexpress class III or class V β-tubulin (Bhattacharya et al., 2011; Ganguly, Yang, Pedroza, et al., 2011) and in cells treated with drugs that are known to inhibit microtubule assembly and block cell division (Yang et al., 2010). Treatment of paclitaxel-dependent cells with concentrations of paclitaxel (or any other drug known to stabilize microtubules) that restores cell division has also been found to suppress microtubule detachment and eliminate the presence of cytoplasmic microtubule fragments. The explanation for these observations is that microtubules that detach from the centrosome expose their minus ends which are seldom, if ever, seen to elongate, but are frequently seen to shorten (Keating et al., 1997; Yang et al., 2010). Because unattached microtubules are able to shorten from both the plus and minus ends, fewer intact microtubules are seen in the cell, but many more fragments are present. These fragments vary in length depending on the initial size of the detached microtubule and how long it has been unattached. In addition, fragments are often seen to translocate toward the outer edges of the cell. This translocation may be powered by treadmilling (addition of subunits at the plus end balanced by loss of subunits at the minus end), but the involvement of microtubule motor molecules cannot be ruled out. These processes result in a relatively uniform distribution of fragments with varying size scattered throughout the cytoplasm. Because of the link between microtubule detachment and fragment formation, it is often possible to gauge the effects of a treatment on microtubule detachment by simply quantifying the number of cells with greater than a threshold number of fragments in the cytoplasm (Yang et al., 2010).

The ability to detect microtubule fragments resulting from detachment can sometimes be problematic because treatments that alter detachment produce varying mitotic defects in different cell types. Some cell lines undergo apoptosis when they are blocked in mitosis and are lost from the cell culture, whereas others slip through the mitotic block and reenter the cell cycle to produce larger and flatter cells that make it easier to detect any fragments that are present. In addition, the use of microtubule fragments as a surrogate for detachment has to be exercised with some caution because other mechanisms can also cause microtubule fragmentation. We have looked for spontaneous fractures due to mechanical stress, but these occur much more rarely and are seldom seen even in microtubules that form hairpin loops. Moreover, the frequency of these events is not cell cycle-dependent nor is it altered by drugs or other perturbations that affect detachment (Bhattacharya et al., 2011; Ganguly et al., 2010; Yang et al., 2010). We have also looked for spontaneous nucleation away from the centrosome and have failed to find any evidence for this mechanism of fragment formation. On the other hand, we have found large numbers of microtubule fragments in cells that overexpress class VI β-tubulin, an isotype whose expression is normally limited to platelets in mammals (Schwer et al., 2001; Yang, Ganguly, Yin, & Cabral, 2011). Unlike the fragments produced by detachment, the formation of fragments due to class VI β-tubulin expression appears to be limited to mitotic cells. Because of their extraordinary stability due to incorporation of this unusual isotype, the fragments are carried forward into subsequent phases of the cell cycle. We do not see increased frequencies of detachment in these cells indicating that they arise by a distinct mechanism (Yang et al., 2011). Similarly, we have seen that overexpression of katanin and other microtubule severing proteins can also produce microtubule fragments, but these fragments tend to be much more uniform in size and are frequently arranged in a linear array as would be expected for multiple severing of a linear filament. A side-by-side comparison of microtubule fragments produced by detachment, class VI β-tubulin overexpression, and katanin

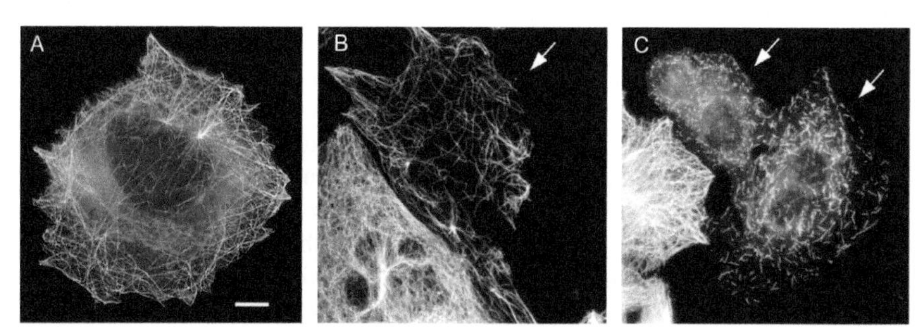

FIGURE 4.3

Mechanisms that generate microtubule fragments. Paclitaxel-dependent CHO β-tubulin mutant Tax 18 grown without the drug (A), CHO cells transfected with class VI β-tubulin (B), and CHO cells transfected with the catalytic subunit of katanin (C) are shown. The cells were stained with an antibody to α-tubulin. Arrows indicate the transfected cells. Note that the pattern of fragments are similar in Tax 18 (A) and class VI β-tubulin-transfected cells (B) even though the mechanisms by which they are generated differ (Ganguly et al., 2010; Yang et al., 2011). On the other hand, the pattern of fragments generated by katanin overexpression is distinct (C). In contrast to (A) and (B), the fragments in panel C are shorter, often follow a linear path due to severing along the length of the microtubule, and long microtubules remaining attached to the centrosome are seldom seen. In addition, the centrosome itself is much less distinct. Scale bar = 10 μm.

overexpression is provided in Fig. 4.3. Although the fragments produced by detachment (Fig. 4.3A) and class VI β-tubulin expression (Fig. 4.3B) have a similar appearance, the fragments produced by severing (Fig. 4.3C) have readily discernable differences. Thus, with a little experience, it is possible to estimate whether a given treatment increases detachment by looking at the pattern of microtubule fragments that are formed.

4.6 MECHANISM OF DETACHMENT

The observation that microtubule detachment increases during mitosis suggests that it is a regulated process facilitated by specific proteins. Given where detachment takes place, it makes sense that proteins involved in the process should be located at centrosomes and spindle poles. Severing proteins such as katanin would seem to be obvious candidates, but as mentioned already, overexpression of these proteins produces a microtubule fragmentation pattern that is quite distinct from the one produced by detachment. Moreover, we have found that mutations in tubulin as well as overexpression of class V or class III β-tubulins that stimulate detachment confer resistance to paclitaxel (Bhattacharya et al., 2011; Ganguly, Yang, Pedroza, et al., 2011; Ganguly et al., 2010), yet overexpression of severing proteins do not (unpublished data). A protein that does appear to be involved in detachment is mitotic centromere-associated kinesin (MCAK). This kinesin family protein is known to cause microtubule disassembly *in vitro* (Desai, Verma, Mitchison, & Walczak, 1999) and is not only associated with mitotic centromeres but also with centrosomes and spindle poles (Ems-McClung & Walczak, 2010). As predicted for a protein involved in detachment, overexpression of MCAK produces a typical pattern of microtubule fragments, increases the detachment frequency, and confers paclitaxel resistance (Ganguly et al., 2011b). Mitotic cells depleted of MCAK produce "hairy" spindles that contain longer and more abundant astral and interpolar microtubules consistent with a decrease in detachment (Kline-Smith & Walczak, 2002). In addition, depletion of MCAK is able to fully or partially correct the abnormally high detachment frequencies associated with class III β-tubulin overexpression and with tubulin mutations in paclitaxel-dependent cell lines (Ganguly, Yang, Pedroza, et al., 2011). Because some paclitaxel-dependent cell lines with abnormally high detachment frequencies can only be partially rescued by MCAK depletion, it is plausible that additional proteins may be involved in the process.

It should be noted that MCAK levels are highest during mitosis, the same time that detachment frequency is highest (Ganguly, Bhattacharya, & Cabral, 2008). At mitotic centromeres, MCAK acts to sever microtubule attachments to misaligned chromosomes, but its role at the spindle poles remains unsettled (Kline-Smith, Khodjakov, Hergert, & Walczak, 2004). We propose that one of its major functions at spindle poles is to facilitate the detachment of microtubules to produce fragments needed for the construction of the mitotic spindle apparatus (Ganguly et al., 2011b).

SUMMARY

Microtubule detachment from centrosomes and spindle poles is a poorly understood process whose importance is only now becoming appreciated. It was initially proposed to play a role in cytoplasmic microtubule turnover (Keating et al., 1997), but more recent studies have demonstrated that it is a cell cycle-regulated process that plays a critical role in cell division and in the mechanism of action of drugs that affect mitotic spindle formation (Ganguly & Cabral, 2011). These drugs either stimulate (e.g., colchicine, vinblastine, nocodazole) or suppress (e.g., paclitaxel, epothilones) microtubule detachment indicating that the frequency of detachment must be maintained at appropriate levels for normal mitotic progression. Tubulin mutations that interfere with cell division often act by stimulating microtubule detachment, and their effects can be counteracted by depletion of MCAK (Ganguly, Yang, Pedroza, et al., 2011). Similarly, toxic effects due to overexpression of MCAK can be counteracted by treatment with paclitaxel (Ganguly, Yang, Pedroza, et al., 2011). Hence, it is likely that kinesin-related proteins such as MCAK play a significant role in the mechanism of detachment, and future studies will be needed to determine what other proteins are involved in the mechanism and how those proteins are arranged and regulated. More work will also be needed to determine how the microtubules are released and the role that microtubule detachment plays during cell division. Is detachment necessary to remodel the interphase microtubules when cells enter mitosis? Does it provide microtubule fragments for mitotic spindle assembly? Does it occur continuously throughout mitosis or only at specific stages? Attempts to answer these and related questions will provide exciting challenges for years to come.

Acknowledgments

We thank Dr. Joanna Olmsted for the EGFP–MAP4 used in our studies. This work was generously supported by NIH Grant CA85955 to F. C.

References

Barlow, S. B., Gonzalez-Garay, M. L., West, R. R., Olmsted, J. B., & Cabral, F. (1994). Stable expression of heterologous microtubule associated proteins in Chinese hamster ovary cells: Evidence for differing roles of MAPs in microtubule organization. *The Journal of Cell Biology*, *126*, 1017–1029.

Bhattacharya, R., Yang, H., & Cabral, F. (2011). Class V β-tubulin alters dynamic instability and stimulates microtubule detachment from centrosomes. *Molecular Biology of the Cell*, *22*, 1025–1034.

Cabral, F. (1983). Isolation of Chinese hamster ovary cell mutants requiring the continuous presence of taxol for cell division. *The Journal of Cell Biology*, *97*, 22–29.

Cabral, F., Wible, L., Brenner, S., & Brinkley, B. R. (1983). Taxol-requiring mutant of Chinese hamster ovary cells with impaired mitotic spindle assembly. *The Journal of Cell Biology*, *97*, 30–39.

Desai, A., Verma, S., Mitchison, T. J., & Walczak, C. E. (1999). Kin I kinesins are microtubule-destabilizing enzymes. *Cell*, *96*(1), 69–78.

Ems-McClung, S. C., & Walczak, C. E. (2010). Kinesin-13s in mitosis: Key players in the spatial and temporal organization of spindle microtubules. *Seminars in Cell & Developmental Biology*, *21*(3), 276–282.

Ganguly, A., Bhattacharya, R., & Cabral, F. (2008). Cell cycle dependent degradation of MCAK: Evidence against a role in anaphase chromosome movement. *Cell Cycle*, *7*, 3187–3193.

Ganguly, A., & Cabral, F. (2011). New insights into mechanisms of resistance to microtubule inhibitors. *Biochimica et Biophysica Acta*, *1816*, 164–171.

Ganguly, A., Yang, H., & Cabral, F. (2010). Paclitaxel dependent cell lines reveal a novel drug activity. *Molecular Cancer Therapeutics*, *9*, 2914–2923.

Ganguly, A., Yang, H., & Cabral, F. (2011a). Class III β-tubulin counteracts the ability of paclitaxel to inhibit cell migration. *Oncotarget*, *2*, 368–377.

Ganguly, A., Yang, H., & Cabral, F. (2011b). Overexpression of mitotic centromere-associated kinesin stimulates microtubule detachment and confers resistance to paclitaxel. *Molecular Cancer Therapeutics*, *10*(6), 929–937.

Ganguly, A., Yang, H., Pedroza, M., Bhattacharya, R., & Cabral, F. (2011). Mitotic centromere associated kinesin (MCAK) mediates paclitaxel resistance. *The Journal of Biological Chemistry*, *286*, 36378–36384.

Ganguly, A., Zhang, H., Sharma, R., Parsons, S., & Patel, K. D. (2012). Isolation of human umbilical vein endothelial cells and their use in the study of neutrophil transmigration under flow conditions. *Journal of Visualized Experiments*, *66*, e4032. http://dx.doi.org/10.3791/4032.

Goshima, G., Mayer, M., Zhang, N., Stuurman, N., & Vale, R. D. (2008). Augmin: A protein complex required for centrosome-independent microtubule generation within the spindle. *The Journal of Cell Biology*, *181*, 421–429.

Hari, M., Yang, H., Zeng, C., Canizales, M., & Cabral, F. (2003). Expression of class III β-tubulin reduces microtubule assembly and confers resistance to paclitaxel. *Cell Motility and the Cytoskeleton*, *56*, 45–56.

Hinchcliffe, E. H., & Sluder, G. (2001). "It Takes Two to Tango": Understanding how centrosome duplication is regulated throughout the cell cycle. *Genes & Development*, *15*, 1167–1181.

Jiang, K., & Akhmanova, A. (2011). Microtubule tip-interacting proteins: A view from both ends. *Current Opinion in Cell Biology*, *23*, 94–101.

Kamath, K., Wilson, L., Cabral, F., & Jordan, M. A. (2005). βIII-Tubulin induces paclitaxel resistance in association with reduced effects on microtubule dynamic instability. *The Journal of Biological Chemistry*, *280*, 12902–12907.

Keating, T. J., Peloquin, J. G., Rodionov, V. I., Momcilovic, D., & Borisy, G. G. (1997). Microtubule release from the centrosome. *Proceedings of the National Academy of Sciences of the United States of America*, *94*(10), 5078–5083.

Kirschner, M., & Mitchison, T. (1986). Beyond self-assembly: From microtubules to morphogenesis. *Cell*, *45*, 329–342.

Kline-Smith, S. L., Khodjakov, A., Hergert, P., & Walczak, C. E. (2004). Depletion of centromeric MCAK leads to chromosome congression and segregation defects due to improper kinetochore attachments. *Molecular Biology of the Cell*, *15*(3), 1146–1159.

Kline-Smith, S. L., & Walczak, C. E. (2002). The microtubule destabilizing kinesin XKCM1 regulates microtubule dynamic instability in cells. *Molecular Biology of the Cell*, *13*, 2718–2731.

Leandro-Garcia, L. J., Leskela, S., Landa, I., Montero-Conde, C., Lopez-Jimenez, E., Leton, R., et al. (2010). Tumoral and tissue-specific expression of the major human beta-tubulin isotypes. *Cytoskeleton (Hoboken)*, *67*(4), 214–223.

Luduena, R. F. (1998). Multiple forms of tubulin: Different gene products and covalent modifications. *International Review of Cytology*, *178*, 207–275.

Mitchison, T., & Kirschner, M. W. (1984). Dynamic instability of microtubules. *Nature*, *312*, 237–242.

Nguyen, H.-L., Charl, S., Gruber, D., Lue, C.-M., Chapin, S. J., & Bulinski, J. C. (1997). Overexpression of full- or partial-length MAP4 stabilizes microtubules and alters cell growth. *Journal of Cell Science*, *110*, 281–294.

O'Connell, C. B., & Khodjakov, A. L. (2007). Cooperative mechanisms of mitotic spindle formation. *Journal of Cell Science*, *120*, 1717–1722.

Piehl, M., Tulu, U. S., Wadsworth, P., & Cassimeris, L. (2004). Centrosome maturation: Measurement of microtubule nucleation throughout the cell cycle by using GFP-tagged EB1. *Proceedings of the National Academy of Sciences of the United States of America*, *101*, 1584–1588.

Rusan, N. M., Fagerstrom, C. J., Yvon, A. M., & Wadsworth, P. (2001). Cell cycle-dependent changes in microtubule dynamics in living cells expressing green fluorescent protein-alpha tubulin. *Molecular Biology of the Cell*, *12*(4), 971–980.

Rusan, N. M., & Wadsworth, P. (2005). Centrosome fragments and microtubules are transported asymmetrically away from division plane in anaphase. *The Journal of Cell Biology*, *168*(1), 21–28.

Saxton, W. M., Stemple, D. L., Leslie, R. J., Salmon, E. D., Zavortink, M., & McIntosh, J. R. (1984). Tubulin dynamics in cultured mammalian cells. *The Journal of Cell Biology*, *99*, 2175–2186.

Schibler, M., & Cabral, F. (1986). Taxol-dependent mutants of Chinese hamster ovary cells with alterations in α- and β-tubulin. *The Journal of Cell Biology*, *102*, 1522–1531.

Schwer, H. D., Lecine, P., Tiwari, S., Italiano, J. E. J., Hartwig, J. H., & Shivdasani, R. A. (2001). A lineage-restricted and divergent β-tubulin isoform is essential for the biogenesis, structure and function of blood platelets. *Current Biology*, *11*, 579–586.

Wang, X. M., Peloquin, J. G., Zhai, Y., Bulinski, J. C., & Borisy, G. G. (1996). Removal of MAP4 from microtubules in vivo produces no observable phenotype at the cellular level. *The Journal of Cell Biology*, *132*(3), 345–357.

Waterman-Storer, C. M., & Salmon, E. D. (1997). Actomyosin-based retrograde flow of microtubules in the lamella of migrating epithelial cells influences microtubule dynamic instability and turnover and is associated with microtubule breakage and treadmilling. *The Journal of Cell Biology*, *139*(2), 417–434.

Yang, H., Ganguly, A., & Cabral, F. (2010). Inhibition of cell migration and cell division correlates with distinct effects of microtubule inhibiting drugs. *The Journal of Biological Chemistry*, *285*, 32242–32250.

Yang, H., Ganguly, A., Yin, S., & Cabral, F. (2011). Megakaryocyte lineage-specific class VI β-tubulin suppresses microtubule dynamics, fragments microtubules, and blocks cell division. *Cytoskeleton (Hoboken)*, *68*(3), 175–187.

Yang, G., Houghtaling, B. R., Gaetz, J., Liu, J. Z., Danuser, G., & Kapoor, T. M. (2007). Architectural dynamics of the meiotic spindle revealed by single-fluorophore imaging. *Nature Cell Biology*, *9*(11), 1233–1242.

CHAPTER

Regulation of Tubulin Expression by Micro-RNAs: Implications for Drug Resistance

Sharon Lobert and Mary E. Graichen

School of Nursing, University of Mississippi Medical Center, Jackson, Mississippi, USA

CHAPTER OUTLINE

Introduction and Rationale **64**
5.1 Methods and Materials **65**
5.1.1 Measuring micro-RNAs 65
5.1.1.1 Choice of Cell Line 65
5.1.1.2 RNA Extraction and qRT-PCR 65
5.1.1.3 PCR Arrays 66
5.1.1.4 Next-Generation Sequencing 66
5.1.2 Upregulating micro-RNAs: Transfections with Precursor PremiRNAs 67
5.1.3 Measuring Activity of Micro-RNAs 67
5.1.3.1 Measuring micro-RNA Targets 67
5.1.3.2 Immunoprecipitation of RISC 67
5.2 Results and Discussion **68**
5.2.1 β-Tubulin Isotypes and miR-100 68
5.2.2 β-Tubulin Isotypes and miR-200c 68
Summary **70**
References **73**

Abstract

In this chapter, we provide an overview of methods for studying micro-RNA regulation of tubulin isotypes. In clinical studies, β-tubulin isotypes were found to be biomarkers for tumor formation. In addition, because changes in the levels of specific β-tubulin isotypes alter the stability of microtubules in mitotic spindles *in vitro*, it has been hypothesized that changes in microtubule protein levels could contribute to chemotherapy resistance. Over the past 15 years, micro-RNAs have been shown

Methods in Cell Biology, Volume 115
ISSN 0091-679X
http://dx.doi.org/10.1016/B978-0-12-407757-7.00005-0

to target mRNAs in signaling pathways involved in tumor suppression, as well as tumorigenesis. Investigating micro-RNA regulation of tubulin isotypes will shed light on the mechanisms underlying the processes that implicate tubulin isotypes as biomarkers for aggressive tumors or chemotherapy resistance. The methods discussed in this chapter include the use of micro-RNA superarrays, next-generation sequencing, real-time PCR experiments, upregulation of micro-RNAs, and immunoprecipitation of RNA-induced silencing complex. We will show examples of data collected using these methods and how these data contribute to understanding paclitaxel resistance.

INTRODUCTION AND RATIONALE

Over the past 15 years, micro-RNAs have been studied as mRNA posttranscriptional silencers (Bartel, 2004; Esquela-Kerscher & Slack, 2006; Heneghan, Miller, & Kerin, 2010; Yu et al., 2010). Micro-RNA transcription and processing has been described in detail (Bartel, 2004). Briefly, micro-RNAs are initially transcribed by RNA polymerase II and folded into hairpin secondary structures (pri-miRNA) that are processed in the nucleus by Drosha and DGCR8 to approximately 75-mers (pre-miRNA). The double-stranded pre-miRNA is transported from the nucleus by XPO5, and the hairpin loop is then cleaved by Dicer1 and further processed in the cytoplasm to a final micro-RNA of 20–25 nucleotides. The double-stranded micro-RNA interacts with RNA-induced silencing complex (RISC) composed of Dicer, double-stranded RNA binding protein TRBP, and Argonaute 2 (Gregory, Chendrimada, Cooch, & Shiekhattar, 2005). One strand of the micro-RNA is released, and the guide strand remains associated with the catalytic protein, Argonaute 2. The micro-RNA can now complex with target mRNAs to promote mRNA degradation. Complementarity of the 5′ nucleotides at positions 2–8 of micro-RNA (seed sequence) with the 3′ untranslated region of the target mRNA is essential for the mRNA degradation processes. Translational silencing is less well understood.

Hundreds of micro-RNAs have been identified that induce tumorigenesis or suppress tumor growth either through degradation of message or by repressing translation (Esquela-Kerscher & Slack, 2006; Yu et al., 2010). For example, one tumor suppressor micro-RNA, miR-100, in ovarian cancer cells (Nagaraja et al., 2010) was implicated in the regulation of cell proliferation and cell survival through interaction with the PI3/Akt/mTOR pathway. We recently showed that miR-100 levels are reduced in MCF7 breast cancer cells by paclitaxel treatment and that miR-100 can regulate β-tubulin isotype classes I, IIA, IIB, and V (Lobert, Jefferson, & Morris, 2011).

Recently, much interest has focused on the miR-200 family, in particular miR-200c. This micro-RNA acts as a tumor suppressor in ovarian cancer cells and tissues (Cittelly et al., 2012). It was shown to regulate β-tubulin isotype class III (*TUBB3*). β-Tubulin class III reduces the stability of microtubules and increases their dynamic instability (Kamath, Wilson, Cabral, & Jordan, 2005). Recent work suggests that β-tubulin class III is a "survival factor" (De Donato et al., 2011). These studies directly link β-tubulin class III protein to the activity of GTPases that contribute to cell proliferation and metastasis. Several studies indicate that increases in β-tubulin class III may contribute to drug resistance in both cell culture and *in vivo* (Cabral, 2008). Furthermore, in ovarian cancer, therapy that includes miR-200c replacement can augment the effectiveness of standard chemotherapy with taxanes (Cittelly et al., 2012), perhaps, in part, because miR-200c targets β-tubulin class III. Thus, the study of micro-RNAs can contribute to our understanding of tubulin regulation and may point the way to novel combination chemotherapy.

In this chapter, we review methods for measuring micro-RNA levels in cell lysates and in RISC. The focus is on studying alterations in micro-RNA levels induced by tubulin-binding drugs and the effects of these alterations on β-tubulin isotypes. We provide examples of data collected using these methods and discuss implications for drug resistance and novel therapies.

5.1 METHODS AND MATERIALS

5.1.1 Measuring micro-RNAs

5.1.1.1 Choice of cell line

Cells and tissues vary in the amounts of specific micro-RNAs (Liang, Ridzon, Wong, & Chen, 2007), and this is reflected in the levels of the primary target mRNAs. For example, we found more than 1000-fold difference in the miR-200c target ZEB1 in two breast cancer cell lines: MDA-MB-231 cells and MCF7 cells. These two cell lines represent very different models for breast cancer. MDA-MB-231 cells lack the HER2, progesterone and estrogen receptors and are considered to be a model for an aggressive triple negative basal type breast cancer phenotype. MDA-MB-231 cells express high levels of the epithelial-to-mesenchymal transition (EMT) protein ZEB1 and low levels of its regulating micro-RNA, miR-200c; whereas, MCF7 cells have estrogen and progesterone receptors and express low levels of ZEB1 and relatively high levels of miR-200c. The basal expression level of the micro-RNA of interest can be determined by quantitative real-time PCR (qRT-PCR), as described below.

5.1.1.2 RNA extraction and qRT-PCR

In common RNA extraction protocols, micro-RNA constitutes about 1% of total RNA extracted from cells and tissues. For micro-RNA cDNA preparation, we find it most efficient to use total RNA rather than attempting to purify micro-RNA. The protocol for preparation of total RNA using organic solvents (TRIzol) has been previously described in detail (Lobert, Hiser, & Correia, 2010). MiScript II RT kits (SABiosciences-Qiagen Corporation, Valencia, CA) are used to prepare miRNA cDNA for PCR experiments. We use the reaction reagents, volumes, and temperatures as recommended by the manufacturer (SABiosciences-Qiagen Corporation, Valencia, CA). Total RNA-containing miRNA is used in the first step, and the 5X miSCript HiSpec buffer is used to prepare the cDNA. We start with 1–2 μg of total RNA, as determined by absorbance at 260 nm using a NanoDrop 1000 UV–Vis spectrophotometer (NanoDrop Technologies, Inc., Wilmington, DE). Using this system, mature miRNAs are polyadenylated and tagged at the 5′ end. The polyadenylation allows oligo DT priming for the reverse transcription. Subsequent PCRs take advantage of the 5′ tag to isolate specific mature miRNAs using a universal primer. Other methods for purifying miRNAs can be found in Redshaw et al. (2013).

Measurement of micro-RNA levels using qRT-PCR with synthesized micro-RNA oligonucleotides for the standard curve yields the micro-RNA copy numbers (Redshaw et al., 2013). The data can be normalized to cell number or microgram total RNA. Alternatively, we use comparative real-time PCR to measure micro-RNA levels in cell culture studies where we have multiple experimental conditions to compare. This method requires the use of preferably two or more housekeeping genes to normalize data. We have found several reliable housekeeping genes for these studies (e.g., *RNU6*, *RNU1A*, *SCA17*, *SNORD44*). Housekeeping genes should be evaluated experimentally for each set of experiments. The housekeeping gene expression should not vary by more than $\pm 0.5\ C_t$ (threshold cycle) with drug treatment or other experimental condition relative to the control samples (Lobert et al., 2010).

5.1.1.3 PCR arrays

We use micro-RNA PCR arrays, Cancer PathwayFinder (SABiosciences-Qiagen Corporation, Valencia, CA), to identify changes in micro-RNA levels associated with paclitaxel treatment. The miScript miRNA PCR arrays are pathway focused and are available in 96- or 384-well formats. Separate plates for control and experimental samples are prepared. There is a free web-based analysis tool that uses the $\Delta\Delta C_t$ method to compare raw data for the experimental and housekeeping genes and calculate relative quantification (http://www.sabiosciences.com/dataanalysis.php). A TRIzol-based protocol is used to purify total RNA from cells, and first strand cDNA synthesis is done according to the superarray manufacturer's instructions. Then template cDNA is used in the qRT-PCR experiments. Further validation of micro-RNA changes associated with paclitaxel treatment is performed using the qRT-PCR validation assays (RT^2 miRNA qPCR Assays, SABiosciences-Qiagen Corporation, Valencia, CA). These kits provide the buffers and primers needed for quantification using real-time comparative PCR.

5.1.1.4 Next-generation sequencing

Next-generation sequencing (massively parallel sequencing) technology for DNA and RNA analysis offers both high-speed data collection and analysis capacity previously unavailable. Several different platforms can be used for micro-RNA data collection and analysis (Brown, 2013; Zhang, Chiodini, Badr, & Zhang, 2011). For example, the Illumina Genome Analyzer can generate 200 giga basepair (Gbp) of short read data per run with accuracy greater than 99.5% (Zhang et al., 2011). The technique permits comparison of thousands of genes with known micro-RNA databases or comparison of unknowns with whole genome sequences. For measurement of drug-induced changes in micro-RNAs, total RNA is purified using a TRIzol-based protocol and processed for analysis. The sample processing and bioinformatics are beyond the scope of this chapter; however, detailed descriptions of next-generation sequencing platforms and software for data analysis can be found in Brown (2013) or Zhang et al. (2011).

5.1.2 Upregulating micro-RNAs: Transfections with precursor premiRNAs

Transfection of double-stranded pre-miRNAs into cells provides supporting data to validate miRNA targets. These experiments can be carried out in six well plates with cells plated at a density of 150,000–250,000 cells per well. Transfections are done using Pre-miR™ miRNA Precursor Starter Kit (Ambion Life Technologies, Grand Island, NY) according to the manufacturer's instructions. After 24 h, the medium is replaced with fresh medium containing DMSO for control cells (treated with a scrambled micro-RNA sequence) or medium with drug. We use siPort™ *Neo*Fx (Ambion Life Technologies, Grand Island, NY) and Optimem medium (ATCC) for the initial transfection reactions with the pre-hsa-miRNA (experimental or negative control scrambled sequence). For drug treatment experiments, we incubate at 37 °C and 5% CO_2 for 24 h and then replace the medium with the appropriate fresh medium. After 24 h, samples can be collected for experimental assays (PCR, Western blotting, etc.).

5.1.3 Measuring activity of micro-RNAs

5.1.3.1 Measuring micro-RNA targets

Several databases are available to assist with the identification of potential and known targets for micro-RNAs (e.g., TargetScan.org, microrna.org, mirdb.org). Measurement of changes in these targets that correlate with levels of micro-RNAs under experimental conditions provides supporting data for changes in micro-RNA activity. For example, we found that mir-200c, a tumor suppressor, is upregulated in MDA-MB-231 breast cancer cells in response to 40-nM paclitaxel treatment (Lobert, Graichen, & Morris, 2013). The upregulation of miR-200c is associated with a significant decrease in the miR-200c target, ZEB1, an inducer of EMT. The transition from epithelial-to-mesenchymal phenotype indicates that cells have reduced anchoring proteins, such as E-cadherin and Claudin 1. These cancer cells thus have the capacity to invade tissues and metastasize. Thus, measurements of ZEB1 inhibition provided evidence for increased amounts and activity of miR-200c.

5.1.3.2 Immunoprecipitation of RISC

Immunoprecipitation of micro-RNAs associated with RISC using Argonaute 2 antibodies is another way to assess the micro-RNA activity. We have used the method to determine whether drug-induced changes of micro-RNA targets correlate with changes in micro-RNA activity. An increased amount of a specific micro-RNA associated with RISC suggests that the activity is increased. In addition, it is expected that increased amounts of target mRNA would also be found associated with RISC if the micro-RNA has increased activity. To obtain sufficient amounts of cell lysate, we grow cells in 300-cm^2 flasks to 80% confluence before initiating drug treatment. For these experiments, we use the reagents and procedure of the Magna RIP™ kit (Millipore, Billerica, MA), except that μMACS™ Protein G Microbeads and μcolumns from Miltenyi Biotech Inc. (Auburn, CA) are substituted for the original beads, and the incubation time of the cell lysate with the Microbeads and the RIPAb+ Ago2 (Argonaute 2) monoclonal antibody (Millipore, Billerica, MA) is 45 min. The RNA is released from the precipitated complex and purified using the TRIzol method (Jedamzik & Eckmann, 2009; Lobert et al., 2010). The micro-RNA and mRNA levels associated with Ago2 are measured using qRT-PCR and normalized to RNA isolated from cell lysate samples that were treated identically except for the immunoprecipitation step.

5.2 RESULTS AND DISCUSSION

5.2.1 β-Tubulin isotypes and miR-100

We showed that paclitaxel treatment of MCF7 breast cancer cells leads to an increase in mRNA for specific β-tubulin isotypes (classes I, IIA, and V). We hypothesized that upregulation of β-tubulin isotypes was due to a reduction in micro-RNAs that regulate β-tubulin mRNA. We found that miR-100 is downregulated by paclitaxel treatment and that transfection of MCF7 cells with miR-100 in the absence and presence of paclitaxel resulted in decrease in mRNA for β-tubulin isotype classes I, IIA, IIB, and V (Lobert et al., 2011; Fig. 5.1A and B).

Because the increase in specific β-tubulin isotype classes with drug treatment is associated with reduction in the tumor suppressor miR-100, which is known to be implicated in cell survival pathways, these data suggest that β-tubulin isotypes could be useful biomarkers for drug resistance. For example, miR-100 downregulates phosphorylated Akt in ovarian cancer cells reducing cell proliferation (Nagaraja et al., 2010). A reduction in miR-100 with paclitaxel treatment could increase phosphorylated Akt in tumor cells. Our data suggest that paclitaxel resistance and increases in specific β-tubulin isotypes (classes I, IIA, and V) could develop as a result of drug-induced reductions in micro-RNA tumor suppressors such as miR-100 (Fig. 5.2).

5.2.2 β-Tubulin isotypes and miR-200c

Triple-negative (basal type) breast cancers that lack estrogen, progesterone, and HER2/neu receptors are frequently the most difficult to treat. Paclitaxel is commonly used in chemotherapy protocols for these aggressive tumors (Anonymous & American Cancer Society, 2012). We studied paclitaxel-induced changes in micro-RNAs in MDA-MB-231 cells that are frequently used as a model for triple negative breast cancers. We found that miR-200c is upregulated about twofold by low dose (40 nM) paclitaxel treatment for 24 h (Fig. 5.3). This was associated with a 40–50% reduction in ZEB1 mRNA, a primary target for miR-200c (Burk et al., 2008). ZEB1 is known to

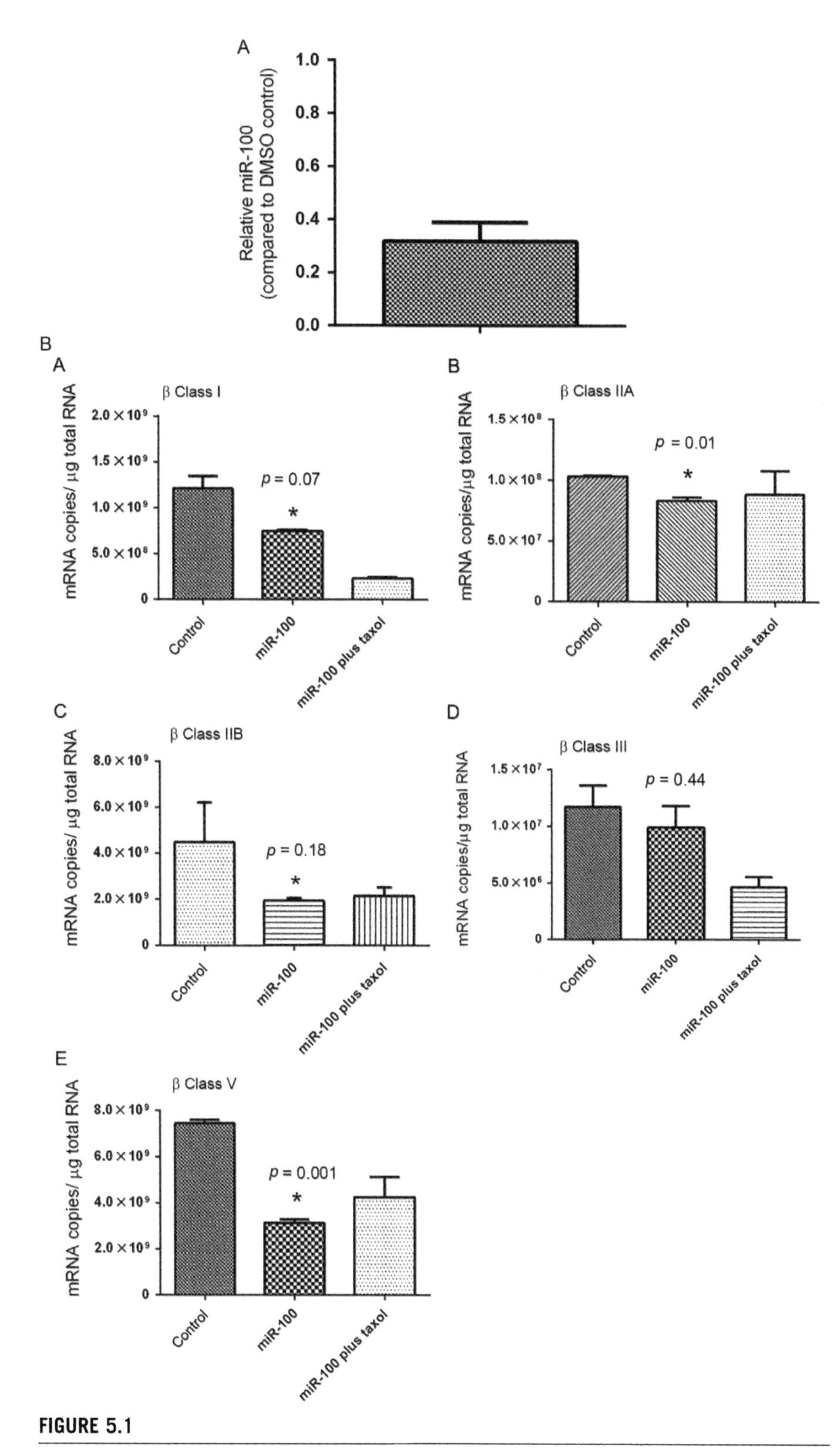

FIGURE 5.1

(A) Relative miR-100 with paclitaxel treatment. MCF7 cells were treated with 400 nM paclitaxel for 24 h. Comparative qRt-PCR was done using *SNORD44* as the housekeeping

play a significant role in EMT, leading to tumor metastasis (Burk et al., 2008). In addition, miR-200c was shown to target β-tubulin class III in ovarian cancer cells and tumors (Cochrane, Spoelstra, Howe, Nordeen, & Richer, 2009; Cittelly et al., 2012). We found that β-tubulin classes I, III, and IVB are downregulated by 40-nM paclitaxel treatment in MDA-MB-231 cells and that these tubulin isotypes appear to be coordinately upregulated with ZEB1 (Fig. 5.4A). MiR-200c likely plays a part in this regulation. Figure 5.4B shows that β-tubulin classes I and III are significantly upregulated when ZEB1 is upregulated threefold by transfection into cells. Smaller increases in β-tubulin classes IVB and V also occur with ZEB1 upregulation. We conclude that the clinical effectiveness of paclitaxel in triple negative breast cancer may be in part due to the drug-induced upregulation of the tumor suppressor miR-200c. In addition, upregulation of β-tubulin isotypes, such as β-tubulin class III, in tumor tissues are likely to be associated with an increase in cell survival factors such as ZEB1 and not a direct effect of β-tubulin class III. Figure 5.5 is a schematic of our findings for paclitaxel-induced changes in miR-200c in MDA-MB-231 cells.

SUMMARY

Over the past 15 years, there has been considerable interest in how noncoding micro-RNAs contribute to mRNA and protein regulation. More than two decades have passed since Cleveland and colleagues showed that one or more cellular cofactors are involved in posttranscriptional regulation of β-tubulins (Pachter, Yen, & Cleveland, 1987; Theodorakis & Cleveland, 1992; Yen, Gay, Pachter, & Cleveland, 1988); however, little is known about the mechanisms underlying differential regulation of these proteins. Using drugs as probes to induce changes in β-tubulin isotypes mRNA associated with changes in micro-RNAs offers a window into the regulation of tubulin genes and the essential cofactors. Furthermore, understanding the regulation of β-tubulin isotypes by micro-RNAs may clarify oncogenic and metastatic pathways and the mechanisms underlying drug resistance.

gene for normalization. The data are plotted as fold difference relative to control cells treated with DMSO. The error bar represents the standard deviation from two independent experiments. (B) Transfection of MCF7 cells with miR-100 in the presence and absence of 400-nM paclitaxel treatment for 24 h. Quantitative RT-PCR data were analyzed using standard curves for each β-tubulin isotype, and mRNA copies were normalized to 1 μg of total RNA. Experiments were done with triplicate individual cell cultures. A representative experiment is shown. The error bars represent the standard deviation from duplicate samples from a single experiment. Panel A: β-tubulin class I; Panel B: β-tubulin class IIA; Panel C: β-tubulin class IIB; Panel D: β-tubulin class III, and Panel E: β-tubulin class V. No changes were found in β-tubulin classes III or IVB, with miR-100 transfection. Student *t*-test was used to compare paired samples. The *p*-values from statistical significance analysis of control and transfected samples (*) are shown.

Figure from Lobert et al. (2011)

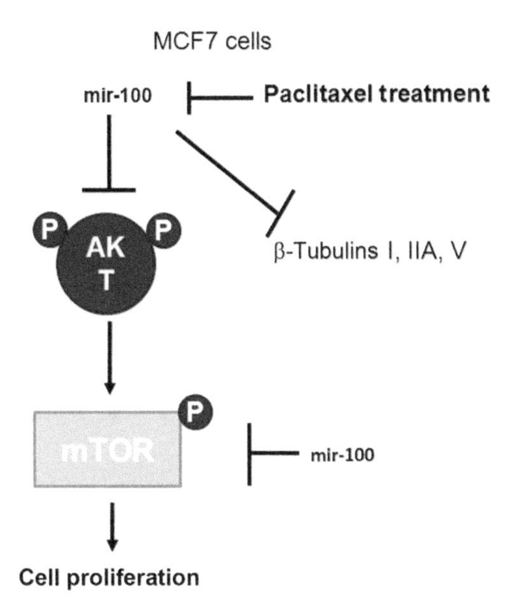

FIGURE 5.2

Schema showing the effects of miR-100 on the Akt/mTOR pathway and β-tubulin isotypes. MiR-100 was shown to reduce levels of phosphorylated Akt and mTOR, signals implicated in tumor progression (Nagaraja et al., 2010). In MCF7 breast cancer cells, miR-100 regulates β-tubulin isotype classes I, IIA, and V (Lobert et al., 2011). Paclitaxel treatment reduces miR-100, and this result has implications for drug resistance and the potential for β-tubulin isotypes to be used as biomarkers for tumor progression. (See color plate.)

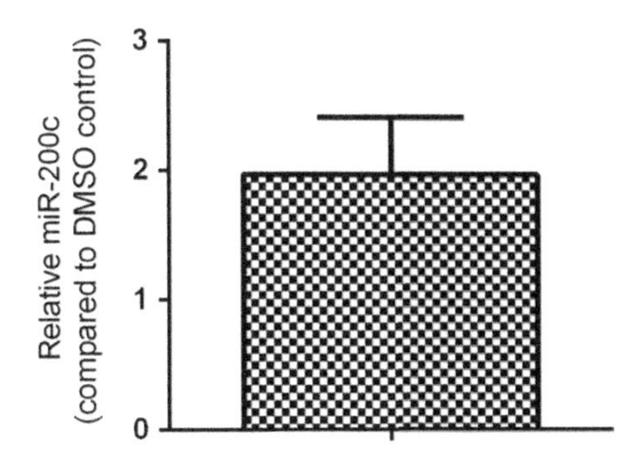

FIGURE 5.3

Comparative qRT-PCR for miR200c. MDA-MB-231 cells were treated with 40 nM paclitaxel for 24 h. Data were normalized to SCA17 as the housekeeping gene.

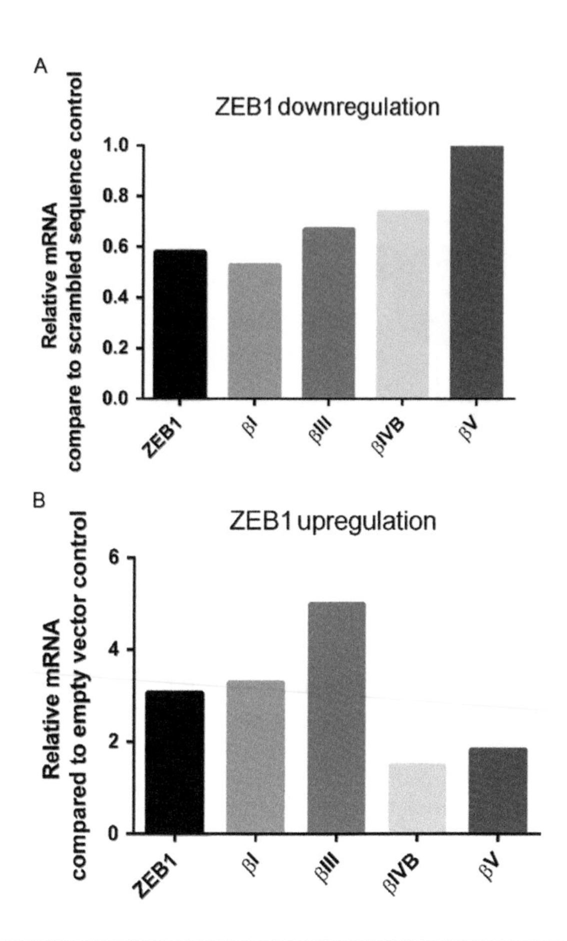

FIGURE 5.4

(A) Downregulation of ZEB1 and relative changes in β-tubulin mRNA in MDA-MB-231 cells. SiRNA technology was used to reduce ZEB1 mRNA levels. Quantitative qRT-PCR was carried out to measure changes in mRNA levels for ZEB1 and β-tubulin isotypes. The bars represent the fold difference in mRNA: ZEB1 SiRNA-treated cells/scrambled sequence-treated cells. The data are representative of three independent experiments. (B) Upregulation of ZEB1 and relative changes in β-tubulin mRNA. A TrueClone™ (OrigGene, Rockville, MD) plasmid expressing human ZEB1 variant 1 (Accession number NM_001128128.1) or empty vector were used to upregulate ZEB1 in cell cultures. Quantitative qRT-PCR was used to measure changes in mRNA for ZEB1 and β-tubulin isotypes. The bars represent the fold difference in mRNA: ZEB1 transfected cell copy number/empty vector transfected cell copy number. The data are representative of three independent experiments.

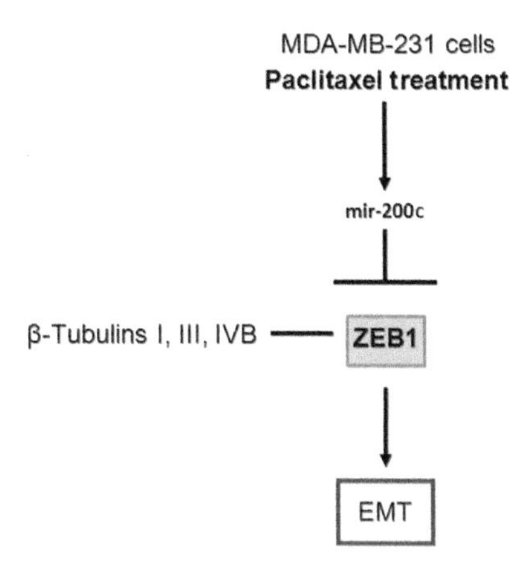

FIGURE 5.5

Schema for paclitaxel treatment of MDA-MB-231 breast cancer cells and survival factors, ZEB1 and β-tubulin isotype classes I, III, and IVB. Paclitaxel treatment increases miR-200c activity. MiR-200c has a reciprocal relationship with the epithelial-to-mesenchymal inducer ZEB1, and miR-200c also regulates β-tubulin isotype classes I, III, and IVB (Lobert et al., 2013). Thus, changes in these β-tubulin isotypes are biomarkers for alterations in cell phenotypes that indicate tumor progression. (See color plate.)

References

Anonymous, American Cancer Society (2012). Chemotherapy for Breast Cancer. http://www.cancer.org/cancer/breastcancer/detailedguide/breast-cancer-treating-chemotherapy, retrieved 4/21/2013.

Bartel, D. P. (2004). MicroRNAs: Genomics, biogenesis, mechanism, and function. *Cell*, *116*, 281–297.

Brown, S. (2013). RNA sequencing with NGS. In S. M. Brown, J. Goecks, & J. Taylor (Eds.), *Next-generation DNA sequencing informatics* (pp. 171–186). New York: Cold Spring Harbor Laboratory Press.

Burk, U., Schubert, J., Wellner, U., Schmalhofer, O., Vincan, E., Spaderna, S., et al. (2008). A reciprocal repression between ZEB1 and members of the miR-200 family promotes EMT and invasion in cancer cells. *EMBO Reports*, *9*(6), 582–589.

Cabral, F. (2008). Mechanisms of resistance to drugs that interfere with microtubule assembly. In T. Fojo (Ed.), *The role of microtubules in cell biology, neurobiology and oncology* (pp. 337–356). New York, NY: Humana Press.

Cittelly, D. M., Dimitrova, O., Howe, E. N., Cochrane, D. R., Jean, A., Spoelstra, N. S., et al. (2012). Restoration of miR-200c to ovarian cancer reduces tumor burden and increases sensitivity to paclitaxel. *Molecular Cancer Therapeutics*, *11*(12), 2556–2565.

Cochrane, D. R., Spoelstra, N. S., Howe, E. N., Nordeen, S. K., & Richer, J. K. (2009). MicroRNA-200c mitigates invasiveness and restores sensitivity to microtubule-targeting chemotherapeutic agents. *Molecular Cancer Therapeutics*, *8*(5), 1055–1066.

De Donato, M., Mariani, M., Petrella, L., Martinelli, E., Zannoni, G. F., Vellone, V., et al. (2011). Class III β-tubulin and the cytoskeletal gateway for drug resistance in ovarian cancer. *Journal of Cellular Physiology*, *227*, 1034–1041.

Esquela-Kerscher, A., & Slack, F. J. (2006). Oncomirs-microRNAs with a role in cancer. *Nature Reviews. Cancer*, *6*, 259–269.

Gregory, R. I., Chendrimada, T. P., Cooch, N., & Shiekhattar, R. (2005). Human RISC couples microRNA biogenesis and posttranscriptional gene silencing. *Cell*, *123*, 631–640.

Heneghan, H. M., Miller, N., & Kerin, M. J. (2010). MiRNAs as biomarkers and therapeutic targets in cancer. *Current Opinion in Pharmacology*, *10*, 543–550.

Jedamzik, B., & Eckmann, C. R. (2009). *Analysis of RNA-protein complexes by RNA coimmunoprecipitation and RT-PCR analysis from Caenorhabditis elegans.* Cold Spring Harb Protoc. *2009*(10):pdb.prot5300. http://dx.doi.org/10.1101/pdb.prot5300.

Kamath, K., Wilson, L., Cabral, F., & Jordan, M. A. (2005). βIII-Tubulin induces paclitaxel resistance in association with reduced effects on microtubule dynamic instability. *The Journal of Biological Chemistry*, *280*(13), 12902–12907.

Liang, Y., Ridzon, D., Wong, L., & Chen, C. (2007). Characterization of microRNA expression profiles in normal human tissues. *BMC Genomics*, *8*, 166. http://dx.doi.org/10.1186/1471-2164-8-166.

Lobert, S., Mary E. Graichen, and Morris, K. (2013). Coordinated regulation of β-tubulin isotypes and epithelial-to-mesenchymal transition protein ZEB1 in breast cancer cells. In press *Biochemistry*.

Lobert, S., Hiser, L., & Correia, J. J. (2010). Expression profiling of tubulin isotypes and microtubule interacting proteins using real-time polymerase chain reaction. In J. J. Correia, & L. Wilson (Eds.), *Methods in cell biology, microtubules,* in vitro. Chennai, India: Elsevier.

Lobert, S., Jefferson, B., & Morris, K. (2011). Regulation of β-tubulin isotypes by micro-RNA 100 in MCF7 breast cancer cells. *Cytoskeleton*, *68*, 355–362.

Nagaraja, A. K., Creighton, D. J., Yu, S., Zhu, H., Gunaratne, P. H., Reia, J. G., et al. (2010). A link between mir-100 and FRAP1/mTOR in clear cell ovarian cancer. *Molecular Endocrinology*, *24*(2), 447–463.

Pachter, J. S., Yen, T. J., & Cleveland, D. W. (1987). Autoregulation of tubulin expression is achieved through specific degradation of polysomal tubulin mRNAs. *Cell*, *51*, 283–292.

Redshaw, R., Wilkes, T., Whale, A., Cowen, S., Huggett, J., & Foy, C. A. (2013). A comparison of miRNA isolation and RT-qPCR technologies and their effects on quantification accuracy and repeatability. *BioTechniques*, *54*(3), 155–164.

Theodorakis, N. G., & Cleveland, D. W. (1992). Physical evidence for cotranslational regulation of β-tubulin mRNA degradation. *Molecular and Cellular Biology*, *12*(2), 791–799.

Yen, T. J., Gay, D. A., Pachter, J. S., & Cleveland, D. W. (1988). Autoregulated changes in stability of polyribosome-bound β-tubulin mRNAs are specified by the first 13 translated nucleotides. *Molecular and Cellular Biology*, *8*(3), 1224–1235.

Yu, Z., Baserga, R., Chen, L., Wang, C., Lisanti, M. P., & Pestell, R. G. (2010). MicroRNA, cell cycle and human breast cancer. *The American Journal of Pathology*, *176*(3), 1058–1064.

Zhang, J., Chiodini, R., Badr, A., & Zhang, G. (2011). The impact of next-generation sequencing on genomics. *Journal of Genetics and Genomics*, *38*(3), 95–109.

CHAPTER

Determining the Structure–Mechanics Relationships of Dense Microtubule Networks with Confocal Microscopy and Magnetic Tweezers-Based Microrheology

Yali Yang* and Megan T. Valentine*,†

**Department of Mechanical Engineering, University of California, Santa Barbara, California, USA*

†The Neuroscience Research Institute, University of California, Santa Barbara, California, USA

CHAPTER OUTLINE

Introduction **76**

6.1 Protocols **78**

6.1.1 Preparation of Samples 78

6.1.1.1 Preparation of Tubulin Proteins 78

6.1.1.2 Preparation of MT Networks Embedded with Magnetic Beads 79

6.1.2 Structure Determination 80

6.1.2.1 Steady-state Network Morphology Determined by Confocal Microscopy 80

6.1.2.2 Dynamic Network Formation Determined by Confocal microscopy 81

6.1.2.3 Analysis of Network Mesh Size 81

6.1.2.4 Analysis of Average Filament Length 82

6.1.3 Magnetic Tweezers Devices for Microscale Manipulation 83

6.1.3.1 Principles of Magnetic Tweezers Technologies 83

6.1.3.2 Calibrations 85

6.1.3.3 Particle Tracking 86

6.1.3.4 Fast Force Switching 87

6.1.3.5 High-force Implementation 88

6.1.3.6 Ring-shaped Magnets Enable Oscillatory Measurements 89

6.1.3.7 Portable Magnetic Tweezers Enables Simultaneous Imaging and Manipulation 90

Methods in Cell Biology, Volume 115

ISSN 0091-679X
http://dx.doi.org/10.1016/B978-0-12-407757-7.00006-2

6.1.4 Analysis of Mechanical Data 90
6.1.4.1 Creep Analysis 91
6.1.4.2 Analysis of Oscillatory Data 93
Summary **93**
Acknowledgments **93**
References **94**

Abstract

The microtubule (MT) cytoskeleton is essential in maintaining the shape, strength, and organization of cells. Its spatiotemporal organization is fundamental for numerous dynamic biological processes, and mechanical stress within the MT cytoskeleton provides an important signaling mechanism in mitosis and neural development. This raises important questions about the relationships between structure and mechanics in complex MT structures. *In vitro*, reconstituted cytoskeletal networks provide a minimal model of cell mechanics while also providing a testing ground for the fundamental polymer physics of stiff polymer gels. Here, we describe our development and implementation of a broad tool kit to study structure–mechanics relationships in reconstituted MT networks, including protocols for the assembly of entangled and cross-linked MT networks, fluorescence imaging, microstructure characterization, construction and calibration of magnetic tweezers devices, and mechanical data collection and analysis. In particular, we present the design and assembly of three neodymium iron boron (NdFeB)-based magnetic tweezers devices optimized for use with MT networks: (1) high-force magnetic tweezers devices that enable the application of nano-Newton forces and possible meso- to macroscale materials characterization; (2) ring-shaped NdFeB-based magnetic tweezers devices that enable oscillatory microrheology measurements; and (3) portable magnetic tweezers devices that enable direct visualization of microscale deformation in soft materials under applied force.

INTRODUCTION

The microtubule (MT) cytoskeleton is essential in maintaining the shape, strength, and organization of cells. Structured MT bundles and networks form the tracks upon which intracellular transport occurs and create the core structure of the mitotic spindle, where they move chromosomes and localize the cleavage furrow (Hirokawa, Noda, Tanaka, & Niwa, 2009). There is increasing evidence that the spatiotemporal organization of the MT cytoskeleton is essential for numerous dynamic biological processes, ranging from neural pathfinding, to the flow of actin in motile and developing cells, to the regulation of protein synthesis (Dent & Kalil, 2001; Kalil & Dent, 2005; Kim & Coulombe, 2010; Rodriguez et al., 2003; Waterman-Storer et al., 2000). For many of these, mechanical stress within the MT cytoskeleton is an important signaling mechanism: for example, tension promotes MT outgrowth at focal adhesion sites, regulates MT turnover, and silences spindle assembly checkpoints to enable cell cycle control (Kaverina et al., 2002; Musacchio & Salmon, 2007; Yvon, Gross, & Wadsworth, 2001). There are increasing efforts to characterize the mechanical properties of the cytoskeleton, and numerous *in vitro* studies have measured the viscoelastic responses of filamentous actin (Broedersz et al., 2010; Chaudhuri, Parekh, & Fletcher, 2007; Gardel, Valentine, Crocker, Bausch, & Weitz, 2003; Gardel et al., 2006; Janmey, Euteneuer, Traub, & Schliwa, 1991; Lee, Ferrer, Lang, & Kamm, 2010; Liu, Koenderink, Kasza, MacKintosh, & Weitz, 2007; Tharmann, Claessens, & Bausch, 2007; Uhde, Ter-Oganessian, Pink, Sackmann, & Boulbitch, 2005). By contrast, relatively little is known of the physical properties of MT networks and how morphology and mechanical interactions give rise to elasticity and force transmission across cellular distances (Janmey et al., 1991; Lin, Koenderink, MacKintosh, & Weitz, 2007; Pelletier, Gal, Fournier, & Kilfoil, 2009; Sato, Schwartz, Selden, & Pollard, 1988). Single MTs are extremely rigid, so the molecular origins of stiffness and stress dissipation are fundamentally different than those of semiflexible actin or flexible polymer systems (Gittes, Mickey, Nettleton, & Howard, 1993; Hawkins, Mirigian, Selcuk Yasar, & Ross, 2010; Head, Levine, & MacKintosh, 2003, 2005; Mickey & Howard, 1995; Taute, Pampaloni, & Florin, 2010; Taute, Pampaloni, Frey, & Florin, 2008; Valdman, Atzberger, Yu, Kuei, & Valentine, 2012; Yang, Bai, Klug, Levine, & Valentine, 2013; Yang, Lin, Kaytanli, Saleh, & Valentine, 2012). This motivates the development and use of new characterization tools that enable direct measurement of structure–property relationships in MT networks.

Here, we describe our development and implementation of a broad set of imaging and microscale manipulation tools to study the properties of reconstituted MT networks. Using custom-built magnetic tweezers devices, we apply calibrated step forces to MT networks using microscale magnetic beads and measure simultaneously the resultant displacement as a function of time (Yang et al., 2012, 2013). The resultant nonuniform deformation field has a characteristic bending radius dictated by the size of the embedded magnetic particle, typically several microns. This length scale is relevant to cargo transport and cellular remodeling. Yet, in this regime, mean-field models that require uniform stretching or bending of filaments no longer apply, limiting the use of traditional continuum mechanics approaches for modeling. Therefore, direct experimental tests are necessary to provide much-needed insight into how the cytoskeleton generates forces, transmits mechanical signals, and maintains cell strength.

The broad class of experimental tools used for microscale manipulation and characterization of polymer materials is collectively known as active microrheology methods. These are distinguished from passive microrheology methods, in which the thermal motions of embedded particles are analyzed. Both forms of microrheology are useful in determining the spatial distributions of stiffness and/or viscosity in heterogeneous materials, or in determining the moduli of precious samples that cannot be obtained in large quantities (Crocker et al., 2000; Gardel, Valentine, & Weitz, 2005; Gardel et al., 2003; Mason & Weitz, 1995; Squires & Mason, 2010). Experimental platforms for active microrheology commonly incorporate an optical microscope to visualize samples and direct the application of force using micron-scale probes.

Atomic Force Microscopy (AFM)-based methods can apply large indentation stresses (~0.1–10 kPa) to soft interfaces and provide simultaneous topographical information, but cannot easily probe the three-dimensional properties of polymer solutions or gels (Kirmizis & Logothetidis, 2010). Optical trapping methods provide nanometer resolution of probe position and can operate at high frequencies. However, optically transparent materials of fairly low index of refraction are required, and since the maximum applied force is typically tens of pico-Newtons, use of optical trapping methods has been limited to fairly soft materials (<50 Pa) (Preece et al., 2011; Valentine et al., 2008). For both optical traps and AFM, the application of constant force requires computer-controlled feedback to compensate for instrument compliance.

Magnetic tweezers devices provide a valuable alternative, allowing for characterization of three-dimensional materials while providing ~nano-Newton forces. However, previous implementations of magnetic tweezers for microrheology have typically relied on the use of electromagnets operating at high current, which can heat samples and exhibit a hysteretic response, or on extremely small distances (~microns) between the pole pieces and magnetic beads, or both (Bausch, Möller, & Sackmann, 1999; de Vries, Krenn, van Driel, & Kanger, 2005; Kollmannsberger & Fabry, 2007; O'Brien, Cribb, Marshburn, Taylor Ii, & Superfine, 2008; Spero et al., 2008). These constraints can lead to unusual experimental geometries with potential interference in high-resolution microscopy, can create steep force gradients within the image plane, and, in the case of devices in which iron pole pieces are submerged into the sample, can create chemically reactive metal ions. By contrast, neodymium iron boron (NdFeB)-based magnetic tweezers are noninvasive and easily provide constant force to the sample plane without the use of feedback control. Because of these advantages, NdFeB magnets have become a standard technology for single-molecule force spectroscopy where femto- to pico-Newton forces are required; however, these have thus far found limited utility in meso- to macroscale materials characterization, which typically requires larger forces to achieve measurable deformations. In this chapter, we describe the design and construction of three new microscope-mounted NdFeB-based magnetic tweezers devices optimized for use with MT networks. Our protocols outline all aspects of instrument assembly and calibration, MT network preparation, structure determination, as well as mechanical data collection and analysis. With these devices, we have successfully characterized many aspects of MT mechanics.

6.1 PROTOCOLS

6.1.1 Preparation of samples

6.1.1.1 Preparation of tubulin proteins

The protocols for unlabeled and rhodamine-labeled tubulin proteins have been previously described in detail. Briefly, unlabeled tubulin is purified from bovine brain by cycles of assembly and disassembly and followed by phosphocellulose chromatography (Miller & Wilson, 2010). Rhodamine-labeled tubulin is prepared by reaction with succinimidyl esters of carboxyrhodamine-6G (C-6157; Invitrogen) (Hyman et al., 1991). We find that using a molar ratio of rhodamine-labeled tubulin to total tubulin of 1:6 is sufficient for visualization using a point-scanning confocal microscope (Olympus Fluoview 500). In some cases, we desire to assess the effects of chemical cross-linking on MT structure and mechanics. For these experiments, we include some fraction of commercial biotinylated porcine brain tubulin (T333P; Cytoskeleton, Inc.) that has been labeled at a ~1:1 ratio of biotin to tubulin heterodimer and free streptavidin (SA20; Prozyme, Inc.). The biotinylated tubulin is delivered as a lyophilate and reconstituted to 10 mg/mL in G-PEM80 buffer (80 mM PIPES, 4 mM $MgCl_2$, 1 mM EGTA, and 1 mM GTP; pH 6.9).

6.1.1.2 Preparation of MT networks embedded with magnetic beads

Entangled MT networks are formed by combining the following reagents on ice: unlabeled and rhodamine-labeled tubulin, 1 mM DTT, 10% (v/v) DMSO, and taxol in G-PEM80 buffer. We typically include taxol (semisynthetic, T7191; Sigma-Aldrich) at a molar ratio of 1:2 taxol to total tubulin to prevent dynamic length changes during the course of the measurement, and we perform the majority of our measurements at room temperature. The typical tubulin concentration ranges from 10 to 50 μM in our measurements. If the effects of dynamic growth and shrinkage on network properties are of interest, then taxol may be omitted; however, in its absence, the polymerized MT solution must be maintained at ~35 °C by a heating stage and/or objective warmer to prevent MT disassembly. The DMSO promotes MT polymerization, and the DTT minimizes disulfide bond formation and photodamage. Tosyl-activated magnetic beads with diameter of 4.5 μm (Dynabeads; Invitrogen) are added to the ice-cold mixture at a final concentration of ~10^6 beads/mL. In some cases, we seek to visualize the long-range deformation field induced by the motion of the magnetic bead and include 2.5-μm latex beads (PS05N; Bangs Laboratories Inc.) at a final concentration of ~5×10^7 beads/mL as fiducial markers.

The ice-cold tubulin solution is then loaded into small rectangular tubes ($0.1 \times 1 \times 50$ mm^3; Friedrich & Dimmock, Inc.) by capillary action; the resultant sample volume is ~5 μL. Prior to loading, the capillary tubes are cleaned by rinsing with 1 M sodium hydroxide, then precoated with reference beads to enable the subtraction of artifactual mechanical or thermal drift, or vibration of the sample, and/or the stage from the real motion of the embedded magnetic particles. To achieve this, ~5.4 μm latex beads (PS06N; Bangs Laboratories, Inc.) are diluted in isopropanol to 5×10^5 beads/mL and loaded into the capillary tubes. The tubes are placed on a flat benchtop for 10 min to promote sedimentation. The tubes are gently dried and baked at 150 °C for 2 min to partially melt the beads onto one side then allowed to cool completely before immediate use.

After introduction to the capillary tubes, entangled networks are immediately sealed with high-vacuum grease and placed in a dry incubator at ~35 °C. It is critical that this process proceed quickly to avoid formation of structures that are shear aligned by loading and to avoid sedimentation of the magnetic beads. The filled capillary tubes are incubated for ~1 h under constant, gentle rotation to prevent the magnetic beads (and latex beads, if included) from settling to the tube surface. This ensures that beads are well dispersed through the depth of the sample. For all mechanical measurements, we select beads that are located in the center of the tube to minimize wall effects. For cross-linked networks, the tubulin solution is first incubated at 35 °C for 3 min in a small microcentrifuge tube, and then streptavidin is added to the desired molar ratio of streptavidin:biotin-labeled tubulin. The solution is well mixed by gently pipetting using a cutoff P20 pipette tip and then immediately loaded into the capillary tube, which is sealed with vacuum grease and incubated at 35 °C for ~1 h under constant rotation, as described earlier. In all cases, after 1 h, the networks are immediately used for imaging and mechanical testing. We have found that the structures are robust for at least several hours, and in some cases as long as 12–18 h, but at longer times, the network structures degrade. Using confocal microscopy, we observe an increase in background fluorescence and fewer and fainter MT filaments, indicating that depolymerization can occur at long times even in the presence of taxol.

6.1.2 Structure determination

6.1.2.1 *Steady-state network morphology determined by confocal microscopy*

In order to ensure quality control, we observe the steady-state structure of every MT network we generate, and we discard any samples that show network collapse, bead aggregation, or other structural irregularities (Yang et al., 2012). We have found this to be particularly important when testing new concentrations or cross-linking conditions; however, given the protocol described above, our sample failure rate is typically ~10% for fresh tubulin samples. This failure rate increases when tubulin aliquots are subjected to multiple freeze–thaw cycles, suggesting that the quality of the tubulin protein is an important determinant of network quality. Additionally, we can determine the network mesh size and the extent of spatial heterogeneity with this technique (see Section 1.2.3). It is possible to image the networks directly through the thin rectangular tubes, which have a wall thickness of 90 μm. We have found that the use of this nonstandard thickness introduces minor spherical aberrations, but these have not caused any significant difficulties in our analysis of the images, and the benefits of using small (~5 μL) volumes greatly outweigh the disadvantages of using nonstandard coverslips. For applications in which such aberrations are unacceptable, use of an objective lens with coverslip thickness correction is recommended.

Confocal microscopy images are obtained using an inverted Fluoview 500 laser scanning system (Olympus) (Fig. 6.1). Steady-state structures are determined by collecting two-dimensional scans (typically 1024 × 1024 pixels) of rhodamine-labeled MT networks using 561 nm laser excitation and a 60 × NA 1.4 oil-immersion objective, with a scan rate of ~ 10 s/scan and magnification of ~200 nm/pixel. In order to achieve the largest dynamic range within the image, PMT gain and offset are manually increased until a small number of pixels are saturated and a small number are

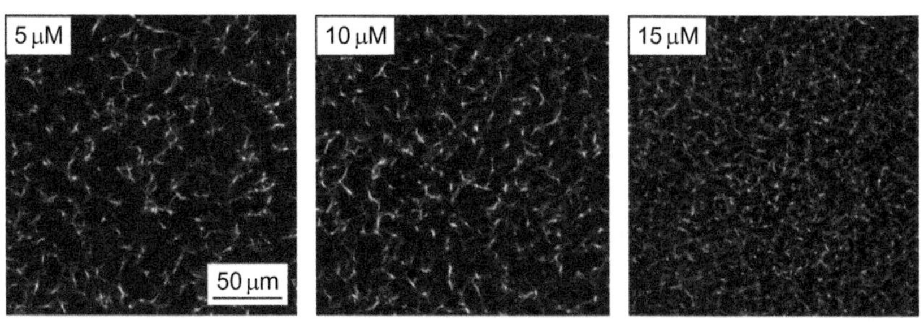

FIGURE 6.1

Representative confocal fluorescence images of entangled MT networks. Using our assembly protocol, the entangled MT networks ranging from 5 to 50 μM are isotropic and homogeneous.

Adapted from Yang et al. (2012). Reproduced by permission of The Royal Society of Chemistry.

black. Based on our typical pinhole diameter of ~100 nm (equivalent to 1 Airy unit) and objective lens, we estimate our *z*-resolution to be ~1 μm.

6.1.2.2 Dynamic network formation determined by confocal microscopy

Time-lapsed confocal fluorescence imaging allows the initial stages of network formation to be observed. To achieve this, we use an upright Fluoview 1000 laser scanning system (Olympus) with an environmental chamber to control the sample temperature. In this case, tubulin proteins are kept ice-cold and loaded into the capillary tubes just prior to imaging. There is no separate incubation step; rather, the environmental chamber is maintained at 30–35 °C, and polymerization is initiated *in situ* and the resulting network growth observed. The lapsed time between the placement of the sample tube on the microscope and the first recorded frame is typically 2–3 min. Since the sample volume is small, temperature equilibration is fast, and we assume that the assembly of the network starts immediately upon placing the tube inside the environmental control box. In detail, we use 559 nm laser excitation through a 25 ×, NA 1.05 water-immersion objective, typically with scan size of 1024 × 1024 pixels^2, scan rate of 10 μs/pixel, and magnification of ~125 nm/pixel. We find a waiting time between each frame of 15–30 s, and total recording time of ~10–20 min is sufficient to capture the dynamics of network formation.

6.1.2.3 Analysis of network mesh size

The average mesh size of the MT network can be determined through analysis of two-dimensional confocal microscopy slices (Yang et al., 2012). First, the raw images are converted to binary images by thresholding to suppress background fluorescence and pixel noise while retaining the gross structural features of the network. Threshold cutoff values are determined by visually comparing the intensities of the brightest pixels of the background to those of the dimmest pixels on the MTs. After thresholding, the distance between nearest-neighbor MT pixels within each row and column is determined. The result is similar to the radial distribution of

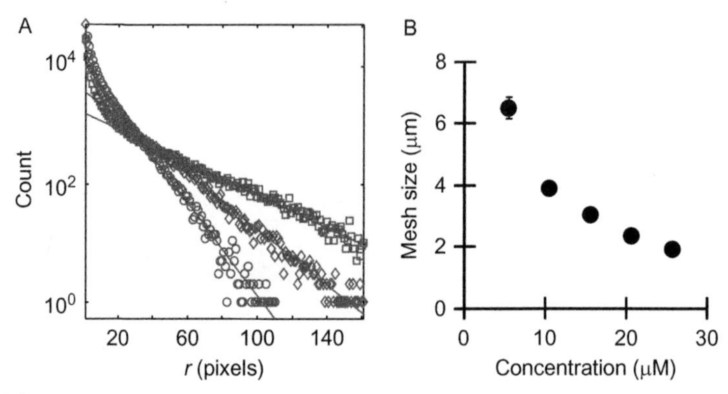

FIGURE 6.2

Determination of mesh size. (A) Distribution of distances between nearest-neighbor MT pixels measured from confocal images of entangled MT networks of 5 μM (squares), 10 μM (diamonds), and 15 μM (circles). The distribution is fitted to an exponential, $P(\xi) = P_0 e^{-r/\xi}$, where r is the nearest-neighbor distance and ξ is the characteristic mesh size. (B) Mesh size of entangled MT networks as a function of tubulin concentration. Each measurement is averaged over 28–57 images. (See color plate.)

Adapted from Yang et al. (2012). Reproduced by permission of The Royal Society of Chemistry.

distances between filament intersections but is implemented in Cartesian coordinates to take advantage of the natural axes of the microscope images. The distribution of distances is plotted and fitted to an exponential $P(\xi) = P_0 e^{-r/\xi}$, where r is the nearest-neighbor distance and ξ is the characteristic mesh size (Fig. 6.2). We find that using an ensemble of 30–60 2D images is sufficient to obtain a robust average value of ξ. Some care is required in the interpretation of ξ. Although ξ varies to some extent with the threshold cutoff, previous studies have demonstrated that ξ provides a good approximation of the average 3D mesh size, but underestimates the maximum pore diameter (Lin et al., 2007).

6.1.2.4 Analysis of average filament length

It is difficult to assess the distribution of MT contour lengths from confocal microscopy images of dense 3D networks. Instead, we determine MT lengths by polymerizing a small volume of MTs (typically at concentrations of 10–50 μM) in a microcentrifuge tube using the same chemical conditions, incubation times, and temperatures that are used in generating networks for mechanical analysis. We find that a molar ratio of 1:6 rhodamine tubulin to total tubulin is sufficiently bright, and we include biotinylated tubulin at a molar ratio of 1:14 biotinylated to total tubulin dimers to enable specific attachment of the MTs to the surface via biotin–streptavidin bonds. After polymerization, we dilute the MTs by a factor of ~100 to a final concentration of ~250 nM into G-PEM80 buffer supplemented with 100 μM taxol to

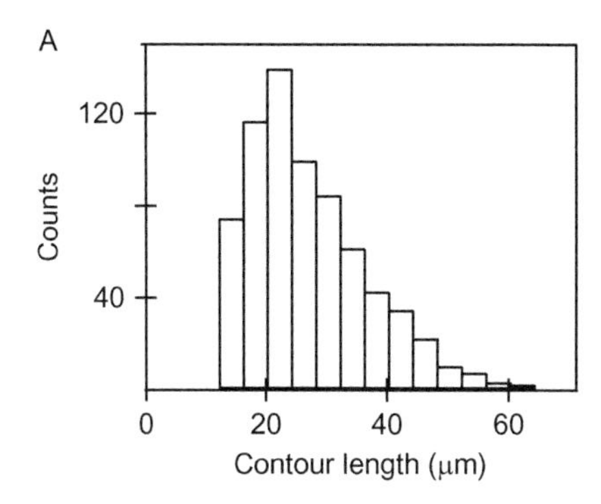

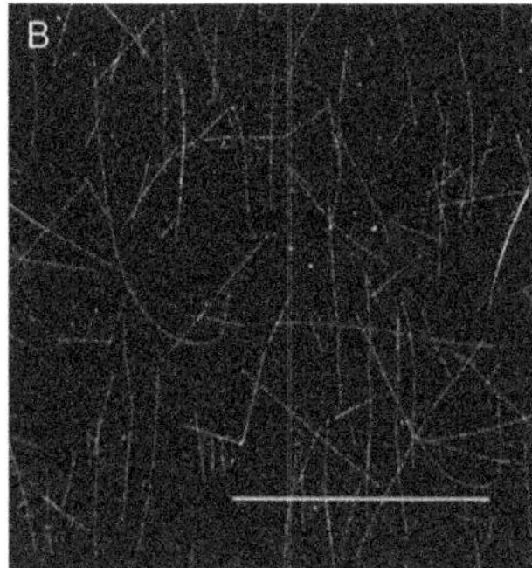

FIGURE 6.3

Determination of MT contour length. (A) The distribution of MT lengths is determined by attaching individual MTs to a coverslip, visualizing them using confocal microscopy, and measuring their lengths manually using built-in measurement tools in ImageJ. Under our polymerization conditions, we find the length distribution is independent of tubulin concentration. The mean length measured from MTs assembled using 50 μM tubulin is 23.3±0.4 μm (SEM; 710 measurements), as shown. (B) Sample image for the length measurement (scale bar is 50 μm; note that image has been adjusted to enhance contrast and brightness prior to printing).

Adapted from Yang et al. (2012). Reproduced by permission of The Royal Society of Chemistry.

preserve MT lengths. We form a microscope flow chamber ($\sim 25 \times 4 \times 0.1\ mm^3$) using two strips of double-sided tape that attach a clean coverslip to a clean glass slide. Prior to introducing the MT solution, we flow in 20 μL of 0.2 mg/mL streptavidin (SA20; Prozyme), incubate 5 min, flow through 100 μL of G-PEM80 with 100 μM taxol to remove any unbound streptavidin, and then flow in 30 μL of diluted MT solution. To minimize photobleaching, we frequently supplement the MT solution with an oxygen scavenging system (glucose, glucose oxidase, and catalase). After a 10-min incubation period during which time nearly all MTs attach to the coverslip, we visualize the surface-bound MTs using Total Internal Reflection Fluorescence (TIRF) or confocal microscopy and measure their lengths manually using built-in measurement tools in ImageJ (Fig. 6.3; Yang et al., 2012).

6.1.3 Magnetic tweezers devices for microscale manipulation

6.1.3.1 Principles of magnetic tweezers technologies

Mechanical measurements are performed using a custom-built magnetic tweezers system that enables precise manipulation of magnetic beads along the optical axis (the z-axis) and simultaneous three-dimensional tracking of bead position (Kim & Saleh, 2008; Manosas et al., 2010; Ribeck & Saleh, 2008). A simple inverted microscope is constructed using an oil-immersion objective, typically 100×, 1.25 NA, to visualize an appropriate field of view around the magnetic particle (Fig. 6.4). The

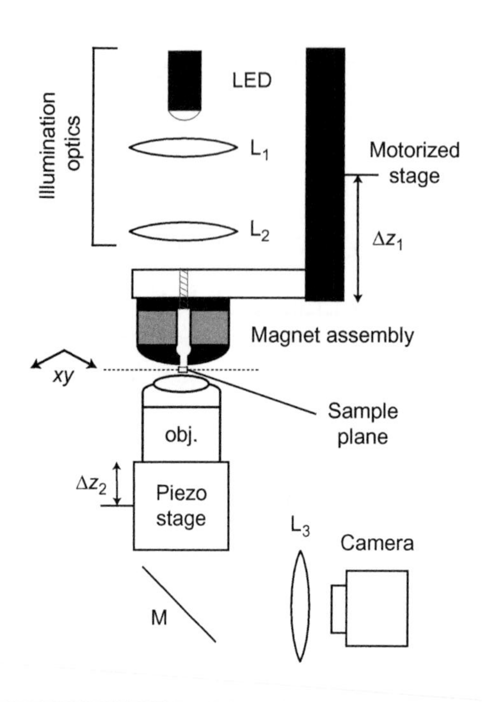

FIGURE 6.4

Schematic of a typical magnetic tweezers device. The sample stage is a manual *x–y* positioning stage with excellent mechanical stability mounted on a heavy-duty platform to minimize the vibrations. Above the sample stage, the magnet assembly is mounted onto a long-distance travel motorized stage which allows the assembly to be moved at a height Δz_1 (typically $\Delta z_1 < 30$ mm) at speeds of up to ~100 mm/s. Above the magnet, the illumination optics consist of the LED housing with built-in collimating lens and a set of lenses (L_1, L_2) that direct light onto the sample. Below the sample stage, the 100× objective lens is mounted on a precision piezoelectric one-axis translation stage. A turning mirror M directs the image to the camera, which is formed by L_3, a simple 150-mm relay lens. The image is captured by a CMOS camera, which is connected to a computer workstation (not shown).

Adapted from Lin and Valentine (2012a). Reprinted with permission from the American Institute of Physics, Copyright 2012.

objective lens is mounted onto a piezoelectric stage (such as P-725; Physik Instrumente) to enable nanopositioning of the focal plane. A 650-nm light-emitting diode (Roithner Lasertechnik) provides semicoherent illumination, and a camera captures brightfield images of the magnetic beads. The sample is placed on a custom-built heavy-duty platform to minimize vibrations. A manual *x–y* positioning stage (i.e., MT25; Marzhauser) with excellent mechanical stability is used to scan through the sample during imaging.

In the simplest case, the applied magnetic field is generated by placing a pair of permanent rare-earth NdFeB magnets (NS-505050; Applied Magnets) oriented such that the alignment of their magnetic moments is antiparallel above the sample stage.

Variation of the magnetic field (and thus magnetic force) at the sample plane is achieved by vertically translating the magnets with a DC-servo motor (such as model M-126.PD1; Physik Instrumente).

Separation distances between the sample plane and magnet range from near contact with the coverslip to tens of millimeters away. The conversion from separation distance to force along the optical axis depends on the magnet type, number, and geometry. Although it is possible to use finite element (FE) modeling to predict with reasonable accuracy the magnetic field generated by a specific magnet array, it is difficult to determine the magnetic moment of commercial superparamagnetic beads, which depends on both the bead properties and the magnitude of the applied magnetic field. Thus, it is typically more reliable to determine experimentally the relationship between separation distance and applied force.

6.1.3.2 Calibrations

Two calibration methods are common (Kim & Saleh, 2008; Manosas et al., 2010; Ribeck & Saleh, 2008). In the first, forces are calibrated by measuring the Brownian motion of a magnetic bead that is tethered to the coverslip by a single DNA molecule, and thus acts as a simple inverted pendulum. The lateral spring constant is given by the ratio of the vertical force to the DNA length. This spring constant is determined by modeling the measured bead trajectory with an overdamped Langevin equation of motion for a particle in a harmonic potential, and fitting the measured power spectrum in position to that predicted from the Langevin model after accounting for issues of finite data sampling rate and instrumental low-pass filtering. The best-fit spring constant, along with the measured length, gives an estimate of the force; this calibration is then repeated at each desired magnet position. This method is particularly useful at small forces, but the DNA tethers tend to break at forces above ~60 pN (Fig. 6.5).

For larger forces, or in cases where purified DNA molecules of known length are not readily available, it is preferable to measure the velocity v at which single magnetic beads of radius R move through a solution of known viscosity η, then to relate this velocity to the force F using Stokes law $F = 6\pi\eta Rv$. The range of velocities that can be reliably determined depends on the camera frame rate and the maximum distance over which particle position can be reliably tracked. At a minimum, ~4–6 particle positions must be determined to accurately measure velocity. We have reliably measured bead velocities over the range of ~0.005–300 μm/s using a pure glycerol solution and a fast CMOS camera with 280 frames per second (fps) for full-frame collection (Lin & Valentine, 2012a). Although tables of glycerol viscosity are available, we recommend independent verification, since viscosity depends sensitively on temperature, and glycerol solutions tend to absorb water in humid environments, which will decrease glycerol concentration and viscosity. We determined viscosity of our glycerol sample to be ~1.15 Pa s, using a strain-controlled rheometer (ARES-LS, TA Instruments) with a cone-plate tool geometry (50 mm diameter, 0.04 rad cone angle, and 0.045 mm gap) at ~19 °C and a strain rate of 1 s^{-1}.

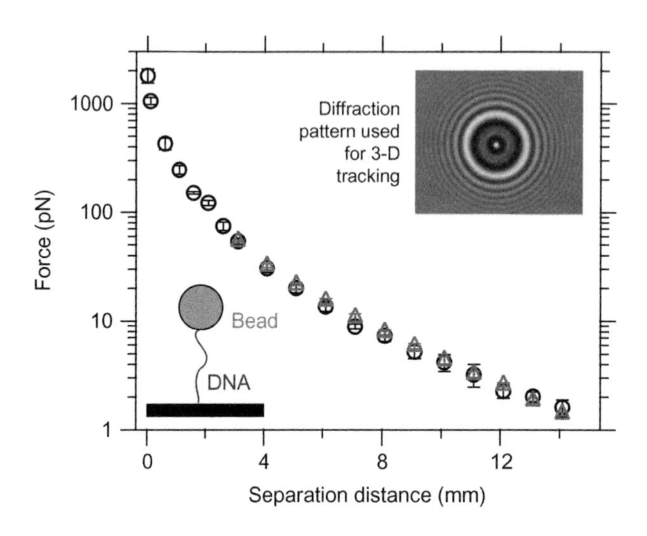

FIGURE 6.5

Calibration for high-force magnetic tweezers device. The main figure shows the calibration curve of force versus the separation distance between the magnet assembly and the sample plane. Data represented by circles are obtained by measuring the velocity at which 4.5-μm diameter beads moved through a glycerol solution of known viscosity under force. Data represented by triangles are obtained by measuring the stretching of a single DNA strand of known length while force is applied to the tethered magnetic bead (schematically depicted in lower inset). Upper inset shows typical diffraction pattern for magnetic bead. (See color plate.)

Adapted from Lin and Valentine (2012a). Reprinted with permission from the American Institute of Physics, Copyright 2012.

6.1.3.3 Particle tracking

Accurate tracking of bead position in three dimensions is essential both for proper calibrations and for analysis of data of beads moving through the MT networks. This tracking is achieved using Fast Fourier Transform-based particle tracking algorithms implemented in LABVIEW (National Instruments), as previously described (Gosse & Croquette, 2002; Ribeck & Saleh, 2008). An autocorrelation of each Fourier image is used to determine the xy centroid position of the bead. Axial position detection is achieved using semicoherent, parallel illumination to generate a diffraction ring pattern that depends sensitively on the distance of the particle from the focal plane. Each image is compared to a lookup table of previously acquired calibration images, and interpolation is used to find the best z coordinate. Because gel microrheology measurements typically involve large particle displacements, this calibration image stack typically consists of ~300 images, where the total calibrated distance is ~30 μm with 10 reference images collected per micrometer in the image plane. We have found that using a computer with a fast, large cache memory is particularly important for the rapid processing of the large image stacks required for 3D tracking of beads. The intrinsic accuracy of the 3D tracking algorithms for beads immobilized in a low-contrast media (such as water) is ~1 nm. When embedded in MT gels, additional noise sources are introduced due to both the thermal fluctuations of the bead and the light scattering from the dense MT network, which can degrade the diffraction of the bead used for 3D tracking. For concentrated entangled networks and cross-linked networks, we estimate our bead tracking accuracy to be <5 nm in all axes. For dilute entangled networks, beads have higher mobility through the network, and our effective resolution is degraded to ~10–20 nm.

6.1.3.4 Fast force switching

Microrheology measurements frequently require fast switching of force amplitudes, either to apply oscillatory stresses to a sample or to apply step stress pulses in a creep measurement protocol. Fast force switching is easily implemented using electromagnets by rapid modulation of the driving current. By contrast, for NdFeB-based magnetic tweezers, force levels are modulated by the physical separation of the magnets and sample. Although the incorporation of a fast linear motor to move the magnets is trivial, the ability to accurately track bead position during the repeated long-distance travel of the magnet array is more challenging. To achieve accurate bead tracking during fast force switching, a very stable long-distance travel motorized stage, a bright illumination source with an intensity that does not vary as the magnet array moves, and a high-speed data acquisition system are required (Lin & Valentine, 2012a).

To ensure mechanical stability, the instrument is typically mounted onto an air-cushioned optical table. A very stable, long-distance travel motorized stage (M414.1PD, Physik Instrumente) is chosen to achieve high pulling force (200 N) and high velocity (100 mm/s). For a typical microrheology experiment, the magnet array is moved at most 15–20 mm to switch the applied force on or off. At maximum motor travel speed, this transition would take ~150–200 ms. An internal PID controller, tuned to maximize acceleration, ensures ~micron-scale repeatability. To damp vibrations, the motor is mounted onto a heavy column (XT95-1000, Thorlabs) filled with ~1 mm steel shot, and the sample is mounted onto a custom-built heavy-duty platform.

To maintain tracking accuracy, the illumination intensity must be constant over the entire magnet travel distance. To achieve this, we use a bright LED light source driven by a stable, current-regulated power supply and focus the light onto the sample using a custom-built optical collection system. This system must meet two critical demands. First, as much LED light must be collected and focused onto the sample as is possible, this will ensure that images can be collected at high frame rates (>100 fps) without substantial contributions of shot noise. Second, the beam waist must be small enough to pass through the magnet array, and the focal depth must be large enough to allow the height of the magnet array to be adjusted without clipping the focused LED beam. To achieve the desired force range for microrheology measurements, the focal depth must be ~20–30 mm. In practice, this requires the illumination arm to be ~1 m long. For convenience, we mount the LED and illumination optics onto the same damped column that supports the long-distance travel motorized stage.

To acquire and process data quickly, we choose a high-resolution camera, fast frame grabber, and high-performance computer workstation for real-time data analysis. We have found particular advantages with the use of a CMOS-based camera (such as the Gazelle model, from Point Grey with a 2048 × 1024 array of 5.5 μm square pixels), which enables a maximum frame rate of 280 fps for full-frame collection. An 8-tap Camera Link frame grabber card (NI PCIe-1433, National Instruments) is used to capture images and relay them to the image processing computer. To maximize the number of beads that can be simultaneously tracked, we select a Dell workstation with a single-chip quad-core processor (Intel® Xeon® Processor X5687) with 3.60 GHz clock speed, 6.4 GT/s QuickPath Interconnect, and a 12 MB Smart Cache. With this data acquisition system, we are able to track ~6 beads at 280 fps in real time using the full 2048 × 1024 pixel array. Tracking more beads is possible but requires a reduction in image size by cropping or binning, decreased frame rate, use of a smaller calibrated image stack size, and/or elimination of other computer demands (such as real-time data display).

6.1.3.5 High-force implementation

Simple magnetic tweezers devices employ an antiparallel pair of NdFeB magnets above the sample stage. This arrangement can produce forces of <100 pN on ~4.5 μm commercial magnetic beads (Lin & Valentine, 2012a). For higher force applications, it is necessary to redesign the magnet assembly to focus and confine the magnetic fields. To optimize the design, FE modeling of the magnetic fields is useful and can provide quantitative predictions of field strengths and gradients for various magnet types, orientations, and shapes. Using this approach, we have found an assembly incorporating two pairs of strong N50 0.5-in. cubic NdFeB magnets and shaped iron "pole pieces" are able to confine and direct the magnetic fields in a manner that produces ~1–2 nN forces at the sample plane on ~4.5 μm magnetic beads (Fig. 6.6). This arrangement not only improves field strength but also increases the field gradient, which, in the limit of saturation of the bead's magnetic moment, is directly proportional to the force applied to the magnetic bead. In practice, we find low-carbon steel to be a good alternative to iron, as it has similar magnetic properties and is easier to machine.

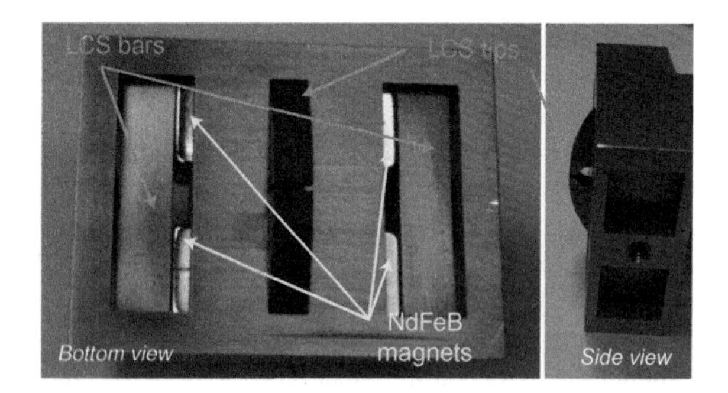

FIGURE 6.6

Magnet assembly optimized for high-force applications. An aluminum housing holds four cubes of NdFeB N50 magnets between two low-carbon steel (LCS) backing bars. Below the magnet array, two LCS tips are press fit into the housing to direct and focus the magnetic fields. (See color plate.)

6.1.3.6 Ring-shaped magnets enable oscillatory measurements

Most NdFeB-based magnetic tweezers employ opposed pairs of cubic magnets that typically allow for only unidirectional pulling, and therefore do not allow for oscillatory measurements around a stress-free state. We have recently shown that for a ring-shaped NdFeB magnet, the gradient of the magnetic field changes sign along the symmetry axis, allowing both pushing and pulling forces to be generated (Lin & Valentine, 2012b). This makes ring magnets ideal for use in oscillatory microrheology measurements, in which frequency-dependent viscoelastic parameters are easily obtained. The overall design is similar to that of simple magnetic tweezers, but the cube magnet array is replaced with a single NdFeB ring magnet of square cross section (i.e., inner radius of 0.125 in., outer radius of 0.375 in., and thickness of 0.125 in., and maximum remanent field $B_r = 1.32$ T). With this geometry, we observe two force reversals as a function of distance along the z-axis (Fig. 6.7). One transition occurs within a millimeter of the front face of the magnet, and a second occurs at an axial distance of ~4 mm. This second, gentler transition is a good candidate for stable manipulation of beads around a zero-force position, a requirement for many applications of oscillatory microrheology. To achieve cyclical actuation, the servo motor that vertically lifts the magnet array is driven using a triangle waveform. We find that bead displacement primarily occurs along the z-axis, with smaller lateral x–y motions. Coupling between

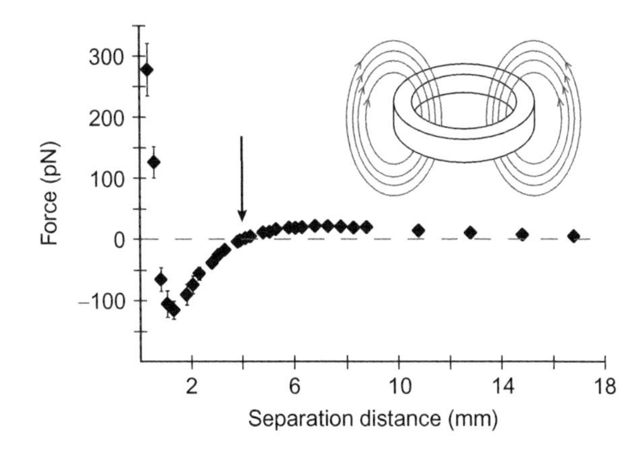

FIGURE 6.7

Calibration curve for ring magnet-based magnetic tweezers device showing force reversals. The main figure shows the calibration curve of force versus separation distance between the magnet assembly and the sample plane. Data are obtained by measuring the velocity at which 4.5 μm diameter beads moved through a glycerol solution of known viscosity under force. Arrow indicates magnet position at which the force gently reverses sign. $F < 0$ indicates pushing forces whereas $F > 0$ indicates pulling. Inset shows the magnetic field lines generated by a ring magnet. (See color plate.)

Adapted from Lin and Valentine (2012b). Reprinted with permission from the American Institute of Physics, Copyright 2012.

z–x or z–y can be minimized by aligning the particle of interest with the center of the ring magnet.

6.1.3.7 Portable magnetic tweezers enables simultaneous imaging and manipulation

In many cases, interpretation of microrheology data requires some knowledge of the MT structure at the surface of the bead. In principle, the magnetic tweezers devices described earlier could be mounted onto an inverted confocal microscope to enable simultaneous imaging and microscope manipulation. In practice, this is a costly solution, which requires a dedicated instrument with substantial structural modifications and custom software and therefore does not allow use of the sophisticated core confocal imaging facilities available at many research institutions. As an alternative, we designed and constructed a custom portable magnetic tweezers device that can be mounted onto any optical microscope (Fig. 6.8; Yang, Lin, Meschewski, Watson, & Valentine, 2011). In this design, two small NdFeB magnets ($0.25 \times 0.25 \times 1$ in.; Applied Magnets) are mounted onto a two-axis translation stage and positioned near the focus of the objective lens in a manner that protects the imaging quality of the microscope. The distance between the magnets and sample can be controllably varied, leading to application of controlled forces to small magnetic particles at the sample plane.

For a typical measurement, the magnets are moved toward the sample until they just touch the coverslip edge, to ensure application of the maximum force. To reduce this force level, the magnets are moved away from the coverslip surface until the separation distance exceeds $\sim$10 mm (at which point the force is nearly zero). The rate of magnet retreat (and therefore the rate of force reduction) can be varied through manual control of the translation stage. In this "sideways-pulling" geometry, beads move perpendicular to the optical axis (the "x" axis) when the force is on. We find that it is most useful to obtain confocal images using both the fluorescence and the transmitted light channels to visualize the MT network and beads, respectively. The magnetic beads available commercially from Invitrogen are slightly autofluorescent under 561 nm excitation, which allows them to be observed in both channels. We find this to be an advantage, since the bead–network interactions can be clearly observed. In this case, particle position is determined using 3D centroid tracking.

6.1.4 Analysis of mechanical data

For microrheology measurements of homogeneous materials (i.e., materials in which the bead radius is much larger than any other network length scale), it is possible to determine the viscoelastic shear moduli directly through analysis of bead motion. This is rarely the case for the MT networks in which the bead size is similar to ξ, and the stiff filaments bend around the bead surface in a manner that does not lead to uniform strain across the sample (e.g., the deformation is nonaffine). Thus, rather than report elastic parameters in terms of modulus, which is a geometry-independent materials property, we instead report network stiffness, which depends on both the modulus and the structure of the network. Similarly, we report bead velocity rather

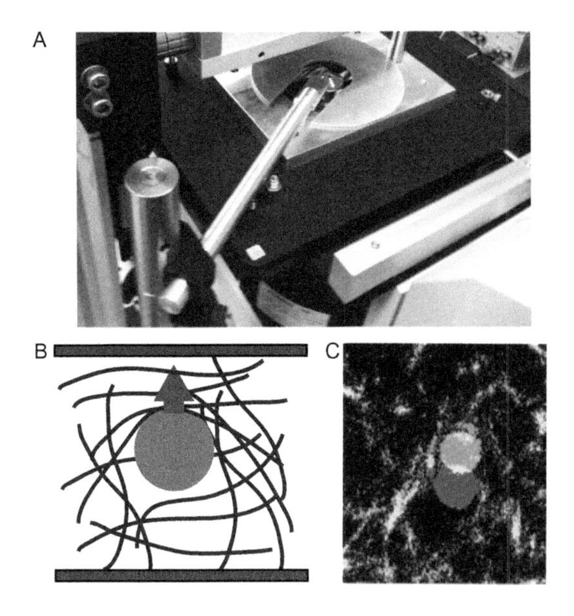

FIGURE 6.8

Design and application of portable magnetic tweezers. (A) Photo of aligned portable magnetic tweezers showing the NdFeB magnets placed directly over the objective and centered with respect to the optical axis; this enables sideways pulling of embedded beads. A sample holder plate is machined to elevate the sample and minimize the distance between the magnets and imaged field of view. (B) Schematic showing direct visualization of cytoskeletal deformation under localized force; magnetic bead (blue) is being pulled in the direction of the red arrow. (C) Dual color image showing force-induced deformation of a 25 μM entangled MT network. Image is an overlay of an image collected under no force (red) and under the application of ~25 pN force (green). In this entangled MT network, the motion of the network is limited to the area just surrounding the bead (4.5 μm diameter), and highly bent filaments can be observed. (See color plate.)

Adapted from Yang et al. (2012). Reproduced by permission of The Royal Society of Chemistry.

than viscosity in the flow regime. Two types of time-dependent loading schemes are used to obtain network response to the sudden application of force (creep response), or to oscillatory loading.

6.1.4.1 Creep analysis

In a creep measurement, an external force is suddenly applied and then maintained at a constant value for a period of time before the load is reduced to zero. We dissect the bead motion into three distinct regimes—a short-time elastic jump, a relaxation transition, and a long-time creep regime (Fig. 6.9A; Yang et al., 2012). Creep velocity v is determined from the slope of a linear fit to bead displacement x versus time t for several (typically <10) seconds prior to the retraction of the magnet pair. We define the start time of the elastic regime by the motion of the magnets toward the sample

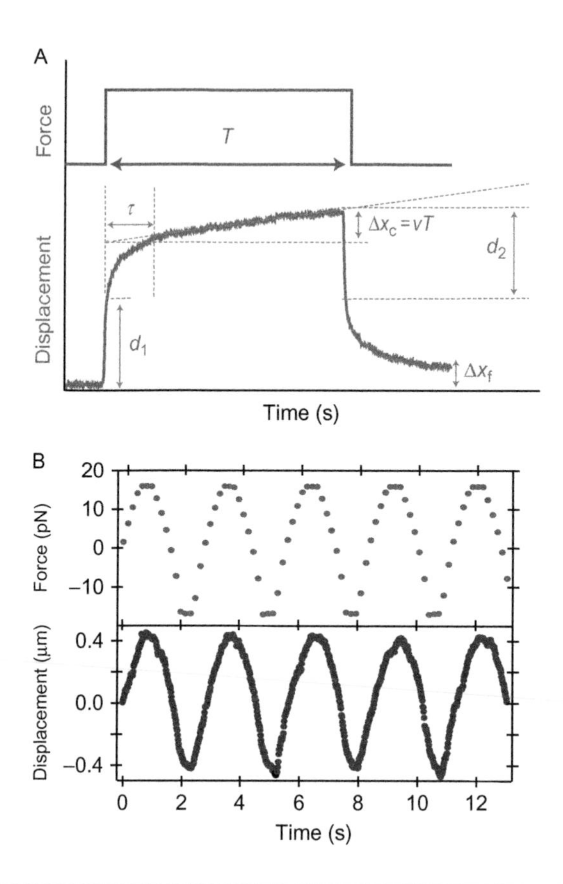

FIGURE 6.9

Representative data traces by magnetic tweezers showing force and bead displacement as a function of time. (A) Schematic of a typical creep response curve under a rectangular force pulse. (B) A typical oscillatory response curve under modulated force. (See color plate.)

Adapted from Yang et al. (2012) and Lin and Valentine (2012b). Reproduced by permission of The Royal Society of Chemistry and the American Institute of Physics, Copyright 2012.

and the end as the time at which the force settles to the commanded value. The "flow regime" follows and ends when the force is reduced at time $t=T$. This can be repeated many times as a function of applied force. After identifying these regimes, the following parameters are calculated: d_1 is the distance the bead travels in the elastic regime and τ is the total time of the relaxation regime. When the magnets are retracted, we identify an elastic recovery regime with jump back distance d_2. The total displacement of the bead due to creep is measured as $\Delta x_c = vT$. The total displacement of the bead after the force pulse is measured as Δx_f. Network stiffness is determined by dividing the applied force F by d_1.

6.1.4.2 *Analysis of oscillatory data*

In oscillatory measurements, the applied force is modulated in time, typically in a sinusoidal or triangle waveform, and the resulting bead displacement measured as a function of time (Fig. 6.9B). This modulation can occur around a stress-free state using a ring magnet geometry, or around an offset force (prestress) using a cube magnet geometry. From these data, we can extract important physical characteristics of the MT network. To measure stiffness, we calculate the ratio of peak force to peak displacement (Lin & Valentine, 2012b). Peak positions and amplitudes are determined through a parabolic fit to those force or position data with amplitudes within ~30% of the peak value. This can be repeated for a range of driving frequencies or forces to directly measure the time- and force-dependent response. The force range is determined by the magnet shape, size, the presence of iron focusing tips, etc., as described earlier. The frequency range is typically limited by the speed range of the motor that drives the magnet position above the sample plane. The ratio of elastic and viscous contributions to gel rheological response can be determined quantitatively by observing the time lag between the peaks in the force and displacement. This time lag is commonly recast in terms of the phase angle δ, defined as the ratio of time lag to oscillation period and normalized such that for elastic gels when $F(t)$ and $\Delta z(t)$ are in phase, $\delta = 0°$, whereas for viscous solutions when $F(t)$ and $\Delta z(t)$ are out of phase, $\delta = 90°$.

SUMMARY

Reconstituted MT networks together with other cytoskeletal networks exhibit a rich set of mechanical properties and provide both a testing ground for fundamental polymer physics and a minimal model of cell mechanics. Microscope-mounted NdFeB-based magnetic tweezers devices enable the study of the relationships between local viscoelastic properties and microstructures. It is possible to vary the magnitude and frequency of applied force, and to simultaneously measure the network structure to determine the molecular origins of MT network strength. We have successfully applied these devices to show that entangled networks are enthalpic viscoelastic solids dominated by filament bending and compression rather than thermal undulations, and that the mechanics of cross-linked networks is dictated primarily by the force-induced failure of chemical bonds (Yang et al., 2012). Future applications will focus on extending these tools to study heterogeneous and enzymatically active cytoskeletal networks of increased chemical complexity to further our understanding of cytoskeleton mechanics and remodeling.

Acknowledgments

Authors thank Jun Lin (UCSB) who with MTV designed and constructed the magnetic tweezers devices outlined herein, Leslie Wilson and Omar Saleh (UCSB) for helpful discussions and for sharing their technical expertise, and Herb Miller (UCSB) for providing purified tubulin proteins. We gratefully acknowledge the support of a Burroughs Wellcome Fund Career Award at the Scientific Interface (to MTV) as well as an NIH/NCRR Shared Instrumentation Grant (1S10RR017753-01) and the MRSEC Program of the NSF under award no. DMR-1121053 for providing core facilities for microscopy and polymer characterization, respectively.

References

Bausch, A. R., Möller, W., & Sackmann, E. (1999). Measurement of local viscoelasticity and forces in living cells by magnetic tweezers. *Biophysical Journal*, *76*(1), 573–579.

Broedersz, C. P., Kasza, K. E., Jawerth, L. M., Munster, S., Weitz, D. A., & MacKintosh, F. C. (2010). Measurement of nonlinear rheology of cross-linked biopolymer gels. *Soft Matter*, *6*(17), 4120–4127.

Chaudhuri, O., Parekh, S. H., & Fletcher, D. A. (2007). Reversible stress softening of actin networks. *Nature*, *445*(7125), 295–298.

Crocker, J. C., Valentine, M. T., Weeks, E. R., Gisler, T., Kaplan, P. D., Yodh, A. G., et al. (2000). Two-point microrheology of inhomogeneous soft materials. *Physical Review Letters*, *85*(4), 888.

de Vries, A. H. B., Krenn, B. E., van Driel, R., & Kanger, J. S. (2005). Micro magnetic tweezers for nanomanipulation inside live cells. *Biophysical Journal*, *88*(3), 2137–2144.

Dent, E. W., & Kalil, K. (2001). Axon branching requires interactions between dynamic microtubules and actin filaments. *The Journal of Neuroscience*, *21*(24), 9759–9769.

Gardel, M. L., Nakamura, F., Hartwig, J. H., Crocker, J. C., Stossel, T. P., & Weitz, D. A. (2006). Prestressed F-actin networks cross-linked by hinged filamins replicate mechanical properties of cells. *Proceedings of the National Academy of Sciences of the United States of America*, *103*(6), 1762–1767.

Gardel, M. L., Valentine, M. T., Crocker, J. C., Bausch, A. R., & Weitz, D. A. (2003). Microrheology of entangled F-actin solutions. *Physical Review Letters*, *91*(15), 158302.

Gardel, M. L., Valentine, M. T., & Weitz, D. A. (2005). Microrheology. In K. Breuer (Ed.), *Microscale diagnostic techniques*. New York: Springer-Verlag.

Gittes, F., Mickey, B., Nettleton, J., & Howard, J. (1993). Flexural rigidity of microtubules and actin filaments measured from thermal fluctuations in shape. *The Journal of Cell Biology*, *120*(4), 923–934.

Gosse, C., & Croquette, V. (2002). Magnetic tweezers: Micromanipulation and force measurement at the molecular level. *Biophysical Journal*, *82*(6), 3314–3329.

Hawkins, T., Mirigian, M., Selcuk Yasar, M., & Ross, J. L. (2010). Mechanics of microtubules. *Journal of Biomechanics*, *43*(1), 23–30.

Head, D. A., Levine, A. J., & MacKintosh, F. C. (2003). Distinct regimes of elastic response and deformation modes of cross-linked cytoskeletal and semiflexible polymer networks. *Physical Review E*, *68*(6), 061907.

Head, D. A., Levine, A. J., & MacKintosh, F. C. (2005). Mechanical response of semiflexible networks to localized perturbations. *Physical Review E*, *72*(6), 061914.

Hirokawa, N., Noda, Y., Tanaka, Y., & Niwa, S. (2009). Kinesin superfamily motor proteins and intracellular transport. *Nature Reviews. Molecular Cell Biology*, *10*(10), 682–696.

Hyman, A., Drechsel, D., Kellogg, D., Salser, S., Sawin, K., Steffen, P., et al. (1991). Preparation of modified tubulins. *Methods in Enzymology*, *196*, 478–485.

Janmey, P. A., Euteneuer, U., Traub, P., & Schliwa, M. (1991). Viscoelastic properties of vimentin compared with other filamentous biopolymer networks. *The Journal of Cell Biology*, *113*(1), 155–160.

Kalil, K., & Dent, E. W. (2005). Touch and go: Guidance cues signal to the growth cone cytoskeleton. *Current Opinion in Neurobiology*, *15*(5), 521–526.

Kaverina, I., Krylyshkina, O., Beningo, K., Anderson, K., Wang, Y.-L., & Small, J. V. (2002). Tensile stress stimulates microtubule outgrowth in living cells. *Journal of Cell Science*, *115*(11), 2283–2291.

Kim, S., & Coulombe, P. A. (2010). Emerging role for the cytoskeleton as an organizer and regulator of translation. *Nature Reviews. Molecular Cell Biology*, *11*(1), 75–81.

Kim, K., & Saleh, O. A. (2008). Stabilizing method for reflection interference contrast microscopy. *Applied Optics*, *47*(12), 2070–2075.

Kirmizis, D., & Logothetidis, S. (2010). Atomic force microscopy probing in the measurement of cell mechanics. *International Journal of Nanomedicine*, *5*, 137–145.

Kollmannsberger, P., & Fabry, B. (2007). High-force magnetic tweezers with force feedback for biological applications. *The Review of Scientific Instruments*, *78*(11), 114301.

Lee, H., Ferrer, J. M., Lang, M. J., & Kamm, R. D. (2010). Molecular origin of strain softening in cross-linked F-actin networks. *Physical Review E*, *82*(1), 4.

Lin, Y.-C., Koenderink, G. H., MacKintosh, F. C., & Weitz, D. A. (2007). Viscoelastic properties of microtubule networks. *Macromolecules*, *40*, 7714–7720.

Lin, J., & Valentine, M. T. (2012a). High-force NdFeB-based magnetic tweezers device optimized for microrheology experiments. *The Review of Scientific Instruments*, *83*(5), 053905.

Lin, J., & Valentine, M. T. (2012b). Ring-shaped NdFeB-based magnetic tweezers enables oscillatory microrheology measurements. *Applied Physics Letters*, *100*(20), 201902.

Liu, J., Koenderink, G. H., Kasza, K. E., MacKintosh, F. C., & Weitz, D. A. (2007). Visualizing the strain field in semiflexible polymer networks: Strain fluctuations and nonlinear rheology of F-actin gels. *Physical Review Letters*, *98*(19), 198304.

Manosas, M., Meglio, A., Spiering, M. M., Ding, F., Benkovic, S. J., Barre, F. X., et al. (2010). Magnetic tweezers for the study of DNA tracking motors. *Methods in Enzymology*, *475*, 297–320.

Mason, T. G., & Weitz, D. A. (1995). Optical measurements of frequency-dependent linear viscoelastic moduli of complex fluids. *Physical Review Letters*, *74*(7), 1250–1253.

Mickey, B., & Howard, J. (1995). Rigidity of microtubules is increased by stabilizing agents. *The Journal of Cell Biology*, *130*(4), 909–917.

Miller, H. P., & Wilson, L. (2010). Preparation of microtubule protein and purified tubulin from bovine brain by cycles of assembly and disassembly and phosphocellulose chromatography. *Methods in Cell Biology*, *95*, 3–15.

Musacchio, A., & Salmon, E. D. (2007). The spindle-assembly checkpoint in space and time. *Nature Reviews. Molecular Cell Biology*, *8*(5), 379–393.

O'Brien, T. E., Cribb, J., Marshburn, D., Taylor Ii, R. M., & Superfine, R. (2008). Chapter 16: Magnetic manipulation for force measurements in cell biology. *Methods in Cell Biology*, *89*, 433–450.

Pelletier, V., Gal, N., Fournier, P., & Kilfoil, M. L. (2009). Microrheology of microtubule solutions and actin-microtubule composite networks. *Physical Review Letters*, *102*(18), 188303.

Preece, D., Warren, R., Evans, R. M. L., Gibson, G. M., Padgett, M. J., Cooper, J. M., et al. (2011). Optical tweezers: Wideband microrheology. *Journal of Optics*, *13*(4), 044022.

Ribeck, N., & Saleh, O. A. (2008). Multiplexed single-molecule measurements with magnetic tweezers. *The Review of Scientific Instruments*, *79*, 094301.

Rodriguez, O. C., Schaefer, A. W., Mandato, C. A., Forscher, P., Bement, W. M., & Waterman-Storer, C. M. (2003). Conserved microtubule-actin interactions in cell movement and morphogenesis. *Nature Cell Biology*, *5*(7), 599–609.

Sato, M., Schwartz, W. H., Selden, S. C., & Pollard, T. D. (1988). Mechanical properties of brain tubulin and microtubules. *The Journal of Cell Biology*, *106*(4), 1205–1211.

Spero, R. C., Vicci, L., Cribb, J., Bober, D., Swaminathan, V., O'Brien, E. T., et al. (2008). High throughput system for magnetic manipulation of cells, polymers, and biomaterials. *The Review of Scientific Instruments*, *79*(8), 083707.

Squires, T. M., & Mason, T. G. (2010). Fluid mechanics of microrheology. *Annual Review of Fluid Mechanics*, *42*, 413–438.

Taute, K. M., Pampaloni, F., & Florin, E.-L. (2010). Chapter 30—Extracting the mechanical properties of microtubules from thermal fluctuation measurements on an attached tracer particle. *Methods in Cell Biology*, *95*, 601–615.

Taute, K. M., Pampaloni, F., Frey, E., & Florin, E.-L. (2008). Microtubule dynamics depart from the wormlike chain model. *Physical Review Letters*, *100*(2), 028102.

Tharmann, R., Claessens, M M A E, & Bausch, A. R. (2007). Viscoelasticity of isotropically cross-linked actin networks. *Physical Review Letters*, *98*(8), 088103.

Uhde, J., Ter-Oganessian, N., Pink, D. A., Sackmann, E., & Boulbitch, A. (2005). Viscoelasticity of entangled actin networks studied by long-pulse magnetic bead microrheometry. *Physical Review E*, *72*(6), 10.

Valdman, D., Atzberger, P. J., Yu, D., Kuei, S., & Valentine, M. T. (2012). Spectral analysis methods for the robust measurement of the flexural rigidity of biopolymers. *Biophysical Journal*, *102*(5), 1144–1153.

Valentine, M. T., Guydosh, N. R., Gutierrez-Medina, B., Fehr, A. N., Andreasson, J. O., & Block, S. M. (2008). Precision steering of an optical trap by electro-optic deflection. *Optics Letters*, *33*(6), 599–601.

Waterman-Storer, C., Duey, D. Y., Weber, K. L., Keech, J., Cheney, R. E., Salmon, E. D., et al. (2000). Microtubules remodel actomyosin networks in xenopus egg extracts via two mechanisms of F-actin transport. *The Journal of Cell Biology*, *150*(2), 361–376.

Yang, Y., Bai, M., Klug, W. S., Levine, A. J., & Valentine, M. T. (2013). Microrheology of highly crosslinked microtubule networks is dominated by force-induced crosslinker unbinding. *Soft Matter*, *9*(2), 383–393.

Yang, Y., Lin, J., Kaytanli, B., Saleh, O. A., & Valentine, M. T. (2012). Direct correlation between creep compliance and deformation in entangled and sparsely crosslinked microtubule networks. *Soft Matter*, *8*(6), 1776–1784.

Yang, Y., Lin, J., Meschewski, R., Watson, E., & Valentine, M. T. (2011). Portable magnetic tweezers device enables visualization of the three-dimensional microscale deformation of soft biological materials. *BioTechniques*, *51*(1), 29–34.

Yvon, A.-M. C., Gross, D. J., & Wadsworth, P. (2001). Antagonistic forces generated by myosin II and cytoplasmic dynein regulate microtubule turnover, movement, and organization in interphase cells. *Proceedings of the National Academy of Sciences of the United States of America*, *98*(15), 8656–8661.

CHAPTER

Studying Mitochondria and Microtubule Localization and Dynamics in Standardized Cell Shapes

7

Andrea Pelikan*, James Sillibourne*, Stephanie Miserey-Lenkei*, Frederique Carlier-Grynkorn*, Bruno Goud* and Phong T. Tran*,†

Cell Biology, Institut Curie, UMR 144 CNRS, Paris, France

†*Cell and Developmental Biology, University of Pennsylvania, Philadelphia, Pennsylvania, USA*

CHAPTER OUTLINE

Introduction **98**
7.1 Design of the Photomask **99**
7.2 Manufacturing of Micropatterned Coverslips **100**
7.3 Seeding and Spreading of Cells **101**
7.4 Mitochondria and Microtubule Visualization **102**
7.4.1 Mitochondria and Microtubule Labeling 102
7.4.2 Live-cell Visualization 103
7.5 Materials **104**
7.5.1 Preparation of Micropatterns 104
7.5.2 Seeding and Spreading of Cell 105
7.5.3 Mitochondria and Microtubule Visualization 105
Conclusion **106**
Acknowledgments **107**
References **107**

Abstract

Mammalian cells show a large diversity in shape and are both shape-changing and mobile when cultured on conventional uniform substrates. The use of micropatterning techniques limits the number of variable parameters, by imposing shape and standardized adhesive areas on the cells, which facilitates analysis. By changing size or

ISSN 0091-679X
http://dx.doi.org/10.1016/B978-0-12-407757-7.00007-4

shape of the micropattern, for example, forcing a polar axis on the cell, it is possible to study how these parameters impact organelle organization, distribution, and dynamics inside the cell. To study the mitochondrial network, which is composed of dynamic tubular organelles dependent on the microtubule cytoskeleton for its distribution, it is important to be able to distinguish between distinct mitochondria. Here, we present a practical method with which we spread the cells on large patterns created with deep UV technique, which not only makes the cells uniform in size and shape as well as immobile, and therefore easier to compare and analyze, but also expands the mitochondrial network and allows for an easier tracking of appropriately labeled individual mitochondria.

INTRODUCTION

The mitochondria are often called the power plants of the cells, generating the ATP the cell needs to carry out its many functions. They are also involved in many other major physiological processes, such as reactive oxygen species biogenesis, calcium homeostasis, and apoptosis. Generally thought of as bean shaped, mitochondria are actually stretched out, tubular structures, which are highly dynamic (Boldogh & Pon, 2007). They constantly undergo fusion, fission, tubulation, and translocation throughout the cell. Mitochondrial dynamics rely on either the actin or microtubule cytoskeleton depending on species and cell type. Knowing the localization and dynamics of mitochondria in healthy and unhealthy cells is crucial for understanding their functions and finding cures for the many mitochondria-based diseases that exist (Chan, 2006).

Individual cultured cells can adopt many different shapes over time, which influence the positions and dynamics of organelles, making analysis difficult. Cells can also be mobile, which makes viewing them over longer periods of time in a region of interest complicated. Historically, mitochondria dynamics have often been studied in neuronal axons (Saxton & Hollenbeck, 2012). This is due to the fact that axons approximate regular 2D structures, where the mitochondria and their motions are rectilinear and parallel to the axonal long axis, making dynamic analysis easily standardized. In other cultured cell types spread on a conventional substrate, where cells adopt diverse shapes and are motile, the organization of mitochondria is not as straightforward to study. To immobilize the cells and standardize cell sizes and shapes, we applied micropatterning techniques using deep UV (Azioune, Storch, Bornens, Théry, & Piel, 2009). This method enables us to image and analyze mitochondria and microtubule dynamics and organization in standardized cells. This technique is straightforward to use in a typical biology lab, requires reasonably simple equipments, and gives robust results. We describe below how the micropatterns are made and how the cells are attached and the mitochondria visualized. This particular protocol is optimized for RPE1 cells but can be modified to study mitochondria in other cell types as well.

7.1 DESIGN OF THE PHOTOMASK

The micropatterns which define different cell sizes and shapes are typically drawn using software such as L-Edit (www.tannereda.com). The files containing the micropatterns are then sent to a company such as Toppan (www.photomask.com) which laser-etches the micropatterns onto a synthetic quartz plate, transparent to deep UV light (wavelength below 200 nm). This is the "photomask" which will be used subsequently in lab to produce the micropatterned substrate to adhere cells. Photomasks vary in size and composition. We find that a typical 12.5 cm × 12.5 cm photomask is useful and can fit many different micropatterns.

The size of individual micropatterns is an important parameter, for example, if the pattern is too big, the cells will not be able to spread completely, and if it is too small, the cells will not spread enough for visualizing the mitochondria properly. Examination of cells spread on a regular coverslip can give reasonable estimates of cell size prior to designing the micropatterns. It is important to keep in mind that cells respond to the size of the micropatterns. Really small or really big patterns change the tension of the underlying actin cytoskeleton, which has been shown to change, for example, centrosome positioning (Rodríguez-Fraticelli, Auzan, Alonso, Bornens, & Martín-Belmonte, 2012). A too small size might even induce apoptosis (Chen, Mrksich, Huang, Whitesides, & Ingber, 1997). It is a good idea to design several sizes to optimize cell spreading for specific studies. Well-spread cells will help visualize the individual mitochondrion (Chevrollier et al., 2012), whereas a more compact spreading may be good for making mitochondria density maps (Schauer et al., 2010).

Another parameter to consider is the shape of the micropatterns, as the adhesion of cell to its substrate profoundly influences cytoskeletal organization, cell polarity, and cell shape (Théry et al., 2006). For example, a round micropattern leads to round cells which have no polarity compared to a crossbow-shaped micropattern, which produces highly polarized cells. In this context, one can design micropatterns where the whole area of the shape will be filled with adhesive substrate, or only the outline of the shape will be adhesive, or any combination of adhesive and non-adhesive regions, etc. The distance separating individual micropatterns should be big enough to prevent a cell from spreading over two patterns or to contact another cell on an adjacent pattern. This distance is dependent on cell size and how mobile the cells are, and therefore on cell types, but 50–100 μm is a good reference.

Finally, the transfer of the micropatterns from the photomask onto individual glass coverslip can be optimized. A typical photomask has dimension 12.5 cm × 12.5 cm, and a typical coverslip has dimension 2 cm × 2 cm. Therefore, the photomask can be sectored into 5 × 5 squared regions, fitting 25 different coverslips simultaneously. Each region can have about 1000 identical micropatterns or 1000 different individual micropatterns, or any other combinations. It is useful to create 1-mm wide borders and to introduce numbering/lettering visible to the eye to distinguish the different regions. Inside each region, further markings may help differentiate different areas of the region which may help subsequent navigation through the coverslip under the microscope, if the patterns are stained with a fluorescent protein.

7.2 MANUFACTURING OF MICROPATTERNED COVERSLIPS

Work in a dedicated clean area, preferably under a hood. Wear gloves and avoid dust by covering the photomask and coverslips. We prefer square coverslips because they are easier to orient; however, round coverslips will also work well. Figure 7.1A summarizes how the micropatterns on photomask is transferred to the coverslip.

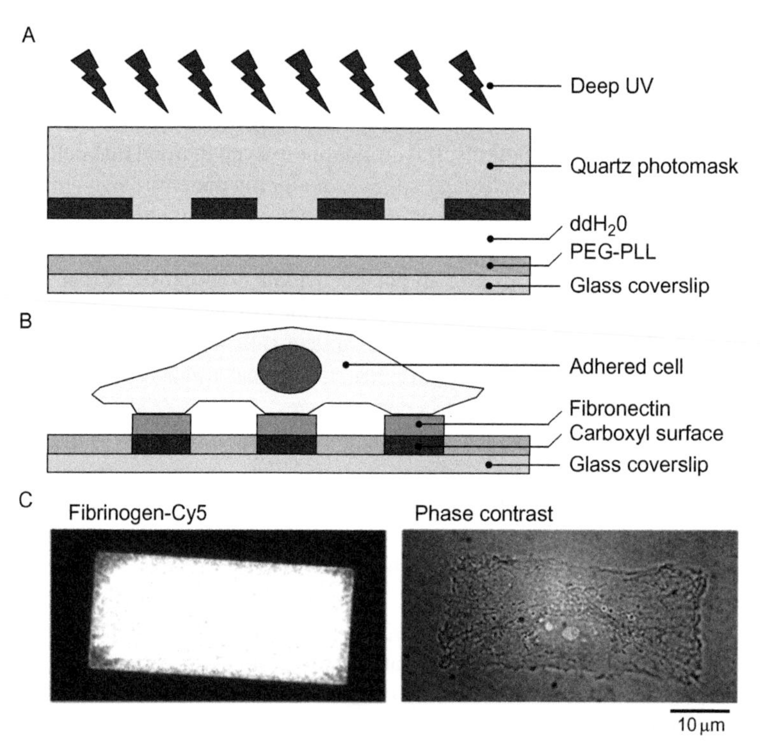

FIGURE 7.1

Deep UV micropatterning technique standardizes cell shape and size for subsequent imaging of mitochondria and microtubules. (A) Schematic of the deep UV technique to transfer the micropatterns from the photomask to the PEG-PLL-coated glass coverslips. The UV irradiation creates carboxyl groups in the exposed areas of the PEG, which subsequently binds to proteins such as fibronectin. (B) Schematic of UV-exposed PEG-PLL-coated coverslip treated with fibronectin then seeded with cells. The fibronectin binds only to the carboxyl surface of the PEG-PLL. Cells adhere only to the fibronectin. (C) Phase-contrast image of an RPE1 cell adhering to the rectangular shape coated with fibronectin. Fibrinogen-Cy5 was added to the fibronectin solution to help visualize the shape of the coated surface. (See color plate.)

1. Clean the coverslips in 70% EtOH for 15 min by sonication. Dry them on KimWipes or with air flow and store in a covered dish to avoid dust.
2. Clean the coverslips inside the plasma chamber for 1 min or by UV irradiation for 5 min.
3. Place the cleaned coverslips upside down on 10–50 μl drops of 0.1 mg/ml PEG-PLL (poly-L-lysine-polyethylene glycol) in 10 mM Hepes, pH 7.4, for 30 min to 1 h on a parafilm in a humid chamber (e.g., a large Petri dish with a wetted Whatman paper under the parafilm). The stock solution of 1 mg/ml in 10 mM Hepes, pH 7.4, is very stable and can be kept at 4 °C for months.
4. Wash the coverslip by dipping it into ddH_2O. Dry with airflow or just in open air under a cover, placed on a KimWipe with the PEG-PLL side up.
5. Rinse the photomask in ddH_2O and 70% EtOH. Dry with airflow or with KimWipes. Expose the brown (reverse) side of the photomask to UV light for 5 min.
6. Place the PEG-PLL side of the coverslip on the brown side of the photomask, on a very small drop of ddH_2O (e.g., 2 μl for a 20-mm square coverslip). Make sure the water spreads evenly to cover the whole coverslip. Press on the coverslip with a plastic pipette tip to remove all air bubbles. Repeat this process if multiple coverslips are needed for multiple regions of the photomask. The coverslips should adhere tightly to the photomask and look iridescent. This is crucial for a sharp final pattern.
7. Place the mask upside down, the silver side facing the UV lamp, and illuminate for 5–10 min. The deep UV irradiation ($\lambda < 200$ nm) effectively removes the carbon C—O—C group from the PEG-PLL which is repellent to proteins, and leaves exposed the carboxyl C—O—H groups which is attractant for proteins.
8. Detach the coverslips by inundating the photomask with ddH_2O and recover the coverslips by gently pushing them to the edge of the photomask using a plastic pipette tip.
9. Dry the coverslips with airflow or by placing on a KimWipe under a cover. The micropatterned coverslips can now be stored for a few months or used at once.

7.3 SEEDING AND SPREADING OF CELLS

Many cell types can be seeded and spread onto the adhesive micropatterns. We typically use RPE1 cells stably expressing fluorescent markers for mitochondria. However, once adhered to the micropatterns, cells can subsequently be fixed for staining with any markers of interest. Below are the steps for seeding and spreading cells onto the micropatterns. Figure 7.1B summarizes how a cell adheres to the UV-exposed fibronectin-coated coverslip. Figure 7.1C shows the rectangular shape coated with fibronectin and fibrinogen-Cy5 and an adhered cell which conforms to the rectangle.

1. Incubate the micropatterned coverslips upside down on drops of 50 μg/ml fibronectin solution diluted in ddH_2O in a humid chamber at room temperature for 1 h. Include in the drops 5 μg/ml fluorescent-labeled fibrinogen to easily

visualize the patterns under the microscope. The proteins will bind tightly to the UV-exposed PEG-PLL regions, serving as substrates of cell adhesion.

2. Put the coverslips into 0% FBS medium in wells in an incubator for 5 min before seeding the cells. High concentrations of serum might impair attaching and spreading.
3. Wash, in PBS, adherent cells that have been growing for a couple of days undisturbed. Detach the cells with trypsin-EDTA at 37 °C for 3–10 min.
4. Carefully but resolutely resuspend the cells in complete medium, for example, by pressing the pipette tip against the bottom of the flask while dispensing them. Use a total of 5 ml for a T25 flask. Add cells to the wells with the coverslips (e.g., 400 μl cell suspension per 3.5 cm well). Alternatively, remove an aliquot for counting while spinning the rest. Seed the cells on the coverslips at a density of approximately 10,000 cells/cm^2.
5. Put the coverslips into the incubator for a brief time (10–20 min for RPE1 cells) before checking the cells. When the cells are attached, ideally there is one cell stuck to each pattern, and the pattern array is easily visible. Wash away unattached excess cells gently. It is important to never remove all medium, since the cells would quickly dry out due to the hydrophobicity of the PEG-PLL coating on the coverslip. HEPES can be used for better buffering outside of the incubator for steps taking longer time, like the washing step.
6. Let the cells spread fully in the incubator (2–4 h for RPE1 cells). Cells are now ready to be visualized live or alternatively be fixed and stained for later visualization. Cells can be fixed with PAF by adding one volume of 6% PAF in PBS to coverslips placed in one volume of medium. If the mitochondria are to be visualized using MitoTracker, incubation (15–30 min) with MitoTracker must take place before fixing.

7.4 MITOCHONDRIA AND MICROTUBULE VISUALIZATION

We typically use RPE1 cells expressing fluorescent markers for mitochondria. There are many protocols for labeling cells with the fluorescent marker of interest. We briefly describe below the methods we use.

7.4.1 Mitochondria and microtubule labeling

There are several ways to fluorescently label the mitochondria and microtubules. For fixed cells, there are a multitude of antibodies against different mitochondrial proteins, and a multitude of antibodies against microtubules. A general method for fixing and staining cells follows:

1. Place the coverslips cell side up in drops of PBS in a humid chamber.
2. Wash the coverslips containing the spread cells in micropatterns with PBS, by alternately aspiring and pipetting 1 ml of PBS, carefully to prevent dehydration.

Wash again with PBSAT and leave a drop for 5 min while preparing the primary antibody solution.

3. Aspirate off the PBSAT and incubate with 100 µl of primary antibody solution for 1 h at room temperature (or overnight at 4 °C). If using several antibodies, they should be recognized by different secondary antibodies. For microtubules, we use anti-α-tubulin B-5-1-2 antibodies.
4. Wash with PBSAT and leave a drop while preparing the secondary antibody solution. Aspirate off the PBSAT and add the secondary antibody for 30 min.
5. Wash in PBSAT and then in PBS. Remove the PBS, add DAPI for 1 min, then wash with ddH_2O, and let air dry under a cover before mounting the coverslips on slides with 20 µl Mowiol.

More commonly used these days to label mitochondria are MitoTracker probes (www.invitrogen.com), which are well adapted for imaging of fixed cells but should not be incubated with the cells for longer than 15–30 min. At longer incubation times, the MitoTracker starts to label other membrane structures and even disturbs the native mitochondrial morphology, which makes it unsuitable to use in live-cell imaging. The MitoTracker must be incubated with the cells prior to fixation.

7.4.2 Live-cell visualization

For live-cell imaging of mitochondria, we use Mito-protein, which is a gene-based marker expressing a 29-amino acid-long polypeptide corresponding to the mitochondrial targeting sequence from human cytochrome *c* oxidase subunit 8. Mito-protein can be tagged with mRFP or EGFP, giving strong signal and low bleaching. Fluorescent Mito-protein can be used for transient transfection or to create stable cell lines. Cells stably expressing fluorescent tubulin can also be made.

For live-cell imaging, the patterned coverslip needs to be mounted in a chamber so that medium can be added onto the cells. Make sure that the size of the coverslips and the chamber are compatible. Different kinds of chambers are commercially available, for example, Ludin chamber (www.lis.ch) or magnetic chamber (www.chamlide.com). We find the magnetic chamber to be very practical due to its quick magnetic closing function and the fact that it exists for different sizes and also for square coverslips, but the gasket might be a bit tricky to clean, which may stress the cells if residual cleaning solvent remains. Letting the gasket incubate in the medium overnight will help dilute any residual solvent.

It is possible to seed and spread cells on the coverslip directly inside the chamber or to move the coverslip with the already spread cells from the well into the chamber. In the latter case, make sure to have a pipette with medium ready before taking the coverslip from the well to minimize the time in open air to prevent dehydration.

The patterned cells inside the chamber are now ready to be imaged live on the microscope. Figure 7.2A compares a fixed cell spread on a nonpatterned coverslip to that of a cell spread on a rectangular shape micropatterned coverslip. Figure 7.2B shows the robustness and repeatability of the micropatterned cells.

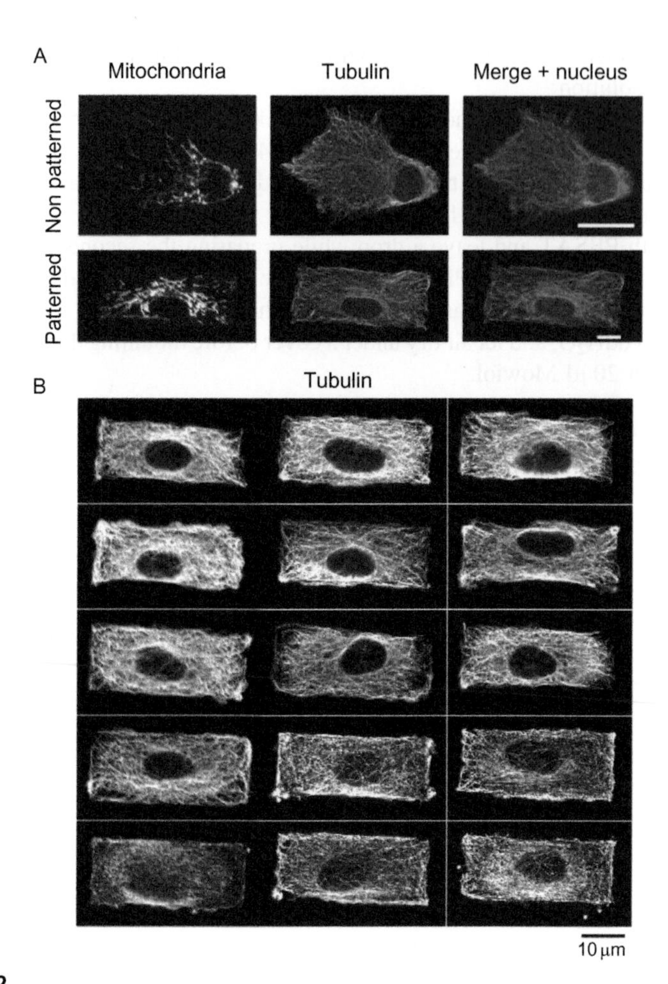

FIGURE 7.2

Micropatterning technique is robust. (A) A comparison between RPE1 cells seeded onto nonpatterned surface and cells seeded onto rectangular patterned surface. Nonpatterned cells can adopt many shapes, making analysis of dynamics subcellular structures such as mitochondria and microtubules challenging. Bar 10 μm. (B) Patterned shapes are repeatable, given rise to standardized cell shapes and size, making analysis of dynamics subcellular structures such as mitochondria and microtubules less challenging. (See color plate.)

7.5 MATERIALS

7.5.1 Preparation of micropatterns

glass coverslips (suitable size for the live-cell chamber)
70% EtOH
large Petri dish

Whatman paper
parafilm
PLL(20)-g[3.5]-PEG(2), stock solution of 1 mg/ml in 10 mM Hepes, pH 7.4 (Surface Solutions GmbH, Switzerland)
10 mM HEPES, pH 7.4
ddH_2O
KimWipes
Plasma cleaner (Harrick Plasma)
synthetic quartz mask with features (Toppan photomasks)
UV lamp, 185 nm (Jelight, UVO Cleaner 42) with an ozone killer (or under a hood).

7.5.2 Seeding and spreading of cell

fibronectin (Sigma, F1141)
fluorescently labeled fibrinogen, resuspended to 1 mg/ml (Invitrogen, e.g., F-13191)
complete culture medium:
  DMEM/F-12 (Invitrogen/Gibco, 21041) complemented with
    1–10% fetal bovine serum (Biowest, S1810)
    2 mM L-glutamine (Invitrogen/Gibco, 25030)
    1% penicillin/streptomycin (Invitrogen/Gibco, 15140)
    2 g/l sodium bicarbonate (Invitrogen/Gibco, 25080) (for CO_2 atmosphere)
culture flasks and dishes (TPP)
PBS
TrypLE (Invitrogen/Gibco, 12605)
coverslips (round or square, fitting the live-cell chamber)
PAF (for fixing)
slides

7.5.3 Mitochondria and microtubule visualization

MitoTracker® Probes (Invitrogen, M7512, M22426, etc.)
fluorescent protein Mito (derived from pEYFP-Mito vector, Clontech, ref. 6115-1)
selection antibiotics for stable cell line
cell culture medium
chamber for live-cell visualization (magnetic from Live Cell Instrument or Cytoo, or Ludin from Life Imaging Services)
Blocking/antibody dilution buffer (PBSAT):
  PBS
  1% BSA fraction V (Sigma, A3059)
  0.5% Triton X-100 (dilute from 25% stock)
Microtubule antibody (e.g., anti-α-tubulin B-5-1-2, Sigma) diluted in PBSAT
Fluorescent secondary antibody diluted in PBSAT
DAPI 1 mg/ml (dilute in H_2O from 5000$\times$ stock)

Mowiol mounting medium:

Add 12 ml of 50% glycerol to 2.4 g of Mowiol 4–88 (Calbiochem, 475904). Mix for 2 h at room temperature. Add 12 ml of 200 mM Tris–HCl, pH 8.5. Incubate at 50 °C (occasionally mixing) until dissolution (~3 h). Pass through a 0.45-μm filter, aliquot, and store at −20 °C.

CONCLUSION

This method for making micropatterns using deep UV light is straightforward to use and does not demand any extravagant instruments. Here, we presented an overview of how to make the micropatterns. Further discussion on troubleshooting and variation of the technique has been published elsewhere (Azioune, Carpi, Tseng, Théry, & Piel, 2010).

Cells on a uniformly protein-coated coverslip are seldom isolated from their neighbors, and often even spread on top of each other, despite seeding reasonably few cells. They adopt a large variation of shapes and are not easily comparable. Forcing a predefined shape and size on cells using micropatterning techniques, on the other hand, has shown to be an efficient way of limiting the number of unknown parameters. Size and shape can be varied rigorously and independently, allowing study of how they influence the cellular organization. Since the cells are normalized, analysis can be standardized and even automated (Chevrollier et al., 2012, Schauer et al., 2010), which would be possible only with difficulty using cells on a conventional substrate. Our current technique confines cells to 2D organization. Future development using 3D micropatterning would be a step further in the direction of recreating the conditions a cell experience in its natural environment (Ochsner, Textor, Vogel, & Smith, 2010).

The current method enables studies of how the shape of the cell and the mechanical forces due to spatially well-defined interactions with the substrate influence the mitochondria and the cytoskeleton. Cells shrink when dying, but there is also a causal link in the opposite direction, that is, cells die when artificially shrunk by adhesion to too small patterns (Chen et al., 1997). From a physiological point of view, this response (cell growth or death) to mechanical deformation of the cell or nucleus is probably important during development and tissue homeostasis, and its deregulation might be implicated in developmental defects and cancer. The mitochondria being a crucial factor in apoptosis (Hoppins & Nunnari, 2012), the method described here should be helpful for better elucidating the mechanisms involved in this process, and the importance of their fragmentation during apoptosis (Frank et al., 2001).

Signaling between cells has been well studied, but less is known of the interactions between cells when it comes to intercellular trafficking (Rustom, Saffrich, Markovic, Walther, & Gerdes, 2004), although there is evidence of transfer of mitochondria or at least their DNA between cells (Spees , Olson, Whitney, & Prockop, 2006). Using cleverly designed micropatterns, on which the cells would contact each other in predefined and controlled ways, should greatly facilitate the study of this phenomenon.

The morphology of the mitochondrial network is not only changing during the cell cycle, but actually has a crucial role during at least the G_1 to S transition (Mitra, Wunder, Roysam, Lin, & Lippincott-Schwartz, 2009). The visualization of the mitochondria in cells spread on micropatterns could help deepening the understanding of how and why their morphology correlates with or even causes the transitions in the cell cycle.

In summary, this method is useful whenever it is important to clearly see and measure mitochondria morphology, localization, and dynamics in cells, fixed or living, under controlled spatial conditions.

Acknowledgments

We thank Nicolas Carpi, Kristin Schauer, and Julie Janvore for helpful technical suggestions. This work was supported by grants from ANR and LaLigue.

References

Azioune, A., Carpi, N., Tseng, Q., Théry, M., & Piel, M. (2010). Protein micropatterns: A direct printing protocol using deep UVs. *Methods in Cell Biology, 97*, 133–146.

Azioune, A., Storch, M., Bornens, M., Théry, M., & Piel, M. (2009). Simple and rapid process for single cell micro-patterning. *Lab on a Chip, 9*, 1640–1642.

Boldogh, I. R., & Pon, L. A. (2007). Mitochondria on the move. *Trends in Cell Biology, 17*, 502–510.

Chan, D. C. (2006). Mitochondria: Dynamic organelles in disease, aging, and development. *Cell, 125*, 1241–1252.

Chen, C. S., Mrksich, M., Huang, S., Whitesides, G. M., & Ingber, D. E. (1997). Geometric control of cell life and death. *Science, 276*, 1425–1428.

Chevrollier, A., Cassereau, J., Ferré, M., Alban, J., Desquiret-Dumas, V., Gueguen, N., et al. (2012). Standardised mitochondrial analysis gives new insights into mitochondrial dynamics and OPA1 function. *The International Journal of Biochemistry & Cell Biology, 44*, 980–988.

Frank, S., Gaume, B., Bergmann-Leitner, E. S., Leitner, W. W., Robert, E. G., Catez, F., et al. (2001). The role of dynamin-related protein 1, a mediator of mitochondrial fission, in apoptosis. *Developmental Cell, 1*, 515–525.

Hoppins, S., & Nunnari, J. (2012). Mitochondrial dynamics and apoptosis—The ER connection. *Science, 337*, 1052–1054.

Mitra, K., Wunder, C., Roysam, B., Lin, G., & Lippincott-Schwartz, J. (2009). A hyperfused mitochondrial state achieved at G1-S regulates cyclin E buildup and entry into S phase. *Proceedings of the National Academy of Sciences of the United States of America, 106*, 11960–11965.

Ochsner, M., Textor, M., Vogel, V., & Smith, M. L. (2010). Dimensionality controls cytoskeleton assembly and metabolism of fibroblast cells in response to rigidity and shape. *PLoS One, 5*, e9445.

Rodríguez-Fraticelli, A. E., Auzan, M., Alonso, M. A., Bornens, M., & Martín-Belmonte, F. (2012). Cell confinement controls centrosome positioning and lumen initiation during epithelial morphogenesis. *The Journal of Cell Biology*, *198*, 1011–1023.

Rustom, A., Saffrich, R., Markovic, I., Walther, P., & Gerdes, H. H. (2004). Nanotubular highways for intercellular organelle transport. *Science*, *303*, 1007–1010.

Saxton, W. M., & Hollenbeck, P. J. (2012). The axonal transport of mitochondria. *Journal of Cell Science*, *125*, 2095–2104.

Schauer, K., Duong, T., Bleakley, K., Bardin, S., Bornens, M., & Goud, B. (2010). Probabilistic density maps to study global endomembrane organization. *Nature Methods*, *7*, 560–566.

Spees, J. L., Olson, S. D., Whitney, M. J., & Prockop, D. J. (2006). Mitochondrial transfer between cells can rescue aerobic respiration. *Proceedings of the National Academy of Sciences of the United States of America*, *103*, 1283–1288.

Théry, M., Racine, V., Piel, M., Pépin, A., Dimitrov, A., Chen, Y., et al. (2006). Anisotropy of cell adhesive microenvironment governs cell internal organization and orientation of polarity. *Proceedings of the National Academy of Sciences of the United States of America*, *103*, 19771–19776.

CHAPTER

Going Solo: Measuring the Motions of Microtubules with an *In Vitro* Assay for TIRF Microscopy

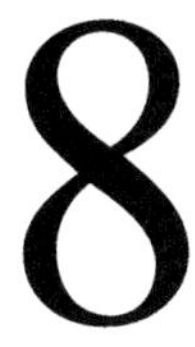

Kris Leslie and Niels Galjart

Department of Cell Biology, Erasmus Medical Center, P.O. Box 2040, Rotterdam, The Netherlands

CHAPTER OUTLINE

Introduction **110**
8.1 Protocols **111**
8.1.1 Preparation of Coverslips and Slides 111
8.1.1.1 Chemical Cleaning of Slides and Coverslips 111
8.1.1.2 Sonication of Slides and Coverslips 111
8.1.2 Assembly of *in vitro* Sample Chambers for the TIRF Microscope 112
8.1.3 Preparation of GMPCCP-stabilized MT Seeds 113
8.1.4 TIRF Microscopy of +TIPs *in vitro* 115
8.1.4.1 Preparation of an Oxygen Scavenger System 116
8.1.4.2 Chemically Functionalizing the Sample Chamber 117
8.1.4.3 Attaching the Seeds and Blocking the Chamber 118
8.1.4.4 Preparation of the Protein Samples 120
8.1.4.5 TIRF Microscopy of the Samples 121
Concluding Remarks **123**
Acknowledgments **123**
References **123**

Abstract

Microtubule (MT) plus-end-tracking proteins (+TIPs) specifically associate with the plus ends of growing MTs, thereby determining, in many different ways, the dynamic behavior of the MTs. Over the past years, a variety of fluorescently tagged +TIPs have been purified. Their reconstitution together with other purified components

ISSN 0091-679X
http://dx.doi.org/10.1016/B978-0-12-407757-7.00008-6

involved in MT plus-end-tracking, and analysis by total internal reflection fluorescence microscopy, has helped to elucidate some of the crucial mechanisms underlying the motion of MTs. For example, +TIP dwell time and association rate, and key MT dynamic instability parameters can be measured in a controlled cell-free environment. In this chapter, we have aimed to describe in an accessible and practical manner how we carry out these assays in our lab. We cover basic steps such as the preparation of glass and sample chambers through to the details of the *in vitro* +TIP assay. When appropriate, we mention common problems providing practical help to overcome potential issues.

INTRODUCTION

The fluorescence microscopy-based *in vitro* assay for measuring microtubule (MT) behavior is essentially the reconstitution of (fluorescent) key components required for MT growth. Besides tubulin, which is virtually always purified from brain (Miller & Wilson, 2010) or purchased (Cytoskeleton Inc.), it also relies upon purified protein components such as EB proteins, which have to be "homemade" and purified (Honnappa, John, Kostrewa, Winkler, & Steinmetz, 2005; Komarova et al., 2009). Furthermore, it also requires the appropriate buffer conditions and essential cofactors for MT growth, such as GTP. The first application of the reconstituted *in vitro* assay, as described here, to study the dynamic instability of MTs in the presence of MT plus-end-tracking proteins (+TIPs) was successfully carried out by Bieling et al. in 2007 on the yeast +TIPs Mal3, Tea2, and Tip1 (Bieling et al., 2007); the proteins were observed to interact on dynamically growing MTs. By labeling the proteins with the fluorescent markers GFP and mCherry, it was possible to follow their binding interactions utilizing two-color total internal reflection fluorescence (TIRF) microscopy. It is the use of this *in vitro* assay, as opposed to studying +TIPs via whole-cell imaging, that has allowed the dissection of the complex interactions of +TIPs with MTs over the past few years (Brouhard et al., 2008; Dixit et al., 2009; Honnappa et al., 2009; Maurer, Bieling, Cope, Hoenger, & Surrey, 2011; Zimniak, Stengl, Mechtler, & Westermann, 2009). Parameters that can be easily measured using fluorescent +TIPs as markers are MT growth rate (+TIPs specifically recognize the ends of growing MTs), the dwell time (or residency time) of the +TIP, and even, when using single-molecule conditions, the association rate (K_{on}) (Bieling et al., 2008; Komarova et al., 2009; Montenegro Gouveia et al., 2010). A major advantage of the fluorescence microscopy-based MT *in vitro* system over that of traditional biochemical assays is that it is possible to visualize and measure the interaction of the proteins on specific sites of the MT, while classical assays measure ensemble MT behavior. By altering biochemical conditions, one can address specific questions such as the effect of ionic strength on +TIP binding (Buey et al., 2011) and investigate the interplay of additional protein factors at the plus end (Montenegro Gouveia et al., 2010). The MT *in vitro* assay is fairly simple to carry out. One complexity lies in the reconstitution of the individual components which sometimes requires troubleshooting. All the proteins involved have to be tagged for visualization in the TIRF microscope and rigorously purified to avoid artifacts due to contaminating proteins.

8.1 PROTOCOLS

8.1.1 Preparation of coverslips and slides

For fluorescence microscopy-based *in vitro* experiments, small glass sample chambers are assembled from microscope slides and coverslips. It is crucial that all glassware is thoroughly cleaned, as the major contaminants for TIRF microscopy, which tend to be organic in nature, exhibit autofluorescence. Cleaning can be done in several ways. If one has access to a plasma cleaner, then this is perhaps the easiest method. However, there are also inexpensive chemical alternatives. We have employed two methods in the lab that successfully remove contaminants.

8.1.1.1 Chemical cleaning of slides and coverslips

One method is to soak the slides and coverslips overnight at room temperature in a saturated KOH solution dissolved in ethanol (~16.5 g of KOH pellets per 100 ml of ethanol, *note*: this produces an extremely basic solution, and hence care should be employed as the solution is highly corrosive and the ethanol is highly flammable). For safety, use the smallest volume required to clean the glass; we use approximately 50 ml to prepare a set of slides and coverslips. The next morning, remove the slides and rinse thoroughly in Milli-Q water six times. The slides and coverslips can then be stored in fresh Milli-Q water for up to 1 week, ensuring that they are protected from dust and other particulate matter. Use slide and coverslip holders made from glass or ceramic material during the overnight soaking step so as to provide efficient cleaning.

8.1.1.2 Sonication of slides and coverslips

The second cleaning method that we have employed is based on sonication. Rest the slides and/or coverslips in an appropriate holder and place them in a clean beaker (100 or 200 ml depending upon the size of the holder). Add a solution of pure isopropanol, 0.2 μm filtered, until it passes the coverslips/slides by approximately 1 cm (see Fig. 8.1). Next, sonicate the mixture for 20 min. After sonication, wash the glass in Milli-Q water three times. Make up a 1 M solution of KOH, 0.2 μm filtered, and add it to the beaker, again making sure that there is at least a 1 cm clearance above the glass, and repeat the sonication for further 20 min. Remove the KOH and wash further three times in Milli-Q water. Store the slides and coverslips in fresh Milli-Q and an appropriate container until use (do not keep them longer than a week). Using this method, a batch of coverslips can be prepared quickly in less than 2 h. The results are comparable to the standard chemical cleaning as outlined above.

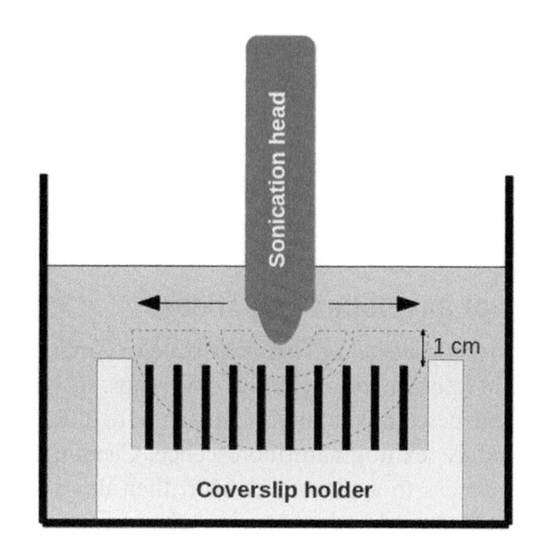

FIGURE 8.1

A schematic of the setup used for sonication of coverslips. Various parts of the apparatus are indicated. Coverslips are shown as vertical black lines. The sonicator head is moved periodically across the coverslips insuring a 1 cm clearance.

8.1.2 Assembly of *in vitro* sample chambers for the TIRF microscope

With the glassware cleaned, you can now proceed to making the *in vitro* sample chamber, a 10-μl chamber where all the components of the MT plus-end-tracking system are assembled.

Required materials

Cleaned coverslips, 18 × 18 mm, rectangular for cell assembly.
Cleaned glass slides.
Double-sided sticky tape (Scotch, 12.7 × 22.8 mm, catalog number 665).
UV curing glue (Norland Optical Adhesive 88, Norland Products, catalog number P/N 8801).[1]
Clean tweezers.

Prior to making the sample cells, draw an outline of the sample chamber on a surface and place a clean slide on top of this template. This allows one to easily position the tape at the correct distance and make sample cells quickly with the same dimensions (see Fig. 8.2). Place the double-sided tape across the slide with a width of 5 mm/0.2″ between each piece of tape. The width is important as it determines the volume (~10 μl) of the sample cell, and subsequent washing steps for the sample chamber are based upon this volume. If you make larger sample cells, then remember to adjust the washing steps accordingly (see Sections 1.4.2 and 1.4.3). However, care must be

[1]Optional.

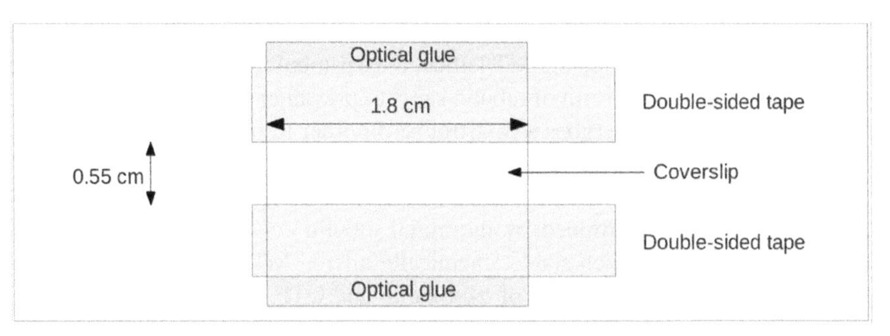

FIGURE 8.2

A diagram of the sample holder with each component highlighted. As indicated in the text, the use of glue is optional. (See color plate.)

taken when washing attached MT seeds as an excess of flow buffer can degrade the seeds (Section 1.4.3) (we have used a maximum of 200 μl without deleterious results).

Once the tape has been applied, dry off the clean coverslip holding it with tweezers. Ideally, drying should be done with nitrogen, but we have also used filtered air without any problems. Alternatively, the coverslips can be dried in a 50 °C oven prior to sample cell assembly. Once the coverslip is ready, place it over the tape as shown in Fig. 8.2. Use a scalpel to remove any excess tape and mold it to the placed coverslip. Make sure that there is good contact between the tape and the mounted coverslip by gently rubbing the surface with the tweezers. With good contact, it will be obvious as the tape sticks to the lower surface. It is very important that this contact is established along the "walls" of the chamber otherwise wash buffers will flow between the coverslip and the tape.

Load a pipette with 10 μl of Norland optical adhesive and apply to both sides of the coverslip, as shown in the figure. Usually, only 3–5 μl per side is required. Use of the glue is not essential but we have found that it can reduce the effects of drift in the microscope, presumably because the glue is more resistant to thermal changes than the double-sided tape and therefore the coverslip remains in place during TIRF experiments. Finally, place the coverslip in a UV box for 5 min to cure the adhesive. Once ready, store the slides in a glass box, precleaned and blown out with nitrogen or filtered air. We usually make fresh slides for 1 day's worth of experiments, but we have stored cleaned coverslips and slides in Milli-Q water for up to a week.

8.1.3 Preparation of GMPCCP-stabilized MT seeds

In cells, as a rule, MTs nucleate in the MT organizing center. For in vitro TIRF experiments we do not have this luxury, and therefore, we have to invent a nucleation point from which fresh MTs can grow. A simple strategy is to make MTs that are chemically stabilized so that they do not undergo catastrophes. From the free ends of the MTs, we can add additional tubulin to promote fresh growth. αβ-Tubulin dimers bound to GTP assemble into MTs, and with this form of tubulin, MTs do not depolymerize. However, once inside the MT lattice, the αβ-tubulin-GTP readily hydrolyzes to αβ-tubulin-GDP. If this form of tubulin starts to prevail at an MT end, then the MT becomes unstable and undergoes a catastrophe. In order to prevent this, we bind αβ-tubulin dimers to a nonhydrolyzable analogue of GTP, which is GMPCCP (GpCpp guanosine-5′-[(α,β)-methyleno]triphosphate). These dimers will associate into MTs to an average length determined by the initial tubulin concentration and they will not depolymerize. From such stable, chemically altered MT "seeds," we can obtain fresh nucleation upon the addition of αβ-tubulin and GTP.

Required materials

- MRB80 buffer: 80 mM piperazinediethanesulfonic acid (PIPES) buffered to pH 6.8 with KOH (room temperature), 1 mM ethylene glycol tetraacetic acid (EGTA), 4 mM $MgCl_2$. Store at 4 °C.
- GpCpp (GMPCCP, guanosine-5′-[(α,β)-methyleno]triphosphate, sodium salt): 10 mM stock. Store at −80 °C. Jena Bioscience, catalog number NU-405S.
- Purified tubulin: 100 μM stock, store at −80 °C. Cytoskeleton Inc., 5 × 1 mg, catalog number T240-B.
- X-Rhodamine tubulin or Tubulin HiLyte 647: 5 μM stock in MRB80 buffer, store at −80 °C. Cytoskeleton Inc., X-Rhodamine: 5 × 20 μg, catalog number TL620M-A or Cytoskeleton Inc., Tubulin HiLyte 647: 5 × 20 μg, catalog number TL670M-A.
- Biotin-tubulin: 50 μM stock in MRB80 buffer, store at −80 °C. Cytoskeleton Inc., catalog number T333P-A (5 × 20 μg).
- Beckman polyethylene tubes, 5 × 20 mm, catalog number 343622.
- Prewarmed airfuge rotor, room temperature.
- Thermal cycler or a tube incubator with a heated lid.

In a 50-μl PCR tube, add the following components: 1.7 μl tubulin, 2.5 μl X-Rhodamine tubulin, 0.7 μl biotin-tubulin, 0.5 μl GMPCCP, and 1.6 μl MRB80 buffer. With these various tubulin concentrations, the seeds will be composed of 78% nonlabeled tubulin, 17% biotin-tubulin, and 5% X-Rhodamine tubulin (*note*: other fluorescently labeled tubulins can also be used). Incubate the solution in a thermal cycler, with a heated lid, for 40 min at a constant temperature of 37 °C. We use a thermal cycler so as not to induce evaporation of the sample, which does happen in a standard 37 °C heating block. After the incubation step, centrifuge the reaction in the airfuge for 10 min in order to pellet the MT seeds. If there is no access to an airfuge, then centrifugation could be carried out in an ultracentrifuge; however, note the very small volumes in use here (7 μl). A colored pellet should be visible. Remove the supernatant, which contains excess GMPCCP, and resuspend gently in 30 μl warm MRB80 buffer. At this stage, we usually check the seeds via TIRF microscopy (see Section 1.4.3). While checking the seeds on the TIRF microscope, keep them warm (room temperature is fine). Once they have been deemed satisfactory, flash freeze the seeds as 4 μl aliquots in liquid nitrogen and store for future use. When using seed aliquots, thaw them at room temperature and keep them warm. Do not attempt to thaw seeds on ice or they will depolymerize. When inspecting seeds, the dye will bleach fairly quickly under standard TIRF conditions (3 mW laser power, 500 ms exposure, 37 °C). A variation in the above protocol that is fairly often used is to add taxol as a seed stabilizing agent. However, we do not recommend this procedure as it may, later on, also interfere with the dynamic behavior of MTs synthesized from the seeds.

8.1.4 TIRF microscopy of +TIPs *in vitro*

The following assay allows the study of the MT system and its associated binding partners in a cell-free manner. One major advantage of this system is that the MT interacting protein of interest is studied independently of complicating *in vivo* factors. In effect, it is possible to dissect the interplay between the binding partner and dynamic MTs *in vitro*, that is, where does the partner bind, for how long, and how does it affect the MT. With the protein of interest purified (and tagged with a fluorophore), all that is required is to have the components for MT growth ready and add them to the prepared sample chambers. TIRF microscopy is then used to image the action of the proteins. With respect to protein tags and fluorophores, it is worth mentioning that a recent study (Zhu et al., 2009) found that the presence of the hexa-HIS tag had an effect on the binding of EB1 to MTs (non-HIS-tagged protein bound with a lower affinity compared to the standard HIS-tagged EB1). These studies were performed using classic (or "bulk") biochemistry approaches and not with the *in vitro* assay described here. Nevertheless, it is worth mentioning as a point of caution when considering the purification of proteins that using the hexa-HIS tag may have an effect on protein binding to the MTs.

Required materials

- MRB80 buffer: 80 mM PIPES buffered to pH 6.8 with KOH (room temperature), 1 mM EGTA, 4 mM $MgCl_2$. On ice.
- MRB80 buffer: 80 mM PIPES buffered to pH 6.8 with KOH (room temperature), 1 mM EGTA, 4 mM $MgCl_2$. Prewarmed to 37 °C (require a minimum of 500 μl per *in vitro* sample).
- Poly-L-lysine PEG biotin (PPL-PEG biotin): 0.2 mg ml^{-1} stock, dissolved in MRB80 buffer. Surface solutions, PPL(20)-g[3.5]-PEG(2)/PEG(3.4)-biotin (50%), 10 mg pack size. A working stock may be stored at 4 °C, the remaining solution can be stored at −80 °C in 200 μl aliquots.
- Streptavidin (available from Sigma–Aldrich) or neutravidin (available from Invitrogen). 1 mg ml^{-1} stock solution in MRB80 buffer. Stored at −80 °C as 20 μl aliquots.
- κ-Casein. Available from Sigma–Aldrich, purified from bovine milk. Catalog number C0406. Stored as a 5 mg ml^{-1} stock solution in MRB80 buffer. 50 μl Aliquots.

- Methyl cellulose, 1% solution. Prepared in MRB80 buffer under sterile conditions. Store at 4 °C.
- Glucose, 1 M in MRB80 buffer. Prepared under sterile conditions, 0.2 μm filtered. Stored at −20 °C as 10 μl aliquots.
- Catalase. Available from Sigma–Aldrich, purified from bovine liver. Catalog number C9322-1G.
- Glucose oxidase. Available from Sigma–Aldrich, purified from *Aspergillus niger*. Catalog number G7141-10KU.
- DTT, 1 M stock.
- GMPCCP-stabilized MT seeds (color as required).
- Tubulin: 100 μM stock, stored at −80 °C. Cytoskeleton Inc., 5 × 1 mg, catalog number T240-B.
- GTP, 50 mM. Stored at −80 °C.
- Purified +TIPs as required.
- Prepared sample chambers.
- Sample tube incubator.
- Beckman 5 × 20 mm polyethylene centrifuge tubes (Beckman, catalog number 343622)
- Vacuum grease or vaseline to seal sample chambers.

8.1.4.1 Preparation of an oxygen scavenger system

In any fluorescent system, dye molecules will undergo photodestruction in a process known as bleaching, and intensity alterations, which is known as blinking. These phenomena are the result of the fluorophores entering the so-called dark state which may be temporary or permanent, and are especially important to consider if the *in vitro* assay is to be used as a single-molecule experiment. If studying an association/disassociation process, such as EB binding to the MT, then abrupt photobleaching can cause a major problem in the interpretation of the results of dwell time. Oxygen scavenging improves fluorophore stability by inhibiting the triplet state and hence reduces the effects of bleaching and blinking (Hubner, Renn, Renge, & Wild, 2001). There are various methods available in the literature for oxygen scavenger systems (Aitken, Marshall, & Puglisi, 2008; Rasnik, McKinney, & Ha, 2006; Swoboda et al., 2012). In our lab, we routinely use the glucose oxidase/catalase system. This scavenger system is made in the following manner:

5 mg catalase.
10 mg glucose oxidase.
15.4 mg DTT.
Dissolved in 500 μl MRB80 buffer.

After components have dissolved, centrifuge the sample for 5 min in a desktop centrifuge at 15,000 rpm to remove any large particulate matter. Usually, we make four 100 μl aliquots and the remaining sample we pipette into 3 μl aliquots. Flash freeze everything in liquid nitrogen and store at −80 °C.

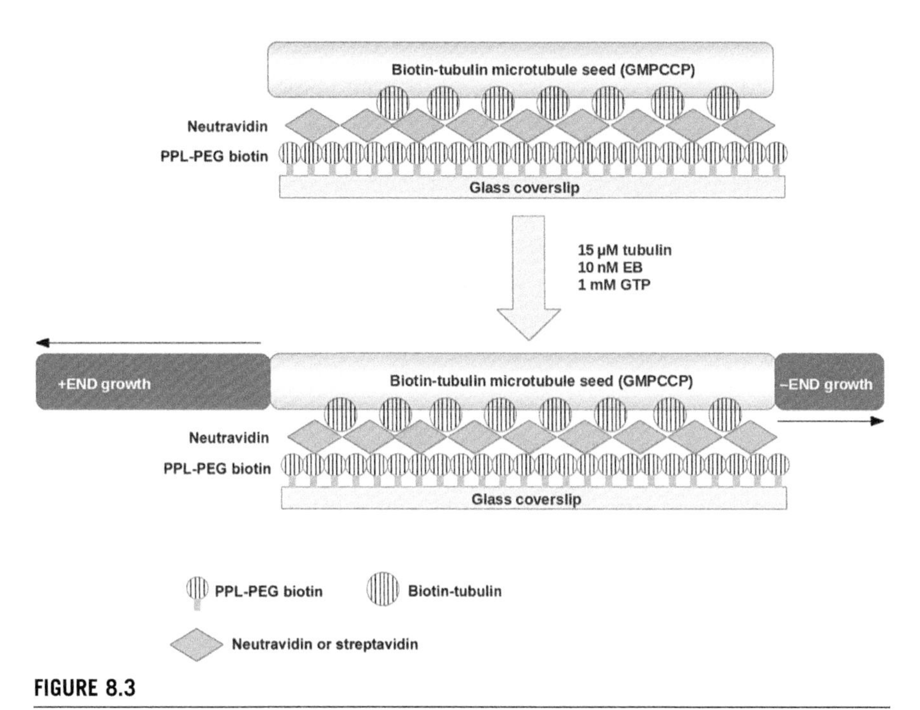

FIGURE 8.3

A schematic of how MT seeds are bound to the coverslip with each component indicated. After coating the glass with PPL-PEG biotin, neutravidin (or streptavidin) is added which reacts with the biotin groups on the surface of the PEG. Finally, the biotin-tubulin-labeled seed is added which binds to the neutravidin layer.

8.1.4.2 ***Chemically functionalizing the sample chamber***

In order to bind the MT seeds to the *in vitro* sample chamber, we need to chemically functionalize the glass, for which we use a simple system involving streptavidin–biotin chemistry. First, the chamber is washed in a solution of PPL-PEG biotin which coats the surface of the glass very efficiently. To this, a streptavidin layer is added and finally the biotin-tagged MT seed (see Fig. 8.3). Given that the K_d of the streptavidin–biotin interaction is in the order of 10^{-14} mol l^{-1}, the binding of the seeds is extremely efficient.

A simple method of adding components to the sample cell is by flowing in solutions using a pipette and removing them with a small piece of tissue paper touching the opposite edge of the coverslip (see Fig. 8.4). The capillary action of the tissue paper draws fluid out of the sample cell. It is important to realize that the sample cell should never become dry so a balance must be achieved between flowing fluid in and drawing fluid out. This can take a little practice to get right, and it is most difficult with the wash steps where larger volumes of buffer are used. Before starting, make sure that the following is prepared:

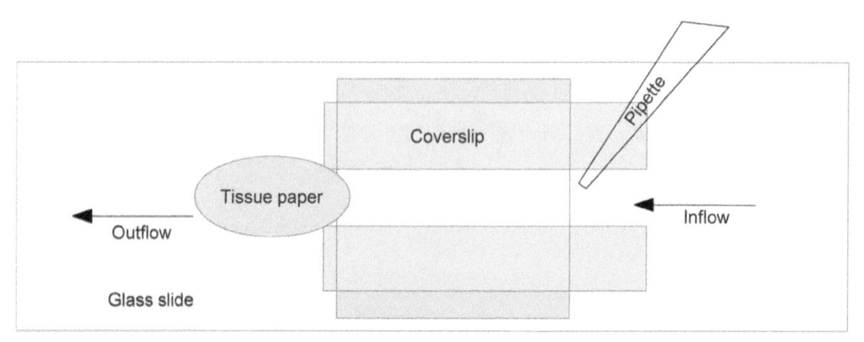

FIGURE 8.4

How to flow components into the sample chamber. The double-sided tape, shown in light gray, is cut so as to have a little extra on the inflow side of the sample chamber. When adding buffer, this makes the process easier by directing the solution toward the coverslip and inhibits buffer from spreading out across the glass when trying to flow it through the sample chamber. (See color plate.)

MRB80 buffer both on ice and warmed to 37 °C.
Streptavidin stock diluted by adding 40 μl of MRB80. Keep at 37 °C.
One κ-casein stock on ice.
An additional κ-casein stock, which is diluted by adding 40 μl of MRB80 and kept at 37 °C.
All other components for the assay are kept on ice.

Another useful addition is some form of humidity chamber so that the sample cell does not dry out. In our lab, we make use of standard 10 cm circular cell culture dishes to which we add a small piece of water-soaked tissue paper. The sample cell is left inside this dish, on a makeshift platform, with the lid closed during incubation steps so as to prevent drying out of the sample chamber. All of the following steps can be carried out at room temperature. With everything prepared and assuming a 10-μl sample cell, carry out the following to prepare the sample chamber:

Flow in 10 μl of PPL-PEG biotin. Allow binding for 5 min.
To remove unbound PEG biotin, flow through 200 μl of warm MRB80 buffer.
Flow through 20 μl of the diluted prewarmed streptavidin. Incubate for 3 min.
To remove excess streptavidin, flow through 50 μl of warm MRB80 buffer.
With these steps complete, the sample chamber is now functional and ready to bind the MT seeds.

8.1.4.3 *Attaching the seeds and blocking the chamber*

While the streptavidin is incubating, retrieve the GMPCCP-stabilized MT seeds from the −80 °C freezer. Under no circumstances place them on ice or they will depolymerize. After removing the aliquot from the freezer, it is best to keep it warm in your hand until placing it in a 37 °C heating block. Given the small volume of the seed stock, it is best to use a heating block with a heated lid so as to avoid evaporation.

If this is the first use of the seed stock, then it will be necessary to check MT integrity. It is also important to test the best seed dilution for *in vitro* experiments. With our TIRF setup, we use a QuantEM 512SC camera (Roper Scientific) with a 512×512 pixel area, 0.065 μm per pixel. We aim to have a seed dilution that gives approximately 3–5 seeds in this field of view. Usually, with a standard seed stock, this is a 1:130 dilution, but depending upon the seed preparation the dilution can vary from 1:100 to 1:160. It is usually best to prepare a few sample chambers to test the best seed dilution prior to experiments. Once the dilution has been obtained, it can be used for the entire stock of frozen seeds, so it is worth investing a little time to be precise. For a TIRF image of MT seeds, see Fig. 8.5.

With the sample chamber prepared, take 1 μl of seeds and dilute as appropriate in warm MRB80 buffer. Flow in 50 μl of the diluted seed stock into the streptavidin-prepared sample chamber. Throw the remaining dilution away. Once the seeds have been diluted, they become unstable so do not use a diluted seed solution for a next round of experiments (it is of course fine to use one seed dilution if carrying out a number of experiments in parallel). Incubate the reaction for 5 min and make sure that the sample chamber is at room temperature (see Section 1.4.4). After 5 min, wash the chamber by flowing in 50 μl of warm MRB80. Finally, flow in 30 μl of warm diluted κ-casein (this acts as a blocking agent to inhibit further binding to the sample chamber surface) and move on to the preparation of the *in vitro* sample.

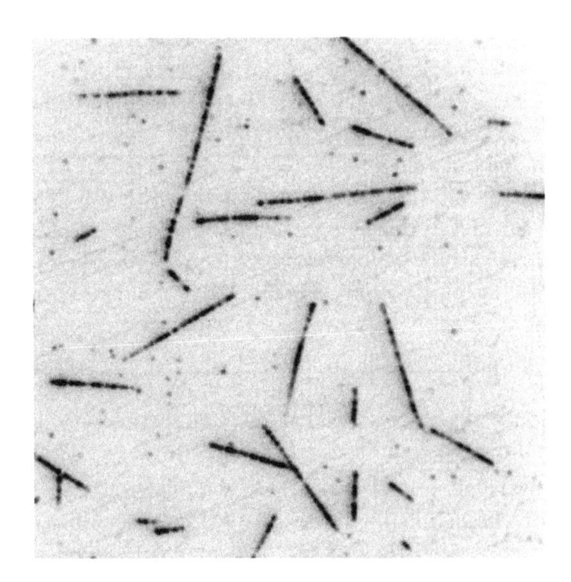

FIGURE 8.5

A TIRF image of GMPCCP-stabilized microtubule seeds labeled with X-Rhodamine tubulin. The seed density in this image would be considered too high but is shown for demonstration purposes. Ideally, one should aim for 3–5 seeds per field of view. The dappled effect in the seeds is due to bleaching of the X-Rhodamine during imaging. The imaging conditions employed were 3 mW 561 nm laser, $100\times$ objective, 500 ms exposure time.

8.1.4.4 Preparation of the protein samples

Now that the sample chambers have been prepared, it is time to make the primary protein mix. It is here that we will add not only the components required for efficient MT growth but also the specific MT-binding proteins under study. For a good work flow, it is recommended to make this mix after the wash step prior to seed attachment. Add the following in a microtube:

6.5 μl MRB80 buffer.
2.0 μl of 1% methyl cellulose.
3.0 μl of 333 mM KCl (50 mM final).
2.0 μl of 5 mg ml^{-1} κ-casein.
0.5 μl of 50 mM GTP (1.2 mM final).
3.0 μl of 100 μM tubulin (15 μM final).
2.0 μl +TIP (~75 nM final concentration for ensemble and ~0.6 nM for single-molecule experiments).
0.5 μl oxygen scavenger mix.
0.5 μl of 1 M glucose.

The concentration of 75 nM final for the +TIP is a good starting point, as with mammalian EB proteins this concentration is known to produce "comets." However, the final concentration used will ultimately depend on the +TIP under study. For single-molecule experiments, we have listed 0.6 nM as a good starting point based on EB3 studies (Montenegro Gouveia et al., 2010). This will provide good coverage of the GDP lattice and plus-end interactions at a density appropriate for analysis. If carrying out a single-molecule experiment, it is a good idea to have some form of tracer to visualize growing MTs as the protein under study will be at too small a concentration to see the MT (end). Previous studies have used a higher concentration (75 nM) of the same protein that is studied in single-molecule "mode," using different fluorescent proteins (and hence channels) to distinguish low and high concentrations (Montenegro Gouveia et al., 2010). This allows one to easily identify the MT in one channel and position single molecules in the other. An alternative method is to label the tubulin instead. This can be done by adding a small tracer amount of, for example, X-Rhodamine or Alexa tubulin (~0.13 μM). The MTs will be faintly visible in the red channel and the +TIP can then be identified in the green channel. Sometimes it is advisable to optimize the concentration of tubulin tracer as too high a concentration will lower the signal-to-noise ratio in the TIRF resulting in a "white out."

Although GFP is widely used, it is not the best choice for single-molecule experiments, primarily due to its blinking behavior. We have seen blinking of GFP in the 20- to 30-ms time scale, which has impeded fast time exposure imaging (e.g., 10 ms) of EB proteins. One way to overcome this is to use alternate fluorophores such as Alexa dyes. However, the preparation of protein samples becomes more complex with such methods.

The components and their concentrations listed above are considered the standard setup for the *in vitro* assay. However, it should be obvious that the possibility to alter a given component and study its subsequent effect provides a unique and powerful tool that is not available for *in vivo* experiments. For example, by altering the KCl concentration, Buey et al. (2011) noted a change in the binding of GFP-EB3 to the lattice of MTs, thereby highlighting the electrostatic nature of EB3 binding to MTs (Buey et al., 2011).

Mix the reaction gently with a pipette and transfer to a 5 × 20-mm Beckman centrifuge tube. Centrifuge the reaction mix in an airfuge for 10 min at 30 psi, room temperature. Failure to centrifuge will usually result in the appearance of aggregates of fluorescent protein, which can be extremely frustrating when collecting data. In case you do not have access to an airfuge, we have had some limited success centrifuging the protein mix in a desktop centrifuge but the final results are far from satisfactory and therefore we highly recommend an airfuge or equivalent centrifugation prior to protein sample loading. With centrifugation complete, flow the reaction mix into the sample chamber and finally seal the chamber using vacuum grease. It might be initially difficult to flow in the sample mix given its higher viscosity, which is primarily due to the methyl cellulose. If this is the case, make sure that a clean dry tissue is used and press the tissue firmly to the coverslip (be careful not to break it). The easiest way to seal the chamber with grease is to take a 2-ml syringe and load a p200 tip at the base. This makes a small exit for the grease which is loaded into the syringe. Use this to seal the sample chambers. It is not recommended to use candle wax as the heat transfer from the molten wax to the sample chamber can be quite high and this may have deleterious effects on the proteins.

8.1.4.5 TIRF microscopy of the samples

Stable heating for the sample, in our system, is provided by a Tokaihit slide holder and additionally from the objective heater. If there is only an objective lens heater, then take note that the sample temperature will be 2–3 °C lower than the temperature set on the objective heater, when the system is at room temperature. We advise to measure the temperature of the *in vitro* cell with a microthermometer prior to experimentation to ensure that the sample is at the required temperature (note that the environment, e.g., the effects of air conditioning, could potentially alter the temperature). Prior to experiments always allow the TIRF system to warm up to operating temperature. If using a stage heater, check the manufacturer's instructions but usually it will take around 40 min to reach working temperature. Failure to do this might result in impeded MT growth (note that the small sample cell size makes it very prone to changes in the environment). The exact opposite problem, that is, too rapid MT growth and a very short sample lifetime, can occur in the summer months with a poorly temperature-regulated TIRF system and environment. Figure 8.6 shows the results we have obtained in-house for MTs in the presence of 75 nM GFP-EB3. This image is from the green channel and hence shows only EB3 interactions, which are clear across the MT seed, lattice, and plus end where there is an obvious comet. Note that there is strong binding to the MT seed. This is expected as the GMPCCP should stabilize the tubulin in a conformation that is a structural mimic of the GTP cap. If using dual color TIRF, one would easily be able to identify the seeds in the red channel. The background in this image is fairly standard for the *in vitro* assay. Usually, the background in the red channel is a little higher. If it is too high, it could be the result of inefficient

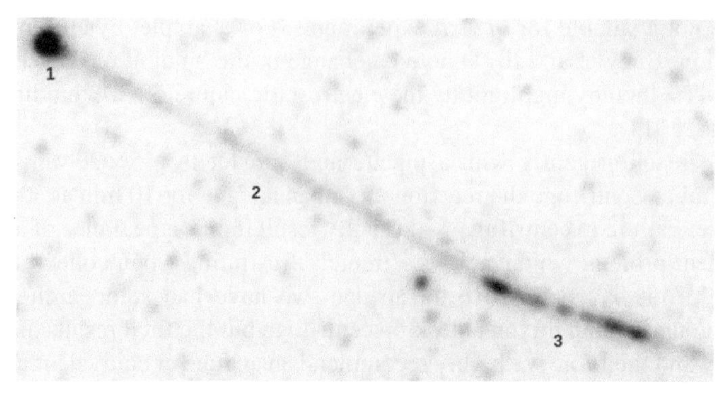

FIGURE 8.6

An *in vitro* experiment with 75 nM 6 × HIS-GFP-EB3 as imaged via TIRF microscopy. The EB comet is clearly visible at the plus end (1), as are lattice interactions along the length of the growing MT (2). The GMPCCP-stabilized seed is visible in (3), where the EB tends to bind more strongly than the GDP lattice, presumably due to the GMPCCP moiety. The imaging conditions were 3 mW 491 nm laser, 100 × objective, 500 ms exposure time, 37 °C.

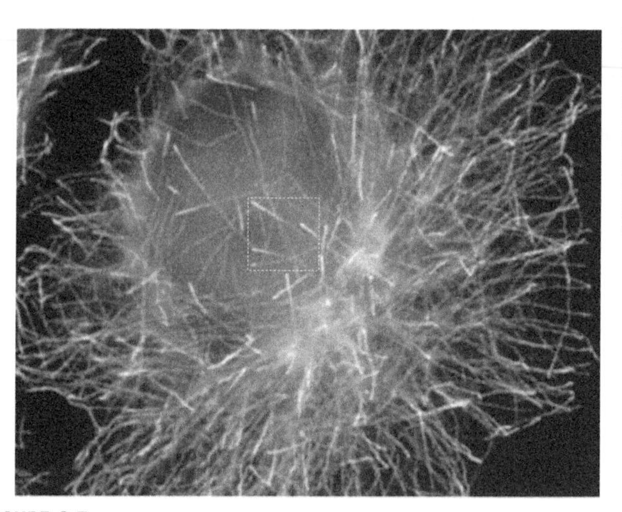

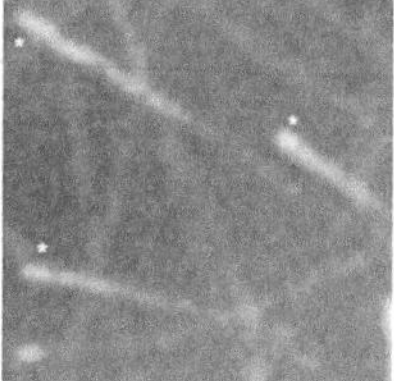

FIGURE 8.7

GFP-EB1 expressed in HeLa cells and imaged via TIRF microscopy. Cells were fixed with methanol and 1 mM EGTA and stained with anti-EB1 (green) and antityrosinated-α-tubulin (red). The EB comet is visible at the plus end (*). (See color plate.)

κ-casein blocking or an issue with either tracer tubulin or the seeds (e.g., lack of centrifugation to a pellet prior to loading, reuse of diluted seeds). As a comparison to the *in vitro* assay, Fig. 8.7 shows EB1 comets produced in a HeLa cell. Again the EB comets are clearly visible and most importantly they are comparable to those produced with the *in vitro* assay.

CONCLUDING REMARKS

The methods described here for the *in vitro* MT assay have been used for a number of years at Erasmus MC. They are easily adaptable for use either in ensemble or single-molecule experiments. In this chapter, we have tried to describe the protocols in enough practical detail so as to make them simple to follow. For theoretical discussions on the molecular mechanisms that govern the motions of MTs and +TIPs, we refer to recent reviews (Akhmanova & Steinmetz, 2010; Galjart, 2010). The key to successful *in vitro* MT experiments is adequate preparation of the reconstituted components. One should spend time ensuring that the tubulins, +TIPs, and other factors that are used in the assay are purified to a high standard and are well characterized; otherwise, the reproducibility of the system will degrade significantly and/or results from the assay may be interpreted incorrectly.

Acknowledgments

This work is financed by the Netherlands Organisation for Scientific Research (NWO, CW-ECHO).

We would like to thank Dr Jeffrey Van Haren for kindly providing us with an image of EB1 comets from HeLa cells.

References

Aitken, C. E., Marshall, A. R., & Puglisi, J. D. (2008). An oxygen scavenging system for improvement of dye stability in single-molecule fluorescence experiments. *Biophysical Journal, 94*, 1826–1835.

Akhmanova, A., & Steinmetz, M. O. (2010). Microtubule +TIPs at a glance. *Journal of Cell Science, 123*, 3415–3419.

Bieling, P., Kandels-Lewis, S., Telley, I. A., van Dijk, J., Janke, C., & Surrey, T. J. (2008). CLIP-170 tracks growing microtubule ends by dynamically recognizing composite EB1/tubulin-binding sites. *Cell Biology, 183*, 1223–1233.

Bieling, P., Laan, L., Schek, H., Munteanu, E. L., Sandblad, L., Dogterom, M., et al. (2007). Reconstitution of a MT plus-end tracking system in vitro. *Nature, 450*, 1100–1105.

Brouhard, G. J., Stear, J. H., Noetzel, T. L., Al-Bassam, J., Kinoshita, K., Harrison, S. C., et al. (2008). XMAP215 is a processive microtubule polymerase. *Cell, 132*, 79–88.

Buey, R. M., Mohan, R., Leslie, K., Walzthoeni, T., Missimer, J. H., Menzel, A., et al. (2011). Insights into EB1 structure and the role of its C-terminal domain for discriminating microtubule tips from the lattice. *Molecular Biology of the Cell, 16*, 2912–2923.

Dixit, R., Barnett, B., Lazarus, J. E., Tokito, M., Goldman, Y. E., & Holzbaur, E. L. F. (2009). Microtubule plus-end tracking by CLIP-170 requires EB1. *Proceedings of the National Academy of Sciences of the United States of America, 106*, 492–497.

Galjart, N. (2010). Plus-end-tracking proteins and their interactions at microtubule ends. *Current Biology, 12*, 528–537.

Honnappa, S., Gouveia, S. M., Weisbrich, A., Damberger, F. F., Bhavesh, N. S., Jawhari, H., et al. (2009). An EB1-binding motif acts as a microtubule tip localization signal. *Cell, 138*, 366–376.

Honnappa, S., John, C. M., Kostrewa, D., Winkler, F. K., & Steinmetz, M. O. (2005). Structural insights into the EB1-APC interaction. *The EMBO Journal*, *24*, 261–269.

Hubner, C. G., Renn, A., Renge, I., & Wild, U. P. (2001). Direct observation of the triplet lifetime quenching of single dye molecules by molecular oxygen. *The Journal of Chemical Physics*, *115*, 9619–9622.

Komarova, Y., De Groot, C. O., Grigoriev, I., Gouveia, S. M., Munteanu, E. L., Schober, J. M., et al. (2009). Mammalian end binding proteins control persistent microtubule growth. *The Journal of Cell Biology*, *184*, 691–706.

Maurer, S. P., Bieling, P., Cope, J., Hoenger, A., & Surrey, T. (2011). GTPgammaS microtubules mimic the growing microtubule end structure recognized by end-binding proteins (Ebs). *Proceedings of the National Academy of Sciences of the United States of America*, *108*, 3988–3993.

Miller, H. P., & Wilson, L. (2010). Preparation of microtubule protein and purified tubulin from bovine brain by cycles of assembly and disassembly and phosphocellulose chromatography. *Methods in Cell Biology*, *95*, 2–15.

Montenegro Gouveia, S., Leslie, K., Kapitein, L. C., Buey, R. M., Grigoriev, I., Wagenbach, M., et al. (2010). In vitro reconstitution of the functional interplay between MCAK and EB3 at microtubule plus ends. *Current Biology*, *19*, 17–22.

Rasnik, I., McKinney, S. A., & Ha, T. (2006). Nonblinking and long-lasting single-molecule fluorescence imaging. *Nature Methods*, *11*, 891–893.

Swoboda, M., Henig, J., Cheng, H. M., Brugger, D., Haltrich, D., Plumeré, N., et al. (2012). Enzymatic oxygen scavenging for photostability without pH drop in single-molecule experiments. *ACS Nano*, *6*, 6364–6369.

Zhu, Z. C., Gupta, K. K., Slabbekoorn, A. R., Paulson, B. A., Folker, E. S., & Goodson, H. V. (2009). Interactions between EB1 and Microtubules dramatic effect of affinity tags and evidence for cooperative behavior. *Journal of Biological Chemistry*, *284*, 32651–32661.

Zimniak, T., Stengl, K., Mechtler, K., & Westermann, S. (2009). Phosphoregulation of the budding yeast EB1 homologue Bim1p by Aurora/Ipl1p. *The Journal of Cell Biology*, *186*, 379–391.

CHAPTER

Analysis of Microtubules in Isolated Axoplasm from the Squid Giant Axon

Yuyu Song*,† and Scott T. Brady†,‡

*Department of Genetics and Howard Hughes Medical Institute, Yale School of Medicine, New Haven, Connecticut, USA

†Marine Biological Laboratory, Woods Hole, Massachusetts, USA

‡Department of Anatomy and Cell Biology, University of Illinois at Chicago, Chicago, Illinois, USA

CHAPTER OUTLINE

9.1 Preparation of Axoplasm 126
9.2 Analysis of Axoplasmic Microtubule Dynamics 131
9.3 Biochemistry of Axoplasmic Microtubules 133
9.4 Immunohistochemistry of Axoplasmic Microtubules 135
References 136

Abstract

Biochemical specialization of cellular microtubules has emerged as a primary mechanism in specifying microtubule dynamics and function. However, study of specific subcellular populations of cytoplasmic microtubules has been limited, particularly in the nervous system. The complexity of nervous tissue makes it difficult to distinguish neuronal microtubules from glial microtubules, and axonal microtubules from dendritic and cell body microtubules. The problem is further compounded by the finding that a large fraction of neuronal tubulin is lost during standard preparations of brain tubulin, and this population of stable microtubules is enriched in axons. Here, we consider a unique biological model that provides a unique opportunity to study axonal microtubules both *in situ* and *in vitro*: isolated axoplasm from the squid giant axon. The axoplasm model represents a powerful system for addressing fundamental questions of microtubule structure and function in the axon.

Microtubules have been studied extensively since the buffer conditions for *in vitro* polymerization were first described. As a result, the biochemistry and biophysics of microtubule dynamics are relatively well understood in the test tube. However, our understanding of microtubules *in situ* is limited at best. In cells, microtubules exhibit

Methods in Cell Biology, Volume 115

ISSN 0091-679X
http://dx.doi.org/10.1016/B978-0-12-407757-7.00009-8

considerable diversity at the molecular level, including tubulin isotypes, posttranslational modifications, and associated proteins. Microtubules in different cell types or even in different subcellular compartments may exhibit strikingly different properties with regard to dynamics, composition, and function. This heterogeneity is particularly striking in neurons, where the bulk of the microtubules are not associated with the microtubule-organizing center, yet may exhibit exceptional stability. The answers to questions about the functional diversity of neuronal microtubules may be critical for understanding many aspects of neuronal development, function, and pathology.

One obstacle to characterizing specific populations of neuronal microtubules is the complexity of nervous tissue. Separating neuronal microtubules from glial microtubules and dendritic microtubules from axonal or cell body microtubules is effectively impossible when using brain tissue as a source, so any studies on the biochemistry and biophysics of neuronal microtubules from brain reflect the properties of a mixed pool. The problem is compounded by the fact that a large fraction of neuronal tubulin is lost during standard preparations of brain tubulin, and this population of stable microtubules has received little attention, despite representing more than 50% of axonal tubulin in mature neurons.

Isolated axoplasm from the squid giant axon provides a unique model system for studying exclusively axonal microtubules both *in situ* and *in vitro*. Although isolated axoplasm has not been widely used, experiments using this model have provided novel insights into the axonal cytoskeleton, and studies on axoplasm have the potential to produce additional insights. Here, we describe the preparation of isolated axoplasms, the use of physiological buffers that more accurately reflect intracellular environments, and examples of experiments that can only be done in this model system (Fig. 9.1).

9.1 PREPARATION OF AXOPLASM

The procedures described here are based on our continuing studies using axons from the Atlantic longfin squid, *Loligo pealeii* and update previous descriptions (Brady, Lasek, & Allen, 1985; Brown & Lasek, 1993; Leopold, Lin, Sugimori, Llinas, & Brady, 1994). Large to medium squid (0.3–0.5 m in length) are preferred as they have axons 300–500 μm in diameter. Suitable squid are seasonally available (April–October typically) at the Marine Biological Laboratory in Woods Hole, MA, but protocols should be readily adapted for other species of squid with suitable axons (>300 μm in diameter). Smaller squid and smaller axons are not suitable as the viscoelastic properties of the axoplasm will lead to disruption of the axoplasm from smaller axons (<150 μm) (unpublished data S. Brady) (Brady, Richards, & Leopold, 1993).

A healthy squid with a translucent body is chosen and decapitated to begin the dissection. The head and tentacles are discarded. The mantle is cut along the dorsal midline to a sheet of muscle that is placed skin side down on the dissecting light table with running seawater. The viscera are removed and the clear pen is carefully pulled away from the mantle, taking care to avoid tearing the nerve fibers that are along each

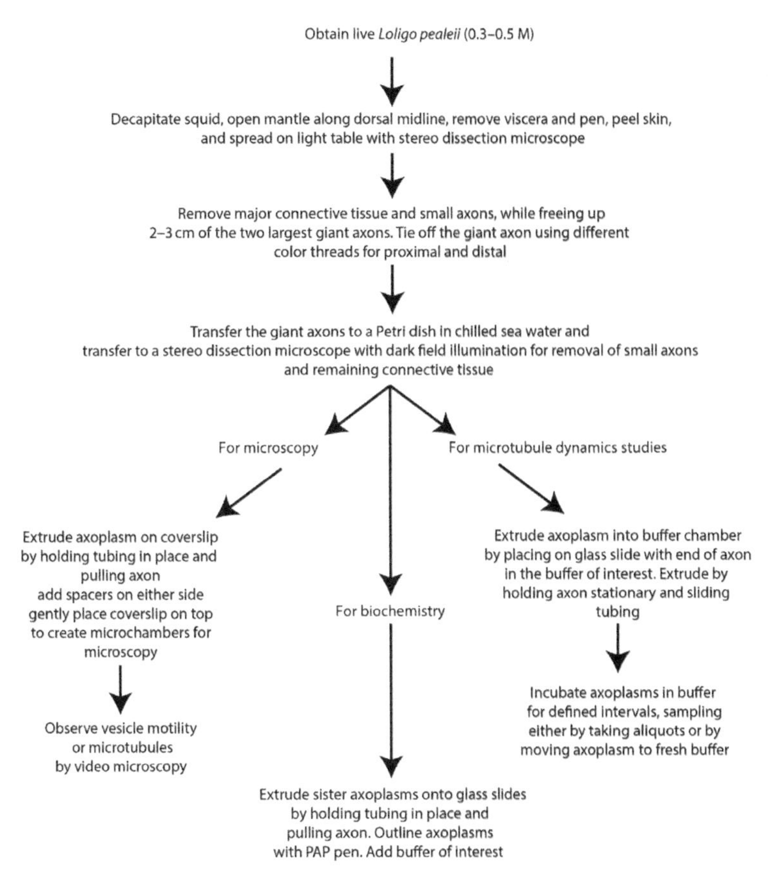

FIGURE 9.1

Flow chart for preparation of axoplasms.

side of the pen. The mantle is turned over and the skin is peeled away to improve visualization of the nerves, then the mantle is pinned to the table with the nerves on the upper surface.

Dissect the nerve bundle containing the largest giant axon (paralleling the pen on both sides) free from the mantle using fine dissection scissors (i.e., 4.5 in. with 10 mm cutting edge, spring action Castroviejo curved scissors, available from George Tiemann or Fine Science Tools) and Dumont 5-45 forceps. Start adjacent to the stellate ganglion and proceed distally to the point where the nerve moves deeper into the muscle, approximately 3 cm (Fig. 9.2). This is rough dissection, taking care to leave connective tissue and smaller axons surrounding the giant axon. These will be removed at the fine dissection step. Leaving them at this stage reduces the chance of damaging the giant axon. When the nerve fibers are freed from the mantle, tie off the two ends with cotton thread to facilitate handling. To ensure that orientation is maintained, proximal and distal ends are ligated with cotton

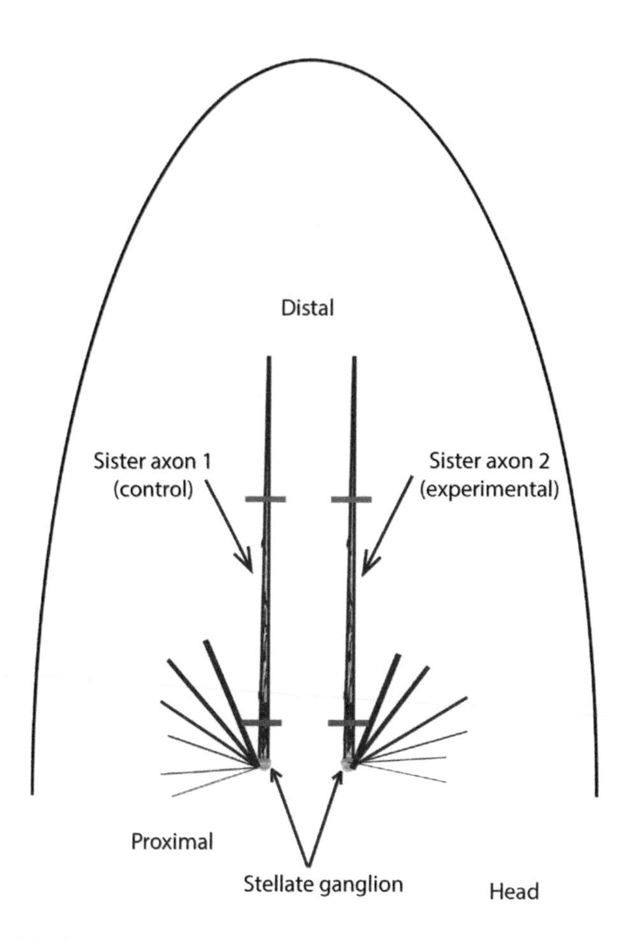

FIGURE 9.2

Dissection of giant axons. A diagram illustrating the position of the squid giant axons after removal of the internal organs and pen. The squid mantle is roughly conical after the head and tentacles are removed. At that point, the mantle is cut along the dorsal midline to create flat sheet of muscle. The viscera can then be removed along with clear stiff pen located in the center of the ventral side of the mantle, between the two largest giant axons. All axons exist in pairs on the right and left side of the midline. The largest diameter axons extend the paralleling of the midline and are the longest axons. The other smaller axons are not used because they are not typically large enough to extrude. The two largest giant axons are dissected free from the surrounding muscle and threads are tied at the distal (green) and proximal ends (red, near the cell bodies in the stellate ganglion). (See color plate.)

thread (black for distal and white for proximal ends). The ganglion is freed from the mantle and the distal end of the nerve beyond the ligation. The nerve should be handled by the threads to avoid damage. Axons from *Loligo* with mantles >0.3 m typically run 400–500 μm in diameter and will produce ~2.5 μl/cm of giant axon.

The dissected nerve is transferred to a 100-mm Petri dish containing cold filtered seawater (use 0.45-μm pore filters and maintain at 0–4 °C) for fine dissection. We prefer natural seawater to reveal axon damage that may affect transport, but some use Ca^{2+}-free artificial seawater for the fine dissection. The nerve bundle is fixed place by wrapping the thread around 18-g hypodermic needles inserted into a small amount of dental wax (Surgident Periphery Wax) placed at opposite edges of the Petri dish.

Using dark-field illumination and a stereo dissection microscope, small axons and connective tissue are gently teased away from the giant axon with Vannas-style iris scissors and No. 5 Dumont forceps, taking care to avoid damaging the membrane of the giant axon. We recommend starting proximal to the stellate ganglion and proceeding distally. Particular caution is needed to avoid cutting small collateral branches that extend from the giant axon occasionally. They may be detectable as slight protrusions on the axon surface and the giant axon may be slight reduced distal to the branch. These collateral branches need to be cut at least 0.1 mm away from the giant axon. After removal of extraneous tissues, the axon is inspected for the presence of small holes, which can be identified as white patches due to the influx of Ca^{2+} from the seawater. Axons with significant white patches should be discarded as the Ca^{2+} activates proteases and may disrupt axoplasmic organization.

Axons to be extruded are removed from the seawater and rinsed briefly in a suitable intracellular buffer (buffer X, see below). Holding the axon by the distal thread, the axon is placed briefly on the filter paper to remove excess fluid and cut adjacent to the proximal thread. The axon is placed on a glass slide or 0 thickness 24 × 60-mm coverslip to extrude for assay of vesicle motility or many biochemical and radiolabeling studies (Brady et al., 1985, 1993; Leopold et al., 1994). To extrude, a section of PE190 tubing is used to compress the distal gently but firmly and the axon is pulled using the distal thread. This leaves a cylinder of axoplasm behind on the coverslip (Fig. 9.3). Typically, a cylinder of 2–2.5 cm in length and roughly 0.4 mm in diameter

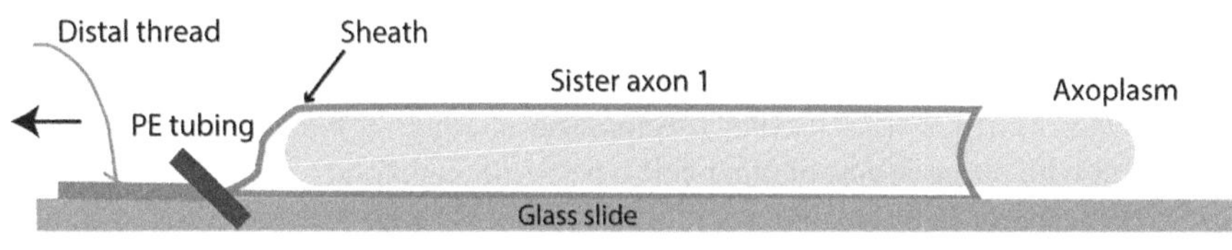

FIGURE 9.3

Extrusion of axoplasm. Once the axon is dissected and cleaned of surrounding small axons/connective tissue, the axoplasm must be extruded. Briefly, the axon is handled by the threads and rinsed in intracellular buffer (buffer X) to remove Ca^{2+} containing sea water. The axon is then blotted on filter paper and cut adjacent to the proximal thread. Using the distal thread, the axon is placed on the coverslip or glass slide. If the goal is to extrude an axoplasm on the surface, one places pressure on the axon close to the thread with a short piece of PE190 tubing and pulls the axon with the thread holding the tubing in place. To extrude into a buffer chamber, the axon is held stationary and the tubing is moved toward the proximal end. (See color plate.)

is used for microscopy, but longer extrusions may be useful for biochemical studies. For imaging, spacers are made by cutting 22 × 22-mm 0 thickness coverslips into 5 × 22-mm sections with a diamond-tipped pen and positioning them on either side of the axoplasm secured by a thin coating of Compound 111 silicon grease (Dow Corning). Compound 111 is stable in the seawater and nontoxic. Do not use high-vacuum silicon grease, which is not designed to seal against aqueous solutions and will be extracted. A top 22 × 22-m 0 thickness coverslip is placed to create a sealed chamber. The top coverslip may be secured with 1:1:1 mix of Vaseline:lanolin:paraffin kept fluid at 50–60 °C, and this creates a microincubation chamber that will maintain the axoplasm for hours. The ends of the chamber are normally kept open to allow perfusion. Chambers prepared in this way have a volume of approximately 25 μl, permitting perfusion of axoplasm with buffers at defined ratio of 1:1 to 1:5, minimizing dilution of axoplasm and allowing introduction of reagents at defined concentrations.

Alternatively, the proximal end may be placed in a small reservoir to extrude into a buffer of choice (Morris & Lasek, 1982). If the extrusion is into a reservoir, the axon is held in place with the proximal end in the fluid and the rest of the axon on a clean glass slide. The tubing is slid from the distal end toward the proximal end. In this case, the axoplasm will extrude into the reservoir while maintaining the form of the axon (Morris & Lasek, 1982).

After extrusion, we routinely incubate the axoplasms in a humidified chamber at 4 °C for 15–30 min, which allows the axoplasm to "rest" after extrusion. Empirically, this produces more reproducible measurements.

Choice of buffers is critical for study of *in situ* microtubule properties. The axoplasm is unique in that the small molecular weight components of the cytoplasm can be directly measured (Table 9.1) (Deffner & Hafter, 1960a, 1960b; Morris & Lasek, 1982). Standard buffers used for study of microtubules *in vitro* such as BRB80 (80 mM PIPES, 1 mM $MgCl_2$, 1 mM EGTA, pH 6.8) and others (Borisy, Marcum, Olmsted, Murphy, & Johnson, 1975; Olmsted & Borisy, 1975; Weisenberg, 1973) diverge significantly from *in vivo* conditions. Analysis of axoplasm shows that *in vivo* conditions have high levels of organic anions (such as amino acids) and K^+ is the major cation (Brady, Lasek, & Allen, 1982; Brady et al., 1985; Morris & Lasek, 1982). Axoplasm is a highly reducing environment, consistent with observations in other cell types. Although total Ca^{2+} is 3.5 mM, free Ca^{2+} levels are much lower, on the order of 100 nM. The axoplasmic cytoskeleton is sensitive to both ionic composition and ionic strength (Brady et al., 1985; Brown & Lasek, 1990, 1993).

An alternative buffer, buffer X (Table 9.2), was developed which retained the major biochemical features of the axoplasmic milieu (Brady et al., 1985). Potassium aspartate serves as the major organic anion with glycine representing the other amino acids. Taurine is the major reducing agent and betaine serves as an organic osmolyte. EGTA is added to buffer the free Ca^{2+} at 50–100 nM. Halides (F, Cl, Br, and I) may be problematic in many studies as they can alter protein–protein interactions significantly (Collins, 2004, 2006; Hearn, Hodder, & Aguilar, 1988; Westh, Kato, Nishikawa, & Koga, 2006) and disrupt the axoplasmic cytoskeleton (Baumgold,

Table 9.1 Nonprotein composition of axoplasm

Class	Component	mM
Amino acids	Nonpolar AA (alanine)	16.14
	Polar AA (glycine)	18.41
	Acidic AA (aspartate)	100.29
	Basic AA (arginine)	6.06
	Betaine	73.7
	Taurine	106.7
	Homarine	20.4
	Cysteic acid	4.9
Organic metabolites	Isethionic acid	164.6
	Glycerol	4.35
	Others	11.02
Carbohydrates	Glucose	0.24
	Mannose	0.92
	Fructose	0.24
	Sucrose	0.24
Inorganic ions	Potassium	344
	Chloride	151.2
	Sodium	35
	Phosphate	17.8
	Magnesium	10
	Sulfate	7.5
	Calcium	3.5
Nucleoside triphosphates	ATP	1

Adapted from Deffner and Hafter (1960a, 1960b) and Morris and Lasek (1982).

Gallant, Terakawa, & Pant, 1981; Brown & Lasek, 1993), so they are minimized in buffer X. Extruded axoplasm placed in buffer X will retain its overall shape and organization for >24 h (Morris & Lasek, 1982) and will maintain fast axonal transport for >3–4 h (Brady et al., 1985, 1993).

9.2 ANALYSIS OF AXOPLASMIC MICROTUBULE DYNAMICS

The stability of axoplasmic organization in buffer X provides a unique model study of microtubule and actin dynamics *in situ* (Morris & Lasek, 1984). Tubulin dimers represent 22% of total axoplasmic protein, with a concentration of approximately 25 μM (Morris & Lasek, 1984). Squid microtubules resemble mammalian microtubules in many ways with regard to its polymerization (although at optimal temperature as 25 °C) and depolymerization (by cold, Ca^{2+} ions, colchicine, etc.)

Table 9.2 Buffer X for squid axoplasm experiments (composition and instructions)

Chemical	MW	M (in stock)	Grams/ 25 ml stock	mM in X	ml stock in 25 ml buffer X
K-Aspartate	171.2	1.0	4.28	350	8.75
Taurine	125.1	Add by wt.	Add by wt.	130	0.407 g
Betaine	135.2	1.0	3.38	70	1.75
Glycine	75.07	1.0	1.88	50	1.25
$MgCl_2 \cdot 6H_2O$	203.31	1.0	5.08	12.9	0.323
$K_2 \cdot$EGTA (adjust stock to pH 7.2 with KOH before bringing to final volume)	380.4	0.1	0.9510	10	2.5
HEPES (adjust stock to pH 7.2 with KOH before bringing to final volume)	238.3	1.0	5.96	20	0.5
$CaCl_2 \cdot 2H_2O$	147.02	1.0	3.68	3	0.075
Glucose	180.2	0.1	0.4505	1	0.25
$K_2 \cdot$ATP (adjust stock to pH 7.2 with KOH before bringing to final volume)	583.4	0.2	2.92 (0.1167 g/ml)	1.0	0.25

Combine aliquots of stock reagents as indicated in the table and check pH. Adjust pH to 7.2 with KOH if needed. Bring to final volume of 25 ml with 20 mM HEPES, pH 7.2 and filter on 45-μm Millipore filter or equivalent. Store in suitable aliquots at –20 °C until ready for use. We routinely prepare aliquots without ATP, bringing the stock to 20 ml rather than the 25 ml final volume and storing as 0.4 ml aliquots. Immediately prior to use, an aliquot is thawed and ATP added from a concentrate stock. Experimental agents can be added at this point and the aliquot brought to a final volume of 0.5 ml.

(Sakai & Matsumoto, 1978). Dilution of extruded axoplasm by 1000-fold in buffer X allows determination of stable polymer, soluble polymer, and free tubulin dimer in axoplasm from a single axon based on the kinetics of extraction for tubulin into the media (Morris & Lasek, 1984). This approach defines a Kinetic Equilibration Paradigm (KEP), which provides a unique perspective on microtubule dynamics *in situ*.

The KEP method allows one to define discrete pools. The amount of free dimer in this assay was significantly less than levels reported from *in vitro* assays of microtubule polymerization, and the level of polymer remaining after 24 h extraction was 15% of the total tubulin (Morris & Lasek, 1984). This is consistent with the observed stability of microtubules in isolated axoplasm with minimal dilution (Weiss, Langford, Seitz-Tutter, & Keller, 1988).

For analysis of depolymerization kinetics, axoplasm should be extruded into a reservoir as described above. The axoplasm will retain its form suspended in the buffer. At this point, the axoplasm can be picked up with Dumont forceps, taking

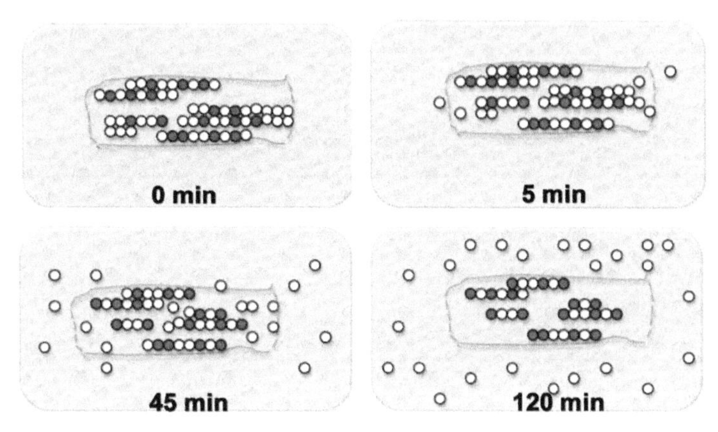

FIGURE 9.4

Kinetic Equilibration Paradigm (KEP). This method was developed originally by Morris and Lasek (1982, 1984) for analysis of monomer–polymer equilibrium of cytoskeletal proteins in the axoplasm. In this diagram, green dots represent modified tubulin with increased stability and white dots represent unmodified labile tubulins (Song et al., 2013). After placing the axoplasms in buffer X, unassembled tubulin dimers would be extracted into the buffer by 5 min. Labile microtubules begin to depolymerize by this time and by 45 min, the tubulin from labile microtubules (soluble polymer) has been extracted. By 120 min, only stable microtubules remain as an axoplasmic "ghost" and do not change over periods as long as 24 h. These ghosts are enriched in modified tubulins. (See color plate.)

care to avoid shearing. Elution of tubulin from the axoplasm can be monitored either by taking aliquots of the media at suitable intervals or by physically transferring the axoplasm into a fresh reservoir of buffer (Fig. 9.4, adapted from Song, 2010).

Alternatively, microtubule behavior can be monitored directly in chambers by imaging with video-enhanced contrast differential interference microscopy (Allen, Allen, & Travis, 1981; Weiss et al., 1988; Weiss, Langford, Seitz-Tutter, & Maile, 1991). This is best done by use of lower ionic strength buffers (i.e., buffer X/2) that promote separation of cytoskeletal elements from the main axoplasm (Brown & Lasek, 1993; Weiss et al., 1988). Exogenous tubulin can be perfused into the chambers and changes in microtubule numbers and stability followed (Weiss et al., 1988). A variant on this approach would perfuse tubulin with a fluorescent tag either chemical (Texas red) or a protein (GFP, mCherry, etc.). Incorporation of fluorescent tubulin could be monitored and the effects of different experimental manipulations evaluated.

9.3 BIOCHEMISTRY OF AXOPLASMIC MICROTUBULES

As microtubules from extruded axoplasm are purely axonal and represent neuronal microtubule properties in this *ex vivo* setting, various biochemical experiments can be carried out to examine different tubulin isotype components, posttranslational

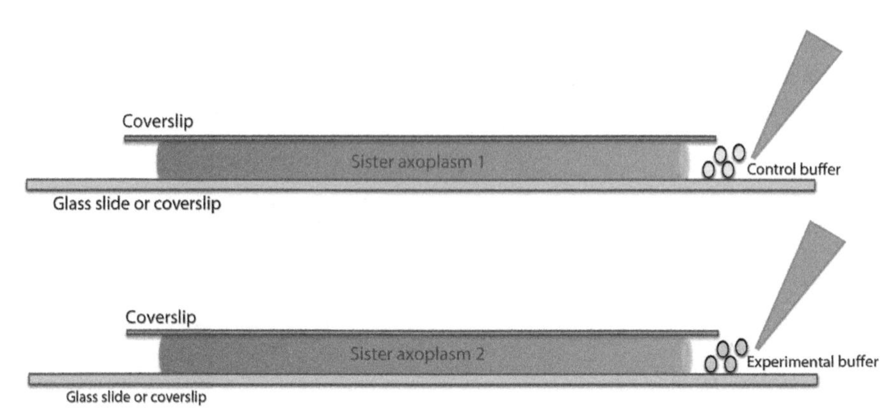

FIGURE 9.5

Perfusion of axoplasms. Buffers of interest can be perfused into axoplasm for studies of microtubule biochemistry or imaging. Ideally, sister axoplasms (two axoplasms from a single squid) are used for comparison to reduce interanimal variability. For biochemistry, the axoplasms are perfused without a top coverslip while imaging studies require the top coverslip. In both cases, buffer volumes are kept small relative to the axoplasm volume to minimize dilution. Typical axoplasms are 5 μl in volume, so buffer volumes are kept at 20–25 μl. This is in contrast to conventional biochemical approaches where dilutions may be 10^3–10^5. (See color plate.)

modifications of tubulins, distribution of microtubule-associated proteins (MAPs), etc. In addition, experimental manipulation of microtubule modifications can be performed to understand how a particular modification may alter microtubule structure, its dynamics and stability, and its affinity for motor proteins and MAPs (Song, 2010).

For biochemical experiments, two "sister" axoplasms of the same length from one squid are extruded on glass slides as described above, one serves as a control while the other is used for experimental manipulations (Fig. 9.5). This ensures that the variability between individual squids is minimized. A circle is drawn around each of the two freshly extruded axoplasms using a PAP Pen (Zymed) or liquid blocker (Super Pap) to delineate the perfusion area before being placed in a humid chamber. Placing the slide on a piece of wet paper towel in the chamber may enhance humidity. After being incubated at 4 °C for 10 min, a control axoplasm is perfused with 30 μl of buffer X/2 + 5 mM ATP alone, while the experimental axoplasm is perfused with the appropriate effector mixed in the same buffer. Both axoplasms are incubated for sufficient amount of time at a temperature determined by specific experimental conditions. Axoplasms are transferred into microcentrifuge tubes where they are homogenized in 30 μl of SDS (1%) or other proper buffers by trituration with a P200 pipette. The samples can be analyzed by SDS-PAGE, followed by immunoblot, autograph (if radio labeled), or direct gel documentation (if fluorescently labeled).

This procedure can also be combined with several other assays such as (1) the axoplasmic microtubule dynamics assay as described above to assess how specific modifications, pharmacological reagents, or MAPs alter microtubule dynamics. (2) Assays for microtubule pelleting, labeling tubulin, tubulin polymerization, etc. (for detailed protocols on those, see http://mitchison.med.harvard.edu/protocols.html). (3) Immunoprecipitation assay to identify interacting proteins or (4) evaluation of pharmacological agents.

The pharmacology of axonal microtubules is of particular interest. For example, antimitotic drugs have been widely used in cancer therapy (Jordan & Wilson, 2004), many of which cause neurological side effects (Windebank & Grisold, 2008), but the underlying mechanism is not well understood. As many of these drugs alter microtubule dynamics and stability (Jordan & Kamath, 2007), it is important to analyze microtubule properties in axons after the drug treatment. The *ex vivo* squid axoplasm system provides well-organized axonal microtubule structure that can be easily analyzed both morphologically (see below) and biochemically. Concurrent analysis of fast axonal transport in response to treatment with different drugs and biochemical reagents (Brady et al., 1993) provides an additional cell biological dimension to studies of microtubules in axoplasm.

9.4 IMMUNOHISTOCHEMISTRY OF AXOPLASMIC MICROTUBULES

A unique feature of the extruded axoplasm is the well-preserved axonal microtubule structure that can be analyzed for morphological changes upon pharmacological treatment, the distribution of other proteins that may or may not interact with microtubules (such as motor proteins, MAPs, and various kinases/phosphatases that modify microtubules and their associated proteins), and the incorporation of exogenous proteins perfused into the system.

Here, we use one example to look at how different tubulin mutations may affect the incorporation of tubulins into microtubules. V5-tagged beta III tubulins harboring various mutations (Tischfield et al., 2010) can be synthesized *in vitro* using a rabbit reticulate system. "Sister" axoplasms are prepared as described in Section 9.3, incubated with 30 μl of buffer X/2 + 5 mM ATP alone as a control or mutant tubulin diluted in the same buffer for 50 min at room temperature. Buffer is removed at the end of incubation and extra solution around the axoplasm is carefully dried by placing a piece of Whatman paper close to the edge of the axoplasm. Fifty microliter of 4% PFA diluted in PBS is used to fix one axoplasm for 50 min at room temperature and then carefully removed as described above. Axoplasms together with the slide are carefully placed in a conical tube filled with 30 ml PBS for 10 min each and repeated three times to wash away the fixative. 1% BSA and 0.1% Triton X-100 in PBS are used to block for at least an hour at room temperature followed by incubation with primary antibodies (anti-V5 and DM1A) diluted in blocking buffer at 4 °C for overnight. Same washing by passing through the conical tubes is repeated three times in PBS before incubation in fluorescent secondary antibodies (Alexa 488 anti-mouse and 594 anti-rabbit) diluted in PBS for an hour at room temperature. After the final washes to remove the secondary antibody, spacers made out of No. 0 coverslips are coated on either side of the axoplasm on the glass with Compound 111 silicon grease, and axoplasms are mounted with mounting medium (such as ProLong Gold Mounting medium) between the two spacers and covered with a No. 1.5 coverslip. For long-term storage, nail polish can be applied to seal the edges. Images are taken and data analyzed by confocal microscopy. These approaches can also be modified to allow analysis of the axoplasmic microtubules by electron microscopy.

References

Allen, R. D., Allen, N. S., & Travis, J. L. (1981). Video-enhanced contrast, differential interference contrast (AVEC-DIC) microscopy: A new method capable of analyzing microtubule related movement in the reticulopodial network of Allogromia laticollaris. *Cell Motility*, *1*, 291–302.

Baumgold, J., Gallant, P., Terakawa, S., & Pant, H. (1981). Tetrodotoxin affects submembranous cytoskeletal proteins in perfused squid giant axons. *Biochemical and Biophysical Research Communications*, *103*(2), 653–658.

Borisy, G. G., Marcum, J. M., Olmsted, J. B., Murphy, D. B., & Johnson, K. A. (1975). Purification of tubulin and associated high molecular weight proteins from porcine brain and characterization of microtubule assembly *in vitro*. *Annals of the New York Academy of Sciences*, *253*, 107–132.

Brady, S. T., Lasek, R. J., & Allen, R. D. (1982). Fast axonal transport in extruded axoplasm from squid giant axon. *Science*, *218*, 1129–1131.

Brady, S. T., Lasek, R. J., & Allen, R. D. (1985). Video microscopy of fast axonal transport in isolated axoplasm: A new model for study of molecular mechanisms. *Cell Motility*, *5*, 81–101.

Brady, S. T., Richards, B. W., & Leopold, P. L. (1993). Assay of vesicle motility in squid axoplasm. *Methods in Cell Biology*, *39*, 191–202.

Brown, A., & Lasek, R. J. (1990). The cytoskeleton of the squid giant axon. In D. L. Gilbert, J. W. J. Adelman, & J. M. Arnold (Eds.), *Squid as experimental animals* (pp. 235–302). New York: Plenum Publishing Corporation.

Brown, A., & Lasek, R. J. (1993). Neurofilaments move apart freely when released from the circumferential constraint of the axonal plasma membrane. *Cell Motility and the Cytoskeleton*, *26*(4), 313–324.

Collins, K. D. (2004). Ions from the Hofmeister series and osmolytes: Effects on proteins in solution and in the crystallization process. *Methods*, *34*(3), 300–311.

Collins, K. D. (2006). Ion hydration: Implications for cellular function, polyelectrolytes, and protein crystallization. *Biophysical Chemistry*, *119*(3), 271–281.

Deffner, G. G., & Hafter, R. E. (1960a). Chemical investigations of the giant nerve fibers of the squid. III. Identification and quantitative estimation of free organic ninhydrin-negative constituents. *Biochimica et Biophysica Acta*, *42*, 189–199.

Deffner, G. J., & Hafter, R. E. (1960b). Chemical investigations of the giant nerve fiber of the squid. *Biochimica et Biophysica Acta*, *42*, 200–205.

Hearn, M. T., Hodder, A. N., & Aguilar, M. I. (1988). High-performance liquid chromatography of amino acids, peptides and proteins. LXXXVI. The influence of different displacer salts on the retention and bandwidth properties of proteins separated by isocratic anion-exchange chromatography. *Journal of Chromatography*, *443*, 97–118.

Jordan, M. A., & Kamath, K. (2007). How do microtubule-targeted drugs work? An overview. *Current Cancer Drug Targets*, *7*(8), 730–742.

Jordan, M. A., & Wilson, L. (2004). Microtubules as a target for anticancer drugs. *Nature Reviews. Cancer*, *4*(4), 253–265. http://dx.doi.org/10.1038/nrc1317.

Leopold, P. L., Lin, J.-W., Sugimori, M., Llinas, R., & Brady, S. T. (1994). The nervous system of Loligo pealei provides multiple models for analysis of organelle motility. In N. J. Abbott, R. Williamson, & L. Maddock (Eds.), *Cephalopod neurobiology: Neuroscience studies in squid, octopus and cuttlefish* (pp. 15–34). Oxford: Oxford University Press.

Morris, J., & Lasek, R. J. (1982). Stable polymers of the axonal cytoskeleton: The axoplasmic ghost. *The Journal of Cell Biology*, *92*, 192–198.

Morris, J., & Lasek, R. J. (1984). Monomer-polymer equilibria in the axon: Direct measurement of tubulin and actin as polymer and monomer in axoplasm. *The Journal of Cell Biology*, *98*, 2064–2076.

Olmsted, J. B., & Borisy, G. G. (1975). Ionic and nucleotide requirements for microtubule polymerization in vitro. *Biochemistry*, *14*(13), 2996–3005.

Sakai, H., & Matsumoto, G. (1978). Tubulin and other proteins from squid giant axon. [Comparative Study]. *Journal of Biochemistry*, *83*(5), 1413–1422.

Song, Y. (2010). *Stabilization of neuronal microtubules by polyamines and transglutaminase: Its roles in brain function (Ph.D)*. Chicago: University of Illinois.

Song, Y., Kirkpatrick, L. L., Schilling, A. B., Helseth, D. L., Chabot, N., Keillor, J. W., et al. (2013). Transglutaminase and polyamination of tubulin: Posttranslational modification for stabilizing axonal microtubules. *Neuron*, *78*, 109–123.

Tischfield, M. A., Baris, H. N., Wu, C., Rudolph, G., Van Maldergem, L., He, W., et al. (2010). Human TUBB3 mutations perturb microtubule dynamics, kinesin interactions, and axon guidance. *Cell*, *140*(1), 74–87. http://dx.doi.org/S0092-8674(09)01558-X [pii] 1016/j.cell.2009.12.011.

Weisenberg, R. C. (1973). Microtubule formation in vitro from solutions containing low calcium concentrations. *Science*, *177*, 1104–1105.

Weiss, D. G., Langford, G. M., Seitz-Tutter, D., & Keller, F. (1988). Dynamic instability and motile events of native microtubules from squid axoplasm. *Cell Motility and the Cytoskeleton*, *10*, 285–295.

Weiss, D. G., Langford, G. M., Seitz-Tutter, D., & Maile, W. (1991). Analysis of the gliding, fishtailing and circling motions of native microtubules. *Acta Histochemica. Supplementband*, *41*, 81–105.

Westh, P., Kato, H., Nishikawa, K., & Koga, Y. (2006). Toward understanding the Hofmeister series. 3. Effects of sodium halides on the molecular organization of H2O as probed by 1-propanol. *Journal of Physical Chemistry A*, *110*(5), 2072–2078. http://dx.doi.org/10.1021/jp055036y.

Windebank, A. J., & Grisold, W. (2008). Chemotherapy-induced neuropathy. *Journal of the Peripheral Nervous System*, *13*(1), 27–46. http://dx.doi.org/10.1111/j.1529-8027.2008.00156.x.

CHAPTER

Imaging GTP-Bound Tubulin: From Cellular to *In Vitro* Assembled Microtubules

10

Hélène de Forges[*,†,1], **Antoine Pilon**[‡,§,1], **Christian Poüs**[‡,¶,2] **and Franck Perez**[*,†,2]

[*]*Institut Curie, Paris Cedex 05, France*
[†]*CNRS UMR144, Paris Cedex 05, France*
[‡]*EA4530, Dynamique des microtubules en physiopathologie, Faculté de Pharmacie, Université Paris-Sud, Châtenay-Malabry, France*
[§]*Unité d'Hormonologie et Immunoanalyse, Pôle de Biologie Médicale et Pathologie, Hôpitaux Universitaires Est Parisien, APHP, Paris, France*
[¶]*Laboratoire de Biochimie-Hormonologie, Hôpitaux Universitaires Paris-Sud, APHP, Clamart, France*

CHAPTER OUTLINE

Introduction **141**
10.1 Imaging GTP Islands in Permeabilized Cells **142**
10.1.1 Permeabilization 143
10.1.2 MB11 Staining 144
10.1.3 Cell Fixation and Colabeling 144
10.2 Imaging GTP Caps and GTP Islands Using Centrosome-Based Microtubule Assembly or Endogenous Microtubule Elongation in Permeabilized Cells **145**
10.2.1 Cell Plating, Culture, and Treatment 145
10.2.1.1 Prior to Assembly of Centrosome-Nucleated Microtubules .. 145
10.2.1.2 Prior to Extracellular Microtubule Elongation 145
10.2.2 Microtubule Growth 145
10.3 Imaging GTP Islands in Microtubules Assembled *In Vitro* **147**
10.3.1 Microtubules Assembled in Suspension with GTP or GMPCPP 147
10.3.1.1 Separate Preparation of GMPCPP and GTP Microtubules ... 147
10.3.1.2 Mixed GMPCPP and GTP Microtubules and Staining with MB11 Antibody 147

[1,2]Equal contributions

ISSN 0091-679X
http://dx.doi.org/10.1016/B978-0-12-407757-7.00010-4

10.3.2 Microtubules Elongated from Taxol Seeds Immobilized on Kinesin-Coated Glass 148
10.3.2.1 Preparation of MAP-free Tubulin 148
10.3.2.2 Fluorescent–Tubulin Labeling 148
10.3.2.3 Recombinant KHC Reconstitution 148
10.3.2.4 Taxol-Stabilized Microtubule Seeds Polymerization 149
10.3.2.5 Preparation of the Incubation Chambers 149
10.3.2.6 Binding of Microtubule Seeds and Microtubule Elongation 149
10.4 Discussion and Future Prospects **150**
Acknowledgments **151**
References **152**

Abstract

Microtubules display a very dynamic behavior, and the presence of the guanosine-triphosphate (GTP) cap at the plus ends of microtubules is essential to regulate microtubule dynamics. Dimitrov et al. (2008) showed that GTP–tubulin is present not only at the plus ends but also in discrete locations along the microtubule lattice. These GTP islands were proposed to contribute to rescue events. Studying the localization of GTP–tubulin in microtubules is essential to better comprehend some core aspects in the regulation of microtubule dynamics. In this chapter, we recapitulate essential tools to study the GTP–tubulin using the recombinant antibody MB11 from permeabilized cells to *in vitro* assays.

Reagents

- AMPPNP (Sigma, Ref: A2647)
- Alexa 488-labeled goat antihuman antibody (Jackson Laboratories, Ref: 709-165-149)
- Anti-tubulin antibody (DM1A; Sigma, Ref: T6199)
- Anti-GTP–tubulin antibody (MB11)
- ATP (Sigma, Ref: A3377)
- Attofluor chamber (Invitrogen)
- β-Mercaptoethanol (Sigma, Ref: M6250)
- Casein (Sigma, Ref: C6905)
- Catalase (Sigma, Ref: C40)
- Dithiothreitol (DTT; Sigma, Ref: D9163)
- D-Glucose (Sigma, Ref: D-3179)
- DMSO (Sigma, Ref: 276855)
- EGTA (EuroMedex, Ref: 1310-B)
- Ethanol (Carlo Erba, Ref: 528151)
- Gelatin (Sigma, Ref: G1393)

Glucose oxidase (Sigma, Ref: G2133)
Glycerol (bidistilled 99.5%; VWR ProLabo, Ref: 24388.295)
GMPCPP (Gena Biosciences, Ref: NU-405L)
GTP (Sigma, Ref: G8877)
Kinesin heavy chain (KHC) motor domain (Cytoskeleton, Inc., KR01)
Methanol (Carlo Erba, Ref: 414819)
NHS ester dyes Cy3 and Cy5 (GE Healthcare, Ref: PA23001 or PA25001)
Nocodazole (Sigma, Ref: M1404)
Paraformaldehyde (EuroMedex, Ref: 15710)
PIPES (EuroMedex, Ref: 1124)
Taxol (Paclitaxel; Enzo Life Sciences, Ref: BML-T104-005)
Triton X-100 (EuroMedex, Ref: 2000-C)

Buffers
PEM: 80 mM PIPES, 1 mM EGTA, 1 mM $MgCl_2$, pH 6.9
PEM-G: PEM + 10% glycerol
Permeabilization buffer: PEM + 10% glycerol + 0.01% Triton X-100
PEM-T: PEM + 1 μM Taxol
PEM-C: PEM + 0.2 mg/mL casein

INTRODUCTION

It is well known that microtubules are highly dynamic polymers, which display dynamic instability (Mitchison & Kirschner, 1984). The role of GTP–tubulin in the regulation of microtubule dynamics has long been known, and it was shown that hydrolysis of GTP at the plus ends of growing microtubules is necessary to trigger depolymerization (Hyman, Salser, Drechsel, Unwin, & Mitchison, 1992). Loss of the GTP cap exposes the unstable GDP-bound tubulin core and leads to depolymerization. Many teams have studied the stabilizing microtubule cap, both its nature and size. Among them, Bayley, Schilstra, and Martin (1990) proposed the lateral cap model based on the hypothesis that the addition of a GTP-tubulin at the end causes the hydrolysis of the GTP of the previous tubulin dimer, which becomes incorporated into the microtubule lattice (Bayley et al., 1990). Drechsel and Kirschner (1994) showed that a minimum of 40 subunits is necessary in the GTP cap to stabilize microtubules. In 1996, Caplow and Shanks proposed, measuring the minimum disassembly rates using guanylyl-(alpha, beta)-methylene-diphosphonate (GMPCPP) microtubules, that a cap composed of only 13 or 14 GTP–tubulin subunits is sufficient to stabilize microtubules (Caplow & Shanks, 1996). Panda, Miller, and Wilson (2002) confirmed this idea and proposed that the stabilizing cap is composed of a monolayer of Tubulin-GDP-Pi.

Once the important role of the GTP–tubulin cap was understood, studies linking the presence of a GTP cap to microtubule-associated proteins (MAPs) arose. For example, in an early study (Severin, Sorger, & Hyman, 1997), kinetochores were found to bind preferentially to GTP microtubules. The authors polymerized chimeric microtubules composed of a GDP segment and GMPCPP ends and found that kinetochores display a higher affinity for microtubules plus ends. This suggested that structural features of the GTP-bound microtubule lattice could be recognizable. In the same study, it was confirmed that Taxol binding to tubulin mimics the GTP-bound conformation of tubulin dimers as previously showed by Arnal and Wade (1995). Indeed, kinetochores bind equally to GMPCPP or GDP portions of microtubules polymerized in the presence of Taxol. In a more recent example, Maurer, Bieling, Cope, Hoenger, and Surrey (2011) showed that EB1, a plus end-tracking protein, and Mal3, its homologue in yeast, bind effectively to GTP– and GTPγS–tubulin, but only weakly to GMPCPP seeds and to the GDP lattice of microtubules polymerized *in vitro*.

Such biochemical studies of the GTP cap were done *in vitro*, but *in vivo* studies have long been missing. To study the GTP cap in cells, we selected a recombinant antibody (MB11) using biotinylated GTPγS–tubulin as an antigen (Dimitrov et al., 2008). Using cosedimentation experiments or immunostaining of *in vitro* polymerized microtubules, MB11 was shown to recognize almost exclusively GTP-loaded microtubules (Dimitrov et al., 2008). Whether MB11 directly recognizes a particular conformation displayed by GTP–tubulin in the polymer (like EB or CLIP170 proteins do) or whether it recognizes some structural defects is still debated. MB11 efficiently stains the plus end of microtubules that are polymerizing and thus allowed to confirm the existence of the GTP cap at the plus ends of growing microtubules in cells. In addition, it led us to propose that GTP–tubulin is present not only at the plus ends but also in discrete dots along the microtubule lattice. We showed in cells that these GTP–tubulin islands are sites of rescue events. We thus proposed that GTP hydrolysis in microtubules might not always be complete and that, upon microtubule depolymerization, a GTP–tubulin island becomes exposed at the plus end and may behave as a GTP cap to promote a rescue event. The mechanisms of rescue involving GTP islands and the mechanisms of catastrophe are summarized and discussed in a recent review (Gardner, Zanic, & Howard, 2013), which points out the importance of rescues in the regulation of microtubule interactions with the cell cortex or with kinetochores.

Studying GTP–tubulin conformation in microtubules using the MB11 antibody is a technical challenge because the MB11 antibody does not recognize GTP–tubulin in microtubules after fixation, probably because fixation alters microtubule conformation. In this chapter, we present four protocols to study the GTP cap at microtubule plus ends and the GTP islands along the microtubule lattice using the anti-GTP–tubulin antibody MB11 in permeabilized cells and in *in vitro* assays.

10.1 IMAGING GTP ISLANDS IN PERMEABILIZED CELLS

MB11 is a conformational antibody. As such, it is very sensitive to antigen denaturation and the best staining results are obtained in permeabilized cells without fixation. To preserve the whole microtubule network, cells should be permeabilized

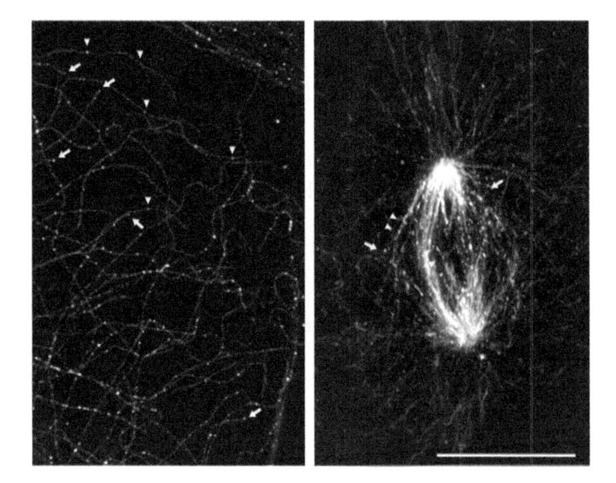

FIGURE 10.1

GTP–tubulin labeled with MB11 antibody (red) on PTK2 cells stably expressing GFP–tubulin (green) imaged on a structured illumination microscope (SIM). Acquisitions were performed in 3D SIM mode, with an n-SIM Nikon microscope before image reconstruction using the NIS-Elements software based on Gustafsson et al. (2008). The system is equipped with an APO TIRF 100 × 1.49 NA oil immersion and an EMCCD DU-897 Andor camera. Arrows point at GTP caps and arrowheads show GTP–tubulin islands, both stained with the MB11 antibody. Signals have been artificially increased for better visualization in the black and white version of the figure. (See color plate.)

using a nonionic detergent in an effective microtubule-stabilizing buffer. Using only a PIPES-based buffer is not sufficient to preserve a large proportion of cellular microtubules for more than a few minutes, thus glycerol and optionally Taxol are also added throughout the labeling operations (Fig. 10.1). To allow multiple labeling, cell fixation can be achieved as soon as incubation with MB11 and consecutive rinses have been performed. GTP–tubulin staining with MB11 is also compatible with the expression of GFP–tubulin (Fig. 10.1).

10.1.1 Permeabilization

Cells are grown in culture medium at 37 °C, 5% CO_2. Replace the culture medium by warm permeabilization buffer. Incubate exactly 3 min at 37 °C. Wash twice 1 s with caution in PEM-G, holding the coverslip.

Optionally, 1 μM Taxol can be added to the extraction, incubation, and washing solutions, especially if labeling is to be correlated with prior *in vivo* imaging. Note that, in contrast with what happens when Taxol is added during microtubule assembly or when using it at high concentration (Dimitrov et al., 2008), low doses of Taxol do not alter the number and length of GTP–tubulin islands *a posteriori*.

Put the coverslip directly on the drop of MB11 antibody.

Tips

Depending on the cell type, adjust the amount of Triton X-100 in the permeabilization buffer, between 0.03% and 1% so that cells remain attached to their substratum. Cells indeed tend to detach very easily after Triton permeabilization. All the steps performed prior to fixation must be made cautiously. As much as possible, do not aspirate or flow medium over the cells. Prefer holding the coverslips between tweezers and process by short immersions.

10.1.2 MB11 staining

Immunofluorescence of GTP–tubulin should be performed in a wet chamber at 37 °C. Drops of antibodies are deposited on a sheet of Parafilm in a closed and humid chamber.

Both MB11 and the antihuman secondary antibody are diluted in PEM-G supplemented with 2 g/L BSA.

After permeabilization, coverslips are put on a drop of MB11 antibody and incubated for 15 min at 37 °C. After incubation, prior to washes, gently add warm PEM-G under the coverslips before picking them up to prevent cells from detaching.

Wash out the primary antibody (3 times, 1 s) in a warm PEM-G bath.

Put the coverslips directly on the secondary antibody drop, on Parafilm, in the humid chamber. Incubate 15 min at 37 °C. Once again, lift slowly the coverslips using PEM-G, wash with caution in PEM-G, and fix the cells.

10.1.3 Cell fixation and colabeling

Cells can then be fixed either with methanol or with paraformaldehyde (PFA) depending on the needs of the experiments.

Fix the cells with cold methanol (−20 °C) for 4 min at −20 °C or with 3–4% PFA for 15 min at room temperature. Wash twice with PBS. Once the cells are fixed, a classical immunofluorescence can be performed for colabeling GTP islands with another staining. Note that cell permeabilization preserves the conformation of microtubules and of tubulin, but plus ends-tracking proteins (+TIPs) are lost from microtubule plus ends.

Tips

- After secondary antibody incubation and washes in PEM-G, put the coverslip on a sheet of Whatman paper to eliminate residual PEM-G. Directly immerse the coverslip in cold methanol.
- After the entire methanol is removed, quickly add PBS in one step. This step is very important to keep cells in good shape and image individual microtubules spread out in the cell.

10.2 IMAGING GTP CAPS AND GTP ISLANDS USING CENTROSOME-BASED MICROTUBULE ASSEMBLY OR ENDOGENOUS MICROTUBULE ELONGATION IN PERMEABILIZED CELLS

After microtubule depolymerization, centrosomes retain their capability to nucleate new microtubules in the presence of free GTP–tubulin once cells are permeabilized and cytosol proteins are extracted. This procedure was adapted from that used by Brinkley et al. (1981). Here, permeabilized cells are used to nucleate oriented microtubules and to control their regime of growth. A similar procedure can be undertaken using permeabilized cells without prior microtubule depolymerization to elongate microtubules in the pericellular region and observe their GTP-bound tubulin domains.

10.2.1 Cell plating, culture, and treatment

10.2.1.1 Prior to assembly of centrosome-nucleated microtubules

HeLa or RPE-1 cells are plated on glass coverslips coated with gelatin and cultured until they reach 50% confluence. Cellular microtubules are depolymerized in two steps: first cells are incubated at 37 °C for 1 h 30 min in 10 μM nocodazole and then put on ice for 2 h in the presence of nocodazole. The drug is then washed out by three rinses in ice-cold culture medium and immediately extracted on ice using PEM supplemented with 0.1% Triton X-100 (three extractions of 1 min). Cells are rinsed twice on ice with PEM to remove Triton.

10.2.1.2 Prior to extracellular microtubule elongation

Cells plated on gelatin and cultured as described above are permeabilized in the permeabilization buffer and rinsed twice in PEM-G without Triton.

10.2.2 Microtubule growth

For both assays, microtubule growth is performed using purified porcine brain tubulin (see Section 3.2.1) diluted at 0.25 g/L in PEM supplemented with 1 mM GTP. To elongate extracellularly endogenous microtubules, the addition of 200–500 μM adenosine triphosphate (ATP) greatly enhances the number of polymerization-competent microtubules (Infante, Stein, Zhai, Borisy, & Gundersen, 2000). For centrosome-nucleated microtubules, extracted cells are kept on ice before polymerization; tubulin assembly starts upon shifting the temperature to 37 °C.

In both assays, assembly is performed for 5–30 min. At the end of the assembly period, soluble tubulin is removed by gently washing coverslips with warm PEM-G and optionally with 1 μM Taxol prior to MB11 labeling, fixation, and colabeling if required. GTP–tubulin labeling is performed as described previously.

Tips

- Cell density should be kept below 50% of confluence to prevent massive cell detachment during the incubations, which is greatly enhanced by the incubation with tubulin.
- Gelatin coating not only helps keeping cell attachment to coverslips but also prevents excessive tubulin adsorption to the substratum and keeps tubulin staining background to reasonable levels.
- Microtubule number and length are limited by the concentration of tubulin added during the elongation phase. Tubulin concentration should be kept lower than 0.5 g/L to prevent excessive protein adsorption on cellular ghosts. To visualize internal GTP islands, growing microtubules to steady state is sufficient. To visualize GTP caps, labeling with MB11 should be performed while microtubules are still in a growth phase. This is usually obtained within the first 10–15 min of incubation, yielding an important amount of short microtubules. Note that some microtubules may also detach (spontaneously or mechanically during incubations and washes) and be observed in various cellular or even extracellular locations (Fig. 10.2).

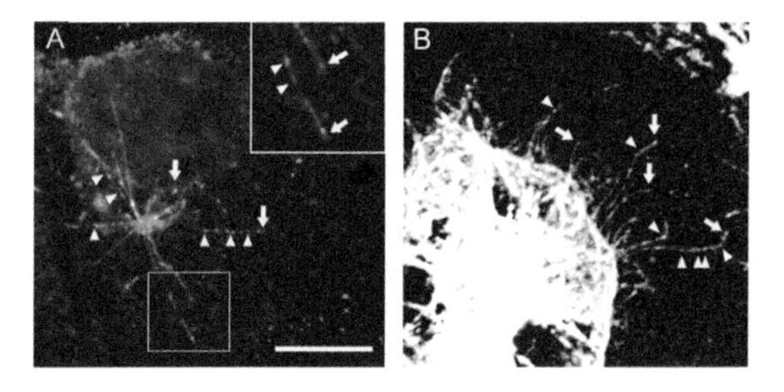

FIGURE 10.2

GTP–tubulin labeling in microtubules nucleated from cellular centrosomes or elongated from cellular microtubules. (A) After complete microtubule disassembly with nocodazole and cold, RPE-1 cells were permeabilized and microtubules were grown from cellular centrosomes with GTP. Note the quite high tubulin staining in cell ghosts and the presence of detached microtubules (inset). Images were acquired on a Leica DMLB microscope. (B) HeLa cells were permeabilized and treated with ATP prior to extracellular microtubule elongation using purified tubulin and GTP. Images are confocal planes acquired using a Zeiss LSM510 microscope. Note that the cell from the right panel detached after fixation, but extracellularly elongated microtubules remained visible. Microtubules were stained for GTP–tubulin using MB11 (red) and total tubulin (green) using anti alpha-tubulin antibody (clone DM1A). GTP caps and internal GTP islands are indicated with arrows and arrowheads, respectively. Scale bar = 10 μm. Signals have been artificially increased for better visualization in the black and white version of the figure. (See color plate.)

10.3 IMAGING GTP ISLANDS IN MICROTUBULES ASSEMBLED *IN VITRO*

In vitro techniques may also be used to visualize GTP–tubulin regions in microtubules. These approaches allow GTP island formation in the absence of MAPs or of insoluble cellular structures after detergent extraction. One of the following protocols also makes use of the GTP analogue GMPCPP, which locks the conformation of tubulin in a GTP-like state. GMPCPP is a slowly hydrolyzable molecule used formerly by Hyman and colleagues (Hyman et al., 1992) who showed that GTP hydrolysis is essential for depolymerization of microtubules and thus for dynamic instability.

10.3.1 Microtubules assembled in suspension with GTP or GMPCPP

10.3.1.1 Separate preparation of GMPCPP and GTP microtubules

Unlabeled and fluorescent tubulins (see Section 3.2) are used to polymerize GMPCPP and GTP microtubules separately. The ratio between fluorescent tubulin and nonfluorescent tubulin is 1:10. Tubulin diluted in PEM supplemented with 10 mM DTT is used at 2.5 g/L (final concentration) and incubated on ice with 1 mM GTP or 1 mM GMPCPP for 5 min. Both mixes are polymerized for 30–45 min at 37 °C. After polymerization, microtubules are diluted at least 50- to 100-fold in warm PEM-T.

10.3.1.2 Mixed GMPCPP and GTP microtubules and staining with MB11 antibody

Microtubules polymerized separately in the presence of each nucleotide are spun down at 30 °C for 10 min at 20,000 × *g* to remove unpolymerized tubulin. Both pellets are resuspended separately in a small volume of PEM-T and then one volume of GMPCPP microtubules is mixed with one volume of GTP microtubules. The MB11 antibody is added to the mixture, which is incubated for 15 min at 37 °C. Microtubules are centrifuged for 15 min at 20,000 × *g* at 30 °C and washed once in PEM-T. They are spun down again; the pellet is resuspended in PEM-T containing the secondary antibody and incubated 15 min at 37 °C.

After final pelleting for 10 min at 20,000 × *g* and resuspension in PEM-T supplemented with antifading agents, a drop of the microtubule suspension is directly observed without fixation at room temperature, with a fluorescence microscope.

Tips

- To pipet polymerized microtubules, use large pipette tips or cut the extremity of the tip so that pipetting does not break polymerized microtubules.
- A small amount of Taxol (1 μM) allows stabilization of the microtubules without changing their conformation as described above. Adding large amounts of Taxol will induce changes in microtubule conformation and such microtubules will be stained all along the lattice by the MB11 antibody.

- The supernatants over pelleted microtubules must be taken off with great care to avoid losing microtubules at each cycle of centrifugation between labelings and washes.

10.3.2 Microtubules elongated from Taxol seeds immobilized on kinesin-coated glass

10.3.2.1 *Preparation of MAP-free tubulin*

MAP-free tubulin is purified from porcine brain as described previously by Walker et al. (1988). MAPs are removed by phosphocellulose chromatography and tubulin aliquots are frozen in liquid nitrogen and stored at −80 °C. For each experiment, an aliquot of purified tubulin is thawed on ice and centrifuged at 20,000 × *g* for 20 min to sediment aggregated tubulin, which is discarded.

10.3.2.2 *Fluorescent–tubulin labeling*

Tubulin is labeled with Cy3 or Cy5 monoreactive NHS ester dyes using a protocol adapted from Peloquin, Komarova, and Borisy (2005). For one labeling, 500 μg of MAP-free tubulin (10 μg/μL) is polymerized with 1 mM GTP for 30 min at 37 °C. One vial of Cy3 or Cy5 monoreactive dye is solubilized in 20 μL DMSO and polymerized microtubules are incubated for 15 min at 37 °C with 5 μL of this solution. Microtubules are then depolymerized for 10 min on ice and labeled tubulin is separated from the excess of unconjugated dye by two cycles of assembly–disassembly. Tubulin-labeled aliquots are frozen in liquid nitrogen and stored at −80 °C.

The advantage of this protocol is that tubulin is labeled in polymerized microtubules and dyes cannot react with primary amines involved in the interaction between tubulin heterodimers during microtubule polymerization. Thus, labeled tubulin is not excluded from microtubules when combined with unlabeled tubulin in the polymerization protocols described above. We could not exclude however that tubulin is labeled on sites important for MAP interactions. In polymerization studies, we used less than 25% of labeled tubulin to avoid the inhibition of MAP interactions with microtubules due to the presence of the fluorochrome.

10.3.2.3 *Recombinant KHC reconstitution*

In these experiments, microtubules are specifically bound on glass coverslips using the interaction with recombinant, active KHC. We use a protocol described for microtubule gliding motility assays (Howard, Hunt, & Baek, 1993), with minor modifications. Classical gliding assay using ATP is first used to determine the polarity of microtubule seeds. Then, microtubule growth is monitored in the absence of movement because of kinesin loading with the nonhydrolyzable ATP analogue, AMPPNP, which allows microtubule/kinesin interaction but prevents nucleotide hydrolysis necessary for microtubule displacement.

Lyophilized KHC is resuspended at 5 μg/μL in kinesin reconstitution buffer composed of 100 mM PIPES pH 7.0, 200 mM KCl, 2 mM $MgCl_2$, 1 mM DTT, 20 μM ATP. 0.5 μL Aliquots are frozen in liquid nitrogen and stored at −80 °C.

For use, one aliquot of kinesin is diluted at 0.5 mg/mL in PEM buffer containing 0.2 mg/mL casein (PEM-C) and 20 μM ATP. Then kinesin is diluted at 10 μg/mL in PEM-C containing 100 μM ATP.

Tips

Casein is used to avoid nonspecific interactions between microtubule seeds or free tubulin with glass coverslips. It is poorly soluble at neutral pH, but it can be rapidly dissolved after effective alkalinization with 10N KOH. After complete solubilization, pH is set back to 6.9 using HCl.

10.3.2.4 Taxol-stabilized microtubule seeds polymerization

To prepare stabilized microtubule seeds as templates to nucleate microtubule growth, we use an alternative method to GMPCPP-stabilized microtubule seeds. Seeds are stabilized using Taxol for easiness and swiftness.

A 4 g/L mix of Cy5-labeled tubulin and nonlabeled tubulin (with a 1:10 ratio of labeled to unlabeled tubulin) is incubated for 15 min at 37 °C with 1 mM GTP and 10% DMSO. These conditions allow the spontaneous polymerization of microtubule seeds. After incubation, microtubule seeds are diluted to 1:100 in PEM containing 10 μM Taxol. These seeds can be stored for 2 days at room temperature.

10.3.2.5 Preparation of the incubation chambers

Sample chambers are prepared using standard glass microscopy coverslips. Coverslips are washed consecutively in 10N HCl, Nanopure water, and 90% ethanol and air-dried. A 2- to 3-mm-width channel is delimited by two strips of adhesive tape on a 25-mm-diameter glass coverslip. A 12-mm-diameter coverslip is clamped on the first one with two additional strips of tape. The solutions are perfused in the incubation chamber using a pipette and a filter paper to aspirate the solutions at the opposite end of the channel. The incubation chamber is then placed in an Attofluor chamber for observation under the microscope.

10.3.2.6 Binding of microtubule seeds and microtubule elongation

10.3.2.6.1 Nonspecific kinesin immobilization

The 10 μg/mL kinesin solution is incubated for 5 min at room temperature in the chamber to allow the adsorption of motor protein on glass. The chamber is then washed three times with PEM containing 0.5 mg/mL casein, and this solution is incubated for 5 min at room temperature to block the nonspecific adsorption of microtubule seeds and of free tubulin added in the experiments. It also prevents kinesin-1 from denaturation (Ozeki et al., 2009).

10.3.2.6.2 Attachment of microtubule seeds and elongation

Microtubule seed suspension is diluted 10 times in PEM containing 10 μM Taxol and 100 μM ATP. Once microtubule seeds are loaded, the chamber is placed at 37 °C on a Zeiss LSM510 confocal microscope (Heidelberg Germany). Microtubule seed attachment and movement are monitored by acquisition of the Cy5 fluorescent signal (63 × 1.4 NA objective).

Seed gliding on kinesin-coated glass allows the unambiguous identification of microtubule plus and minus ends: because immobilized kinesin walks toward the plus ends, the seeds move with minus ends in front. To stop seeds from gliding prior to elongation, the chamber is washed twice with PEM-C supplemented with 10 μM Taxol and 1 mM AMPPNP.

To elongate microtubule seeds, a 1 g/L mix of unlabeled and Cy3-labeled tubulin (with a 1:4 ratio of labeled to unlabeled tubulin) in PEM supplemented with 1 mM GTP is perfused twice in the chamber to eliminate Taxol and to avoid stabilization of growing microtubules. To monitor microtubule elongation, time-lapse sequences are acquired during 20 min. Microtubules that elongate at the plus and minus ends of the seeds interact with the kinesin adsorbed on glass and polymerize parallel to the confocal plane. At the end of microtubule growth, free tubulin is removed by washing the chamber with PEM-G containing 1 μM Taxol.

Tips

- Imaging of microtubule polymerization is done at 37 °C. Placing a humid lid over the incubation chamber so that liquid in the channel does not evaporate from its extremities is strongly recommended.
- We also recommend using an oxygen scavenger cocktail (20 mM D-glucose, 0.02 mg/mL glucose oxidase, 0.08 mg/mL catalase, and 0.5% β-mercaptoethanol). Oxygen scavenging protects dyes from photobleaching and prevents tubulin oxidation, which can cause the spontaneous breaking and depolymerization of microtubules. This cocktail is added to every solution injected in the chamber before imaging. It has to be used within 1 h after preparation.

To label GTP–tubulin, we use the protocol described above (see Section 1.2). After labeling, microtubules can be imaged directly or fixed with methanol for further analysis (Fig. 10.3).

10.4 DISCUSSION AND FUTURE PROSPECTS

The MB11 conformational antibody specifically binds to GTP–tubulin. It is a unique tool to study the distribution of GTP–tubulin islands along microtubules and their implication in microtubule dynamics. Although the exact structure that the antibody recognizes in tubulin is not yet known, MB11 may prove to be a powerful tool to study modifications in the microtubule network depending on changes in the cell status in pathophysiological conditions (e.g., differentiation, division, migration).

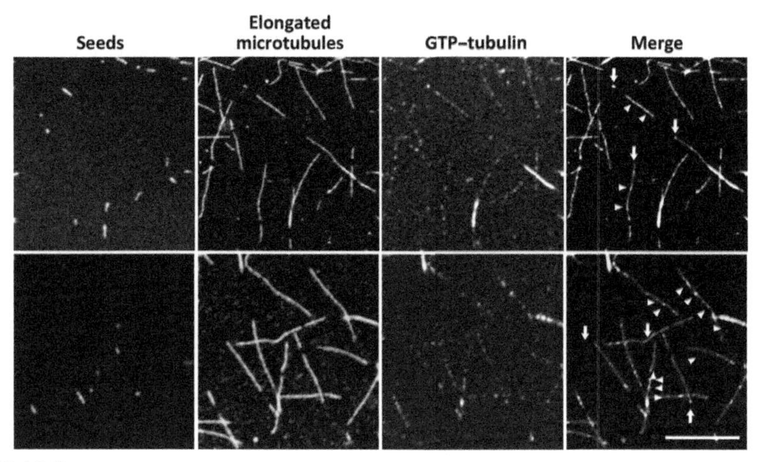

FIGURE 10.3

GTP–tubulin labeling in microtubules elongated from Taxol-stabilized seeds attached to kinesin heavy chain. Taxol-stabilized microtubule seeds prepared using Cy5-labeled tubulin were attached to kinesin. After seed immobilization with AMPPNP, microtubules were elongated with Cy3-labeled tubulin and then subjected to MB11 staining revealed with A488-labeled secondary antibody. Fields from two experiments are shown. Images are confocal planes acquired with a Zeiss LSM510 microscope. GTP caps and internal GTP islands are indicated with arrows and arrowheads, respectively. Scale bar = 10 μm. Signals have been artificially increased for better visualization in the black and white version of the figure. (See color plate.)

Certain microtubule-stabilizing MAPs, like the Von Hippel–Lindau antioncogene (Thoma et al., 2010), may modulate the occurrence of microtubule internal GTP islands. It suggests that cells may finely regulate rescue events by controlling the local density of GTP islands. Understanding how GTP islands occur in microtubules and if other MAPs facilitate their persistence is an important challenge. Conversely, understanding whether GTP-bound tubulin in the internal regions of microtubules or in the GTP cap of growing microtubules can be readily recognized by +TIPs to control microtubule elongation and rescues will be of great interest to get a more comprehensive view of rescue events. However, colabelings of GTP–tubulin and +TIPs in living cells are still an open challenge. Gaining access to such information will be valuable, for example, to measure the actual size of the GTP cap or to evaluate whether GTP–tubulin conformation is immobile or can propagate along microtubules.

Acknowledgments

We thank A. Dimitrov (Institut Curie, Paris) who contributed to the setup of some of the protocols described here. We thank L. Sengmanivong for support in SIM microscopy and acknowledge the Nikon Imaging Centre at Institut Curie. The authors also thank V. Nicolas (Institut Fédératif de Recherche IPSIT, Univ. Paris-Sud 11) for helpful advice in imaging *in vitro* assembled microtubules on the LSM510 microscope. F. P. team is supported by the Centre National de la Recherche Scientifique and by the Institut Curie and by grants from the Agence National de la Recherche, the Fondation pour la Recherche Médicale, and the Institut National du Cancer. H. F. is supported by the Ministère de l'Enseignement Supérieur et de la Recherche and the Fondation ARC pour la Recherche sur le Cancer.

References

Arnal, I., & Wade, R. H. (1995). How does taxol stabilize microtubules? *Current Biology*, *5*, 900–908.

Bayley, P. M., Schilstra, M. J., & Martin, S. R. (1990). Microtubule dynamic instability: Numerical simulation of microtubule transition properties using a Lateral Cap model. *Journal of Cell Science*, *95*, 33–48.

Brinkley, B. R., Cox, S. M., Pepper, D. A., Wible, L., Brenner, S. L., & Pardue, R. L. (1981). Tubulin assembly sites and the organization of cytoplasmic microtubules in cultured mammalian cells. *The Journal of Cell Biology*, *90*, 554–562.

Caplow, M., & Shanks, J. (1996). Evidence that a single monolayer tubulin-GTP cap is both necessary and sufficient to stabilize microtubules. *Molecular Biology of the Cell*, *7*, 663–675.

Dimitrov, A., Quesnoit, M., Moutel, S., Cantaloube, I., Pous, C., & Perez, F. (2008). Detection of GTP-tubulin conformation in vivo reveals a role for GTP remnants in microtubule rescues. *Science*, *322*, 1353–1356.

Drechsel, D. N., & Kirschner, M. W. (1994). The minimum GTP cap required to stabilize microtubules. *Current Biology*, *4*, 1053–1061.

Gardner, M. K., Zanic, M., & Howard, J. (2013). Microtubule catastrophe and rescue. *Current Opinion in Cell Biology*, *25*, 14–22.

Gustafsson, M. G., Shao, L., Carlton, P. M., Wang, C. J., Golubovskaya, I. N., Cande, W. Z., Agard, D. A., & Sedat, J. W. (2008). Three-dimensional resolution doubling in wide-field fluorescence microscopy by structured illumination. *Journal of Biophysics*, *94*, 4957–4970.

Howard, J., Hunt, A. J., & Baek, S. (1993). Assay of microtubule movement driven by single kinesin molecules. *Methods in Cell Biology*, *39*, 137–147.

Hyman, A. A., Salser, S., Drechsel, D. N., Unwin, N., & Mitchison, T. J. (1992). Role of GTP hydrolysis in microtubule dynamics: Information from a slowly hydrolyzable analogue, GMPCPP. *Molecular Biology of the Cell*, *3*, 1155–1167.

Infante, A. S., Stein, M. S., Zhai, Y., Borisy, G. G., & Gundersen, G. G. (2000). Detyrosinated (Glu) microtubules are stabilized by an ATP-sensitive plus-end cap. *Journal of Cell Science*, *113*(Pt. 22), 3907–3919.

Maurer, S. P., Bieling, P., Cope, J., Hoenger, A., & Surrey, T. (2011). GTPgammaS microtubules mimic the growing microtubule end structure recognized by end-binding proteins (EBs). *Proceedings of the National Academy of Sciences of the United States of America*, *108*, 3988–3993.

Mitchison, T., & Kirschner, M. (1984). Dynamic instability of microtubule growth. *Nature*, *312*, 237–242.

Ozeki, T., Verma, V., Uppalapati, M., Suzuki, Y., Nakamura, M., Catchmark, J. M., et al. (2009). Surface-bound casein modulates the adsorption and activity of kinesin on SiO_2 surfaces. *Biophysical Journal*, *96*, 3305–3318.

Panda, D., Miller, H. P., & Wilson, L. (2002). Determination of the size and chemical nature of the stabilizing "cap" at microtubule ends using modulators of polymerization dynamics. *Biochemistry*, *41*, 1609–1617.

Peloquin, J., Komarova, Y., & Borisy, G. (2005). Conjugation of fluorophores to tubulin. *Nature Methods*, *2*, 299–303.

Severin, F. F., Sorger, P. K., & Hyman, A. A. (1997). Kinetochores distinguish GTP from GDP forms of the microtubule lattice. *Nature*, *388*, 888–891.

Thoma, C. R., Matov, A., Gutbrodt, K. L., Hoerner, C. R., Smole, Z., Krek, W., et al. (2010). Quantitative image analysis identifies pVHL as a key regulator of microtubule dynamic instability. *The Journal of Cell Biology*, *190*, 991–1003.

Walker, R. A., O'Brien, E. T., Pryer, N. K., Soboeiro, M. F., Voter, W. A., Erickson, H. P., et al. (1988). Dynamic instability of individual microtubules analyzed by video light microscopy: Rate constants and transition frequencies. *The Journal of Cell Biology*, *107*, 1437–1448.

CHAPTER

11

Tubulin-Specific Chaperones: Components of a Molecular Machine That Assembles the α/β Heterodimer

Guoling Tian and Nicholas J. Cowan

Department of Biochemistry and Molecular Pharmacology, New York University Langone Medical Center, New York, New York, USA

CHAPTER OUTLINE

Introduction **156**

11.1 Methods **159**

11.1.1 cDNA Clones and Vectors 159

11.1.2 Expression of TBCA, TBCB, and TBCC 161

11.1.2.1 Bacterial Cell Growth and Induction 161

11.1.2.2 Preparation of Bacterial Lysates 162

11.1.3 Purification of TBCA, TBCB, and TBCC 162

11.1.3.1 Chromatography on Either Sepharose Q (TBCA and TBCB) or Sepharose S High-Performance (TBCC) Ion Exchange Resins 162

11.1.3.2 Anion Exchange Chromatography on MonoQ 163

11.1.3.3 Gel Filtration Chromatography (optional) 163

11.1.4 Expression of TBCE 163

11.1.4.1 Construction of Recombinant Baculovirus 163

11.1.5 Purification of TBCE 164

11.1.5.1 Preparation of Insect Cell Lysates 164

11.1.5.2 Anion Exchange Chromatography 164

11.1.5.3 Chromatography on Hydroxylapatite 165

11.1.5.4 Size Exclusion Chromatography (optional) 165

11.1.6 Expression of TBCD 165

11.1.7 Purification of TBCD 166

11.1.7.1 Preparation of HeLa Cell Lysates 166

11.1.7.2 Anion Exchange Chromatography on Q Sepharose High Performance 166

11.1.7.3 Anion Exchange Chromatography on MonoQ 166

Methods in Cell Biology, Volume 115

ISSN 0091-679X
http://dx.doi.org/10.1016/B978-0-12-407757-7.00011-6

11.1.7.4 Chromatography on Hydroxylapatite 166
11.1.7.5 Size Exclusion Chromatography 167
11.2 Discussion **167**
References **169**

Abstract

The tubulin heterodimer consists of one α- and one β-tubulin polypeptide. Neither protein can partition to the native state or assemble into polymerization competent heterodimers without the concerted action of a series of chaperone proteins including five tubulin-specific chaperones (TBCs) termed TBCA–TBCE. TBCA and TBCB bind to and stabilize newly synthesized quasi-native β- and α-tubulin polypeptides, respectively, following their generation via multiple rounds of ATP-dependent interaction with the cytosolic chaperonin. There is free exchange of β-tubulin between TBCA and TBCD, and of α-tubulin between TBCB and TBCE, resulting in the formation of TBCD/β and TBCE/α, respectively. The latter two complexes interact, forming a supercomplex (TBCE/α/TBCD/β). Discharge of the native α/β heterodimer occurs via interaction of the supercomplex with TBCC, which results in the triggering of TBC-bound β-tubulin (E-site) GTP hydrolysis. This reaction acts as a switch for disassembly of the supercomplex and the release of E-site GDP-bound heterodimer, which becomes polymerization competent following spontaneous exchange with GTP. The tubulin-specific chaperones thus function together as a tubulin assembly machine, marrying the α- and β-tubulin subunits into a tightly associated heterodimer. The existence of this evolutionarily conserved pathway explains why it has never proved possible to isolate α- or β-tubulin as stable independent entities in the absence of their cognate partners, and implies that each exists and is maintained in the heterodimer in a nonminimal energy state. Here, we describe methods for the purification of recombinant TBCs as biologically active proteins following their expression in a variety of host/vector systems.

INTRODUCTION

The α/β-tubulin heterodimer was originally thought to assemble spontaneously via association of the two constituent polypeptides, with a binding constant in the micromolar range (Detrich & Williams, 1978). More recent measurements based on plasmon resonance suggest a dissociation constant in the range of 10^{-11} M (Caplow & Fee, 2002). In any event, it has never proved possible to purify α- or β-tubulin in native form free from its counterpart. Moreover, expression of α- or β-tubulin in *Escherichia coli*, either alone or together, leads to their deposition within the host cells as completely insoluble inclusion bodies. Attempts to recover native tubulin from these insoluble materials have consistently proved futile, in spite of the well-established principle that all the information required for a given protein to assume its correct three-dimensional structure is contained within its primary structure (Anfinsen, 1973). However, *in vitro* translation in a eukaryotic cell extract (such as that derived from rabbit reticulocyte lysate) of the same sequences that yield insoluble material in *E. coli* results in the generation of soluble tubulin that is functional in terms of its ability to polymerize into microtubules (Cleveland, Kirschner, & Cowan, 1978). This posed the following paradox: tubulin translated in a prokaryotic cell context does not fold and leads to the production of inclusion bodies, while translation of the identical sequences in eukaryotic cells leads to the generation of functional tubulin heterodimers.

The deposition of insoluble α- and β-tubulin in *E. coli* cells has been successfully exploited in order to develop an *in vitro* folding assay for these proteins (Cowan, 1998). The method depends on the ability to label the recombinant protein in the prokaryotic host without labeling any host cell proteins. This is done using a vector in which the expression of recombinant sequences is driven by a T7 promotor: in the presence of ^{35}S-methionine and rifampicin (a drug that inhibits *E. coli* RNA polymerase, but not T7 polymerase), only the recombinant protein is labeled (Studier, Rosenberg, Dunn, & Dubendorff, 1990). The labeled inclusion bodies can be relatively easily purified because of their extreme insolubility, and the recombinant proteins unfolded in 8 M urea. This procedure yields probes of sufficiently high purity and specific activity (i.e., $>10^6$ cpm/μg) that they can be used in *in vitro* folding assays to identify factors that are required for productive folding. The products of such *in vitro* folding reactions can be readily identified by their characteristic mobility on native polyacrylamide gels.

Our development and use of this assay led to the discovery and purification of the cytosolic chaperonin (Gao, Thomas, Chow, Lee, & Cowan, 1992) (termed CCT, for *C*ytosolic *C*haperonin containing *T*-complex polypeptide 1; also termed TriC, for *T-ri*ng *C*omplex). This is a large, ribosome-sized multimolecular complex assembled from eight different (though related) polypeptides into a structure that is readily visible in the electron microscope as a double toroid. CCT polypeptides are distantly related to those of GroEL, the chaperonin that is present in *E. coli* and that functions in the facilitated folding of a significant proportion (estimated to be at least 5%) of newly synthesized proteins (Hartl & Hayer-Hartl, 2002; Lorimer, 1996; Young, Agashe, Siegers, & Hartl, 2004). All chaperonins, including CCT, function by providing a sequestered environment within the toroidal cavity where folding can take place in the absence of nonproductive interactions with other proteins. Cycles of ATP hydrolysis and ADP/ATP exchange result in allosteric changes in the chaperonin that govern the binding and release of the target protein (Spiess, Meyer, Reissmann, & Frydman, 2004; Valpuesta, Martin-Benito, Gomez-Puertas, Carrascosa, & Willison, 2002). In the case of α- and β-tubulin, interaction with CCT is an obligatory part of the folding reaction (Cowan & Lewis, 2001). Moreover, while GroEL participates in the facilitated folding of a wide range of proteins in *E. coli* cells, it cannot facilitate the productive folding of actins or tubulins, the two principle targets of CCT (Tian, Vainberg, Tap, Lewis, & Cowan, 1995b). This explains why it is not possible to produce soluble tubulins via their expression in *E. coli.*

Unlike actin and other obligate targets of CCT, neither α- nor β-tubulin can partition to the native state as a result of ATP-dependent interaction with CCT alone (Gao, Vainberg, Chow, & Cowan, 1993). Rather, tubulin subunits released from CCT are assembled into α/β heterodimers by interaction with several proteins known as cofactors or tubulin-specific chaperones (TBCs) in a reaction that depends on GTP hydrolysis by the chaperone-bound β-tubulin (Bhamidipati, Lewis, & Cowan, 2000; Cowan and Lewis, 2001; Lewis, Tian, & Cowan, 1997; Tian et al., 1997). The overall pathway whereby tubulin heterodimers are formed *de novo* in higher eukaryotes is depicted in Fig. 11.1. Quasi-native intermediates (defined as containing a native GTP-binding pocket) generated as a result of multiple cycles of ATP-dependent interaction with CCT (Tian, Vainberg, Tap, Lewis, & Cowan, 1995a) are captured and stabilized by TBCA or TBCD (in the case of β-tubulin) or by TBCB and TBCE (in the case of α-tubulin). There is free exchange of β-tubulin between TBCA and TBCD and of α-tubulin between TBCB and TBCE. TBCE/α and TBCD/β interact with each

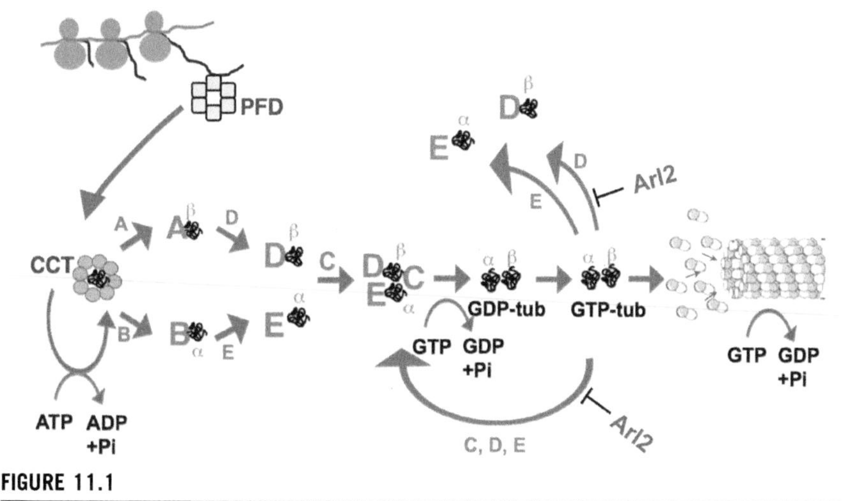

FIGURE 11.1

The chaperone-dependent tubulin folding and heterodimer assembly pathway. Nascent α- and β-tubulin polypeptides are bound by the chaperone protein prefoldin (blue) (Vainberg et al., 1998) and transferred to the cytosolic chaperonin CCT (orange). As a result of multiple rounds of ATP-dependent interaction with the chaperonin, the tubulin target proteins partition to a quasi-native state defined by the acquisition of a native GTP-binding pocket. Quasi-native folding intermediates are captured and stabilized by TBCB (in the case of α-tubulin) or TBCA (in the case of β-tubulin). The tubulin target proteins are transferred by free exchange to TBCD and TBCE, forming TBCD/β and TBCE/α; these complexes associate to form a supercomplex. Entry of TBCC into this supercomplex triggers hydrolysis of GTP in the E-site, destabilizing it and releasing newly formed α/β-tubulin heterodimers. Following free exchange of E-site guanine nucleotide, these heterodimers are competent for entry into the microtubule polymer. Note that TBCD and TBCE are each capable of disrupting the heterodimer in the back-reaction (shown as purple arrows in this figure); in the presence of TBCC and GTP, this results in a perpetual cycling of tubulin polypeptides through the supercomplex. The activity of TBCD is modulated via its interaction with the small Ras family GTPase Arl2 (green). Based on Tian et al. (1997), Tian, Bhamidipati, Cowan, and Lewis (1999), Vainberg et al. (1998), Tian, Huang, Parvari, Diaz, and Cowan (2006), Bhamidipati et al. (2000), Cowan and Lewis (2001). (For interpretation of the references to color in this figure legend, the reader is referred to the online version of this chapter.)

other, forming a multimolecular complex; TBCC enters this complex, forming a supercomplex that releases native tubulin heterodimers upon E-site GTP hydrolysis. The hydrolysis of GTP by β-tubulin in the supercomplex may be thought of as a switch for the release of heterodimers because cofactors have a much lower affinity for GDP-tubulin than for GTP-tubulin (Tian et al., 1999). Importantly, in addition to participating in the generation of *de novo* tubulin heterodimers, TBCC, TBCD, and TBCE can act together as a GTPase activator (GAP) for native tubulin. The biological significance of this activity has not been clearly established *in vivo*, but it may serve as a quality-control mechanism that continually checks for the ability of native heterodimers to hydrolyze GTP. It may also contribute toward modulating microtubule dynamics by influencing the size of the pool of GTP-bound tubulin (Tian et al., 1999).

Not surprisingly, the tubulin-specific chaperone-dependent reaction that assembles the heterodimer can be driven in reverse: incubation *in vitro* of native heterodimers with a molar excess of TBCD or TBCE results in heterodimer disruption and (at least in the case of TBCD) the formation of the TBCD/β complex (the TBCE/α complex does not appear to exist as a stable entity). These reactions (which we refer to as the back reaction, shown as purple arrows in Fig. 11.1) are also apparent *in vivo*: when cells are transfected with plasmids engineered for the overexpression of TBCD or TBCE, their microtubule network is destroyed (Fig. 11.2). A noteworthy feature of tubulin destruction by TBCD is that it is modulated by interaction of TBCD with the small Ras family member GTPase Arl2 (Bhamidipati et al., 2000).

This chapter describes methods for the preparation and purification of the tubulin-specific chaperones. All five of these proteins were originally isolated from a tissue source (bovine testis) via multiple chromatographic dimensions in which the protein of interest was assayed via CCT-driven *in vitro* tubulin folding reactions. These laborious and time-consuming protocols have now been superseded by the production of the corresponding biochemically active recombinant proteins in a variety of host/vector systems, and it is these methods that are described below. The sequences of TBCA-E are well conserved among mammals, so the protocols we provide should be applicable for their purification via expression using cloned cDNAs from any mammalian species. The methods we describe are for the proteins in unmodified form. This is because the effect on biological activity (if any) of the addition of a tag to facilitate affinity purification is uncertain. The yield of recombinant protein in the host systems we have developed is variable (depending on the chaperone in question), but the chromatographic methods we describe are for the most part straightforward. The *in vitro* CCT-dependent folding assay used to demonstrate TBC activity has been described in detail elsewhere (Cowan, 1998).

11.1 METHODS

11.1.1 cDNA clones and vectors

Full-length cDNA clones encoding TBCA, TBCB, TBCC, TBCD, and TBCE are available a number of commercial sources (e.g., Origene Inc., Cambridge Biosciences Inc., Invitrogen Inc., GeneCopoeia Inc.). For expression, the vector of choice

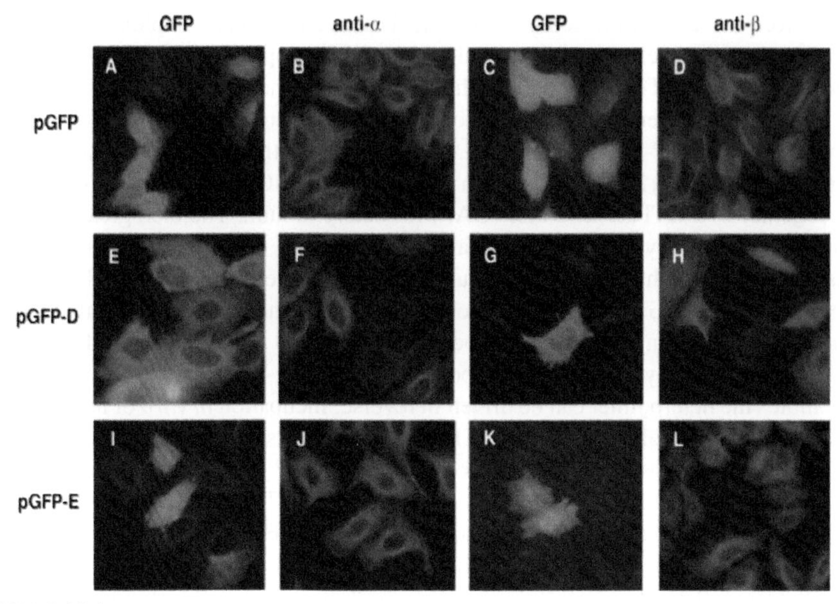

FIGURE 11.2

Microtubule destruction by overexpression of TBCD or TBCE in cultured cells. HeLa cells were transfected with constructs engineered for the expression of GFP alone (as a control; A–D), GFP-TBCD (E–H), or GFP-TBCE (I–L). Microtubules are shown in red, detected with either an anti-α-tublin antibody (B, F, and J) or an anti-β-tubulin antibody (D, H, and L). (For interpretation of the references to color in this figure legend, the reader is referred to the online version of this chapter.)

From Bhamidipati et al. (2000)

depends on the tubulin-specific chaperone: TBCA, TBCB, and TBCC, for example, can be expressed in *E. coli* in good yield as biologically active proteins. The preferred expression vector for these cloned cDNAs is one belonging to the pET series (Invitrogen Inc.), in which expression is driven from an inducible T7 promoter. Unfortunately, although TBCD and TBCE can also be expressed in *E. coli*, the recombinant protein is completely insoluble and cannot be recovered as biologically active material. Resort must therefore be made to eukaryotic host/vector systems. In the case of TBCE, the biologically active protein can be expressed and purified from insect sf9 cells via infection with recombinant baculovirus. A number of commercially available vectors and kits are available for this purpose. Although TBCD can also be expressed as a soluble protein in insect sf9 cells, the yield is relatively poor and we have not found it possible to purify the protein in active form. The reason for this finding is not clear, but it may reflect the absence of some posttranslational modification that is required for activity. In any event, we have found that expression of TBCD in cultured human cells via the use of recombinant adenovirus yields material that can be purified with relative ease and is biologically active (Tian, Thomas, & Cowan, 2010).

11.1.2 Expression of TBCA, TBCB, and TBCC

TBCA, TBCB, and TBCC can be expressed as soluble, biologically active recombinant proteins in good yield in *E. coli* host cells. Note that in the case of TBCC, the protein exists as two isoforms: short (TBCC) and long (TBCC-L). The biological significance of these two isoforms is unknown; both are equally active in *in vitro* heterodimer assembly assays. The protocol below is for the expression and purification of the long form.

11.1.2.1 Bacterial cell growth and induction

The following protocol is for the preparation of bacterial host cells expressing either TBCA, TBCB, or TBCC:

a. Transform the T7 expression plasmid into *E. coli* BL21DE3 cells using antibiotic selection appropriate for the pET vector in use. Incubate at 37 °C until the resulting bacterial colonies are easily visible to the naked eye (typically overnight).

b. Transfer the colonies en masse into a flask containing 1 l of sterile Luria broth containing the selection antibiotic. This can be conveniently done by pipetting 5 ml of broth directly onto the plate, scraping up the colonies with a sterile spreader, and transferring the bacterial suspension to the growth flask. It is not necessary to pick a single colony; indeed, in addition to the greatly extended time required for growth, doing so may lead to diminished yield because of the larger number of division cycles with an accompanying enhanced chance of plasmid loss.

c. Grow with shaking at 37 °C until $A_{600} = 1.0$–1.2. Remove 1 ml of culture as an uninduced control. Induce expression in the bulk culture by the addition of isopropyl thiogalactoside to 1 mM. Continue growth with shaking at 37 °C for a further 2.5 h; remove a 1 ml aliquot of the culture for use in determining the efficiency of expression of the recombinant protein. Harvest the bacteria in these small aliquots by centrifugation in 1.5 ml Eppendorf tubes, discard the supernatant, and suspend directly in SDS-PAGE loading buffer. If necessary, the pellet can be efficiently dispersed by brief sonication using a micro tip powered by an ultrasonic generator.

d. Harvest the bulk of the bacterial culture by centrifugation at $10{,}000 \times g$.

e. Resuspend the bacterial pellet in 10 ml 10 mM Tris–HCl, pH 7.2, 1 mM EDTA ($T_{10}E_1$) by vigorous vortexing. Dilute to 500 ml with $T_{10}E_1$ and repeat step d. At this stage, the bacterial pellet may be stored frozen at −20 °C.

f. Analyze the bacterial protein content on a 10% SDS-PAGE gel. In each case, the recombinant protein should be readily visible as a Coomassie-stained band that is absent from the corresponding uninduced control.

11.1.2.2 ***Preparation of bacterial lysates***

a. Resuspend the washed bacterial pellet in 10 ml of $T_{10}E_1$ containing a cocktail of protease inhibitors and transfer to a French Pressure Cell. Lyse the bacteria under 1500 psi with minimally two passes or until the lysate has no detectable viscosity when discharged drop-wise from a Pasteur pipet.

b. Centrifuge the lysate at 30,000 × *g* for 15 min at 4 °C. Decant the supernatant and centrifuge at 200,000 × *g* for 15 min at 4 °C in an ultracentrifuge.

c. The particle-free supernatant can either be applied directly to the first chromatographic dimension as described below or stored by flash freezing. The latter is most conveniently accomplished by adding the cleared supernatant in a drop-wise stream from a pipet directly into a small Dewar flask containing liquid nitrogen. The resulting frozen pellets can be stored indefinitely in a −70 °C freezer.

11.1.3 Purification of TBCA, TBCB, and TBCC

11.1.3.1 ***Chromatography on either Sepharose Q (TBCA and TBCB) or Sepharose S High-Performance (TBCC) ion exchange resins***

a. Thaw the particle-free extracts prepared as described in Section 1.2.2c in a 37 °C water bath, taking care not to let the temperature of the melt rise above 4 °C. Filter the extract through a 0.4 μ membrane to ensure removal of all particulate material.

b. Measure the total protein using, for example, the Bio-Rad Protein Assay reagent (Bio-Rad Inc.). The purification procedures described below are for yields of less than 200 mg of total protein in the starting material. Larger yields require a prorated increase in the size of the ion exchange columns used.

c. Apply the filtrate to a 10/10 (10 mm × 10 cm) column of Sepharose Q High-Performance anion exchange (QHP) resin equilibrated in 20 mM Tris–HCl, pH 7.2, 0.5 mM $MgCl_2$, 1 mM EDTA (in the case of TBCA and TBCB) or an equivalent column of Sepharose S High-Performance cation exchange resin equilibrated in 10 mM $NaPO_4$, pH 6.8, 1 mM EDTA (in the case of TBCC). In either case, this can be done using either the 10 ml or 50 ml FPLC superloop. Wash the column with at least 2 column volumes of equilibration buffer or until the absorbance at 280 nm declines to a value of 0.2 or less.

d. Develop the column with a 100-ml linear gradient of equilibration buffer containing 0.5 M $MgCl_2$ (in the case of TBCA and TBCB) or equilibration buffer containing 0.25 M $NaPO_4$, pH 6.8 (in the case of TBCC), collecting fractions of 2 ml. The columns can be run at room temperature, but the fractions should be stored on ice as soon as they emerge.

e. Assay an aliquot (2–5 μl) of the fractions emerging from the ion exchange column by SDS-PAGE (12%). The approximate expected conductivity ranges for the elution positions of TBCA, TBCB, and TBCC are 7.8–9.0, 10.4–12.8, and 9.3–10.0 mS/cm, respectively. The recombinant proteins should be readily visible as prominent Coomassie-stained bands migrating at 12, 40, and 38 kDa, respectively.

f. Pool the fractions identified as containing the recombinant protein of interest and concentrate to 2 ml using a Millipore Centriprep device (Millipore Inc.) or equivalent. The pooled concentrated material can be stored at −70 °C following the addition of glycerol to 10%.

11.1.3.2 ***Anion exchange chromatography on MonoQ***

a. Exchange the protein obtained by the ion exchange steps described earlier into 10 mM $NaPO_4$ buffer, pH 7.5, 0.1 mM $MgCl_2$, 1 mM DTT using a PD10 desalting column (GE Healthcare Inc.).
b. Apply the sample to a 0.5 × 5.0 cm (5/5) MonoQ column (GE Healthcare Inc.) equilibrated in desalting buffer.
c. Wash the column with 2 column volumes of desalting (start) buffer and develop with a 50-ml linear gradient containing start buffer plus 0.25 M NaO_4 buffer, pH 7.5. TBCB and TBCC emerge from this dimension as major peaks in the range of 13.7–15.6 and 4.3–5.6 mS/cm, respectively.
d. Pool the fractions contained in this peak and concentrate to 0.5 ml in a Millipore Centricon 10 concentration device.

11.1.3.3 ***Gel filtration chromatography (optional)***

a. Apply the material obtained in Section 1.3.2d to a 22-ml Superdex 75 gel filtration column (GE Healthcare Inc.) equilibrated in 20 mM Tris–HCl, pH 7.5, 0.15 M NaCl, 1 mM EDTA, 1 mM DTT. TBCA, TBCB, and TBCC emerge as symmetrical peaks with apparent molecular masses of approximately 30, 50, and 55 kDa, respectively.
b. Pool the fractions contained in the peak, concentrate them as described in Section 1.3.2d and store as multiple small aliquots at −70 °C after flash freezing in liquid nitrogen.

11.1.4 Expression of TBCE

11.1.4.1 ***Construction of recombinant baculovirus***

We have found the BakPak kit sold by Clontech Inc. to be relatively straightforward and convenient for the purpose of recombinant virus production; similar kits are available from other manufacturers (e.g., Invitrogen Inc.). As complete materials and comprehensive instructions are given with these commercial products, including the provision of vectors for the insertion of the sequences to be expressed, no details are provided here. We recommend that both the titer of the amplified recombinant viruses and the time of infection be optimized for maximum yield of recombinant protein production, as described in the commercially provided user's manuals.

11.1.5 Purification of TBCE

11.1.5.1 Preparation of insect cell lysates

a. Dislodge cells that remain attached to dishes containing infected sf9 cells with a polypropylene scraper. Pool the suspensions into 500 ml conical polypropylene centrifuge bottles and centrifuge at 2000 × *g* in a Beckman J6 centrifuge or equivalent. Remove the supernatant by aspiration with a Pasteur pipet linked to a vacuum flask, taking care not to disturb the cell pellet.
b. Resuspend the cell pellet by gently tapping the centrifuge bottle.
c. Gently add 50 ml of cold $T_{10}E_1$, swirl to generate a homogeneous suspension, centrifuge at 2000 × *g* at 4 °C, and remove and discard the supernatant as before.
d. Resuspend the cell pellet by gently tapping the centrifuge bottle.
e. Add 5 ml of ice-cold $T_{10}E_1$ containing protease inhibitor cocktail (Roche Inc.).
f. Transfer to a 15-ml glass Dounce homogenizer with a type A (tight fitting) pestle. Lyse the cells with 20 full strokes of the homogenizer at 0 °C.
g. Centrifuge the lysate at 500 × *g* at 4 °C for 5 min. Carefully aspirate the supernatant and transfer to a clean centrifuge tube.
h. Resuspend the nuclear pellet in half the original volume of ice-cold $T_{10}E_1$ by persistent tapping, centrifuge at 500 × *g* as before, and combine the supernatant with that obtained in step g.
i. Centrifuge the combined supernatants at 200,000 × *g* at 4 °C for 15 min in an ultracentrifuge (such as a Beckman Optima).
j. Collect the supernatants and store flash frozen in liquid nitrogen as described in Section 1.2.2c.

11.1.5.2 Anion exchange chromatography

a. Thaw the flash frozen sf9 cell extracts prepared as described in Section 1.5.1 in a 37 °C water bath, taking care that the melt does not warm above 4 °C.
b. Filter through a 0.4-μ Millipore membrane to ensure removal of any aggregated material that may have formed.
c. Measure the total protein content of the sample using, for example, the Bio-Rad Protein Assay Reagent. The total soluble protein recovered from 15 to 20 dishes (20 cm × 20 cm) of infected sf9 cells should be in the range of 80–100 mg.
d. Apply the mixture to an FPLC column of Q High-Performance anion exchange resin (GE Healthcare Inc.) (1 cm × 10 cm) equilibrated in 10 mM Tris–HCl, pH 8.0, 1 mM EDTA, and 1 mM DTT. For volumes in the range of 5–10 ml, it is most convenient to use the 10 ml superloop for this purpose.
e. Wash the column after sample loading with 2 column volumes of equilibration buffer and develop with a 100-ml linear gradient of equilibration buffer containing 0.25 M NaCl, collecting fractions of 2 ml. The column may be run at room temperature, but it is advisable to transfer the fractions to an ice bucket as soon as they emerge from the column.
f. Analyze a small proportion (e.g., 5 μl) of the relevant fractions by SDS-PAGE. As a guide, TBCE emerges from the column in the conductivity range 7.3–10.6 mS/cm.

The recombinant protein should be visible as a conspicuous band migrating with an apparent mass of 62 kDa.

11.1.5.3 Chromatography on hydroxylapatite

a. Pool the fractions containing TBCE from the anion exchange column and concentrate them to 2 ml using a Millipore Centriprep device or equivalent.
b. Exchange the protein into 10 mM $NaP0_4$ buffer, pH 7.5, 0.1 mM $MgCl_2$, 1 mM DTT using a PD10 desalting column (GE Healthcare Inc.).
c. Apply the desalted sample to an FPLC column of hydroxylapatite (Pentax, American Chemical Co., Inc.) (0.5×2.0 cm) equilibrated in 10 mM $NaP0_4$ buffer, pH 7.5, 0.1 mM $MgCl_2$, 1 mM DTT using the 2 ml loop.
d. Wash the column after sample loading with 2 ml of equilibration buffer and develop with a 20 ml linear gradient of equilibration buffer containing 0.25 M $NaP0_4$, pH 7.5. Collect fractions of 1 ml. The column may be run at room temperature, but it is advisable to transfer the fractions to an ice bucket as soon as they emerge from the column.
e. TBCE emerges from the Pentax column as a symmetrical peak in the conductivity range 2.4–4.7 mS/cm.
f. Analyze a small proportion (e.g., 2.5 μl) of the samples contained within this conductivity range by SDS-PAGE to determine the purity and yield of the product.

11.1.5.4 Size exclusion chromatography (optional)

a. Pool the fractions containing TBC from the hydroxylapatite column and concentrate them to 0.5 ml at 4 °C using a Millipore Centricon device or equivalent.
b. Apply this material to a Superdex 200 gel filtration column (22 ml) equilibrated in 10 mM Tris–HCl, pH 8.0, 0.15 M NaCl, 1 mM EDTA, 1 mM DTT. Collect 0.5 ml fractions.
c. TBCE emerges from the column as a symmetrical peak at 13–14 ml and should have a purity of 98% or better as judged by SDS-PAGE. This material can be stored flash frozen in small aliquots at −70 °C.

11.1.6 Expression of TBCD

TBCD expressed in *E. coli* is insoluble and cannot be recovered by renaturation. We have also attempted to express TBCD as a recombinant protein in insect sf9 cells. While this results in a modest yield of recombinant protein that is soluble, we found this material to be unstable (for a reason that is unclear) in the sense that attempts to purify it by any chromatographic procedure resulted in massive losses. We therefore adopted a procedure that depends on the use of recombinant adenovirus engineered for the expression of TBCD in HeLa cells. We have expressed both human and bovine TBCD via adenovirus vectors, and both are biologically active, although they have distinctive properties upon expression in cultured cells (Tian et al., 2010). We have found the AdEasy kit sold by Stratagene Inc. to be relatively straightforward and convenient for the purpose of recombinant adenovirus production; similar kits are available from other manufacturers. As complete and detailed instructions are given with these commercial products, including the provision of vectors for the insertion of the sequences to be expressed and the cell lines required for recombinant viral construction and propagation, no description is provided here. However, as in the case of recombinant baculovirus, we recommend that both the titer of the amplified recombinant viruses and the length of time of infection be optimized for production of recombinant TBCD. Even under optimal conditions, the yield of recombinant TBCD is rarely better than 1–5% of the total protein. Nonetheless, even at the low end of this range, it is possible to purify this material to homogeneity as described below.

11.1.7 Purification of TBCD

11.1.7.1 *Preparation of HeLa cell lysates*

The preparation of a particle-free lysate of adenovirus-infected HeLa cells is identical to that used for the preparation of lysates from infected insect sf9 cells. Follow the steps described in Section 1.5.1a–j.

11.1.7.2 *Anion exchange chromatography on Q Sepharose High Performance*

Purification of recombinant TBCD on Q Sepharose High Performance (QHP) follows the same procedure as that described for TBCE (Section 1.5.2). TBCD emerges from the QHP column in the conductivity range 17.5–21.5 mS/cm. The relatively low yield of the recombinant protein in adenovirus-infected HeLa cells make it advisable to locate it as a 116 kDa band by Western blotting using an anti-TBCD antibody.

11.1.7.3 *Anion exchange chromatography on MonoQ*

a. Pool the fractions identified as containing TBCD and concentrate to 2 ml using a Millipore Centricon device (Millipore Inc.) or equivalent.
b. Exchange the protein into 20 mM NaP04 buffer, pH 7.5 using a PD10 desalting column (GE Healthcare Inc.).
c. Apply the sample to a 0.5 × 5.0 cm (5/5) MonoQ column (GE Healthcare Inc.). Wash and elute the column as described in Section 1.3.2c. TBCD emerges from this column in the conductivity range 19.0–22.0 mS/cm and can be detected either by Western blotting with an anti-TBCD antibody or as a Coomassie stained band migrating at 116 kDa.

11.1.7.4 *Chromatography on hydroxylapatite*

a. Pool the fractions identified as containing TBCD and concentrate to 2 ml using a Millipore Centricon device (Millipore Inc.) or equivalent.
b. Exchange the protein into $NaP0_4$ buffer, apply this material to a hydroxylapatite column, wash and develop as described in Section 1.5.3b–d. TBCD emerges as an identifiable absorbance peak in the conductivity range 1.5–6.5 mS/cm.

c. Analyze a small proportion of the fractions contained within this conductivity range by 8% SDS-PAGE to determine the purity and yield of the product.

11.1.7.5 Size exclusion chromatography

A final purification step is done as described in Section 1.5.4, except that the optimal gel filtration column for TBCD is Superose 6 rather than Superdex 200. TBCD emerges from this column as a symmetrical peak with an apparent molecular mass of about 120 kDa. This material can be concentrated using a Millipore Centricon 30 concentration device, flash frozen in small aliquots and stored at −70 °C.

11.2 DISCUSSION

The ultimate test of the quality of the tubulin-specific chaperones prepared by the methods described here is their performance in *in vitro* folding reactions. In these reactions, highly labeled unfolded α- or β-tubulins made by expression in *E. coli* are presented by sudden dilution from denaturant into an aqueous reaction containing CCT, ATP, GTP, and one or more of the tubulin-specific chaperones prepared as described earlier. Detailed protocols describing the preparation of labeled tubulin probes, the purification of CCT, and the assembly of the folding reaction are described in Cowan (1998). Alternatively, in the case of TBCD and TBCE, the activity of the purified proteins may be tested via their ability to disrupt the native tubulin heterodimer in the back-reaction. In either case, the reaction products are analyzed by electrophoresis on nondenaturing (ND) polyacrylamide gels, a procedure described in detail by Fanarraga, Carranza, et al. (2010). An example of the complexes formed by TBCA, TBCC, TBCD, and TBCE either alone or in combination and their migration properties is shown in Fig. 11.3. Such analyses (among others) were essential for the elucidation of the overall tubulin heterodimer assembly pathway shown in Fig. 11.1, but they have also proved useful in the analysis of spontaneously occurring human mutations in *TUBA1A* and *TUBB2B*. These mutations result in devastating defects in the neuronal migration events that accompany normal brain development during late embryogenesis. For example, some of the disease-causing mutations result in defective interactions of CCT-generated folding intermediates with either TBCB (in the case of *TUBA1A*) (Tian et al., 2008) or TBCA (in the case of *TUBB2B*) (Jaglin et al., 2009). Mutations in the human gene encoding TBCE are also known to cause the rare inherited disease HRD (hypoparathyroidism, mental retardation, and facial dysmorphism) (Parvari et al., 2002; Tian et al., 2006).

It is likely that the tubulin-specific chaperone-dependent reactions that participate in the *de novo* assembly of the α/β-tubulin heterodimer occur *in vivo* in free solution in a manner that does not depend on microtubules or microtubule dynamics. This is because the heterodimer is assembled upon transcription/translation of α- or β-tubulin in rabbit reticulocyte lysate, a cell-free cocktail in which the endogenous heterodimer concentration (about 0.1 mg/ml) is well below the critical concentration (C_C) required for polymerization into microtubules. However, there is accumulating

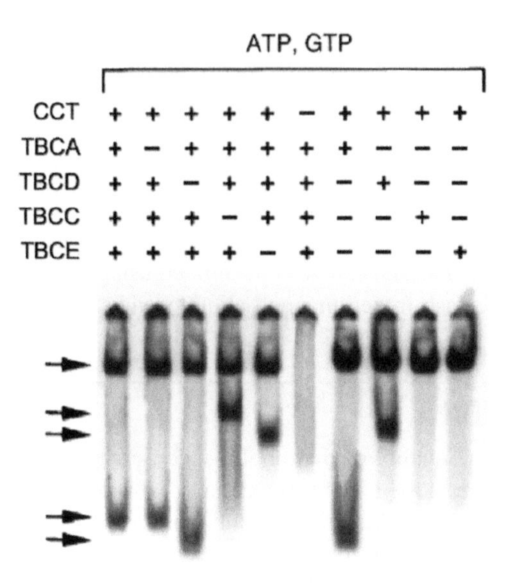

FIGURE 11.3

An example of *in vitro* folding reactions containing purified tubulin-specific chaperones used to determine the necessary and sufficient conditions for assembly of the α/β-tubulin heterodimer. ^{35}S-labeled, unfolded β-tubulin was suddenly diluted from 8 M urea into buffer containing CCT, ATP, GTP, and the tubulin-specific chaperones shown in this figure. Purified native tubulin heterodimer was included as a source of the α-tubulin subunit. Reaction products were resolved by electrophoresis on nondenaturing PAGE. Arrows (top to bottom) denote the migration positions of the CCT/β-tubulin binary complex, the TBCDβ/TBCEα supercomplex, the TBCD/β complex, native α/β-tubulin heterodimers, and the TBCA/β complex, respectively. Note that TBCA and TBCB are not essential for *de novo* heterodimer assembly *in vitro.*

From Tian et al. (1996)

evidence that at least some of the TBC proteins can modulate microtubule behavior in addition to their involvement in heterodimer assembly. For example, TBCB plays a role in determining microtubule behavior in the neuronal growth cone (Lopez-Fanarraga et al., 2007) and regulates microtubule density in microglia during their transition to reactive states (Fanarraga, Villegas, Carranza, Castano, & Zabala, 2009). Moreover, TBCD is not only distributed throughout the cytosol, but it also localizes at centrosomes and the midbody and is required for proper spindle organization, cell abscission, centriole formation, and ciliogenesis (Cunningham & Kahn, 2008; Fanarraga, Bellido, Jaen, Villegas & Zabala, 2010). Both TBCB and TBCE contain conserved CAP-Gly (cytoskeleton-associated protein glycine-rich) domains (Tian et al., 1997), which include a motif (GKNDG) responsible for targeting to the C-terminal EEY/F sequence motifs of CLIP170, EB proteins, and microtubules (Weisbrich et al., 2007). Thus, it is conceivable that TBCB and TBCE also influence dynamic behavior by interacting either directly

Table 11.1 Accession numbers for structures of human (*) or mouse (**) tubulin-specific chaperones in the PDB database (http://www.rcsb.org/pdb/home/home.do)

TBCA*	Whole molecule	1H7C
TBCB**	N-terminal Ubl-domain	1V6E
TBCB**	C-terminal CAP-Gly domain	1WHG
TBCC*	N-terminal domain	2L3L
TBCC*	C-terminal domain	2YUH
TBCE**	C-terminal Ubl-domain	1WJN

with microtubule plus ends or indirectly via other +TIPs. No clear evidence for such interactions currently exists, but this may reflect equilibrium dissociation constants that are too weak for the formation of stable complexes.

The ability to generate tubulin-specific chaperones as recombinant proteins has been essential for the acquisition of structural data (Table 11.1), although (with the exception of TBCA) this information is incomplete in the sense that it is currently limited to individual domains, and there is as yet no structural data on TBCD. Nonetheless, the available structures represent an important step toward a mechanistic understanding of the molecular machinery that assembles the α/β heterodimer.

References

Anfinsen, C. B. (1973). Principles that govern the folding of protein chains. *Science*, *181*(96), 223–230.

Bhamidipati, A., Lewis, S. A., & Cowan, N. J. (2000). ADP ribosylation factor-like protein 2 (Arl2) regulates the interaction of tubulin-folding cofactor D with native tubulin. *The Journal of Cell Biology*, *149*(5), 1087–1096.

Caplow, M., & Fee, L. (2002). Dissociation of the tubulin dimer is extremely slow, thermodynamically very unfavorable, and reversible in the absence of an energy source. *Molecular Biology of the Cell*, *13*(6), 2120–2131.

Cleveland, D. W., Kirschner, M. W., & Cowan, N. J. (1978). Isolation of separate mRNAs for alpha- and beta-tubulin and characterization of the corresponding in vitro translation products. *Cell*, *15*(3), 1021–1031.

Cowan, N. J. (1998). Mammalian cytosolic chaperonin. *Methods in Enzymology*, *290*, 230–241.

Cowan, N. J., & Lewis, S. A. (2001). Type II chaperonins, prefoldin, and the tubulin-specific chaperones. *Advances in Protein Chemistry*, *59*, 73–104.

Cunningham, L. A., & Kahn, R. A. (2008). Cofactor D functions as a centrosomal protein and is required for the recruitment of the gamma-tubulin ring complex at centrosomes and organization of the mitotic spindle. *The Journal of Biological Chemistry*, *283*(11), 7155–7165.

Detrich, H. W., 3rd., & Williams, R. C. (1978). Reversible dissociation of the alpha beta dimer of tubulin from bovine brain. *Biochemistry*, *17*(19), 3900–3907.

Fanarraga, M. L., Bellido, J., Jaen, C., Villegas, J. C., & Zabala, J. C. (2010). TBCD links centriogenesis, spindle microtubule dynamics, and midbody abscission in human cells. *PLoS One*, *5*(1), e8846.

Fanarraga, M. L., Carranza, G., Castano, R., Nolasco, S., Avila, J., & Zabala, J. C. (2010). Nondenaturing electrophoresis as a tool to investigate tubulin complexes. *Methods in Cell Biology*, *95*, 59–75.

Fanarraga, M. L., Villegas, J. C., Carranza, G., Castano, R., & Zabala, J. C. (2009). Tubulin cofactor B regulates microtubule densities during microglia transition to the reactive states. *Experimental Cell Research*, *315*(3), 535–541.

Gao, Y., Thomas, J. O., Chow, R. L., Lee, G. H., & Cowan, N. J. (1992). A cytoplasmic chaperonin that catalyzes beta-actin folding. *Cell*, *69*(6), 1043–1050.

Gao, Y., Vainberg, I. E., Chow, R. L., & Cowan, N. J. (1993). Two cofactors and cytoplasmic chaperonin are required for the folding of alpha- and beta-tubulin. *Molecular and Cellular Biology*, *13*(4), 2478–2485.

Hartl, F. U., & Hayer-Hartl, M. (2002). Molecular chaperones in the cytosol: From nascent chain to folded protein. *Science*, *295*(5561), 1852–1858.

Jaglin, X. H., Poirier, K., Saillour, Y., Buhler, E., Tian, G., Bahi-Buisson, N., et al. (2009). Mutations in the beta-tubulin gene TUBB2B result in asymmetrical polymicrogyria. *Nature Genetics*, *41*(6), 746–752.

Lewis, S. A., Tian, G., & Cowan, N. J. (1997). The alpha- and beta-tubulin folding pathways. *Trends in Cell Biology*, *7*, 479–485.

Lopez-Fanarraga, M., Carranza, G., Bellido, J., Kortazar, D., Villegas, J. C., & Zabala, J. C. (2007). Tubulin cofactor B plays a role in the neuronal growth cone. *Journal of Neurochemistry*, *100*(6), 1680–1687.

Lorimer, G. H. (1996). A quantitative assessment of the role of the chaperonin proteins in protein folding in vivo. *The FASEB Journal*, *10*(1), 5–9.

Parvari, R., Hershkovitz, E., Grossman, N., Gorodischer, R., Loeys, B., Zecic, A., et al. (2002). Mutation of TBCE causes hypoparathyroidism-retardation-dysmorphism and autosomal recessive Kenny-Caffey syndrome. *Nature Genetics*, *32*(3), 448–452.

Spiess, C., Meyer, A. S., Reissmann, S., & Frydman, J. (2004). Mechanism of the eukaryotic chaperonin: Protein folding in the chamber of secrets. *Trends in Cell Biology*, *14*(11), 598–604.

Studier, F. W., Rosenberg, A. H., Dunn, J. J., & Dubendorff, J. W. (1990). Use of T7 RNA polymerase to direct expression of cloned genes. *Methods in Enzymology*, *185*, 60–89.

Tian, G., Bhamidipati, A., Cowan, N. J., & Lewis, S. A. (1999). Tubulin folding cofactors as GTPase-activating proteins: GTP hydrolysis and the assembly of the alpha/beta-tubulin heterodimer. *The Journal of Biological Chemistry*, *274*, 24054–24058.

Tian, G., Huang, M. C., Parvari, R., Diaz, G. A., & Cowan, N. J. (2006). Cryptic out-of-frame translational initiation of TBCE rescues tubulin formation in compound heterozygous HRD. *Proceedings of the National Academy of Sciences of the United States of America*, *103*(36), 13491–13496.

Tian, G., Huang, Y., Rommelaere, H., Vandekerckhove, J., Ampe, C., & Cowan, N. J. (1996). Pathway leading to correctly folded beta-tubulin. *Cell*, *86*(2), 287–296.

Tian, G., Kong, X. P., Jaglin, X. H., Chelly, J., Keays, D., & Cowan, N. J. (2008). A pachygyria-causing alpha-tubulin mutation results in inefficient cycling with CCT and a deficient interaction with TBCB. *Molecular Biology of the Cell*, *19*(3), 1152–1161.

Tian, G., Lewis, S. A., Feierbach, B., Stearns, T., Rommelaere, H., Ampe, C., et al. (1997). Tubulin subunits exist in an activated conformational state generated and maintained by protein cofactors. *The Journal of Cell Biology*, *138*(4), 821–832.

Tian, G., Thomas, S., & Cowan, N. J. (2010). Effect of TBCD and its regulatory interactor Arl2 on tubulin and microtubule integrity. *Cytoskeleton*, *67*, 706–714.

Tian, G., Vainberg, I. E., Tap, W. D., Lewis, S. A., & Cowan, N. J. (1995a). Quasi-native chaperonin-bound intermediates in facilitated protein folding. *The Journal of Biological Chemistry*, *270*(41), 23910–23913.

Tian, G., Vainberg, I. E., Tap, W. D., Lewis, S. A., & Cowan, N. J. (1995b). Specificity in chaperonin-mediated protein folding. *Nature*, *375*(6528), 250–253.

Vainberg, I. E., Lewis, S. A., Rommelaere, H., Ampe, C., Vandekerckhove, J., Klein, H. L., et al. (1998). Prefoldin, a chaperone that delivers unfolded proteins to cytosolic chaperonin. *Cell*, *93*(5), 863–873.

Valpuesta, J. M., Martin-Benito, J., Gomez-Puertas, P., Carrascosa, J. L., & Willison, K. R. (2002). Structure and function of a protein folding machine: The eukaryotic cytosolic chaperonin CCT. *FEBS Letters*, *529*(1), 11–16.

Weisbrich, A., Honnappa, S., Jaussi, R., Okhrimenko, O., Frey, D., Jelesarov, I., et al. (2007). Structure-function relationship of CAP-Gly domains. *Nature Structural and Molecular Biology*, *14*(10), 959–967.

Young, J. C., Agashe, V. R., Siegers, K., & Hartl, F. U. (2004). Pathways of chaperone-mediated protein folding in the cytosol. *Nature Reviews. Molecular Cell Biology*, *5*(10), 781–791.

CHAPTER

Heterotrimeric G Proteins and Microtubules

12

Witchuda Saengsawang* and Mark M. Rasenick†,‡

*Department of Physiology, Faculty of Science, Mahidol University, Bangkok, Thailand

†Departments of Physiology & Biophysics and Psychiatry, University of Illinois at Chicago, Chicago, Illinois, USA

‡The Jesse Brown VA Medical Center, Chicago, Illinois, USA

CHAPTER OUTLINE

Introduction 174

12.1 General Protocols 174

12.1.1 *In Vitro* Determination of the Association Between Tubulin and Gsα 174

12.1.1.1 Protein Purification 175

12.1.1.2 Pull-down Assays 177

12.1.1.3 Binding Analysis By SPR 177

12.1.1.4 Peptide Array Membrane Analysis 178

12.1.2 Functional Consequence of Tubulin–Gsα Interaction 179

12.1.2.1 Steady-state Tubulin GTPase Assay 179

12.1.2.2 Single-turnover Tubulin GTPase Assay 180

12.1.3 Functional Consequences of Gsα-mediated Activation of Tubulin GTPase: Gsα and Gsα-derived Peptides Increase Microtubule Dynamics 181

12.1.3.1 Microtubule Polymerization Assay 181

12.1.3.2 Microtubule Dynamics Assay By Video Microscopy 181

12.1.4 Cellular Consequences of Gsα Interaction With Tubulin/microtubules 183

12.1.4.1 Binding Studies in Cells 183

12.1.4.2 Microtubule Dynamics Assay in Cells 185

12.1.4.3 Cell Morphology Assay 187

12.2 Buffer Compositions 187

Concluding Comments 188

Acknowledgments 188

References 188

Methods in Cell Biology, Volume 115

ISSN 0091-679X
http://dx.doi.org/10.1016/B978-0-12-407757-7.00012-8

Abstract

Microtubules, major components of the cytoskeleton, play important roles in a variety of cellular functions including mitosis, intracellular transport, and the modulation of cell morphology. Several studies have demonstrated that specific G-protein alpha subunits bind to tubulin with a high affinity (~130 nM) and elicit various functional effects on tubulin and microtubules. In this chapter, we present a description of the protocols for several methods that are used to determine the interaction between heterotrimeric G proteins and tubulin, as well as functional consequences of the interactions including protocols for protein purification, binding assays, tubulin GTPase assays, microtubule dynamics assays, and assays for cytoskeletal consequences of G-protein-coupled receptor signaling.

INTRODUCTION

Microtubules have been implicated in G-protein signaling for more than 30 years, and a direct binding between select G-protein α subunits (Gsα, Gi1α, and Gi2α) and tubulin was observed a decade later (Wang, Yan, & Rasenick, 1990). This was followed by demonstration of tubulin interaction with Gqα for regulation of phospholipase signaling (Popova, Garrison, Rhee, & Rasenick, 1997). Initial studies suggested that microtubules were instrumental in G-protein signaling and that their disruption facilitated signaling, by liberating G proteins from some unidentified constraint. Subsequent studies revealed direct transactivation of Gα subunits due to transfer of GTP between tubulin and Gα (Popova et al., 1997; Yan, Popova, Moss, Shah, & Rasenick, 2001), although a structural basis for this was not revealed until recently (Dave et al., 2011; Layden et al., 2008). During the course of these studies, it became apparent that the interactions between tubulin and Gα were bidirectional and that Gα was capable of activating tubulin GTPase (Roychowdhury & Rasenick, 1994) and this increased dynamic behavior of microtubules (Roychowdhury, Panda, Wilson, & Rasenick, 1999). While Gsα had long been observed in the cytosol, lending credibility to the possibility that it could regulate microtubules (Rasenick et al., 1984), the demonstration that this accompanied the activation of Gsα (Wedegaertner, Bourne, & von Zastrow, 1996) and the observation of this phenomenon in living cells (Yu & Rasenick, 2002) gave rise to the observations that activated Gsα could, indeed, regulate microtubule-based neurite outgrowth (Yu, Dave, Allen, Sarma, & Rasenick, 2009) and microtubule dynamics (Dave et al., 2011). These observations and the methods that made them possible will constitute the basis of this chapter.

12.1 GENERAL PROTOCOLS

12.1.1 *In vitro* determination of the association between tubulin and Gsα

Several *in vitro* binding assays including an overlay method using ^{125}I tubulin and G proteins immobilized nitrocellulose (Wang & Rasenick, 1991), a pull-down assay (Yu et al., 2009), and surface plasmon resonance (SPR) (Dave et al., 2011) have been

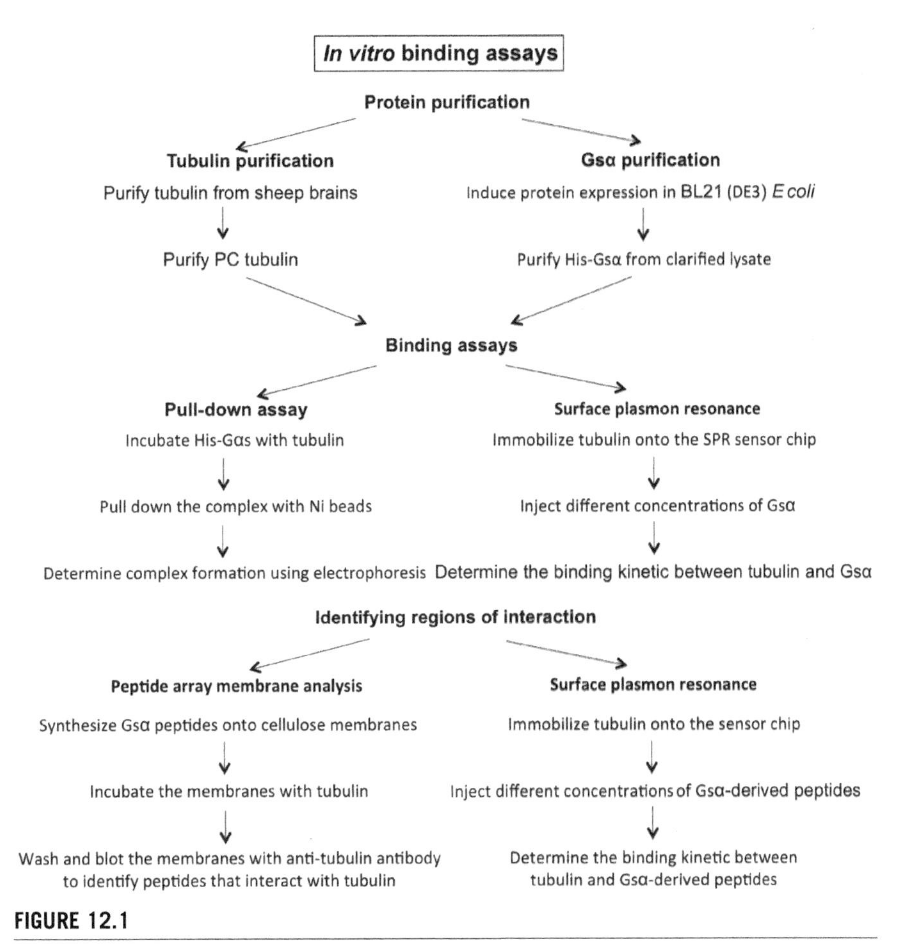

FIGURE 12.1

Flowchart illustrating the steps in the *in vitro* binding assays including protein purification, binding assays, and identifying regions of interaction.

used to demonstrate that Gα forms stable complexes with tubulin. Here, we will describe details for the two latter methods used to determine the association between Gsα and tubulin. Flowcharts illustrating the purification and *in vitro* binding assay protocols are shown in Fig. 12.1.

12.1.1.1 ***Protein purification***

For all *in vitro* binding and functional studies, we use purified tubulin and purified Gsα. The purity of tubulin and the Gsα is crucial for determining the binding properties and function of Gsα–tubulin complexes. The methods for purification of these two proteins have been described by many investigators (Linder, Ewald, Miller, & Gilman, 1990; Miller & Wilson, 2010; Shelanski, Gaskin, & Cantor, 1973). Below, we briefly describe the methods used in this lab.

12.1.1.1.1 Tubulin

Brains are obtained from freshly killed sheep from a local slaughterhouse and are placed on ice immediately upon removal from animals. The gray matter is separated from the white matter and blended in cold PEM buffer within 1 h. Particulate matter and cellular debris are removed at 100,000 × *g* for 1 h at 4 °C. Supernatants are collected and microtubule polymerization is induced by adding 125 μM GTP, 500 μM ATP, and 30% glycerol (in PEM buffer) at 37 °C for 1 h with gentle agitation. Microtubules are then pelleted by centrifugation at 100,000 × *g* for 30 min at 37 °C. The microtubule pellet is depolymerized by resuspending in 100 ml PEM buffer at 4 °C and incubating on ice for 30 min. Microsomes, protein aggregates, and other polymers are removed by centrifugation at 4 °C at 100,000 × *g* for 30 min. The second cycle of microtubule polymerization is performed by adding1 mM GTP and 30% glycerol for 1 h at 37 °C, and microtubules are pelleted by centrifugation at 100,000 × *g* for 1 h at 37 °C. Microtubule pellets are resuspended in cold PEM buffer and stored at −80 °C as "non-PC tubulin" (tubulin containing microtubule-associated proteins, MAPs).

MAPs are removed from tubulin using a phosphocellulose column to obtain purified tubulin (called PC-tubulin). Tubulin-containing MAPs are loaded on a PEM buffer (100 mM PIPES, 1 mM EDTA and 1 mM $MgCl_2$, pH 6.9)-equilibrated phosphocellulose column (Whatman P-11) at 4 °C. PC-tubulin is eluted with 30 ml PEM, concentrated to 10 mg/ml and stored at −80 °C. Purity of the PC-tubulin, as determined by SDS gel electrophoresis, should be >99% on a Coomassie-stained gel.

12.1.1.1.2 Gsα

His-Gsα or the mutationally-activated His-Gsα^{Q227L} is purified from BL21 (DE3) *Escherichia coli* (Novagen) (Linder et al., 1990). The frozen glycerol stock of bacteria is cultured overnight in 10 ml LB at 37 °C. Four milliliters from this culture are diluted into 1 l of 2YT medium (16 g/l Bacto-Tryptone, 10 g/l Bacto-yeast extract, and 5 g/l NaCl) and cultured to 0.4–0.6 OD. Protein expression is induced with 30 μM isopropyl β-D-1-thiogalactopyranoside for 20 h at 15 °C. After induction, the bacterial pellets are resuspended in TSG buffer, lysed with lysosyme (10 μg/ml), and sonicated 10 times for 30 s at 2-min intervals (Fisher Scientific Sonic Dismembrator 550). The crude lysate is then clarified by ultracentrifugation at 100,000 × *g* for 90 min at 4 °C, and the buffer is adjusted to GAT buffer (see Section 2). The lysate is loaded on a Co^{3+}superflow resin (Talon, BD Bioscience) equilibrated in GAT buffer and allowed to flow through at 2 ml/min. The column is washed with GAT buffer containing 0 and 20 mM imidazole and eluted with 30 and 150 mM imidazole. Finally, the GsαQL is concentrated using a Pierce ICON concentrator and dialyzed into storage buffer. As a rule, Gsα protein purified in this manner is >85% pure on a Coomassie-stained SDS gel. Proteins are further purified by HPLC using Gel filtration (Superdex 75, Amersham, England) if higher purity is needed. To test functionality, wild-type Gsα is incubated with NaF (10 mM) and $AlCl_3$ (50 μM). Tryptophan fluorescence is determined with excitation at 280 nm and emission at 340 nm before and after the addition of fluoride to monitor the extent to which Gsα is converted to the activated state. Note that the Q227L variant is the mutationally activated form of Gsα, as the endogenous GTPase is inactivated.

12.1.1.2 Pull-down assays

To determine the binding of Gsα to tubulin, we use pull-down assays with purified 6His-Gsα and purified tubulin. 6His-Gsα (described above) is loaded with GDPβS (GDP-form) or GTPγS (GTP form) and purged of free nucleotide by concentration/ dialysis incubated with tubulin and incubated with a one- or threefold molar ratio of PC-tubulin/Gα for 1 h at room temperature. Complexes are then incubated with 50 μl of Ni^{2+}-nitrilotriacetic acid-agarose beads for 2.5 h at 4 °C. The samples are centrifuged at 13,000 × *g* for 15 min and washed with 50 mM Tris buffer. These beads bind 6-His, thus binding Gsα and any proteins with which it complexes. The samples are resuspended in 1 × SDS sample buffer and separated on 12% SDS-polyacrylamide gels. The gels are stained with Coomassie Blue before drying. The data show a strong preference for tubulin binding to the GTP form (Q227L) of Gsα (Fig. 12.2A).

12.1.1.3 Binding analysis by SPR

The real-time interaction between Gsα and tubulin is measured using surface plasmon resonance (SPR). The information achieved from this technique can be used to determine the kinetics of bi-molecular interactions (Hahnefeld, Drewianka, & Herberg, 2004; Jason-Moller, Murphy, & Bruno, 2006).

12.1.1.3.1 Tubulin immobilization on SPR sensor chip

Tubulin is immobilized in HBS-P buffer pH 7.4 at a flow rate of 10 μl/min on a carboxymethyl dextran-coated sensor chip (CM5) using amine coupling according to the manufacturer's instructions. *N*-Hydroxysuccinimide (NHS) and 1-ethyl-3-(3-dimethylaminopropyl) carbodiimide hydrochloride (EDC) are used to link the free amino acids on tubulin to the sensor chip. Flow cells are injected with 50% NHS/ EDC at a flow rate of 10 μl/min followed by 100 μl of 1 μM tubulin in PEM and 1 M ethanolamine (pH 8.5) to prevent further coupling. This resulted in 200–1000 RU (response unit) of tubulin bound to the sensor chip (1 RU ≈ 1 pg/mm^2 protein). A reference flow cell is injected with PEM instead of tubulin. All reactions are performed at 25 °C. All buffers are freshly prepared using milli-Q water, filtered, and degassed before use.

12.1.1.3.2 Detecting the binding between Gsα and tubulin using SPR

Different concentrations of $Gs\alpha^{Q227L}$ and GsαWT (60, 100, and 200 nM) are prepared freshly in HBS-P buffer and injected onto the flow cell at a flow rate of 30 μl/min. The real-time interaction between tubulin and Gsα is then monitored as RU. Figure 12.2B shows sensorgrams of real-time interaction between injected $Gs\alpha^{Q227L}$ or GsαWT and immobilized tubulin. The sensorgram from buffer blank injections and the reference flow cell (without ligand) are subtracted in order to correct for the buffer shifts created during injection and the nonspecific binding, respectively. The final curves are then fitted to a 1:1 Langmuir association model with drifting baseline to obtain the binding kinetics. $Gs\alpha^{Q227L}$ interacts with tubulin in a concentration-dependent manner. The affinity of $Gs\alpha^{Q227L}$ binding to tubulin is measured as 100 nM. After each injection, the surface is regenerated by injection of 1 M NaCl, 0.5% triton X-100 in HBS-P buffer to removed bound protein. Gsα are also injected onto the reference flow cell (no

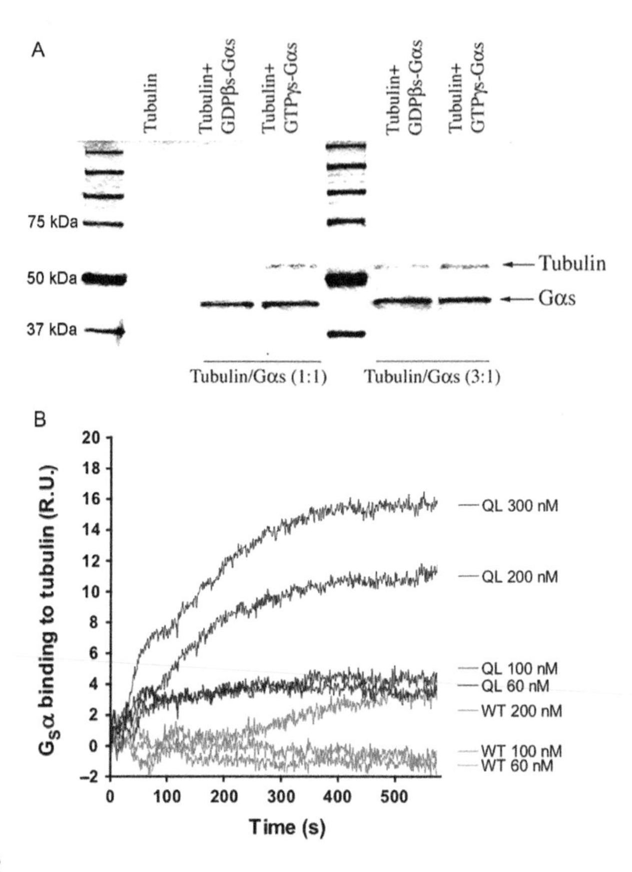

FIGURE 12.2

In vitro determination of the association between tubulin and Gsα. (A) Coprecipitation of Gsα and tubulin. His-Gαs-GTPγS (5 μg) or His-Gαs-GDPβS (5 μg) was incubated with tubulin for 2 h at room temperature followed by pull down using nickel-agarose beads. After separation by SDS-PAGE, the gel was stained with Coomassie Blue. (B) Sensorgrams representing the surface plasmon resonance analysis of the interaction between immobilized tubulin and GsαQL or GsαWT. The concentration series used were 60, 100, and 200 nM. GsαQL binds to tubulin with $k_a = 5 \times 10^4\ M^{-1}\ s^{-1}$, $k_d = 5 \times 10^{-3}\ s^{-1}$, and $K_D = 100$ nM. (See color plate.)

This research was originally published by Dave et al. (2011)

ligand) and buffer is also injected into each flow cell. In addition, BSA injected into the flow cells failed to detectably bind tubulin.

12.1.1.4 Peptide array membrane analysis

To identify the Gsα regions that specifically interact with tubulin, we use peptide array analysis. Peptides corresponding to the primary Gsα sequence as well as the Gtα sequence are covalently attached to a cellulose-based membrane. As Gtα (transducin) has been shown to not interact with tubulin despite significant sequence and structural similarities with Gsα (Chen, Yu, Skiba, Hamm, & Rasenick, 2003; Wang et al., 1990), it serves as a good control to determine which regions on Gsα bind specifically to tubulin. Peptides are synthesized onto a cellulose membrane with a proprietary PEG spacer (AIMS Scientific Products, Braunschweig, Germany) via the C-terminal amino acid in sequential spots by the use of a SPOT synthesis kit (SIGMA genosys, St. Louis, MO, USA) (Frank, 2002). Peptides from Gsα and Gtα are divided into overlapping peptides (15 amino acids in length with 10 amino acid overlap between sequential peptide, 70 total spots).

The peptide array membranes are blocked with TBS-containing 0.1% Tween-20 (TBS-T) with 2.5% milk for 1 h, washed with TBS-T, and incubated overnight at 4 °C with 150 nM tubulin in TBS-T buffer. Next, the membranes are washed three times with RIPA buffer and incubated with anti α-tubulin antibody (Sigma, St. Louis, MO, USA), followed by the horseradish peroxidase conjugated secondary antibody (1 h each at room temperature in the RIPA buffer containing 1% milk), and developed with enhanced chemiluminescence (ECL) Western blotting detection reagents (Amersham Biosciences). Following the tubulin binding determination, the membranes are stripped by using a stripping buffer and then reprobed with ECL to verify that residual protein–antibody complexes are removed before reuse. Also, controls are done to verify that there is no nonspecific binding of the primary or secondary antibody to the membranes and that no residual tubulin remains bound to the peptides following the stripping procedure. These peptides identified from peptide array are then determined for their affinities to immobilized tubulin using the SPR technique described above.

12.1.2 Functional consequence of tubulin–Gsα interaction

Gsα binding to tubulin has several functional consequences including activation of tubulin GTPase, which can be evaluated using a tubulin GTPase assay, and transfer of GTP between tubulin and Gsα, which can be determined by a transactivation assay (Popova et al., 1997; Yan et al., 2001). Only the former will be discussed in this chapter.

The nucleotide state on tubulin is an important factor that determines microtubule dynamics. Both the α- and β-subunits of tubulin bind GTP, but only the GTP on the β-subunit can be hydrolyzed to GDP and exchanged for a new GTP (E-site). This hydrolysis occurs due to tubulin's intrinsic GTPase activity, which is normally activated during the process of polymerization upon addition of another tubulin dimer (Carlier & Pantaloni, 1983). A flowchart demonstrating the tubulin GTPase assay protocols is shown in Fig. 12.3.

12.1.2.1 *Steady-state tubulin GTPase assay*

PC-tubulin is allowed to bind GTP (Roychowdhury & Rasenick, 1994). The samples are then incubated with $Gs\alpha^{Q227L}$ (which lacks intrinsic GTPase activity) or ovalbumin at 30 °C for 30 min and treated with 1% SDS at room temperature for 15 min. Nucleotide analysis is done by thin-layer chromatography on polyethyleneimine-cellulose plates. Two microliters of a 10 mM solution of GTP and GDP is spotted 1.5 cm apart on a polyethyleneimine-cellulose thin-layer plate, followed by 2 μl of each samples. The spots containing GTP or GDP are visualized with a UV lamp,

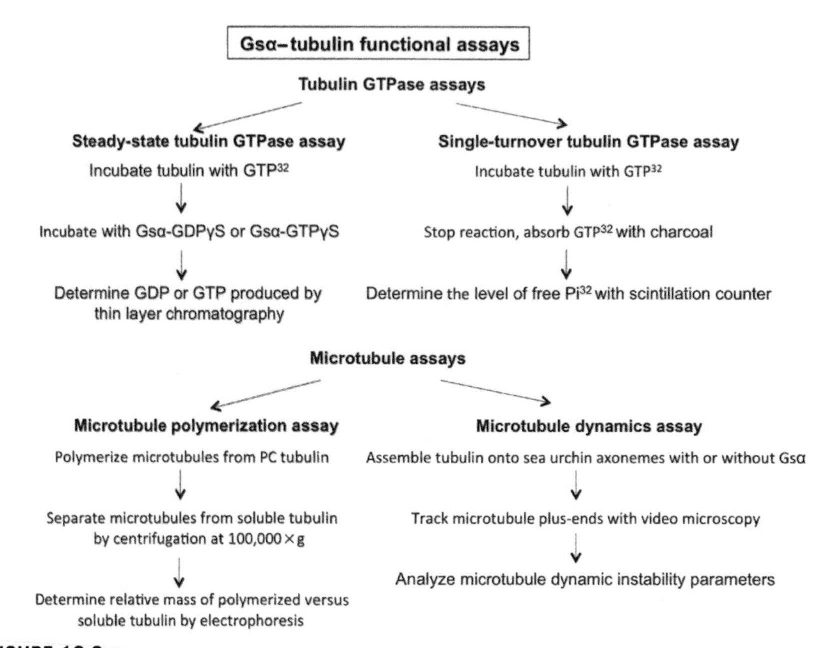

FIGURE 12.3

Flowchart illustrating the steps in the Gsα–tubulin functional assays including tubulin GTPase assays and microtubule assays.

and plates are exposed to film for autoradiography or the ^{32}P nucleotide contained in those spots. Quantitative analysis is done by measuring the integrated optical density of each GDP spot. The integrated optical density represents the GTPase activity as measured by GDP formation.

12.1.2.2 *Single-turnover tubulin GTPase assay*

To determine the effect of Gsα on tubulin GTPase activity, independent of possible rebinding of new GTP to tubulin, a single turnover GTPase activity is performed. The method is similar to a previously described protocol with some modifications (Mejillano, Shivanna, & Himes, 1996). Briefly, tubulin is preloaded with ^{32}P-γ-GTP and separated from free nucleotide. Tubulin is then incubated with different concentrations of Gsα, and released ^{32}Pi is quantified.

12.1.2.2.1 Preparation of tubulin loaded with ^{32}P-GTP

Tubulin is stripped of nucleotide by incubating with charcoal–dextran on ice for 10 min with periodic mixing using a vortex mixer. The charcoal is removed by two centrifugations at 16,000 × *g* for 15 min. Next, tubulin is incubated with [^{32}P]-γ-GTP for 30 min on ice, and the unbound GTP is removed using a protein-desalting column (Pierce). A sample of this tubulin is counted to determine the labeling efficiency (mol GTP^{32}/mol tubulin), which is typically 0.6–0.9 mol GTP/mol tubulin.

12.1.2.2.2 GTPase assay

Tubulin is incubated with Gsα QL for 20 min at 37 °C in PEM buffer. The reaction is stopped with 5% cold perchloric acid followed by mixing (vortexing). Reactions are extracted by incubating with charcoal–phosphate (5% Norit-A-activated charcoal in 50 mM Na_3PO_4, pH 3.0) for 15 min on ice with periodic vortexing followed by centrifugation two times at 16,000 × *g* for 30 min to remove the charcoal. Charcoal absorbs GTP^{32} and leaves free Pi^{32} in the supernatant. The supernatant is counted in a scintillation counter (we use a Beckman LS6000I), and the tubulin GTPase activity is expressed as millimole $^{32}P_i$ released/min/mol tubulin. The result, shown in Fig. 12.4A, reveals that Gsα increases tubulin GTPase activity in a concentration-dependent manner with EC50 = 1.2 μM, V_{max} = 20 mmol of GTP hydrolyzed/min/mol of tubulin. Using this method, we also have demonstrated that Gsα-peptide (270–284) corresponding to the interface between Gsα and tubulin also increases tubulin GTPase activity in a concentration-dependent manner with V_{max} = 0.070 pmol P_i released/min/μg tubulin and EC50 = 24 μM (Dave et al., 2011).

12.1.3 Functional consequences of Gsα-mediated activation of tubulin GTPase: Gsα and Gsα-derived peptides increase microtubule dynamics

Microtubules are stabilized by the presence of GTP at the plus-end. As a result of Gsα-activating tubulin GTPase, Gsα also increases microtubule dynamic instability. We determine the ability of microtubule to polymerize in the presence of Gsα using a microtubule polymerization assay and evaluate microtubule dynamics using real-time measurement by video microscopy. A flowchart demonstrating the microtubule assays protocol is shown in Fig. 12.3.

12.1.3.1 Microtubule polymerization assay

It is hypothesized that, for a given population of tubulin, the percentage present as microtubules is diminished in the presence of Gsα or peptides corresponding to the Gsα tubulin binding site. To test this, microtubule polymer mass is determined using 15 μM tubulin in PEM buffer with 200 μM GTP, pH 6.9 (G-PEM). Tubulin is polymerized for 1 h at 37 °C, $Gs\alpha^{Q227L}$ (exchanged into G-PEM buffer using a Microcon spin concentrator), added to microtubules for 1 h at 37 °C, and the microtubules are then separated from soluble tubulin at 100,000 × *g* for 1 h at 37 °C (Beckman TL-100). The final reaction volume, including Gsα, is 20 μl. The pelleted protein is resuspended in 20 μl of water at 4 °C. Two microliters of each fraction is run on a 10% SDS-polyacrylamide gel (125 V, 2 h), followed by Coomassie Blue staining, to determine the relative mass of polymerized versus soluble tubulin (Fig. 12.4B).

12.1.3.2 Microtubule dynamics assay by video microscopy

To study the effect of Gsα on microtubule dynamic instability, purified bovine brain tubulin (15 μM) is assembled onto sea urchin (*Strongylocentrotus purpuratus*) axonemes in PMEM buffer in the presence of 2 mM GTP. The reaction mixture is incubated

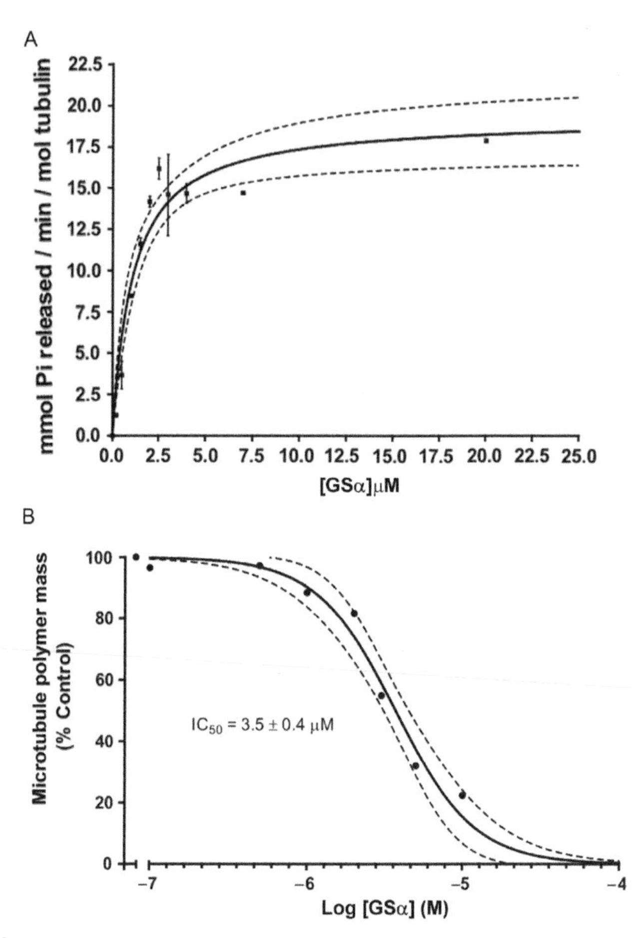

FIGURE 12.4

Activated Gsα increases the intrinsic GTPase activity of tubulin and promote microtubule depolymerization. (A) Single-turnover GTPase. Mutationally activated (GTPase-deficient) Gsα (Q227L) was incubated with tubulin GTP (200 nM) for 30 min, and the tubulin GTPase rate was determined. $EC_{50}=1.2$ μM, $V_{max}=20$ mmol of GTP hydrolyzed/min/mol of tubulin, and $n_H=1.0$. Dashed lines: 95% confidence interval for a hyperbolic fit. Tubulin (23 μM) was incubated with GsαQL for 1 h at 30 °C in the presence of 1 mM GTP. Microtubules were pelleted by centrifugation, and the amount of tubulin in supernatant and pellet fractions were quantified. GsαQL inhibited microtubule assembly with an IC_{50} of 3.5 ± 0.4 μM.

This research was originally published by Dave et al. (2011)

at 30 °C for 40 min in the presence or absence of different concentrations of GsαQL- or Gsα-derived peptide, or the control peptide Gtα. Tracking of microtubule plus ends is carried out at 30 °C by video-enhanced differential interference contrast microscopy using an Olympus IX71 inverted microscope with a ×100 (numerical aperture = 1.4) oil immersion objective. The end of an axoneme that possesses more, faster growing, and longer microtubules than the opposite end is designated as the plus-end. The real-time, 10-min videos are analyzed using Real Time Measurement (RTM II) software, and the data are collected using IgorPro (MediaCybernetics, Bethesda, MD, USA). Microtubules are considered to be growing if they increase in length by >0.3 μm at a rate of ≥ 0.3 μm/min. Shortening events are identified by a >1-μm length change at a rate of ≥ 2 μm/min. We calculate the catastrophe frequency by dividing the total number of catastrophes (transitions to shortening) by the time the microtubules are growing and in the attenuated state. The rescue (transition from shortening to growing) frequency is calculated as the total number of rescue events divided by the total time shortening. Dynamicity is calculated as the sum of the total growth length and the total shortening length divided by the total time. Table 12.1 shows the effects of GsαQL and Gsα-derived peptide (270–284) on microtubule dynamics.

12.1.4 Cellular consequences of Gsα interaction with tubulin/microtubules

Several forms of evidence obtained from *in vitro* binding assays clearly show that tubulin directly interacts with Gsα. In cells, this interaction and function of tubulin–Gsα complex also occur (Yu et al., 2009). In PC12 cells, activated Gsα interacts with tubulin and increases microtubule dynamic instability. In addition, the ability of Gsα to induce dynamic instability results in neurite-like process outgrowth in PC12 cells. A flowchart illustrating the cellular functional assays is shown in Fig. 12.5.

12.1.4.1 *Binding studies in cells*

In order to show that Gsα and tubulin are complexed in cells, HA-tagged Gsα is expressed in PC12 cells. The Gsα is then activated by either adding cholera toxin (which inactivates GTPase of Gsα) or mutational activation (Q227L). The Gsα–tubulin complexes are determined by coimmunoprecipitation using anti-HA antibody to pull down HA-tagged Gsα and anti-tubulin antibody to detect the tubulin presence in the complex. Immunofluorescence techniques also reveal colocalization of Gsα and tubulin in PC12 cells after cholera toxin activation (Yu et al., 2009).

12.1.4.1.1 Immunoprecipitation

PC12 cells infected with denoviruses coding for GFP, Ad/GFP, Ad/Gsα, or Ad/Gsα^{Q227L} are cultured for 40 h and then washed twice in PBS. Cells are lysed in 500 μl of lysis buffer on ice for 30 min. The lysate is collected and cleared by centrifuging at 12,000 × *g* for 20 min at 4 °C. After adjusting protein concentration to equal amounts for each sample, the supernatant (450 μl) is transferred to 1.5-ml microcentrifuge tubes and incubated with agarose beads coated with antimouse IgG for 1 h at 4 °C with continuous gentle inversion. The agarose beads are pulled down by centrifuging at room temperature and discarded. The lysate is then incubated with 5 μl of monoclonal antibody against HA for 20 h at 4 °C, and then the antibody/lysate mixture is incubated with agarose beads coated with antimouse IgG for 2 h at

Table 12.1 Effects of GsαQL, and Gsα peptide on microtubule dynamic instability

Dynamic instability parameters	Control	GsαQL (2 μm)	Gsα peptide (20 μm)
Growing rate (μm/min)	1.6±0.1	2.6±0.1[a]	2.6±0.3[a]
Shortening rate (μm/min)	8.9±0.7	12.4±1.3[a]	13.6±1[a]
Time growing (%)	39	35	21.5
Time shortening (%)	11	23	15.6
Time attenuated (%)	50	42	62.9
Catastrophe frequency (per min)	0.26±0.02	0.45±0.02[a]	0.75±0.1[b]
Rescue frequency (per min)	1.42±0.2	0.98±0.1[c]	1.36±0.3
Dynamicity	1.44	2.35	2.42

Microtubules were polymerized to steady state at the ends of axoneme seeds in the absence and presence of GsαQL or Gsα peptide, and the dynamic instability parameters were determined. Data are mean ± S.E.

[a]*$p<0.01$ as observed in a t test with respect to control.*
[b]*$p<0.001$ as observed in a t test with respect to control.*
[c]*$p<0.05$ as observed in a t test with respect to control.*
This research was originally published by Dave et al. (2011).

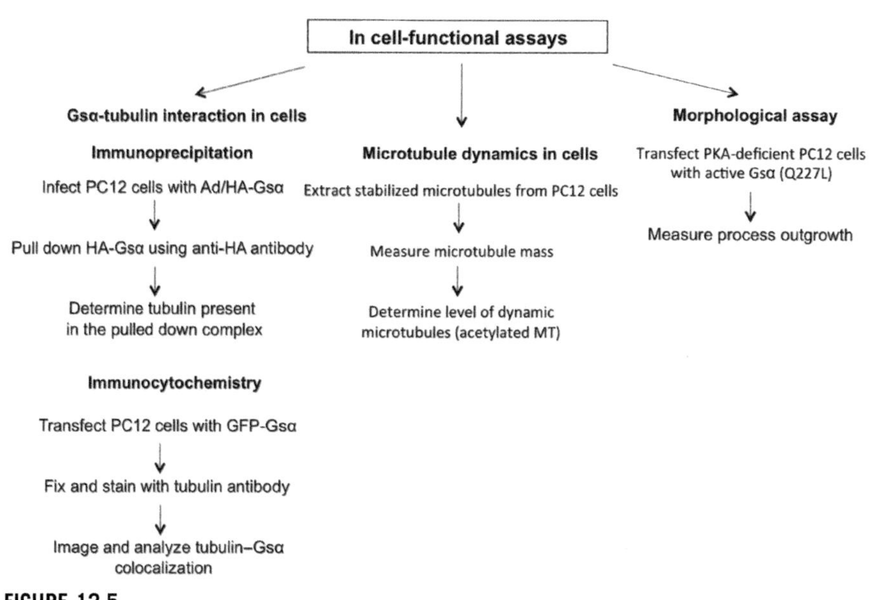

FIGURE 12.5

Flowchart illustrating the steps in the cellular functional assays including Gsα–tubulin interaction, microtubule dynamics in cells, and morphological assays.

4 °C with continuous gentle inversion. After the agarose beads are washed with lysis buffer three times, the 50 μl of SDS-PAGE sample buffer is added to the agarose beads. Fifteen microliters of supernatant is applied onto 5–12% gradient SDS-PAGE, and the resolved proteins are analyzed on a Western blot using the antibody against α-tubulin. The membrane is stripped and reblotted with antibody against Gαs to show sample loading (Fig. 12.6A).

12.1.4.1.2 Immunocytochemistry

To reveals colocalization between activated Gsα and tubulin. PC12 cells are grown in Dulbecco's-modified Eagle's medium (Invitrogen) supplemented with 10% fetal bovine serum, 5% horse serum, and 1% antibiotic (penicillin and streptomycin). Cells are maintained in a 5% CO_2 incubator at 37 °C. Medium is changed every 3 days, and cells are passaged once per week. PC12 cells in 12-well culture plates are transfected with Gαs-GFP using GenePORTERTM transfection reagent (Gene Therapy Systems, Inc., San Diego, CA, USA). The cells are treated with cholera toxin or vehicle control, washed twice with PBS, and fixed with cold 100% methanol for 4 min after extraction with 0.2% (w/v) saponin in microtubule-stabilizing buffer. The coverslips are then incubated with PBSS buffer (PBS plus 0.01% saponin) containing 10% bovine serum albumin for 20 min and then incubated in a 1:1000 dilution of anti-α-tubulin in PBSS buffer for 3 h. Subsequently, the coverslips are washed with PBSS four times and incubated with a 1:180 dilution of secondary antibodies labeled with TRITC in PBSS buffer for 40 min. These coverslips are washed with PBSS buffer four times and mounted on the slide with mounting medium. The slides are air-dried and examined by deconvolution microscopy. Images are captured with the Applied Precision, Inc. (Seattle, WA, USA) DeltaVision system built on an Olympus IX-70 base. Z-stacks are then deconvolved using the Softworx software. To quantify the colocalization between Gαs and microtubules, images are imported to Volocity (Improvision Inc., Waltham, MA, USA) for colocalization analysis. The extent of overlap between Gαs and microtubules is defined with Pearson's correlation.

12.1.4.2 ***Microtubule dynamics assay in cells***

By using detergent-extracted cytoskeletons under microtubule-stabilizing conditions, cellular microtubule mass can be measured in the absence of unassembled tubulin. Activated Gsα decreases the stable pool of polymerized microtubules in intact cells. In addition, using antibodies specific to different type of tubulin posttranslational modification demonstrated that activated Gsα also reduces the level of dynamic microtubules (acetylated MT) in the cells.

12.1.4.2.1 Measurement of detergent-insoluble microtubules

To measure microtubule mass, detergent-extracted cytoskeletons, free of unassembled tubulin, are prepared under microtubule-stabilizing conditions (Drubrin, Feinstein, Shooter, & Kirschner, 1985; Solomon, Magendantz, & Salzman, 1979). The tubulin content of the cytoskeletons is measured with immunoblotting. In brief, cells in a 25-ml flask are washed once with 37 °C PBS and once with extraction buffer. Cells

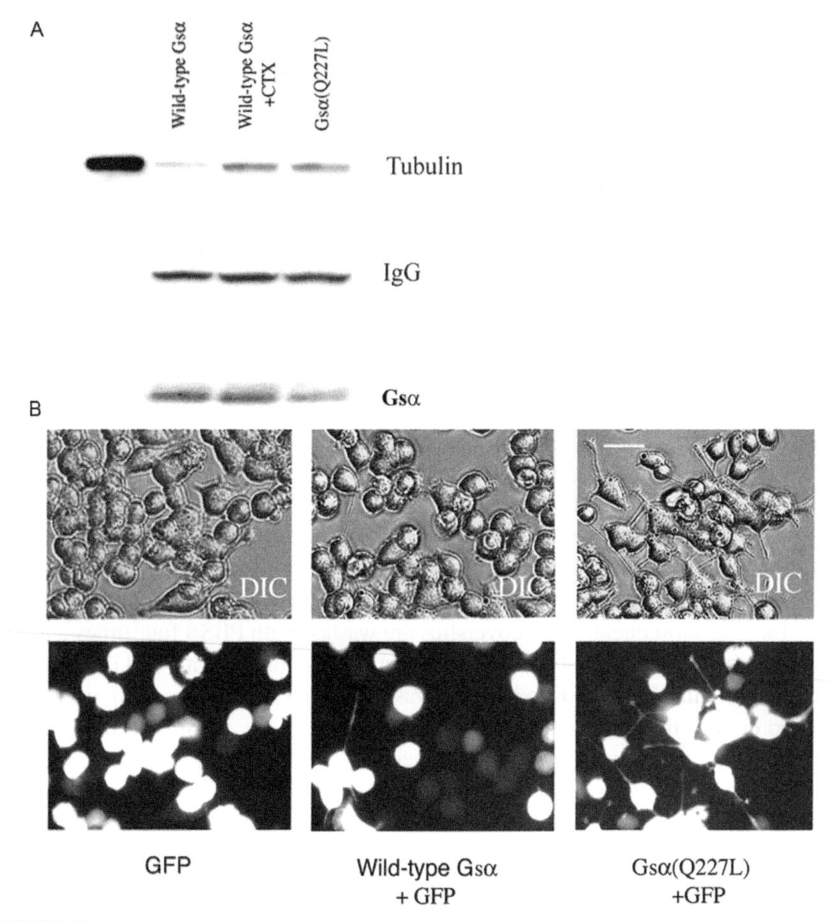

FIGURE 12.6

Cellular consequences of Gsα–tubuiln interaction. (A) Coimmunoprecipitation of Gsα with tubulin in cells. PC12 cells were infected with adenovirus containing GFP alone, HA-Gsα^{Q227L}-GTP, or HA-Gsα (wild-type) cDNA. Thirty-six hours after infection, wild-type Gsα-infected cells were treated with cholera toxin (*CTX*) for 1.5 h. Cells were lysed, and Gsα was immunoprecipitated using an HA antibody. Western blots with anti-tubulin antibody reveal the extent of tubulin complexed with the two forms of Gsα. (B) Activated Gsα promotes neurite outgrowth in PKA-deficient PC12 cells (PC12–123.7). Cells were infected with adenovirus containing both GFP and Gsα or Gsα^{Q227L}. Images were acquired after cells were infected with indicated viruses. Lower: Fluorescence of GFP indicated infected cells expressing Gsα. Upper: Differential interference contrast. Bar: 20 μm.

This research was originally published by Yu et al. (2009)

are subsequently extracted twice for 8 min with 0.5 ml of extraction buffer containing 0.1% Triton X-100 and protease inhibitors. After excess extraction buffer is drained from each flask, 0.5 ml of lysis buffer is added for 3–5 min to solubilize the detergent-extracted cytoskeletons. In addition, the 0.5 ml of extraction buffer used to extract PC12 cells is centrifuged for 1 min to collect insoluble material that came off of the culture flask during extraction. This material is added back to the lysis mixture in lysis buffer. The viscous cytoskeletal lysate is boiled for 3 min and then centrifuged for 10 min at 2000 × *g*, and the DNA-containing pellet is removed. The protein concentration of the extracted and cytoskeletal fractions is determined by the Lowry assay. Equal amounts of cytoskeletal protein fraction samples are loaded onto SDS-polyacrylamide gels, and the tubulin contents are determined by immunoblotting.

12.1.4.3 *Cell morphology assay*

Studies in PC12 cells have demonstrated that activated Gsα (Q227L) promotes morphological changes by inducing neurite outgrowth (Fig. 12.6B). This effect of Gsα is cAMP/PKA independent as the same result was observed in PKA-deficient PC12 cell line (PC12–123.7). In addition, treatment of PKA-deficient cells with cAMP analog and specific Epac agonists did not induce neurite outgrowth (Yu et al., 2009). To determine the consequence of Gsα, increasing microtubule dynamics on cell morphology Gsα or Gsα (Q227L) expressing PKA-deficient PC12 cell line (PC12–123.7) are grown in the condition mentioned above. The extent of neurite outgrowth is quantitated in living PC12–123.7 cells. The morphological differentiation of cells is determined by the percentage of neurite-bearing cells. One individual, blinded to experimental conditions, scores a cell as neurite-bearing if a cell contains at least one slender projection that exceeded the cellular diameter in length. As would be surmised from the assays showing the considerable preference tubulin shows for activated Gsα that conformation is far more effective in stimulating neurite outgrowth than wild-type Gsα.

12.2 BUFFER COMPOSITIONS

1. PEM buffer: 100 mM PIPES, 1 mM $MgCl_2$, and 1 mM EDTA, pH 6.9
2. TSG buffer: 20 mM Tris, 500 mM NaCl, 5 mM $MgCl_2$, 50 μM GDP, 200 μM PMSF, and 2 mM β-mercaptoethanol, pH 8.0
3. GAT buffer: 20 mM Tris, 500 mM NaCl, 20% glycerol, 10 mM $MgCl_2$, 30 mM KCl, 2 mM β-mercaptoethanol, 200 μM PMSF, and 50 μM GTP, pH 8.0
4. Gsα storage buffer: 50 mM HEPES, 1 mM DTT, and 1 mM EDTA, pH 8.0
5. HBS-P buffer: 10 mM Hepes, 150 mM NaCl, with 0.005% (v/v) surfactant P20, pH 7.4
6. HEPES buffer: 10 mM Hepes, 5 mM $MgCl_2$, 150 mM NaCl, and 1 mM β-mercaptoethanol, pH 8.0
7. PMEM buffer: 87 mM PIPES, 36 mM MES, 1 mM EGTA, and 2 mM $MgCl_2$, pH 6.8
8. Lysis buffer: 25 mM Na_2HPO_4, 400 mM NaCl, and 0.5% SDS, pH 7.2

9. Microtubules-stabilizing buffer: 80 mM PIPES/KOH, 1 mM $MgCl_2$, 1 mM EGTA, 30% (v/v) glycerol, and 1 mM GTP, pH 6.8
10. Extraction buffer: 100 mM PIPES, 1 mM $MgSO_4$, 2 mM EGTA, 100 μM EDTA, and 2 M glycerol, pH 6.75
11. Stripping buffer: 625 mM Tris–HCl, 100 mM β-mercaptoethanol, and 2% (w/v) SDS, pH 6.7

CONCLUDING COMMENTS

Activated Gsα, translocated to the cytosol from the plasma membrane, appears to directly interact with tubulin and acts as a direct modulator of microtubule dynamics, due primarily to the activation of GTPase by Gsα. This provides a mechanism for the modulation of microtubule-based cellular structures by the activation of G-protein-coupled receptors and does so in a manner independent of second messenger generation. Hopefully, this chapter has delineated a number of methods that will allow further elucidation of this underexplored aspect of cell biology.

Acknowledgments

Much of the science described in this chapter was supported by NIH (MH 39595, DA020568, MH 07800) and the Veteran's Administration (BX11049).

References

Carlier, M. F., & Pantaloni, D. (1983). Taxol effect on tubulin polymerization and associated guanosine 5′-triphosphate hydrolysis. *Biochemistry*, *22*, 4814–4822.

Chen, N. F., Yu, J. Z., Skiba, N. P., Hamm, H. E., & Rasenick, M. M. (2003). A specific domain of Gialpha required for the transactivation of Gialpha by tubulin is implicated in the organization of cellular microtubules. *The Journal of Biological Chemistry*, *278*, 15285–15290.

Dave, R. H., Saengsawang, W., Lopus, M., Sonya, R. H., Wilson, L., & Rasenick, M. M. (2011). A molecular and structural mechanism for G-protein mediated microtubule destabilization. *The Journal of Biological Chemistry*, *286*, 4319–4328.

Drubrin, D. G., Feinstein, S. C., Shooter, E. M., & Kirschner, M. W. (1985). Nerve growth factor-induced neurite outgrowth in PC12 cells involves the coordinate induction of microtubule assembly and assembly-promoting factors. *The Journal of Cell Biology*, *101*, 1799–1807.

Frank, R. (2002). The SPOT-synthesis technique. Synthetic peptide arrays on membrane supports—Principles and applications. *Journal of Immunological Methods*, *267*, 13–26.

Hahnefeld, C., Drewianka, S., & Herberg, F. W. (2004). Determination of kinetic data using surface plasmon resonance biosensors. *Methods in Molecular Medicine*, *94*, 299–320.

Jason-Moller, L., Murphy, M., & Bruno, J. (2006). Overview of Biacore systems and their applications. *Current Protocols in Protein Science*, *Chapter 19*, Unit 19 13.

Layden, B. T., Saengsawang, W., Donati, R. J., Yang, S., Mulhearn, D. C., Johnson, M. E., et al. (2008). Structural model of a complex between the heterotrimeric G protein, Gsalpha, and tubulin. *Biochimica et Biophysica Acta, 1783*, 964–973.

Linder, M. E., Ewald, D. A., Miller, R. J., & Gilman, A. G. (1990). Purification and characterization of Go alpha and three types of Gi alpha after expression in Escherichia coli. *The Journal of Biological Chemistry, 265*, 8243–8251.

Mejillano, M. R., Shivanna, B. D., & Himes, R. H. (1996). Studies on the nocodazole-induced GTPase activity of tubulin. *Archives of Biochemistry and Biophysics, 336*, 130–138.

Miller, H. P., & Wilson, L. (2010). Preparation of microtubule protein and purified tubulin from bovine brain by cycles of assembly and disassembly and phosphocellulose chromatography. *Methods in Cell Biology, 95*, 3–15.

Popova, J. S., Garrison, J. C., Rhee, S. G., & Rasenick, M. M. (1997). Tubulin, Gq, and phosphatidylinositol 4,5-bisphosphate interact to regulate phospholipase Cbeta1 signaling. *The Journal of Biological Chemistry, 272*, 6760–6765.

Rasenick, M. M., Wheeler, G. L., Bitensky, M. W., Kosack, C. M., Malina, R. L., & Stein, P. J. (1984). Photoaffinity identification of colchicine-solubilized regulatory subunit from rat brain adenylate cyclase. *Journal of Neurochemistry, 43*, 1447–1454.

Roychowdhury, S., Panda, D., Wilson, L., & Rasenick, M. M. (1999). G protein alpha subunits activate tubulin GTPase and modulate microtubule polymerization dynamics. *The Journal of Biological Chemistry, 274*, 13485–13490.

Roychowdhury, S., & Rasenick, M. M. (1994). Tubulin-G protein association stabilizes GTP binding and activates GTPase: Cytoskeletal participation in neuronal signal transduction. *Biochemistry, 33*, 9800–9805.

Shelanski, M. L., Gaskin, F., & Cantor, C. R. (1973). Microtubule assembly in the absence of added nucleotides. *Proceedings of the National Academy of Sciences of the United States of America, 70*, 765–768.

Solomon, F., Magendantz, M., & Salzman, A. (1979). Identification with cellular microtubules of one of the co-assembling microtubule-associated proteins. *Cell, 18*, 431–438.

Wang, N., & Rasenick, M. M. (1991). Tubulin-G protein interactions involve microtubule polymerization domains. *Biochemistry, 30*, 10957–10965.

Wang, N., Yan, K., & Rasenick, M. M. (1990). Tubulin binds specifically to the signal-transducing proteins, Gs alpha and Gi alpha 1. *The Journal of Biological Chemistry, 265*, 1239–1242.

Wedegaertner, P. B., Bourne, H. R., & von Zastrow, M. (1996). Activation-induced subcellular redistribution of Gs alpha. *Molecular Biology of the Cell, 8*, 1225–1233.

Yan, K., Popova, J. S., Moss, A., Shah, B., & Rasenick, M. M. (2001). Tubulin stimulates adenylyl cyclase activity in C6 glioma cells by bypassing the beta-adrenergic receptor: A potential mechanism of G protein activation. *Journal of Neurochemistry, 76*, 182–190.

Yu, J. Z., Dave, R. H., Allen, J. A., Sarma, T., & Rasenick, M. M. (2009). Cytosolic Galpha s acts as an intracellular messenger to increase microtubule dynamics and promote neurite outgrowth. *The Journal of Biological Chemistry, 284*, 10462–10472.

Yu, J. Z., & Rasenick, M. M. (2002). Real-time visualization of a fluorescent G(alpha)(s): Dissociation of the activated G protein from plasma membrane. *Molecular Pharmacology, 61*, 352–359.

CHAPTER

13

Purification and Biophysical Analysis of Microtubule-Severing Enzymes *In Vitro*

Juan Daniel Diaz-Valencia*, Megan Bailey† and Jennifer L. Ross*

**Department of Physics, University of Massachusetts Amherst, Amherst, Massachusetts, USA*

†Molecular and Cellular Graduate Program, University of Massachusetts Amherst, Amherst, Massachusetts, USA

CHAPTER OUTLINE

Introduction **192**

Importance of *in Vitro* Biochemical and Biophysical Studies 194

13.1 Methods **194**

13.1.1 Purification From Sf9 Cells 195

13.1.1.1 Amplifying Virus 195

13.1.1.2 Protein Expression 196

13.1.1.3 Protein Purification 196

13.1.1.4 Concentration Determination 198

13.1.1.5 Special Considerations for Severing Enzymes 198

13.1.2 Required Elements for Biophysical Assays 200

13.1.2.1 Microtubules 200

13.1.2.2 Experimental Chamber 202

13.1.3 Biophysical Assays 203

13.1.3.1 Epifluorescence and Single Molecule TIRF Imaging methods 204

13.1.3.2 Two-color Colocalization 205

13.1.3.3 Severing and Depolymerization 207

13.1.3.4 Binding and Diffusion 207

13.1.3.5 Photobleaching 208

Discussion and Summary **211**

Acknowledgments **211**

References **211**

Methods in Cell Biology, Volume 115

ISSN 0091-679X
http://dx.doi.org/10.1016/B978-0-12-407757-7.00013-X

Abstract

Microtubule-severing enzymes are a novel class of microtubule regulators. They are enzymes from the ATPases associated with various cellular activities family (AAA+) that utilize ATP to cut microtubules into smaller filaments. Discovered over 20 years ago, there are still many open questions about severing enzymes. Both cellular and biochemical studies need to be pursued to fully understand how these enzymes function mechanistically in the cell. Here, we present methods to express, purify, and test the biophysical nature of these proteins *in vitro* to begin to address the biochemical and biophysical mechanisms of this important and novel group of microtubule destabilizers.

INTRODUCTION

The microtubule cytoskeleton is a network of microtubule filaments required for cell morphology and intracellular transport. This network is organized and remodeled to perform cellular activities such as differentiation and cell division. Various microtubule-associated proteins (MAPs) interact with microtubules to regulate the network. Stabilizing and nucleating MAPs are positive regulators. Other important regulators are required to destabilize the network for the purpose of remodeling the location and dynamics of the filaments. For instance, maintenance of flagella, axons, and interphase networks requires remodeling and destabilization of specific microtubules or portions of the microtubule lattice.

Recent cell and biochemical work has begun to highlight the importance and novel mechanisms behind negative microtubule regulators or destabilizers. Several microtubule depolymerizing enzymes derived from the kinesin super family promote disassembly in an adenosine 5′-triphosphate (ATP)-dependent manner including kinesin-8 (Gupta, Carvalho, Roof, & Pellman, 2006; Varga et al., 2006), the mitotic centromere-associated kinesin or kinesin-13 (Desai, Verma, Mitchison, & Walczak, 1999; Helenius, Brouhard, Kalaidzidis, Diez, & Howard, 2006; Newton, Wagenbach, Ovechkina, Wordeman, & Wilson, 2004), and kinesin-14 (Cai, Weaver, Ems-McClung, & Walczak, 2009; Sproul, Anderson, Mackey, Saunders, & Gilbert, 2005; Troxell et al., 2001).

Nature developed an exquisite way to remodel microtubules in the form of another class of diffusible factors called microtubule-severing enzymes. These ATPases negatively regulate microtubules and belong to the meiotic clade of the AAA+ protein superfamily (Frickey & Lupas, 2004; Ogura & Wilkinson, 2001). Amidst these proteins are katanin (McNally & Vale, 1993), spastin (Roll-Mecak & Vale, 2005), and fidgetin (Cox, Mahaffey, Nystuen, Letts, & Frankel, 2000; Mukherjee et al., 2012a; Zhang, Rogers, Buster, & Sharp, 2007). These proteins contain the AAA+ domain on their C terminal halves. The 230 amino acid AAA+ segment is highly conserved and contains all the elements involved in ATP binding and hydrolysis (Fig. 13.1A, C, D) (Hanson & Whiteheart, 2005).

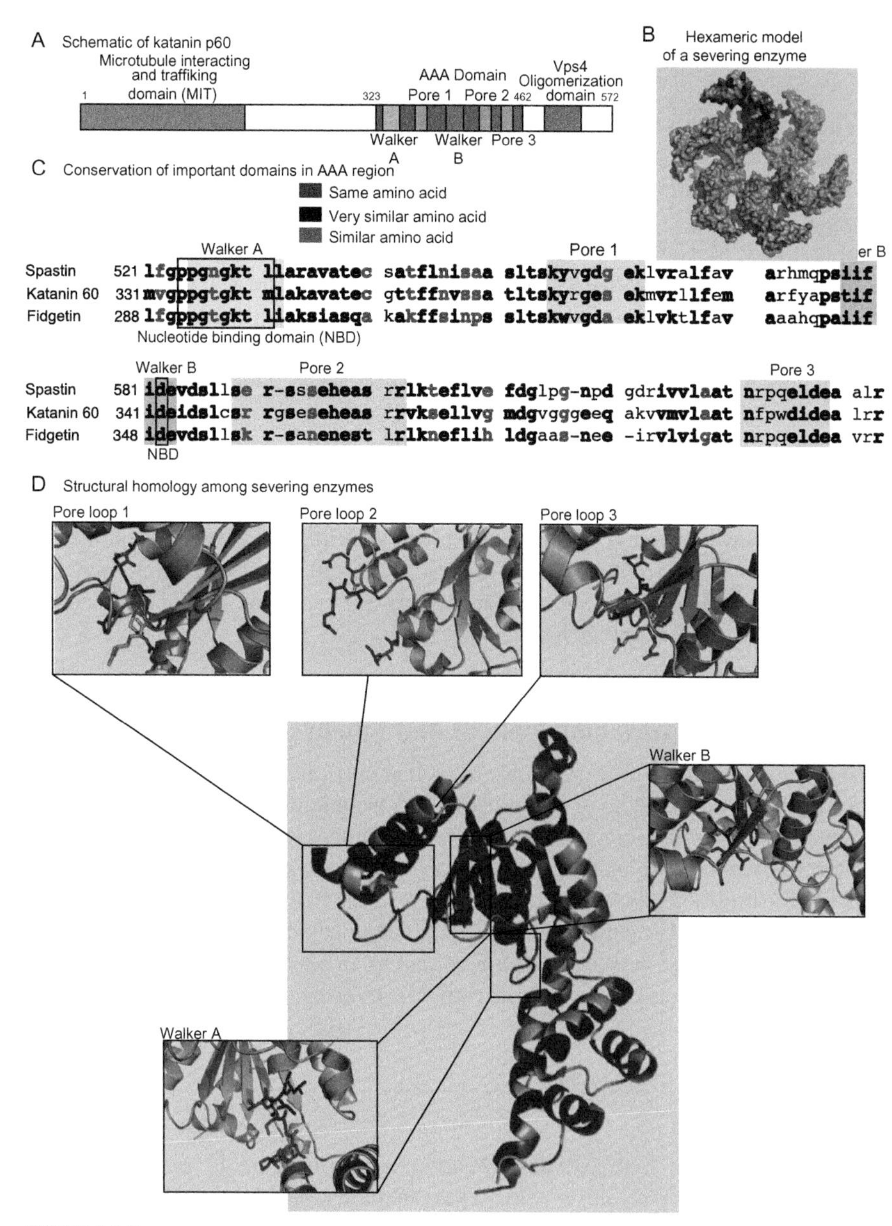

FIGURE 13.1

Conservation of domains in all known severing enzymes. (A) Schematic of katanin p60 showing the microtubule interacting and trafficking (MIT) domain (pink), ATPases associated with diverse cellular activities (AAA) domain (purple) with the Walker A domain (yellow) and Walker B domain (blue), the regions associated with the pore loops (green), and the Vps4 oligomerization domain (orange). Spastin and katanin have similar domain structure, although it is important to note that fidgetin's MIT domain has not been identified yet. (B) A model of the hexameric structure that severing enzymes are thought to adopt. Conserved residues (pink), very similar residues (dark blue), and similar residues (blue) are denoted in the shading comparing *Drosophila melanogaster* proteins. Note that this model only includes the AAA domain of spastin. (C) *D. melanogaster* amino acid sequences

Continued

The microtubule-severing enzymes katanin (Hartman J.J. and Vale R.D. 1999) and spastin (Roll-Mecak & Vale, 2008) assemble into biologically active hexamers driven by ATP binding (Fig. 13.1B). The severing activity relies on the carboxy-terminal tail (CTT) of tubulin (Evans, Gomes, Reisenweber, Gundersen, & Lauring, 2005; McNally & Vale, 1993; Roll-Mecak & Vale, 2005, 2008; White & Lauring, 2007). These enzymes are proposed to interact with the surface of the microtubule lattice via an amino-terminal microtubule interacting and trafficking domain. They also specifically bind to the CTT of tubulin via an active pore region of the AAA+ domain, the active site of the enzyme. The AAA+ domains are thought to translocate the CTT of the tubulin through the pore via ATP hydrolysis. This action pulls the CTT of the tubulin subunit away from the lattice, possibly locally unfolding the tubulin dimer, loosening the interdimer bonds, and generating a break in the microtubule.

In vivo, microtubule severing has been shown to be important for a vast number of cellular processes, nevertheless, the location and regulation of severing enzymes may vary depending on the subcellular region, cell type, and organism (recently reviewed by Sharp & Ross, 2012).

Importance of *in vitro* biochemical and biophysical studies

Recapitulating a cellular process outside the cell greatly facilitates the development of a mechanistic understanding of that process. The technique of *in vitro* reconstitution enables one to control the components for a cellular process, with nothing superfluous, to determine what is absolutely required for a given phenomenon to occur. Numerous reductionist studies of this sort have been used over the past 30 years to learn much of what we know about basic microtubule mechanics and dynamics. Using the methods outlined in this chapter, we hope that more research can be conducted to illuminate the underlying mechanisms controlling microtubule severing.

13.1 METHODS

Here, we present an itemized description of our procedure to purify and test microtubule-severing enzymes *in vitro*. These methods were employed in our recent publications (Diaz-Valencia et al., 2011; Mukherjee et al., 2012b; Zhang et al., 2011).

FIGURE 13.1—Cont'd from the AAA domain of the three known severing enzymes. This sequence shows conservation in all three enzymes of the important domains for the enzyme function. The color scheme is the same as in B. (D) A structural representation of the AAA region of spastin (PDB 3B9P). The zoomed-in views show the important residues in the Walker A, Walker B, and pore loop regions. The color scheme is the same as in B. (See color plate.)

13.1.1 Purification from Sf9 cells

13.1.1.1 *Amplifying virus*

13.1.1.1.1 Buffers, reagents, and equipment

13.1.1.1.1.1 ***Reagents***

- Cellfectin transfection kit (Invitrogen, 10362-100, Carlsbad, CA)
- 6-well plate (Thermo Fisher Scientific, 353224, Agawam, MA)
- Petri Dish, 150 × 15 mm Polystyrene (Thermo Fisher Scientific, 08-757-148, Agawam, MA)
- Sf9 insect cell cultures
- Serum-free media—SFM-900 II media (Gibco, 10902, Carlsbad, CA)
- HI fetal bovine serum (FBS) (Gibco, 16140, Carlsbad, CA)
- Bacmid

13.1.1.1.1.2 ***Equipment***

- Shaking incubator
- Tabletop centrifuge for 15 ml conicals

13.1.1.1.2 Detailed procedures

Recombinant baculovirus is prepared according to Bac-to-Bac system protocol (Invitrogen, Carlsbad, CA). All information included here is also found in the manual.

13.1.1.1.2.1 ***Transfecting Sf9 cells***

1. First, following the kit protocol, transfect 1×10^6 Sf9 cells in a 6-well plate with ~2 μg of the bacmid DNA using Cellfectin or equivalent Sf9 transfection reagent. There should be no FBS or antibiotics in the media during the transfection.
2. Once cells have been transfected, let the culture incubate for 3–5 h at 27 °C.
3. Change the transfection media on the cells to fresh SFM media with 5% FBS.
4. Incubate the cells for 72 h or until you see the cells lysing and dying. If you are not sure if the cells are infected, wait for 5–7 days and then harvest the virus. Wrap the 6-well plate in a damp towel and cover with a plastic wrap to keep moist and prevent drying out (this is a humid chamber).

13.1.1.1.2.2 ***Harvesting P1 Recombinant Baculovirus stock***

1. Transfer media from transfected cells to a 15-ml conical with a sterile 10-ml pipette.
2. Centrifuge gently at 1500 × *g* in a clinical centrifuge for 5 min at room temperature to remove any remaining cells from the media.
3. Transfer the supernatant to a new 15-ml conical with a sterile 10-ml pipette.
4. Wrap the conical in tin foil to protect from the light and store at 4 °C. This is the P1, which is also referred to as the first titer of virus.

13.1.1.1.2.3 ***Making P2 and P3 Recombinant Baculovirus stocks***

1. Plate 3×10^6 cells in 25 ml in a 150 × 15-mm Petri Dish.

2. Wait for 10 min or until the cells are attached to the plate and add 20 μl of P1 recombinant baculovirus stock to cells. This is going to be the P2 recombinant baculovirus stock.
3. Incubate the cells in a humid chamber for 5–7 days at 27 °C.
4. Harvest P2 recombinant baculovirus stock by centrifugation the same way the P1 recombinant baculovirus stock was harvested.
5. Repeat the same steps for a P3 recombinant baculovirus stock.

13.1.1.1.2.4 ***Notes.*** It is important to remember that not every baculovirus will work in exactly the same way. For some baculoviruses, the P2 recombinant baculovirus stock works best, and for others, the P3 recombinant baculovirus stock is best. The length of incubation (45–96 h) makes a significant difference. You should determine these factors using small cultures for testing protein expression. You can always reinfect the cells, or to save time, even just go back to a previous P1 recombinant baculovirus stock to reinfect cells and make more P2 and P3 recombinant baculovirus stocks. This is useful because it takes several days to complete this process.

13.1.1.2 ***Protein expression***

13.1.1.2.1 Buffers, reagents, and equipment

- P2 or P3 (second or third titer virus) to infect cells and produce functional protein
- 0.5-l Erlenmeyer flasks (Sigma-Aldrich, CLS430422, St. Louis, MO)
- 250 ml of SFM-900 II media (Gibco, 10902, Carlsbad, CA)
- HI FBS (Gibco, 16140, Carlsbad, CA)
- 250 ml of cells at 1×10^6 cells/ml
- Refrigerated shaking incubator to grow cells

13.1.1.2.2 Detailed procedures

Use P3 Recombinant 6 × His-GFP-katanin baculovirus stock to infect Sf9 cells at a multiplicity of infection (MOI) of one baculovirus particle infecting each cell for expression and purification of the severing enzymes. Higher MOI or number of insect cells is not recommended because severing enzymes when highly overexpressed tend to form aggregates which are inactive.

1. Sf9 insect cells are infected with 15 μl of P3 recombinant baculovirus stock and grown in 0.5-l Erlenmeyer flasks using 250 ml of SFM-900 II media supplemented with 5% HI FBS serum at a density of 1×10^6 cells/ml.
2. Let the infected culture grow at 27 °C, 130 rpm in a refrigerated shaking incubator for 72 or 96 h depending on the virus.

13.1.1.3 ***Protein purification***

13.1.1.3.1 Buffers, reagents, and equipment

13.1.1.3.1.1 ***Buffers: (All buffers should be made fresh on the day of the experiment)***

- Lysis buffer: 50 mM Tris (pH 8.5), 250 mM NaCl, 5 mM $MgCl_2$, 50 μM ATP, 1 mM PMSF, 7 mM 2-mercaptoethanol, 20 mM Imidazole, 10% sucrose.

- Wash Buffer-1: 50 mM Tris (pH 8.5), 250 mM NaCl, 5 mM $MgCl_2$, 50 μM ATP, 1 mM PMSF, 7 mM 2-mercaptoethanol, 20 mM Imidazole, 10% sucrose.
- Wash Buffer-2: 50 mM Tris (pH 8.5), 250 mM NaCl, 5 mM $MgCl_2$, 50 μM ATP, 1 mM PMSF, 7 mM 2-mercaptoethanol, 40 mM Imidazole, 10% sucrose.
- Elution buffer: 50 mM Tris (pH 8.5), 250 mM NaCl, 5 mM $MgCl_2$, 50 μM ATP, 1 mM PMSF, 7 mM 2-mercaptoethanol, 500 mM Imidazole, 10% sucrose.
- Severing Buffer I: 10% sucrose, 20 mM HEPES, 300 mM NaCl, 3 mM $MgCl_2$, 5 mM DTT, 50 μM ATP, pH 7.0.
- Coomassie Blue stain and destain.

13.1.1.3.1.2 ***Reagents***

- Infected insect cell culture
- Nickel beads (Qiagen, 30210, Valencia CA)
- Chromatography column (Bio-Rad 0.8 × 4 cm Poly-Prep Chromatography Columns, 731-1550, Hercules, CA)
- Spin column (Millipore, UFC801024, Billerica, MA)
- Funnel
- Kimwipes

13.1.1.3.1.3 ***Equipment***

- Microfluidizer (Avestin Inc., Ontario, Canada)
- Ultracentrifuge
- RC-6 Sorvall centrifuge with rotors that can hold 50 and 15 ml conicals
- Access to a cold room or a refrigerator to place column

13.1.1.3.2 Detailed procedures

1. Harvest the cells by centrifuging for 10 min at 4 °C at 1000 × *g*. Save samples of supernatants to troubleshoot using SDS-PAGE gel.
2. Gently resuspend the pellet in cold lysis buffer. At this point, all steps should be done on ice or in a cold room.
3. In order to homogenize the insect cells and recover the recombinant proteins, pass the cells through a microfluidizer with nitrogen under 15,000–20,000 PSU twice. The lysate should be clear after lysis. A high-pressure homogenizer (Avestin EmulsiFlex-B30) can be used as well. It is important to note that lysing the cells by sonication does not result in functional protein.
4. To remove cell debris, the lysed cells are centrifuged for 30 min at 91,500 × *g* at 4 °C.
5. During the centrifugation of the protein, prepare the Ni^{2+}-NTA beads for use in affinity purification. First, dilute with 1 ml of ddH_2O into 1 ml bed volume of beads. Centrifuge at 1500 × *g* for 5 min, remove ddH_2O from beads, and resuspend in ddH_2O. Repeat three times. Resuspend in lysis buffer after the last centrifugation step.
6. Discard the pellet of cell debris and use the supernatant for the subsequent steps. Pass the cell lysate through a funnel with a kimwipe to remove all cell

debris in the supernatant that may clog up the column. Save samples of supernatants to troubleshoot using SDS-PAGE gel.

7. Mix the supernatant of the lysed cells in batch with Ni^{2+}-NTA beads equilibrated in lysis buffer. Agitate the beads and lysate mixture for 120 min at 4 °C to bind protein to the beads.
8. Wash buffers can be made ahead of time but add the ATP right before washing the beads. Remove unbound protein by resuspending the beads in Wash Buffer-1 three times and centrifuging at 1500 × *g* for 5 min at 4 °C. Make sure the beads settle at the bottom of the tube completely before removing the wash buffer. Save samples of supernatants to troubleshoot using SDS-PAGE gel.
9. In between the steps to remove excess protein, prepare the spin column by centrifuging with ddH_2O four times.
10. Wash the beads with bound protein three times with Wash Buffer-2 by centrifuging at 1500 × g for 5 min at 4 °C. Save samples of supernatants to troubleshoot using SDS-PAGE gel.
11. After the washes are complete, pour the beads into a chromatography column.
12. To elute the protein, incubate the beads with 1 ml of elution buffer for 10 min and then allow it to flow through the column.
13. Repeat this step three times to recover as much protein as possible.
14. To identify which fractions the protein eluted in, place 2 μl of each elution fraction on filter paper and stain with Coomassie blue. Destain for 30 s.
15. Pool all fractions that contain the protein in a spin column to buffer exchange into Severing Buffer I and concentrate into 250 μl final volume. See Fig. 13.2 for a schematic depiction of purification process.

13.1.1.4 Concentration determination

13.1.1.4.1 Buffers, reagents, and equipment

- 100 μl of Severing Buffer I
- Pierce BCA protein assay kit (Thermo Fisher Scientific, 23227, Agawam, MA)
- Microplate Reader (BioTek, Synergy 2 Multi-Mode, Winooski, VT)
- 96-well plate with flat bottom wells (Thermo Fisher Scientific, 12565501, Agawam, MA)

13.1.1.4.2 Detailed procedures

The concentration of GFP-katanin is determined by using the Pierce BCA protein assay kit following the microplate protocol. Once we know the concentration of protein, the protein remains on ice and we continue with the microscope assays.

13.1.1.5 Special considerations for severing enzymes

1. Microtubule-severing enzymes are very difficult to work with, and it can be frustrating when they are not functional every time. In our experience, we purify functional protein one out of every three times. Previous research reported ability to drop freeze in LN2 and store in LN2. We have been unable to replicate this, so we use fresh protein. While this can be a challenge, it is useful to know that you did not do anything wrong specifically. If the

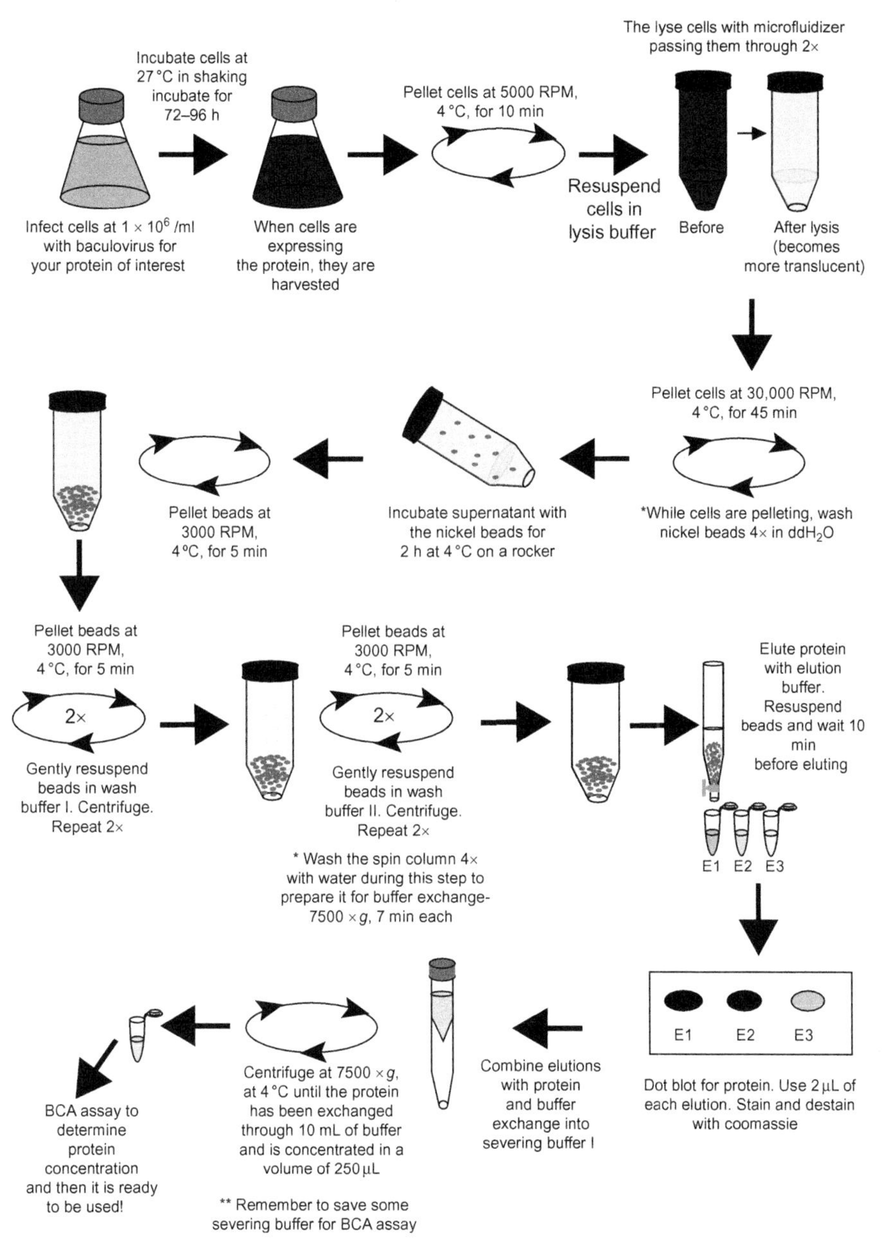

FIGURE 13.2

Purification diagram. This figure outlines the purification process for severing enzymes with a 6 × His tag. We have highlighted steps with important notes. It should be noted that the amount of expression of the protein depends on the virus expression. (See color plate.)

protein purification is consistently not working, you should try changing buffers and baculovirus. We have found removing the unbound protein in batch results in better protein than column purification. Perhaps, this is due to the length of time it takes to run a column.

13.1.2 Required elements for biophysical assays

13.1.2.1 Microtubules

Microtubules are the substrate of microtubule-severing enzymes. We prepare several different types of microtubules in order to determine if these AAA+ enzymes recognize differences in the nucleotide state of the polymer, shifts in the number of protofilaments, or have a preference for the plus- or minus-end. To do this, we make microtubules stabilized with Taxol or guanalyl-(α,β)-methylene diphosphate (GMPCPP). We also use polarity-marked microtubules and microtubules with protofilament defects that are visualizable via fluorescence.

We purify and prepare our own tubulin from pig brains following the protocol from Shelanski (Shelanski, Gaskin, & Cantor, 1973) and Hyman (Hyman et al., 1991). We use TRITC rhodamine-labeled tubulin (Cytoskeleton, Catalog number TL590M, Denver, CO) or in-house-labeled tubulin. In-house labeling is done as described in Hyman et al. (1991). We use DyLight 650, DyLight 550, or DyLight 488 with NHS ester (Pierce/Thermo Scientific, 46402, Rockford, IL).

13.1.2.1.1 Buffers, reagents, and equipment

13.1.2.1.1.1 Buffers

- PEM-100: 100 mM pipes, 2 mM EGTA, 2 mM $MgCl_2$, pH 6.9 with KOH.

13.1.2.1.1.2 Reagents

- Unlabeled tubulin (5 mg/ml)
- Labeled tubulin (5 mg/ml)
- Paclitaxel (Taxol) (Sigma-Aldrich, T7402, St. Louis, MO)
- Guanosine 5′-triphosphate sodium salt hydrate (GTP) (Sigma-Aldrich, G8877, St. Louis, MO)
- GMPCPP (Sigma-Aldrich, M3170, St. Louis, MO)

13.1.2.1.1.3 Equipment

- Micro ultracentrifuge
- Micro ultracentrifuge tubes
- Tabletop temperature-controlled centrifuge

13.1.2.1.2 Detailed procedures

13.1.2.1.2.1 Taxol-stabilized microtubules

1. Mix unlabeled tubulin (5 mg/ml) with labeled tubulin at any percentage from 5% to 20% labeling. If using lyophilized rhodamine tubulin from cytoskeleton, we recommend hydrating in PEM-100 and incubating for 10 min on ice before combining in order to dissolve the protein fully.

2. Clarify the tubulin by centrifuging at 366,000 × *g* for 10 min at 4 °C in a micro ultracentrifuge or airfuge to remove any aggregated tubulin.
3. Collect the supernatant and add 1 mM GTP.
4. Incubate at 37 °C for 30 min to nucleate and polymerize the tubulin into filaments.
5. Add 40 μM Taxol.
6. Incubate at 37 °C for 30 min to equilibrate Taxol.
7. Spin down the microtubules at 16,000 × *g* in a tabletop centrifuge for 10 min at 27 °C.
8. Resuspend the pellet in PEM-100 with 40 μM Taxol.

13.1.2.1.2.2 ***GMPCPP microtubules***

1. Mix unlabeled tubulin (5 mg/ml) with labeled tubulin at any percentage from 5% to 20% labeling. Treat lyophilized tubulin as above.
2. Clarify the tubulin by centrifuging at 366,000 × *g* for 10 min at 4 °C in ultracentrifuge to remove any aggregated tubulin.
3. Collect the supernatant and add 1 mM GMPCPP.
4. Incubate at 37 °C for 30 min to nucleate and polymerize the tubulin into filaments.
5. Spin down the microtubules at 16,000 × *g* in a tabletop centrifuge for 10 min at 27 °C.
6. Resuspend the pellet in PEM-100.

13.1.2.1.2.3 ***Polarity-marked microtubules.*** Polarity-marked microtubules are created by growing GTP-tubulin microtubules from GMPCPP-microtubule seeds. In order to differentiate the seed from the elongation, they must have a ~10-fold difference in labeling intensity in the same color channel, or be labeled with different fluorophores. We do not grow GMPCPP seeds with Taxol, but we do use it to stabilize elongated extensions after polymerization.

1. For both the GMPCPP seeds and the GTP-tubulin elongation, combine the labeled tubulin with the unlabeled to the desired percentages. We recommend 20% labeling for seed and 3% labeling for elongations, if using the same fluorophore.
2. Assemble GMPCPP seeds as you made GMPCPP microtubules above. Clarify the elongation tubulin and keep it on ice.
3. After incubating the GMPCPP tubulin at 37 °C, shear the GMPCPP microtubules by passing them through a Hamilton syringe (gauge 22S) three times to create short seeds.
4. Add 1 mM GTP to the elongation tubulin at 5 mg/ml and leave on ice.
5. Incubate the elongation tubulin for 1 min at 37 °C to bring the solution up in temperature, but not to polymerize yet.
6. Add 2 μl of the GMPCPP seeds to the elongation tubulin and incubate at 37 °C for 20 min to polymerize GTP extensions from GMPCPP seeds.
7. Add Taxol to 40 μM final concentration to stabilize elongated microtubules. Use this type of microtubule within 12 h to prevent end-to-end annealing and loss of polarity marker.

13.1.2.1.2.4 ***Protofilament defect microtubules.*** Microtubules polymerized with GTP and stabilized with Taxol display 12–13 protofilaments (Arnal & Wade, 1995), while microtubules polymerized with GMPCPP contain mostly 14 protofilaments (Hyman, 1995). In order to create microtubules with defects, we end-to-end anneal these two types of microtubules together after each has formed. This occurs spontaneously at high concentration, as evidenced by striped microtubules in solutions of older polarity-marked microtubules.

1. Create GMPCPP and Taxol-stabilized microtubules as described above. If using the same fluorophore on both, we recommend a 10-fold difference in labeling concentrations for the two types.
2. Shear each set of microtubules by passing the microtubules through a Hamilton syringe (gauge 22S) three times.
3. Combine equal parts of GMPCPP and Taxol-stabilized microtubules at high concentration (5 mg/ml).
4. Incubate the microtubules together for 2–24 h at room temperature to enable end-to-end annealing.

13.1.2.2 Experimental chamber

We perform all biophysical assays in microscope flow chambers with microtubules immobilized on the surface of a coverslip. This description is similar to that found in Dixit and Ross (2010).

13.1.2.2.1 Buffers, reagents, and equipment

13.1.2.2.1.1 Buffers

- PEM-100

13.1.2.2.1.2 Reagents

- Monoclonal antibeta tubulin clone TUB 2.1 (Sigma, T5201-200UL, St. Louis, MO)
- 5% Pluronic F127 (Sigma, P2443-250G, St. Louis, MO)
- Silanized coverslip (22 × 30 × 1.5 mm, Thermo Fisher Scientific, Catalog number 12-544-A, Agawam, MA)
- Cover glass (25 × 75 × 1 mm, Thermo Fisher Scientific, Catalog number 12-544-4, Agawam, MA)
- Double-stick tape (Scotch 3M, St. Paul, MN)
- Kimwipes or filter paper

13.1.2.2.2 Detailed procedures

1. Flow chambers are built by attaching parallel double-sided tape strips to a glass slide. A silanized glass coverslip is attached to the glass slide with double-sided tape. The procedure to create silanized coverslips is detailed in Dixit and Ross (2010). In this way, a channel 0.1 mm deep, 3 mm wide, and 30 mm long is created with a volume of approximately 13 μl (Fig. 13.3). To

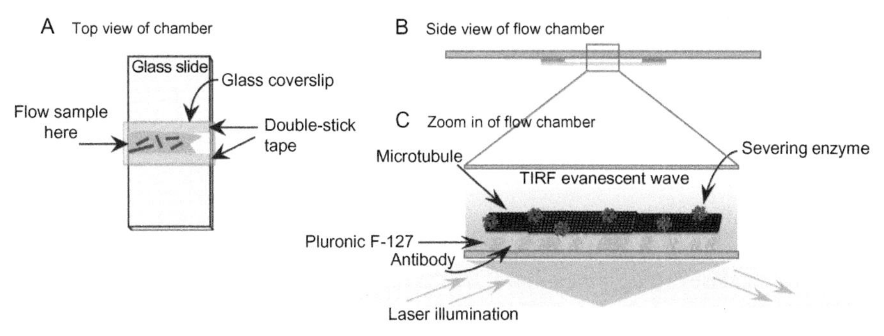

FIGURE 13.3

Flow chamber diagram. (A) A diagram of the top view of flow chamber. The coverslip is closest to the objective. It is stuck to the glass coverslip with double-stick tape to make a chamber you can flow into. (B) A side view of the chamber. (C) A zoomed-in view of the chamber showing the surface preparation. Antitubulin antibody (yellow) is bound and Pluronic F-127 (orange) is added to block the glass. Microtubules are added to the chamber to bind to the tubulin antibodies. Finally, a solution of severing enzymes, ATP, glucose, DTT, and an oxygen scavenging system are flowed into the chamber. The severing enzymes are depicted by the green hexamers on the microtubule. We use the evanescent field of the TIRF illumination to visualize the severing enzymes. (For interpretation of the references to color in this figure legend, the reader is referred to the online version of this chapter.)

avoid any leakage, gently press the coverslip that is in contact with the double-sided tape.

2. A 1:100 dilution of monoclonal anti-alpha tubulin antibody (2%, v/v) is prepared in PEM-100 and flowed in to fill the chamber.
3. The antibody is incubated at room temperature for 5 min.
4. Flow 5% Pluronic F-127 prepared in PEM-100. Centrifuge the F-127 solution prior to use each day. Use a filter paper to absorb the liquid from the other end to create flow. The chamber is again incubated at room temperature for 5 min to bind F-127 to open hydrophobic surfaces.
5. Flow in a 1:100 dilution of microtubules. See previous section on microtubules. Incubate the chamber at room temperature for 5 min.
6. To remove any nonattached microtubules, flow in a chamber wash buffer made of PEM-100 with or without 40 μM Taxol depending on the type of microtubules used. At this point, the chamber is ready to be used.

13.1.3 Biophysical assays

We have published results using several biophysical assays to study the activity of katanin severing (Diaz-Valencia et al., 2011; Mukherjee, 2012; Zhang, 2011). In all cases, we use fluorescence imaging to examine severing of microtubules. Here, we describe how to perform these experiments in detail.

13.1.3.1 Epifluorescence and single molecule TIRF imaging methods

13.1.3.1.1 Buffers, reagents, and equipment

13.1.3.1.1.1 Buffers

- Severing Buffer II: 10% sucrose, 20 mM HEPES, 100 mM NaCl, 3 mM $MgCl_2$, 5 mM DTT, 50 μM ATP, pH 7.0
- PEM-100
- Imaging buffer: Severing Buffer II, 20 μM Taxol, 0.1% Pluronic F-127, 0.5 mg/ml bovine serum albumin (BSA), 10 mM DTT, 40 mM glucose, 40 μg/ml glucose oxidase, 16 μg/ml catalase, 8 mM phosphocreatine, 0.16 mg/ml creatine phosphokinase with 2 mM ATP

13.1.3.1.1.2 Reagents

- Taxol
- Pluronic F-127 (Sigma-Aldrich, P2443, St. Louis, MO)
- BSA (Fraction V, Fisher Scientific, BP1605, Fair Lawn, NJ)
- DL-Dithiothreitol (DTT) (Amresco, 0281, Solon, OH)
- D-(+)-Glucose (Glucose) (Sigma-Aldrich, G5767, St. Louis, MO)
- Glucose oxidase from *Aspergillus niger* (Sigma-Aldrich, G2133, St. Louis, MO)
- Catalase from bovine liver (Sigma-Aldrich, C3155, St. Louis, MO)
- Phosphocreatine (Sigma-Aldrich, P7936, St. Louis, MO)
- Creatine phosphokinase from rabbit muscle (Sigma-Aldrich, C3755, St. Louis, MO)
- ATP magnesium salt (Sigma-Aldrich, A9187, St. Louis, MO)
- Adenosine 5′-(β,γ-imido)triphosphate lithium salt hydrate (AMPPNP) (Sigma-Aldrich, A2647, St. Louis, MO)
- Hexokinase from *Saccharomyces cerevisiae* (Sigma-Aldrich, H4502, St. Louis, MO)
- Stock of severing enzyme (active)
- Mutant severing enzyme (inactive) as a negative control

13.1.3.1.1.3 Equipment. Fluorescent species were visualized using epifluorescence and total internal reflection fluorescence (TIRF) on a microscope built around a Nikon Eclipse Ti microscope, as described previously (Ross & Dixit, 2010). Epifluorescence is excited using a Xe–Hg lamp (Nikon). TIRF excitation of GFP molecules used a 488-nm 125 mW Argon-ion air laser (Melles Griot) or a 50-mW 488 nm solid state (SpectraPhysics/Newport). The system has a high numerical aperture objective (60 ×, NA = 1.49, Nikon, Melville, NY). An additional 4 × magnification is placed before the camera to give 67.5 nm/pixel. Images are acquired with an electron multiplier CCD Cascade II camera (Photometrics, Tucson, AZ) or an IXON EM-CDD camera (Andor, South Windsor, CT).

13.1.3.1.2 Experimental procedures

1. Using the flow chamber with microtubules, as described above, image the microtubules to ensure that they are adhered to the surface of the cover glass and easily detected.

2. Establish the best exposure time for the microtubules. We typically take time-lapse movies with 5–20 s between frames.
3. Perform a control experiment without severing enzymes to test that buffers and imaging conditions do not cause microtubule break down. Create a chamber with microtubules, record for 3 min, and then flow through with imaging buffer. Add the imaging buffer slowly and carefully so that you do not change the position or lose focus (we use the Perfect Focus System of our Ti-E microscope). If the microtubules go out of focus, stop recording and refocus before beginning to record again. Record a 20-min movie to determine that the microtubules are not photodamaged due to imaging alone. If photodamage occurs, readjust the imaging parameters (exposure time or time between frames) so that you are not damaging the microtubules.
4. To test that the severing enzyme is functional, repeat the same experimental procedure as with the control chamber, except include severing enzyme in the imaging buffer that contains the ATP and oxygen scavenging system. Begin with a low concentration (50 nM) and work up to higher concentrations to find the concentration that enables active severing.

13.1.3.2 Two-color colocalization

13.1.3.2.1 Experimental procedures

For two-color imaging, we perform the same experiment described above, except we set up the microscope to record sequential images in red and green fluorescence. Otherwise a dual-view imaging system can be employed for simultaneous imaging.

13.1.3.2.2 Analysis

1. The data are recorded as 16-bit images using NIS Elements AR software (Nikon). The data are exported at 16-bit tiff files for each color.
2. All movies were aligned using the ImageJ plug-in stackreg in translation mode (Thévenaz, Ruttimann, & Unser, 1998).
3. The motion of single GFP-katanin molecules and the fluorescence signal of microtubules are analyzed using kymographs (space–time plots) generated with the Multiple Kymograph plug-in for ImageJ (J. Rietdorf and A. Seitz, European Molecular Biology Laboratory, Heidelberg, Germany). The kymographs are graphical representations of spatial position over time in which the *y*-axis represents the progression of time as going downward (Fig. 13.4). To generate kymographs from a stack of tiffs, draw a segmented line on each microtubule. Restore the selected line on the green channel stack of images. Use the kymograph plug-in on both channels using the same selected microtubule region of interest. Kymographs are a useful tool for looking at both the microtubules and the labeled severing enzyme because you can evaluate when the enzyme is found at a severing event.

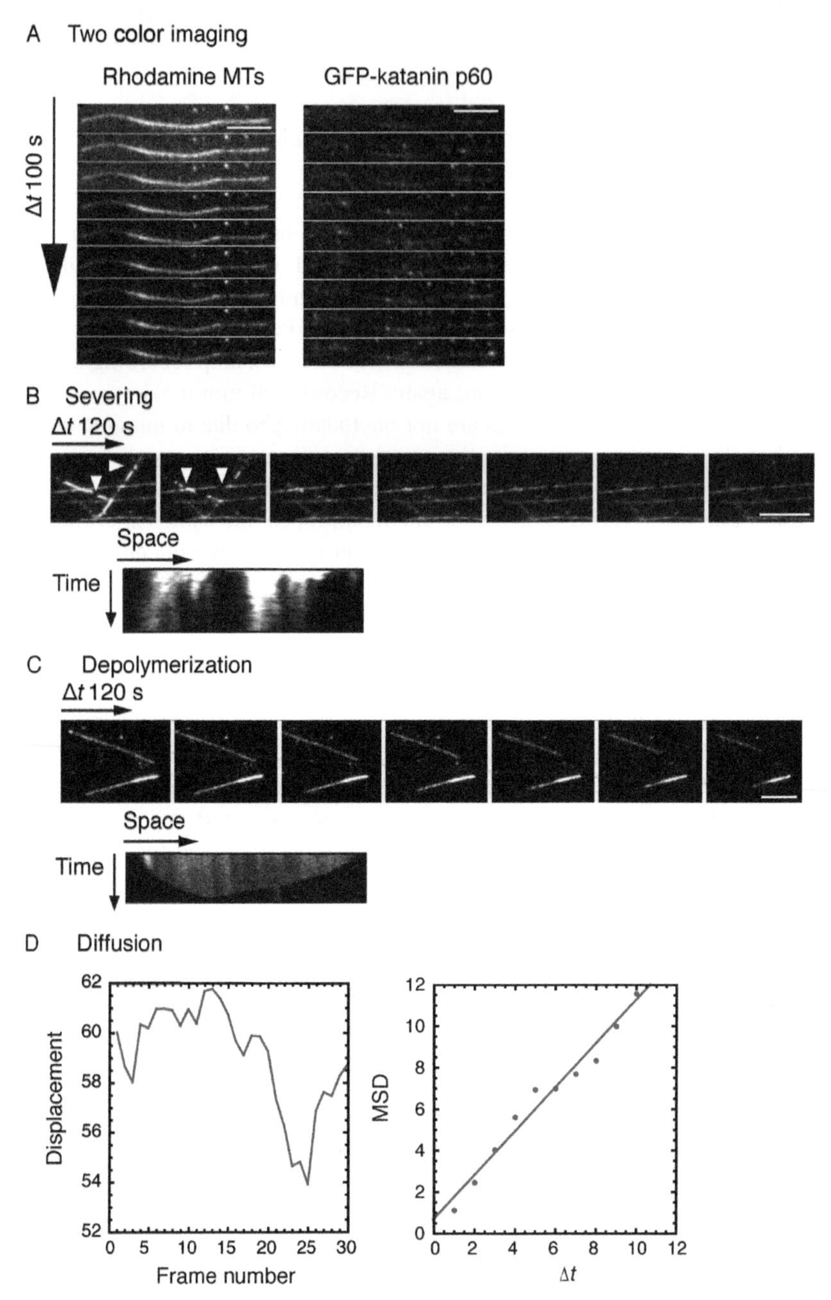

FIGURE 13.4

Visualizing severing and depolymerization. (A) Two-color time series showing severing and depolymerization of microtubules (left) and binding of katanin (right). (B) Time series of katanin-severing microtubules. A severing event is when a gap appears in the microtubule (white arrows). The kymograph shows the microtubule over the course of the movie, where the *x*-axis is space and the *y*-axis is time. (C) Time series of katanin-depolymerizing microtubules. The kymograph shows an example of depolymerization. All scale bars are 5 μm. (D) Plot of the displacement of a single katanin complex over 30 frames (left) and a mean squared displacement (MSD) over shifted time, Δt (right). (For color version of this figure, the reader is referred to the online version of this chapter.)

13.1.3.3 ***Severing and depolymerization***

13.1.3.3.1 Experiment

The experiment is executed in the same way as the two-color colocalization experiments except that it is only essential to use labeled microtubules. The enzyme can be either unlabeled or labeled.

13.1.3.3.2 Analysis

13.1.3.3.2.1 ***Depolymerization rate analysis.*** When microtubules depolymerize in the presence of katanin, the ends shrink and depolymerization is observed as the loss of signal at the ends of microtubules (Fig. 13.4A). Also, when microtubules are severed in the lattice, new ends are created and they also disassemble at these new ends. To measure the rate of decay of the microtubule ends, create a kymograph of the microtubule over time, use the angle tool in ImageJ to draw two intersecting lines and measure the average angle at each end (Fig. 13.4C). The rate of depolymerization is determined as one over the tangent of a given angle in distance pixel per time pixel. We convert the rate data using the known time interval and distance scale, which depends on the pixel conversion factor for the camera. Then, using graphing software, we plot the average rate of depolymerization for the plus-ends and minus-ends of microtubules if the polarity of the microtubules is known.

13.1.3.3.2.2 ***Severing frequency analysis.*** First, determine the length of all microtubules present at the beginning of the time series of images. Then, count by hand the number of severing events along microtubules using either the movie or the kymographs. Microtubule breaks due to severing look like an interruption in signal along the microtubule. Sometimes, many breaks occur close together, so it is important to be careful to count individual breaks instead of a big break made of several tiny breaks close to each other. Given the resolution limit of fluorescence microscopy, approximately half the wavelength of the emitted light, we require that over 200 nm of filament is removed in order to detect a break. A length of 200 nm is composed of over 300 dimers. Since our microtubules are Taxol stabilized, they do not depolymerize by having dimers removed, so the severing enzyme is most likely actively removing these dimers.

In order to quantify the density of severing events per unit time in a microtubule, count the number of clear severs observed and divide by the length of microtubule and the time of imaging. To quantify the frequency of severing at interfaces, use "Protofilament Defect Microtubules" from above (Section 1.2.1.2.4) and count the number of severing events at every interface in individual microtubules. Divide by the total number of frames to acquire a severing rate. As a control, compare the frequency of severing at interfaces with the frequency of severing 1–2 μm far away from interfaces.

13.1.3.4 ***Binding and diffusion***

13.1.3.4.1 Experiment

1. First, take an image of microtubules in epifluorescence.
2. Image GFP-katanin at low concentrations using TIRF such that individual molecules are visible. Take time-lapse movies with 100 ms frame rates, or faster,

with no delay between frames. Assays can be performed with the enzyme in trapped kinetic states using AMPPNP to mimic ATP state, ATP, ADP, or hexokinase to mimic Apo (no-nucleotide) state. The experiments elucidate the effect of the nucleotide on the binding and motility of the microtubule-severing enzyme.

13.1.3.4.2 Analysis

1. The quantitative measurement of binding of GFP-katanin is determined by counting the number of binding events manually.
2. The duration of association for each event is measured from kymographs by recording the number of pixels (frames) each molecule is bound. The molecule must associate and dissociate within the movie to be included. We do not include molecules that are already associated before the start of the movie. The number of pixels needs to be converted to seconds using the frame rate, typically 100 ms/frame.
3. Diffusion is determined by tracking individual and well-separated fluorescent enzymes using the SpotTracker plug-in in ImageJ. First, rotate the frames of the movie so that the microtubule is horizontal. This creates diffusion in one dimension and is easier to analyze.
4. Run the SpotTracker plug-in and save the results.
5. Analyze the results in Excel by calculating the mean squared displacement and plotting it over elapsed time shift, Δt, as follows. For each time shift of $\Delta t = 1, 2, 3, \ldots, N$ frames, calculate the difference between two positions $(x_1, x_2, x_3, \ldots x_N)$ as Δx_i, where i represents the index of the displacement. For instance, for a frame shift of one frame, Δt is 1, and one calculates $\Delta x_1 = x_2 - x_1$, $\Delta x_2 = x_3 - x_2$, $\Delta x_3 = x_4 - x_3, \ldots$. This is the "displacement" part of the mean-squared displacement.
6. Next, you must square these displacements: Δx_i^2.
7. Finally, the squares of the displacements must be averaged for each time shift so that you get the average over all i of Δx_i^2 for each time shift, Δt.
8. Finally, plot the averages of the squared displacements as a function of Δt. Fit a line to the resulting data. The slope equals the displacement squared per unit time, which is a diffusion coefficient. Make sure to convert the data from pixels to $\mu m^2/s$. See Table 13.1 for an example data set and the plot is featured in Fig. 13.4.

*13.1.3.5 **Photobleaching***

13.1.3.5.1 Experiment

To quantify the oligomeric state of GFP-katanin, we used photobleaching assays. These assays can be performed by flowing in 0.01 mg/ml anti-GFP antibodies in a flow chamber to fix GFP-katanin on the surface of the coverslip. After incubation, we flow GFP-katanin with a nucleotide to test the oligomerization state in the presence of ATP, or AMPPNP. Then, we image a field with many spots continuously with the laser until they become dark.

Table 13.1 Data from Spot Tracker Plug-in

		Δt (frames)									
Frame #	**x spot [px]**	**1**	**2**	**3**	**4**	**5**	**6**	**7**	**8**	**9**	**10**
1	60	**Ex: (x2−x1) ^2**									
2	58.66	1.8	**Ex: (x3−x1) ^2**								
3	58.01	0.42	3.96	**Ex: (x4−x1) ^2**							
4	60.33	5.38	2.79	0.11	**Ex: (x5−x1) ^2**						
5	60.18	0.02	4.71	2.31	0.03	**Ex: (x6−x1) ^2**					
6	60.95	0.59	0.38	8.64	5.24	0.9	**Ex: (x7−x1) ^2**				
7	60.96	0	0.61	0.4	8.7	5.29	0.92	**Ex: (x8−x1) ^2**			
8	60.89	0	0	0.5	0.31	8.29	4.97	0.79	**Ex: (x9−x1) ^2**		
9	60.28	0.37	0.46	0.45	0.01	0	5.15	2.62	0.08	**Ex: (x10−x1) ^2**	
10	60.94	0.44	0	0	0	0.58	0.37	8.58	5.2	0.88	**Ex: (x11−x1) ^2**
11	60.36	0.34	0.01	0.28	0.36	0.35	0.03	0	5.52	2.89	0.13
12	61.67	1.72	0.53	1.93	0.61	0.5	0.52	2.22	1.8	13.4	9.06
13	61.75	0.01	1.93	0.66	2.16	0.74	0.62	0.64	2.46	2.02	13.99

Continued

Table 13.1 Data from Spot Tracker Plug-in—cont'd

		Δt (frames)									
Frame #	***x* spot [px]**	**1**	**2**	**3**	**4**	**5**	**6**	**7**	**8**	**9**	**10**
14	61.36	0.15	0.1	1	0.18	1.17	0.22	0.16	0.17	1.39	1.06
15	60.74	0.38	1.02	0.86	0.14	0.04	0.21	0.02	0.05	0.04	0.31
16	59.67	1.14	2.86	4.33	4	0.48	1.61	0.37	1.49	1.66	1.64
17	59.1	0.32	2.69	5.11	7.02	6.6	1.59	3.39	1.39	3.2	3.46
18	59.87	0.59	0.04	0.76	2.22	3.53	3.24	0.24	1.14	0.17	1.04
19	59.85	0	0.56	0.03	0.79	2.28	3.61	3.31	0.26	1.19	0.18
20	59.26	0.35	0.37	0.03	0.17	2.19	4.41	6.2	5.81	1.21	2.82
21	57.34	3.69	6.3	6.4	3.1	5.43	11.56	16.16	19.45	18.75	9.12
22	56.24	1.21	9.12	13.03	13.18	8.18	11.76	20.25	26.21	30.36	29.48
23	54.63	2.59	7.34	21.44	27.25	27.46	19.98	25.4	37.33	45.29	50.69
24	54.8	0.03	2.07	6.45	19.89	25.5	25.7	18.49	23.72	35.28	43.03
25	53.9	0.81	0.53	5.48	11.83	28.73	35.4	35.64	27.04	33.29	46.79
26	56.86	8.76	4.24	4.97	0.38	0.23	5.76	8.94	9.06	5.02	7.9
27	57.61	0.56	13.76	7.9	8.88	1.88	0.07	2.72	5.02	5.11	2.22
28	57.45	0.03	0.35	12.6	7.02	7.95	1.46	0.01	3.28	5.76	5.86
29	58.26	0.66	0.42	1.96	19.01	11.97	13.18	4.08	0.85	1	2.53
30	58.73	0.22	1.64	1.25	3.5	23.33	15.44	16.81	6.2	1.93	0.28

Calculate the average of the column, this is the Mean Squared Displacement (MSD)

Δt (*x*-axis, Figure 13.4D, ii)	**1**	**2**	**3**	**4**	**5**	**6**	**7**	**8**	**9**	**10**
MSD (Δt) (*y*-axis, Figure 13.4D, ii)	1.12	2.46	4.03	5.62	6.94	6.99	7.7	8.34	9.99	11.58

13.1.3.5.2 Analysis

Data are analyzed as described previously (Ross & Dixit, 2010).

DISCUSSION AND SUMMARY

Obtaining active enzyme from every purification is a great challenge. Purifying katanin has proved to be a hard task as inferred by the dearth of publications that undertake their biochemical and biophysical characterization. The lack of stability is common in some other AAA+ proteins since dynein and CLPX are also known to lose activity.

The protocol and the assays explained here provide the first steps in the process of further understanding the biochemical and biophysical properties of microtubule-severing enzymes. These techniques will be useful in clarifying the role that this novel class of enzymes has on regulating microtubules.

Acknowledgments

We would like to thank David Sharp, Dong "Ray" Zhang, and Suranjana Mukherjee for their collaboration, advise, and initial preparation of the bacmid and virus for the Drosophila GFP-katanin. This work was supported by March of Dimes Basil O'Connor Starter Grant #5-FY09-46 and NSF CMMI-0928540 to J. L. R. J. L. R. was also supported by NSF MRI Grant DBI-0923318, and a Cottrell Scholars Award from Research Corporation for Science Advancement #20031. M. B. and J. D. D.-V. were supported by NSF Grant DMR-1207783.

References

Arnal, I., & Wade, R. H. (1995). How does Taxol stabilize microtubules? *Current Biology: CB*, *5*, 900–908.

Cai, S., Weaver, L. N., Ems-McClung, S. C., & Walczak, C. E. (2009). Kinesin-14 family proteins HSET/XCTK2 control spindle length by cross-linking and sliding microtubules. *Molecular Biology of the Cell*, *20*, 1348–1359.

Cox, G. A., Mahaffey, C. L., Nystuen, A., Letts, V. A., & Frankel, W. N. (2000). The mouse fidgetin gene defines a new role for AAA family proteins in mammalian development. *Nature Genetics*, *26*, 198–202.

Desai, A., Verma, S., Mitchison, T. J., & Walczak, C. E. (1999). Kin I kinesins are microtubule-destabilizing enzymes. *Cell*, *96*, 69–78.

Diaz-Valencia, J. D., Morelli, M. M., Bailey, M., Zhang, D., Sharp, D. J., & Ross, J. L. (2011). Drosophila Katanin-60 Depolymerizes and Severs at Microtubule Defects. *Biophysical Journal*, *100*, 2440–2449.

Dixit, R., & Ross, J. L. (2010). Studying plus-end tracking at single molecule resolution using TIRF microscopy. *Methods in Cell Biology*, *95*, 543–554.

Evans, K. J., Gomes, E. R., Reisenweber, S. M., Gundersen, G. G., & Lauring, B. P. (2005). Linking axonal degeneration to microtubule remodeling by Spastin-mediated microtubule severing. *The Journal of Cell Biology, 168*, 599–606.

Frickey, T., & Lupas, A. N. (2004). Phylogenetic analysis of AAA proteins. *Journal of Structural Biology, 146*, 2–10.

Gupta, M. L., Carvalho, P., Roof, D. M., & Pellman, D. (2006). Plus end-specific depolymerase activity of Kip3, a kinesin-8 protein, explains its role in positioning the yeast mitotic spindle. *Nature Cell Biology, 8*, 913–923.

Hanson, P. I., & Whiteheart, S. W. (2005). AAA+ proteins: Have engine, will work. *Nature Reviews. Molecular Cell Biology, 6*, 519–529.

Hartman, J. J., & Vale, R. D. (1999). Microtubule Disassembly by ATP-dependent oligomerization of the AAA enzyme katanin. *Science, 286*, 782–785.

Helenius, J., Brouhard, G., Kalaidzidis, Y., Diez, S., & Howard, J. (2006). The depolymerizing kinesin MCAK uses lattice diffusion to rapidly target microtubule ends. *Nature, 441*, 115–119.

Hyman, A. A. (1995). Microtubule dynamics: Kinetochores get a grip. *Current Biology: CB, 5*, 483–484.

Hyman, A., Drechsel, D., Kellogg, D., Salser, S., Sawin, K., Steffen, P., et al. (1991). Preparation of modified tubulins. *Methods in Enzymology, 196*, 478–485.

McNally, F. J., & Vale, R. D. (1993). Identification of katanin, an ATPase that severs and disassembles stable microtubules. *Cell, 75*, 419–429.

Mukherjee, S., Diaz Valencia, J. D., Stewman, S., Metz, J., Monnier, S., Rath, U., et al. (2012a). Human fidgetin is a microtubule severing the enzyme and minus-end depolymerase that regulates mitosis. *Cell cycle (Georgetown, Tex.), 11*, 2359–2366.

Mukherjee, S., Diaz-Valencia, J. D., Stewman, S., Monnier, S., Rath, U., Asenjo, A. B., et al. (2012b). Human fidgetin is a microtubule severing enzyme and minus-end depolymerase that regulates mitosis. *Cell Cycle, 11*, 1–8.

Newton, C. N., Wagenbach, M., Ovechkina, Y., Wordeman, L., & Wilson, L. (2004). MCAK, a Kin I kinesin, increases the catastrophe frequency of steady-state HeLa cell microtubules in an ATP-dependent manner in vitro. *FEBS Letters, 572*, 80–84.

Ogura, T., & Wilkinson, A. J. (2001). AAA+ superfamily ATPases: Common structure–diverse function. *Genes to Cells: Devoted to Molecular & Cellular Mechanisms, 6*, 575–597.

Roll-Mecak, A., & Vale, R. D. (2005). The Drosophila homologue of the hereditary spastic paraplegia protein, spastin, severs and disassembles microtubules. *Current Biology: CB, 15*, 650–655.

Roll-Mecak, A., & Vale, R. D. (2008). Structural basis of microtubule severing by the hereditary spastic paraplegia protein spastin. *Nature, 451*, 363–367.

Ross, J. L., & Dixit, R. (2010). Multiple color single molecule TIRF imaging and tracking of MAPs and motors. *Methods in Cell Biology, 95*, 521–542.

Sharp, D. J., & Ross, J. L. (2012). Microtubule-severing enzymes at the cutting edge. *Journal of Cell Science, 125*, 2561–2569.

Shelanski, M. L., Gaskin, F., & Cantor, C. R. (1973). Microtubule assembly in the absence of added nucleotides. *Proceedings of the National Academy of Sciences of the United States of America, 70*, 765–768.

Sproul, L. R., Anderson, D. J., Mackey, A. T., Saunders, W. S., & Gilbert, S. P. (2005). Cik1 targets the minus-end kinesin depolymerase kar3 to microtubule plus ends. *Current Biology: CB, 15*, 1420–1427.

Thévenaz, P., Ruttimann, U. E., & Unser, M. (1998). A pyramid approach to subpixel registration based on intensity. *IEEE transactions on image processing: A publication of the IEEE Signal Processing Society*, *7*, 27–41.

Troxell, C. L., Sweezy, M. A., West, R. R., Reed, K. D., Carson, B. D., Pidoux, A. L., et al. (2001). pkl1(+)and klp2(+): Two kinesins of the Kar3 subfamily in fission yeast perform different functions in both mitosis and meiosis. *Molecular Biology of the Cell*, *12*, 3476–3488.

Varga, V., Helenius, J., Tanaka, K., Hyman, A. A., Tanaka, T. U., & Howard, J. (2006). Yeast kinesin-8 depolymerizes microtubules in a length-dependent manner. *Nature Cell Biology*, *8*, 957–962.

White, S. R., & Lauring, B. (2007). AAA+ ATPases: Achieving diversity of function with conserved machinery. *Traffic*, *8*, 1657–1667.

Zhang, D., Grode, K. D., Stewman, S. F., Diaz-Valencia, J. D., Liebling, E., Rath, U., et al. (2011). Drosophila katanin is a microtubule depolymerase that regulates cortical-microtubule plus-end interactions and cell migration. *Nature Cell Biology*, *13*, 361–369.

Zhang, D., Rogers, G. C., Buster, D. W., & Sharp, D. J. (2007). Three microtubule severing enzymes contribute to the "Pacman-flux" machinery that moves chromosomes. *The Journal of Cell Biology*, *177*, 231–242.

CHAPTER

Measurement of *In Vitro* Microtubule Polymerization by Turbidity and Fluorescence

14

Matthew Mirigian*, Kamalika Mukherjee†, Susan L. Bane† and Dan L. Sackett*

**Program in Physical Biology, Eunice Kennedy Shriver National Institute of Child Health and Human Development, NIH, Bethesda, Maryland, USA*

†*Department of Chemistry, Binghamton University, State University of New York, Binghamton, New York, USA*

CHAPTER OUTLINE

14.1 Background and Theory 216
14.1.1 Background 216
14.1.2 Critical Concentration 217
14.1.3 Polymerization Curves and the "Turbidity Coefficient" 219
14.1.4 Polymerization Promoters 222
14.1.5 Measurement of Drug Effects—Polymerization Inhibitors 223
14.2 Materials and Equipment 223
14.2.1 Equipment 224
14.2.2 Materials 225
14.3 Methods 225
14.3.1 Preparation of Tubulin 225
14.3.2 Turbidity Assay in Cuvettes 225
14.3.3 Turbidity Assay in Multiwell Plates 226
14.3.4 Analysis of Turbidity 227
14.3.5 Fluorescence Assay in Multiwell Plates 227
References 228

Abstract

Tubulin polymerization may be conveniently monitored by the increase in turbidity (optical density, or OD) or by the increase in fluorescence intensity of diamidino-phenylindole. The resulting data can be a quantitative measure of microtubule (MT) assembly, but some care is needed in interpretation, especially of OD data.

Methods in Cell Biology, Volume 115
2013 Published by Elsevier Inc.
ISSN 0091-679X
http://dx.doi.org/10.1016/B978-0-12-407757-7.00014-1

Buffer formulations used for the assembly reaction significantly influence the polymerization, both by altering the critical concentration for polymerization and by altering the exact polymer produced—for example, by increasing the production of sheet polymers in addition to MT. Both the turbidity and the fluorescence methods are useful for demonstrating the effect of MT-stabilizing or -destabilizing additives.

14.1 BACKGROUND AND THEORY

14.1.1 Background

This chapter presents protocols for the assay of *in vitro* microtubule (MT) polymerization from purified proteins in defined buffer solutions. Purified tubulin will polymerize into MT and other polymers under correct conditions. This was discovered years ago, partially as a result of using a buffer solution that chelated calcium (Weisenberg, 1972). This reaction has been widely used, not only to understand MT polymerization *per se* but also as a means to check the action of other proteins that alter MT properties or polymerization. Perhaps, its widest application has been in evaluating the action of MT-active drugs (Hamel, 2003).

Measuring MT polymerization depends on measuring the change in some physical property that differs between a solution of dimeric tubulin and a solution of polymers like MT. The most obvious of these properties is size. As dimers assemble into polymers, the solution scatters more light, which can be detected by the increased optical density measurable by turbidity assay in a spectrophotometer (Gaskin, Cantor, & Shelanski, 1974). The difference between dimer and polymer size also allows for the use of sedimentation and filtration assays (Bollag et al., 1995). In addition to size differences, polymerization also changes binding properties of the protein so that fluorescent probes such as diamidino-phenylindole (DAPI) that bind to tubulin increase their emission intensity upon polymerization of tubulin (Bonne, Heuséle, Simon, & Pantaloni, 1985; Heusele, Bonne, & Carlier, 1987).

Turbidity is the most widely used method for following tubulin polymerization because it is simple, requires no unusual equipment, and is quantifiable. A number of papers have described the theory behind these measurements and how to maximize the information they may yield (Berne, 1974; Detrich, Jordan, Wilson, & Williams, 1985; Gaskin, 2011; Hall & Minton, 2005). It is readily adapted to multiwell plate assays and indeed is commercially available in this form (Davis, Martinez, Nelson, & Middleton, 2010). All that is required is a spectrophotometer and an optically clear vessel (cuvette or multiwell plate) to hold the tubulin sample in the light beam. With moderate care, the resulting turbidity readings can yield quantitative measurements of assembly kinetics and steady-state polymer concentrations. Slightly more complicated protocols and analyses will also yield information on polymer form (are these tubulin polymers really MT?) and reversibility. Some of these are described below.

The drawbacks of a turbidity assay are not always appreciated but may lead to considerable misinterpretation. MT formation is not the only thing that can cause an increase in turbidity in a tubulin solution. Addition of HCl can do the same thing, by precipitating the tubulin in aggregates. It is important to check that the observed turbidity is due to formation of MT. Other non-MT polymers of tubulin can also form, and interpretation depends on understanding the nature of the product. These can be distinguished by careful measurement of the turbidity (OD) yield per polymer concentration, and by the turbidity as a function of wavelength (see below).

Fluorescence enhancement upon polymerization provides a simple and sensitive method for measuring polymerization that is less sensitive to different polymer forms than is turbidity. A drawback is the need for equipment that is less widespread than spectrophotometers: spectrofluorometers or fluorescence-based plate readers.

A number of compounds have been described whose fluorescence increase can be used to monitor MT polymerization; DAPI has been most widely applied for this purpose (Bane, Ravindra, & Zaydman, 2007; Barron et al., 2003). DAPI fluorescence increases on binding to tubulin, and the affinity of DAPI for tubulin increases upon tubulin polymerization. Thus, the fluorescence emission of a given concentration of DAPI will increase upon tubulin addition and increase even more so when that tubulin polymerizes. This relatively straightforward effect allows assay of polymerization that is less affected by polymorphic assembly than is turbidity (Heusele et al., 1987). A small cautionary note is that addition of DAPI may influence the polymer form being measured (Vater, Böhm, & Unger, 1993).

Although, in principle, a fluorescence assay might be expected to be more sensitive (requiring less of the assayed reagents, i.e., tubulin here) than one based on optical density, in this application that may not be so. This is because the amount of tubulin required is driven by the critical concentration of tubulin polymerization and the volume of the assay rather than by the sensitivity of the optical method.

14.1.2 Critical concentration

An important parameter to consider in the design of any *in vitro* tubulin polymerization experiment is the critical concentration (C_C). This is the total tubulin concentration below which no polymerization occurs, or equivalently, the concentration of dimeric (nonpolymerized) tubulin (C_D) in steady-state equilibrium with MT polymer. For any experiment designed to measure polymerization, the initial (total) tubulin concentration (C_T) must be higher than the critical concentration (C_C). The critical concentration is not an intrinsic property of tubulin, but rather is a system parameter that is dependent on the solution composition (buffer, pH, ionic strength, divalent cations, etc.) and temperature. It is equivalent to the K_d ($=1/K_a$) for tubulin at the MT ends and therefore gives thermodynamic information about the system.

The most thorough way to measure the critical concentration for a given set of conditions is to prepare multiple samples with increasing C_T, allow them to polymerize to steady state, measure some parameter that is proportional to polymerized tubulin, such as OD or fluorescence, and plot that parameter versus C_T. When the line through the data is extrapolated to zero polymer, the intersection on the C_T axis is the C_C. Since the system is in steady-state equilibrium, an equivalent procedure is to allow polymerization to occur at high C_T and then dilute to several lower C_T, allow those to come to steady state, and plot the result. Performing both a series of increasing initial (unpolymerized) concentrations and dilution-based decreasing concentrations (from a polymerized sample) should yield the same C_T and overlapping data sets. This is illustrated

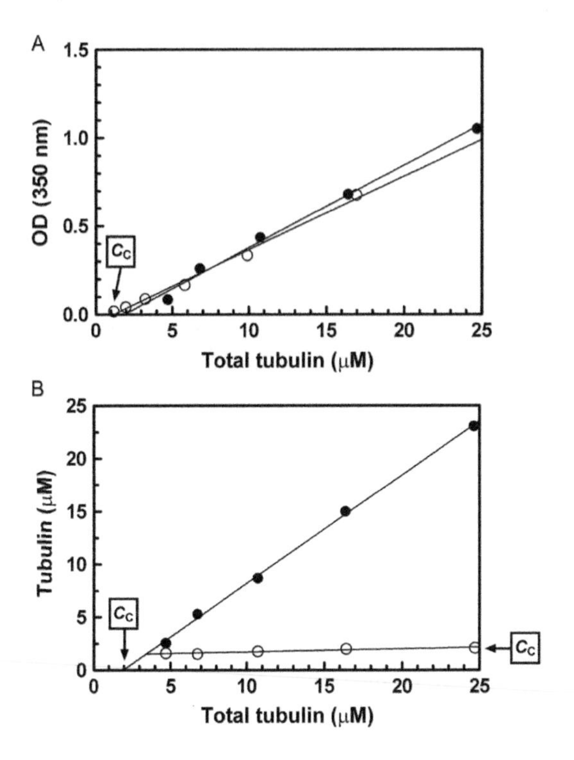

FIGURE 14.1

Critical concentration for polymerization determined by OD and by pelleting. Polymerization in Pipes–Mg–TMAO was monitored as a function of total tubulin concentration by OD measurements as well as by a pelleting assay. (A) Polymerization monitored by OD. The filled circles represent reactions in which unpolymerized tubulin at the indicated concentrations was allowed to polymerize and OD recorded. The open symbols represent an experiment in which tubulin was initially polymerized at 1.7 mg/ml (17 μM) and then sequentially diluted with prewarmed buffer to the indicated concentrations. At each point, the sample was allowed to equilibrate for ~8 min, OD recorded, and additional buffer added. Both data sets indicate a C_C of 1–1.2 μM (see arrow). (B) Polymerization determined by a pelleting assay. Samples were removed at steady state from the reactions indicated by the solid symbols in panel (A), and centrifuged for 5 min at room temperature in a Beckman Airfuge at 30 psi (150,000 × *g*). Protein contents were determined for the supernatant and the pellet, and plotted as concentration in the sample. The filled symbols are the concentrations of polymer pelleted and indicate a critical concentration of about 2 μM. The open symbols are the concentration of unpolymerized protein in the supernatant and indicate a C_C of about 1.5 μM. The low slope of this line shows that essentially all of the tubulin was polymerization competent.

in Fig. 14.1 for tubulin polymerization promoted by the natural organic osmolyte trimethylamine oxide (TMAO) (Sackett, 1997), see Section 1.4.

It is worth the time to determine the critical concentration of tubulin in the system that will be used in the assay. This can be done by plotting the steady-state OD versus C_T, as just described, or by measuring the extent of assembly by OD followed by centrifugation of the same sample. This will also reveal how much of the tubulin is polymerization competent. Once steady state is reached in the spectrophotometer, remove an aliquot from each sample (to determine total tubulin concentration, C_T) and a second aliquot for centrifugation. Pellet the polymer from this aliquot by centrifugation at ~200,000 × *g* for 5 min in an Airfuge (or at ~20,000 × *g* for 15 min in a microcentrifuge). Remove an aliquot of the supernatant and determine protein concentration. This is the concentration of the soluble dimer, C_D. The concentration of the polymer (C_P) can now be calculated, since $C_T = C_P + C_D$. Plot the data as shown in Fig. 14.1. The measurement of C_D and C_P, plotted as a function of C_T, should show C_D increasing up to C_C and then remaining constant above that. If the slope of C_D as a function of C_T is greater than 0 above C_C, this is an indication that some fraction of the tubulin dimers are not competent to polymerize (dead tubulin). If the slope is 0, that is, if C_D remains nearly constant at $C_T > C_C$, then this may be taken as evidence that essentially all of the tubulin is polymerization competent.

A simpler way to determine C_C is to centrifuge a polymerized sample at a *single* value of C_T, yielding C_D from the supernatant, which is equal to the C_C as long as there is no significant incompetent tubulin. This condition can be shown by a (previous) experiment with the same tubulin preparation, using the procedure described earlier, or in quick form, by repeated centrifugation on a second sample prepared at a higher C_T. If the supernatants yield the same C_D, then this may be taken to be the C_C.

14.1.3 Polymerization curves and the "turbidity coefficient"

An important but often overlooked component of a polymerization assay monitored by OD is the shape of the polymerization curve. Tubulin assembly is characterized by a lag time, a period of net growth, and a steady state (plateau). An increase in the amount of polymerized tubulin is accompanied by an increase in the maximum slope as well as the plateau of the curve. Figure 14.2 shows assembly curves in a few critical concentration experiments, quantitated in Table 14.1. Note how the maximum slope increases as the amount of assembled tubulin (plateau) increases. The three parameters of the curve to examine are the lag period (the time before the OD begins to increase), the maximum rate of OD increase, and the OD reached at steady state. These parameters are often highly correlated, but need not be. Note that the lag time may be relatively long or barely observable. The lag time generally reflects the number of nucleation events, which affects the *number* of MT (an increased number of nucleation events produces an increased number of shorter MT), but may not affect the total mass of polymer produced. Thus, in comparing two solutions, a shorter lag phase in one could indicate an increased number of nucleation events producing an increased number of shorter MT, even if the total mass of MT at plateau is the same between the two solutions.

The relationship between steady-state OD and polymer mass is a very useful parameter. The turbidity coefficient, which we will refer to as ε^* (by analogy to ε, the extinction coefficient for absorbance), is the turbidity (OD) of the polymer per concentration, thus OD/C_P. The value of ε^* is different for different polymers. A solution containing pure MT at a concentration of 1 mg/ml (C_P) will have an OD of not more

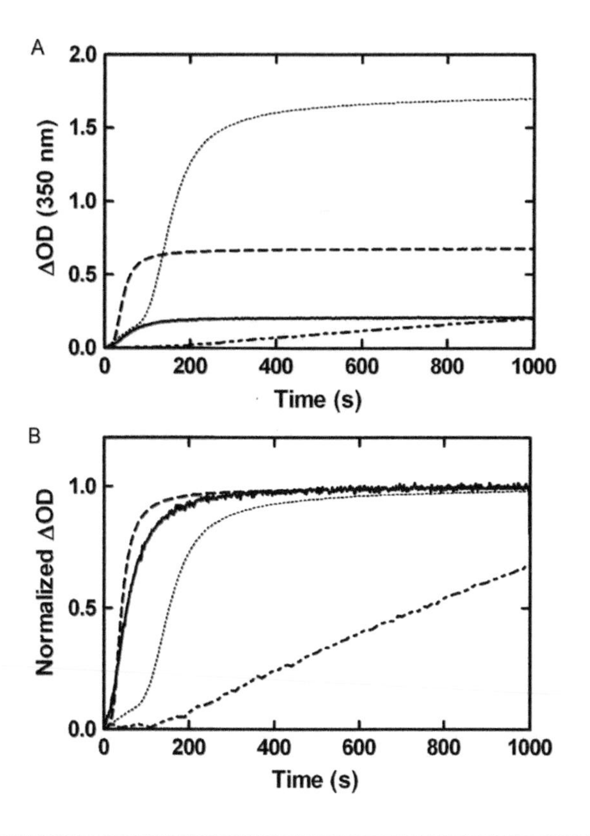

FIGURE 14.2

Polymerization timecourses in four different buffers. The first 1000 s of polymerization, monitored by OD_{350}, is shown for four buffers from Table 14.1. Pipes–Mg + 1 M glycerol is indicated by the dashed and dotted line, Pipes–Mg + 1 M TMAO by the dashed line, 1 M glutamate by the dotted line, and Pipes–Mg + 10 μM paclitaxel by the solid line. (A) The ΔOD, determined by subtracting the initial OD corresponding to the buffer and soluble tubulin from all subsequent readings, is plotted versus time showing polymerization kinetics of the systems. (B) The same data are plotted as ΔOD normalized to the steady state (plateau) OD for each buffer, which is $\Delta OD/\Delta OD_{max}$. The OD_{max} was taken at 30 min (1800 s) of polymerization and is not shown on these graphs. The normalized graph shows more clearly the differences in lag times as well as differences in the rates at which polymerization occurs as each system approaches its maximum amount of polymer under the different buffer conditions.

than 0.24 at 350 nm and 1 cm pathlength (Correia & Williams, 1983). A solution with sheet polymers of tubulin will have a higher OD than this (as well as characteristic changes in a plot of OD vs. wavelength (Detrich et al., 1985; Hall & Minton, 2005)). Thus simply calculating ε* will allow a quick check of whether the polymerization reaction is yielding MT or, for example, a mixture of MT and sheets. An example of this is the polymerization of tubulin in 1 M glutamate shown in Fig. 14.2 and

Table 14.1 Buffer effects on polymerization reactions

Buffer	Critical concentration C_C (g/l)	A_{350}/(mg/ml polymer) $= A/C_P = \varepsilon^*$
Pipes/Mg (0.1 M/0.5 mM) pH 7.0	2.51	0.23
Pipes/Mg (0.1 M/0.5 mM) + 1 M Glycerol pH 7.0	1.41	0.22
Pipes/Mg (0.1 M/0.5 mM) + 1 M TMAO pH 7.0	0.16	0.24
1 M Glutamate pH 6.6	0.22	0.65
Pipes/Mg (0.1 M/0.5 mM) pH 7.0 + 10 μM paclitaxel	0.07	0.50

Critical concentrations and turbidity coefficients are presented for polymerization in several buffers. Critical concentrations *in the different buffer conditions were determined by pelleting assay. Buffers were prepared beforehand except for the paclitaxel-containing buffer where the paclitaxel was added immediately before the reaction. In addition, DMSO was kept at 1% for all buffers to maintain consistency due to its necessity in the case of paclitaxel. Reactions of 50 μl were carried out directly in centrifuge tubes so as not to lose protein in transfer. Reactions were prepared by adding to the buffer 0.5 μl of 0.1 M GTP to a final concentration of 1 mM, followed by paclitaxel in that one case, then 6 μl of 25 mg/ml purified tubulin (3 mg/ml final concentration), except in the case of paclitaxel-containing buffer where 1 μl tubulin was added (0.5 mg/ml final concentration). The solutions were placed in a 30°C water bath for 30 min. The tubes were removed from the bath and centrifuged in the Airfuge for 5 min. The supernatant was assayed with BioRad Protein Assay to determine soluble tubulin concentration (C_D), corresponding to C_C.*
Turbidity coefficients *were measured using optical density measurements carried out in a spectrophotometer and reactions prepared in 50 μl cuvette. Reactions were prepared as similarly as possible to the reactions performed for critical concentration measurements. The components of the polymerization solution were added to the cuvette, buffer, GTP, paclitaxel (in the one case), and tubulin. These were mixed quickly, but thoroughly by pipetting up and down, and then placed into the spectrophotometer at 30°C. Optical density at 350 nm was monitored for 30 min at 2 s intervals (this short interval is only necessary for conditions that give a short lag time, i.e., paclitaxel). The ΔOD is determined by subtracting the initial OD value in the lag phase from the plateau OD in the steady-state phase. The ΔOD divided by the calculated polymer concentration gives ε*.*

quantitated in Table 14.1. The high OD seen in the upper plot (OD vs. time) in glutamate is not due to an increased mass of MT produced, but rather is due to production of some fraction of sheet polymers, indicated by the ε^* value of 0.65. All that is required to check that the OD is due to production of MT is to measure the OD at steady state, measure C_P as described earlier, and divide OD by C_P. It should be noted that production of non-MT polymers does not mean an invalid assay. Indeed, the increased OD yield by glutamate polymerization gives a stronger signal of polymerization, and the polymers have the expected sensitivity to temperature and tubulin-binding drugs as do MT (Hamel, 2003). It simply means that changes in OD can be influenced by polymer *type* as well as by polymer *quantity*, and a simple control will check this.

This is an important control because often the addition of a test compound may result in increased or decreased OD, relative to a control solution, and this will be interpreted (and reported) as an increase or decrease in the quantity of MT produced. Often it can be that the compound altered the *type* of polymer produced, rather than the *quantity*. This simple procedure can avoid such an incorrect interpretation.

Additional information may be obtained about the nature and shape of the polymers produced in a polymerizing solution of tubulin by collecting the OD at multiple wavelengths and determining the wavelength dependence of OD. This is often a valuable procedure and has been described in several publications, but is beyond the scope of this paper (see, e.g., Andreu & Timasheff, 1982; Detrich et al., 1985).

We have determined the critical concentration, C_C, and the turbidity coefficient ε^*, for a number of buffer conditions, using a single preparation of purified tubulin and a single temperature. These results are presented in Table 14.1 for the conditions shown in the curves in Fig. 14.2 as well as other buffer conditions. These data can provide part of the information for designing a polymerization experiment.

Several points should be made about the data in Table 14.1. It is clear that the critical concentration is strongly influenced by the buffer chosen for the polymerization. A fairly high concentration of tubulin is required for assembly to occur in Pipes/Mg buffer alone, though addition of 1 M glycerol lowers this by about twofold. Addition of 1 M TMAO or 10 μM paclitaxel lowers the required tubulin concentration by a factor of 10, as does adding 1 M glutamate (not shown). Use of 1 M glutamate alone has a similar result. A slightly less obvious result of buffer choice is revealed by the turbidity coefficient ε^*. Pipes/Mg buffer alone, or with 1 M glycerol or with 1 M TMAO, promote assembly that has a ε^* of ~0.23, consistent with polymers that are mostly MT. 1 M glutamate alone, or addition of 1 M glutamate to Pipes/Mg buffer (not shown) or addition of 10 μM paclitaxel to Pipes/Mg buffer, promotes assembly that has a ε^* of 0.5 or greater, indicating that many sheet polymers are also present in the solution.

A further point concerns the temperature. The values of C_C shown in Table 14.1 were determined at 30 °C. Note that in some cases, especially paclitaxel-promoted assembly, increasing the temperature to 37 °C is not likely to cause much increase in polymer yield, because the C_C is already very low at 30 °C.

14.1.4 Polymerization promoters

Tubulin will polymerize in a simple buffer that contains Mg^{2+} and GTP and is near neutral pH. However, the C_C can be rather high (see Table 14.1), so it is not unusual to add components to enhance polymerization by lowering the C_C. These fall into two categories: "tubulin-specific" and "thermodynamic." Tubulin-specific agents require concentrations that are comparable to tubulin (tens of μM) and have more-or-less specific sites of interactions. Examples are paclitaxel (Taxol™) and proteins such as microtubule-associated proteins (MAPs) or polyamines. Nontubulin-specific or "thermodynamic" promoters require concentrations of hundreds of mM to 1 M or higher and favor polymerization by "molecular crowding" or water-exclusion effects. These are often natural organic osmolytes and include glycerol, glutamate, and TMAO (Hamel & Lin, 1981; Lee & Timasheff, 1975; Sackett, 1997), but also include dimethylsulfoxide (DMSO), often used to promote polymerization and simultaneously to introduce sparingly soluble test compounds (Himes, Burton, & Gaito, 1977). Glutamate is an unusual example of a promoter, since it can be both the buffer and promoter—1 M Na glutamate is a complete solution in which tubulin–GTP will polymerize.

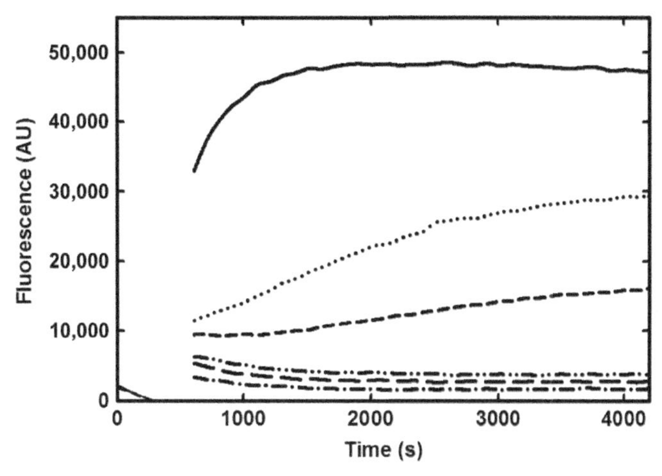

FIGURE 14.3

Polymerization timecourses monitored by the change DAPI fluorescence, performed as described in the text. Solutions of bovine brain tubulin (8 μM), DAPI (10 μM), and GTP (1 mM) in PME buffer were incubated with varying concentrations of podophyllotoxin (1 μM (·····); 2 μM (----); 4 μM (-·-·-·-); 6 μM (---); 8 μM (-·-·-)) at room temperature for 35 min. The control (—) had no podophyllotoxin. The multiwell plate was incubated at 37 °C in the SynergyMx microplate reader for 5 min, after which the fluorescence at 450 nm (360 nm excitation) was recorded to establish a baseline. Microtubule assembly was then induced by paclitaxel in DMSO to a final concentration of 8 μM. No data were collected in the interval between 5 and 10 min because the assay plate was removed for addition of paclitaxel during that period. The final concentration of DMSO in each sample was 8% (v/v).

14.1.5 Measurement of drug effects—Polymerization inhibitors

The effect of MT-destabilizing, polymerization-inhibiting drugs can be measured by following the inhibition of polymerization as a function of drug concentration. This is illustrated in Fig. 14.3 using DAPI fluorescence to monitor MT polymerization.

14.2 MATERIALS AND EQUIPMENT

There are three components of the experiment that require decisions: (1) the buffer, (2) the protein, and (3) the instrument. We will discuss these in sequence.

- The buffer used here is based on Na Pipes, but Na glutamate is a very attractive alternative for turbidity assay since it allows the reaction to be "tuned" simply by changing the concentration of glutamate (Hamel, 2003). In addition, glutamate promotes production of sheet polymers as well as MT, and these produce higher OD per milligram polymer than do MT, increasing the signal of polymerization. Often polymerization-promoting compounds are included in the buffer, as discussed in Section 1.3. Here, we suggest Pipes/Mg/TMAO because it produces a low C_C and hence requires lower amount of tubulin, and because it produces mostly MT (see Table 14.1), and these are reversible (see Fig. 14.1). Additionally, it shares with glutamate and other polymerization promoters the ability to "tune" the reaction.

- The protein used here is purified tubulin from mammalian brain, usually bovine or rat. A useful alternative is known as "microtubule protein" (MTP), consisting of tubulin plus MAPs. MTP will usually polymerize at lower concentrations than tubulin alone due to the assembly promoting properties of the MAPs.
- The instrument used for turbidity measurements is a spectrophotometer, here monochrometer-based, capable of reading cuvettes or multiwell plates. A filter-based unit can also give quite useful results, however. Fluorescence-based measurements are here performed in a monochrometer-based instrument for multiwell plates, but again filter-based instruments would also be suitable.

Assays are most often read in cuvettes, but multiwell plates are easy to use. There are advantages and disadvantages to each. Cuvette-based assays have the advantage that the light path for every sample is the same, so quantitative measurements may be better done this way. Multiwell plates have the advantage that many samples may be run simultaneously as easily as one. Both media can accept rather small samples—cuvettes that require 50 μl are readily available, and multiwell plates with half the usual diameter in a 96-well format are readily available and also require only 50 μl samples (although we use 60 μl in both). Thermostatting the samples is often available on plate- and cuvette-based instruments and makes possible experiments not readily accomplished without it, but room-temperature experiments are still very possible and valuable.

14.2.1 Equipment

1. Spectrophotometer (cuvette or plate reading, such as SpectraMax Plus, Molecular Devices, Sunnyvale, CA), preferably with control of the temperature of the sample chamber (though useful measures can be made at room temperature).
2. Spectrofluorometer (such as Synergy Mx multimode microplate reader (BioTek) (Software used: Gen 5)).
3. Cuvettes, small volume, such as 50 μl self-masking cuvettes, usable for either turbidity or fluorescence (Starna Cells, Atascadero, CA; Hellma Analytics, Plainview, NY).
4. Multiwell plates, 96-well half-area plates allow turbidity or fluorescence assays with 50–60 μl volumes (Costar 3695, Corning Life Sciences, Corning, NY).
5. Microcentrifuge (room temperature).
6. Microcentrifuge (4 °C).
7. Table-top ultracentrifuge (optional)—a Beckman Optima Max, or similar, with a rotor such as a TLA-100 or equivalent can be very useful, and equivalent to an Airfuge for pelleting MT.
8. Airfuge (optional). The Beckman Airfuge is an extremely useful centrifuge for quickly pelleting small samples (<100 μl) at high g (150–200,000 × g).

14.2.2 Materials

1. Tubulin (either commercial or lab purified). Tubulin is available commercially (e.g., Sigma-Aldrich, St. Louis, MO; Cytoskeleton, Boulder, CO), but purification of MTP or tubulin from bovine or rat brain is not difficult and protocols are available (Andreu, 2007; Hamel & Lin, 1981; Sackett, Knipling, & Wolff, 1991), but not discussed here. Tubulin can also be readily purified from nonneural sources for these assays (Sackett, Werbovetz, & Morrissette, 2010). Whatever the source, tubulin is usually stored at −80 °C in aliquots in order to minimize the numbers of freeze–thaw cycles. We purify our tubulin in-house, store it at high concentration (25 mg/ml) in GTP-free PM buffer, and limit freeze–thaw cycles to two thaws.
2. PM buffer = 0.1 M Pipes, 1 mM $MgCl_2$, pH 6.9. Pipes free acid is not very soluble in water, so is titrated into solution, and then to pH 6.9, with NaOH. A PM buffer with 1.0 M TMAO should be used when assaying depolymerizing agents, and 0.8 M TMAO should be used for stabilizing agents.
3. GTP stock solution = 0.1 M in PM buffer.
4. DMSO.
5. Podophyllotoxin: Stock solution = 4.1 mg/ml DMSO gives 10 mM. Working solution = 1 mM in DMSO. This is an inexpensive and effective inhibitor of polymerization.

14.3 METHODS

14.3.1 Preparation of tubulin

Tubulin may be commercial or lab prepared. Once an aliquot is thawed for an experiment, a working stock is prepared at 2.5 mg/ml in PM buffer and kept on ice. A brief high-speed spin (5 min at 20,000 × *g* or top speed in a 4 °C microcentrifuge or table-top ultracentrifuge) is performed to remove any aggregates, and the top 90% of the solution is removed to a new tube. With good tubulin preparations, this results in the loss of very little material and, for many purposes, can be skipped once this is verified. The tube is kept on ice.

14.3.2 Turbidity assay in cuvettes

The cuvette of choice is a microvolume, self-masking cuvette with a working volume of 50 μl, though we usually use 60 μl, to avoid a meniscus in the light path. The final sample will be 60 μl, 1.25 mg/ml tubulin, 0.5 mM GTP, 5% (v/v) DMSO, 1.0 M TMAO, in PM buffer. Tubulin and GTP concentrations may be altered as desired for particular experiments. Lowering the TMAO concentration (e.g., to 0.8 M) will raise the C_C and make the reaction more suitable for assay of polymerization-promoting compounds (such as paclitaxel). The sample is prepared in two halves: tubulin plus any test drug in one half (both at twice the desired final concentration) and buffer plus GTP at twice its desired final concentration. The two halves are then combined. The steps in this procedure are the following:

1. Prewarm the spectrophotometer to 30 °C (setting this will depend on the instrument—with the SpectraMax Plus, it is simply entered into the software setup).
2. Prepare the sample in a small microcentrifuge tube on ice. The sample consists of 30 μl of the tubulin solution at 2.5 mg/ml (twice the desired final concentration) in PM buffer with 1.0 M TMAO. If a drug is to be added, it can be added in 3 μl of DMSO to the tubulin solution and preincubated in a small microcentrifuge tube on ice. Control reaction would have 3 μl of DMSO alone.
3. The final 27 μl of solution consisting of PM buffer with 1.0 M TMAO, 1 mM GTP. This is mixed by pipetting up and down, and the total transferred to the cuvette. In this sort of experiment with multiple samples, we would prepare a larger volume of the PM–TMAO–GTP solution and just add 27 μl of the mixture per sample.
4. After adding the sample to the cuvette (being careful to avoid air bubbles), place it in the spectrophotometer and record OD (350 nm) versus time. A total time of 20 min and a 20-s interval between readings is usually sufficient.
5. (Optional). At the end of the recording period, the sample is removed with a micropipet and centrifuged for 5 min at full speed in an Airfuge ($\sim$200,000 $\times$ g), or 15 min at full speed ($\sim$20,000 $\times$ g) in a room-temperature microcentrifuge. The top 30 μl is removed and used for protein assay of the supernatant, in order to determine ε^* (see 14.3.4 Analysis of Turbidity).

14.3.3 Turbidity assay in multiwell plates

The sample preparation is very similar to that in the cuvette assay, but modified for multiwell plates.

1. Set the reading chamber of the plate reader to 30 °C.
2. Prepare two small rectangular styrofoam boxes with openings sufficient to hold a multiwell plate. Fill one with ice and the other with ~1–2 cm of room-temperature water.
3. Place the half-area, 96-well plate on the ice.
4. Place test compounds in 3 μl DMSO into each well. Also perform the following controls: 3 μl of DMSO alone, to assess the effect of the vehicle DMSO on the polymerization, and 3 μl of 1 mM podophyllotoxin as a negative (no polymerization) control.
5. Pipet the tubulin solution (30 μl, 2.5 mg/ml) into each well and mixed by stirring with the pipet tip.
6. Keep the plate on ice until all wells have samples, and longer if a preincubation with test compounds is desired.
7. Carefully add 27 μl of the PM–GTP–TMAO solution carefully to each well and mix the sample by up and down pipetting, being careful to avoid bubbles (this takes practice).
8. Move the plate to the styrofoam water box and float it on the water for 1 min. Timing this step carefully will make the experiment quite reproducible.

9. Quickly wipe the bottom of the plate, dry with a paper towel, and then again with a Kimwipe under the wells to be read.
10. Put the plate into the instrument and record OD (350 nm) at 20 s intervals for a total of 20 min.
11. (Optional). At the end of the recording, 50 μl samples are removed from each well and centrifuged as in the cuvette assay, and supernatant removed for protein assay.

14.3.4 Analysis of turbidity

The most important point to remember in interpreting turbidity is that turbidity does not only arise from MTs, and that changes in turbidity yield may originate in changes in polymer *type* as well as in polymer *amount*. For this reason, it can be very valuable to measure the protein concentration on the supernatant of the samples taken and centrifuged at the end of the OD readings, as discussed in Section 1.2. If the OD at the end of the readings differs between two samples, but the supernatant protein concentration is the same, it may be concluded that the polymers in the two samples are not the same, and further study, for example, electron microscopy, might be considered. Similarly, differences in lag time or maximum slope may be clues to suggest further experiments.

Beyond this point, there are three parts of the timecourse that are useful in the interpretation of turbidity assays of polymerization: the lag time, the rate of increase, and the steady-state level of OD. As solution conditions more strongly favor polymerization (higher tubulin concentration, increased temperatures (up to 37 °C), addition of polymerization promotors) the lag phase shortens, the rate of increase steepens, and the final steady-state OD value increases. Detailed discussions of these points may be found in Gaskin et al. (1974), Correia and Williams (1983), Hall and Minton (2005), and Gaskin (2011).

14.3.5 Fluorescence assay in multiwell plates

14.3.5.1 Introduction

This protocol uses a final volume of 200 μl in each well. Although it consumes more tubulin than the previously described assay, it is easier for less-experienced experimentalist (such as an undergraduate student) to perform. Paclitaxel-induced assembly is used because the critical concentration of tubulin in the presence of paclitaxel is very low; therefore, almost all of the tubulin in each sample will assemble into MTs. Nevertheless, this particular procedure uses about twice as much tubulin as the turbidity assays described. The total volume in each well can be decreased by reducing each solution volume proportionally. Note that, while this protocol uses a temperature of 37 °C, the polymer yield in the control would not be much different at lower temperature (e.g., 30 °C) as discussed in Section 1.3 and shown in Table 14.1.

The protocol described below is to assay for inhibitors of MT assembly. All buffer solutions contain DMSO at a final concentration of 4%. If ligands to be tested are potential inducers of MT assembly, omit paclitaxel and adjust the DMSO volumes accordingly.

14.3.5.2 Material

1. PME buffer (0.1 M Pipes, 1 mM $MgSO_4$, 2 mM EGTA, pH 6.90).
2. GTP stock solution (~65 mM in double-distilled water).
3. DMSO.
4. DAPI stock solution (~1.5 mM in double-distilled water).
5. Paclitaxel stock solution (2 mM in DMSO); paclitaxel working solution (200 μM in DMSO).
6. Ligand to be tested in DMSO.
7. FLUOTRAC 200, 96-well microplate, black/clear bottom (Grenier Bio-One, Germany).

14.3.5.3 Method

1. Set the reading chamber of the plate reader to 37 °C.
2. Set the excitation at 360 nm and the emission at 450 nm.
3. Prepare ligand solutions in DMSO. We prepare individual samples that are 25×, the desired final concentration. In this way, the same volume of ligand solution is added to each well.
4. Prepare tubulin stock solution that contains GTP and DAPI in PME buffer to the following final concentrations: 8.7 μM tubulin, 1.09 mM GTP, and 10.9 μM DAPI.
5. Place 184 μl tubulin stock solution in each well.
6. Add 8 μl of the ligand sample in DMSO to the well.
7. Mix by stirring the solution with a P20 pipet tip while pipetting up and down. Incubate the plate at room temperature to allow the ligand to bind to the protein. We usually perform the incubation for about 30–40 min.
8. Place the plate in the reading chamber and incubate for 5 min.
9. Record the fluorescence at 450 nm to obtain a baseline.
10. Remove the plate from the chamber.
11. Add 8 μl of the paclitaxel working solution in DMSO (to a final paclitaxel concentration of 8 μM) and mix carefully using the pipet. Try to avoid bubble formation at this point.
12. Return the plate to the plate reader.
13. Monitor the fluorescence at 450 nm over time until a satisfactory plateau is reached.

References

Andreu, J. M. (2007). Large scale purification of brain tubulin with the modified Weisenberg procedure. *Methods in Molecular Medicine*, *137*, 17–28.

Andreu, J. M., & Timasheff, S. N. (1982). Tubulin bound to colchicine forms polymers different from microtubules. *Proceedings of the National Academy of Sciences of the United States of America*, *79*, 6753–6756.

Bane, S. L., Ravindra, R., & Zaydman, A. A. (2007). High-throughput screening of microtubule-interacting drugs. *Methods in Molecular Medicine*, *137*, 81–88.

Barron, D. M., Chatterjee, S. K., Ravindra, R., Roof, R., Baloglu, E., Kingston, D. G., et al. (2003). A fluorescence-based high-throughput assay for antimicrotubule drugs. *Analytical Biochemistry*, *315*, 49–56.

Berne, B. J. (1974). Interpretation of the light scattering from long rods. *Journal of Molecular Biology*, *89*, 755–758.

Bollag, D. M., McQueney, P. A., Zhu, J., Hensens, O., Koupal, L., Liesch, J., et al. (1995). Epothilones, a new class of microtubule-stabilizing agents with a taxol-like mechanism of action. *Cancer Research*, *55*, 2325–2333.

Bonne, D., Heuséle, C., Simon, C., & Pantaloni, D. (1985). 4′,6-Diamidino-2-phenylindole, a fluorescent probe for tubulin and microtubules. *The Journal of Biological Chemistry*, *260*, 2819–2825.

Correia, J. J., & Williams, R. C., Jr. (1983). Mechanisms of assembly and disassembly of microtubules. *Annual Review of Biophysics and Bioengineering*, *12*, 211–235.

Davis, A., Martinez, S., Nelson, D., & Middleton, K. (2010). A tubulin polymerization microassay used to compare ligand efficacy. *Methods in Cell Biology*, *95*, 331–351.

Detrich, H. W., III, Jordan, M. A., Wilson, L., & Williams, R. C., Jr. (1985). Mechanism of tubulin assembly. Changes in polymer structure and organization during assembly of sea urchin tubulin. *The Journal of Biological Chemistry*, *260*, 9479–9490.

Gaskin, F. (2011). Analysis of microtubule assembly kinetics using turbidimetry. *Methods in Molecular Biology*, *777*, 99–105.

Gaskin, F., Cantor, C. R., & Shelanski, M. L. (1974). Turbidimetric studies of the in vitro assembly and disassembly of porcine neurotubules. *Journal of Molecular Biology*, *89*, 737–755.

Hall, D., & Minton, A. P. (2005). Turbidity as a probe of tubulin polymerization kinetics: A theoretical and experimental re-examination. *Analytical Biochemistry*, *345*, 198–213.

Hamel, E. (2003). Evaluation of antimitotic agents by quantitative comparisons of their effects on the polymerization of purified tubulin. *Cell Biochemistry and Biophysics*, *38*, 1–22.

Hamel, E., & Lin, C. M. (1981). Glutamate-induced polymerization of tubulin: Characteristics of the reaction and application to the large-scale purification of tubulin. *Archives of Biochemistry and Biophysics*, *209*, 29–40.

Heusele, C., Bonne, D., & Carlier, M. F. (1987). Is microtubule assembly a biphasic process? A fluorimetric study using 4′,6-diamidino-2-phenylindole as a probe. *European Journal of Biochemistry*, *165*, 613–620.

Himes, R. H., Burton, P. R., & Gaito, J. M. (1977). Dimethyl sulfoxide-induced self-assembly of tubulin lacking associated proteins. *The Journal of Biological Chemistry*, *252*, 6222–6228.

Lee, J. C., & Timasheff, S. N. (1975). The reconstitution of microtubules from purified calf brain tubulin. *Biochemistry*, *14*, 5183–5187.

Sackett, D. L. (1997). Natural osmolyte trimethylamine N-oxide stimulates tubulin polymerization and reverses urea inhibition. *The American Journal of Physiology*, *273*, R669–R676.

Sackett, D. L., Knipling, L., & Wolff, J. (1991). Isolation of microtubule protein from mammalian brain frozen for extended periods of time. *Protein Expression and Purification*, *2*, 390–393.

Sackett, D. L., Werbovetz, K. A., & Morrissette, N. S. (2010). Isolating tubulin from nonneural sources. *Methods in Cell Biology*, *95*, 7–32.

Vater, W., Böhm, K. J., & Unger, E. (1993). Effects of the fluorescence dye DAPI on microtubule structure in vitro: Formation of novel types of tubulin assembly products. *Acta Histochemica*, *94*, 54–66.

Weisenberg, R. C. (1972). Microtubule formation in vitro in solutions containing low calcium concentrations. *Science*, *177*, 1104–1105.

CHAPTER

Live-Cell Imaging of Microtubules and Microtubule-Associated Proteins in *Arabidopsis thaliana*

15

Jessica Lucas

Department of Biology, Santa Clara University, Santa Clara, California, USA

CHAPTER OUTLINE

Introduction ... **232**

15.1 Protocols ... **234**

15.1.1 Building Transgenic *A. Thaliana* ... 234

15.1.1.1 Transgene Construction ... 235

15.1.1.2 Plant Transformation ... 236

15.1.1.3 Transgenic Plant Recovery ... 237

15.1.1.4 Screening T2 Seed Lines ... 238

15.1.2 Sample Preparation ... 239

15.1.3 Imaging ... 240

15.2 Materials ... **242**

15.2.1 Sterile Media for Sterile Plant Growth ... 242

15.2.2 *A. Tumefaciens* Suspension for Plant Transformation ... 242

15.2.2.1 Virulence Buffer ... 242

Concluding Comments ... **243**

Acknowledgment ... **243**

References ... **243**

Abstract

Microtubules and microtubule-associated proteins (MAPs) play fundamental roles in plant growth and morphogenesis. The ability to observe microtubules and MAPs in living cells using fluorescent protein fusions has propelled plant scientists forward and given them the opportunity to answer longstanding biological questions. In combination with the genetic resources available in the model plant *Arabidopsis*

Methods in Cell Biology, Volume 115

ISSN 0091-679X
http://dx.doi.org/10.1016/B978-0-12-407757-7.00015-3

thaliana, our mechanistic understanding of how the microtubule cytoskeleton affects plant life has dramatically increased. It is a simple process to construct transgenic *A. thaliana* plants that express fluorescent protein fusions by using the disarmed plant pathogen *Agrobacterium tumefaciens*. Several screening steps are necessary to ensure that the fusion protein accurately mimics the native protein because transgenes are inserted randomly into the *A. thaliana* genome. To image the fluorescent proteins *in planta*, confocal microscopy is used to alleviate issues caused by specimen thickness and autofluorescence.

INTRODUCTION

Plants are essential for human livelihood because they produce food, fuel, shelter, and many medicines. Because plant products are important to our society, there is much interest in understanding the complex mechanisms that govern plant growth and development in different environmental contexts. Several lines of evidence indicate that the microtubule cytoskeleton is a key component of the growth machinery in all organisms, including plants (Chan, 2012; Lloyd, 2011; Lucas & Shaw, 2008; Nick, 2012; Wasteneys, 2004). However, we know relatively little about the molecular relationship between plant growth and the microtubule cytoskeleton (Crowell, Gonneau, Stierhof, Hofte, & Vernhettes, 2010; Smith & Oppenheimer, 2005; Szymanski & Cosgrove, 2009; Wasteneys & Fujita, 2006). The ability to label and image proteins of interest with fluorescent proteins in living cells has allowed researchers to build a deeper mechanistic understanding of biology. In this review, we will focus on techniques used to image fluorescent protein reporters in living seedlings of *Arabidopsis thaliana*. *A. thaliana* is a model flowering plant for genetics, development, and cell biology. Specifically, we will discuss imaging microtubules and microtubule-associated proteins (MAPs).

In most plant cells, four types of microtubule arrays cyclically appear throughout the cell cycle: the interphase cortical array, the preprophase band, the spindle, and the cytokinetic phragmoplast (Muller, Wright, & Smith, 2009; Van Damme & Geelen, 2008). Genetic or chemical disruption of any array severely disrupts growth and morphogenesis (Baskin, Wilson, Cork, & Williamson, 1994; Corson et al., 2009). As in other organisms, the spindle segregates genetic material during M-phase. The preprophase band and phragmoplast are unique to plants, and both are required for the accurate positioning of new cell walls (Rasmussen, Humphries, & Smith, 2011; Smith, 2001). The precise placement of new cell walls is an important patterning step in plant development, and the mechanisms in which the preprophase band and phragmoplast mediate this process are largely unknown (De Smet & Beeckman, 2011; Muller et al., 2009; Van Damme, Vanstraelen, & Geelen, 2007). The microtubules of the preprophase band and interphase array are linked along their lengths to the plasma membrane by currently unknown proteins (Hardham & Gunning, 1978). The interphase array affects properties of the cellulosic cell wall, which affects cell expansion and therefore growth and morphology (Chan, 2012; Lloyd, 2011; Lucas & Shaw, 2008; Nick, 2012; Wasteneys, 2004). Depending upon environmental conditions and the developmental context, interphase arrays are organized into different configurations (Fig. 15.1;

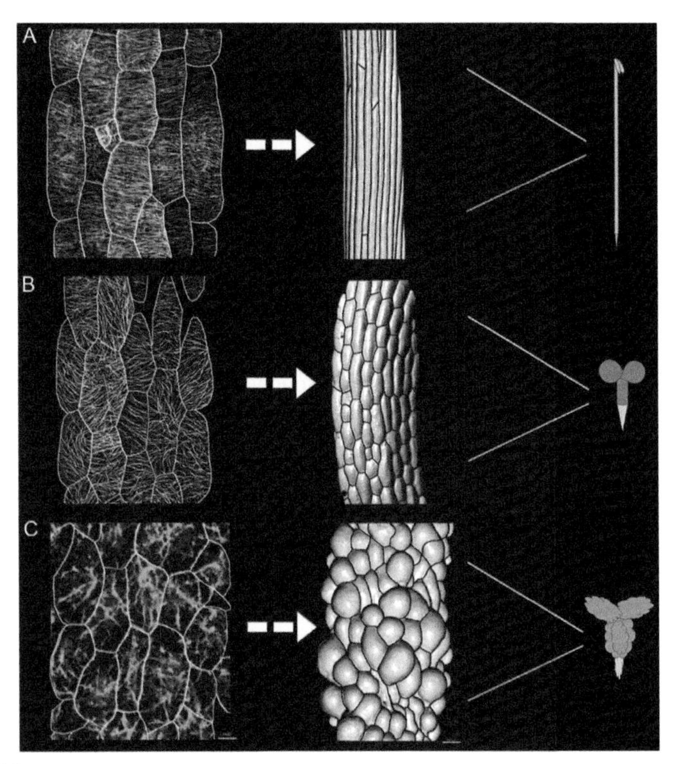

FIGURE 15.1

Correlation between interphase array organization and cell growth in the *A. thaliana* seedling stem. The left images are confocal micrographs showing interphase microtubules labeled with GFP:alpha-tubulin6 driven from a constitutive 35S promoter in epidermal cells of young seedling stems. The boundaries between cells have been drawn in as white lines. The center images are surface renderings of the outer surface of the seedling stem. Black lines represent the cell boundaries. Cartoon drawings of entire seedlings are presented on the right. (A) Seedlings grown in complete darkness exhibit transverse microtubule arrays (left), cell elongation (center), and axial growth of the stem (right). (B) When grown in the light, cell and seedling growth are restrained (center and right), and microtubules are organized into mesh arrays (left). (C) Cells grow isotropically in the presence of oryzalin, a chemical that inhibits microtubule polymerization (middle). When the interphase array is depolymerized (left), the growth of the seedling is uncontrolled (right). For full microscopy methods, see Lucas, Nadeau, and Sack (2006). (See color plate.)

Chan, 2012; Lloyd, 2011; Lucas & Shaw, 2008; Nick, 2012; Wasteneys, 2004). These differences in array organization lead to cell walls that constrain expansion in different ways. In this respect, interphase microtubules are an important regulatory component of the machinery that controls growth and morphology in plant cells.

Given the important role of the interphase array, much effort has been invested to determine how this array organizes without a centrosome (Ehrhardt, 2008;

Ehrhardt & Shaw, 2006; Pastuglia & Bouchez, 2007; Wasteneys & Ambrose, 2009). Although the dominant life phase of most plants lacks a discrete microtubule organizer, gamma-tubulin and the associated proteins of the core ring complex are present in plants (Nakamura, Ehrhardt, & Hashimoto, 2010). Live-cell imaging studies using fluorescent transgenes in *A. thaliana* have revealed that gamma-tubulin localizes transiently to the cell cortex and nucleates new polymers adjacent to the plasma membrane (Murata et al., 2005). At the resolution of light microscopy, the gamma-tubulin nucleation complexes localize primarily on existing microtubule lattices, and new polymers are formed parallel or at 40° angles to existing microtubules (Murata et al., 2005).

Microtubule organization emerges from the combination of the dynamic behavior of microtubules, the action of various MAPs, and the geometry of cell (Ehrhardt, 2008; Ehrhardt & Shaw, 2006; Pastuglia & Bouchez, 2007; Wasteneys & Ambrose, 2009). Plant genomes encode MAPs similar to those found in other organisms such as kinesins, katanins, MAP65's, EB1's, XMAP215, and CLASP (Ambrose, Shoji, Kotzer, Pighin, & Wasteneys, 2007; Burk, Liu, Zhong, Morrison, & Ye, 2001; Chan, Calder, Doonan, & Lloyd, 2003; Hamada, 2007; Jiang & Sonobe, 1993; Kirik et al., 2007; Reddy & Day, 2001; Whittington et al., 2001). The function of several of these proteins is similar to the animal or yeast counterpart. For example, the *A. thaliana* catalytic subunit of katanin severs microtubules from cortical nucleation sites, releasing the polymers for hybrid treadmilling (Nakamura et al., 2010). However, other plant MAPs appear to have acquired plant-specific functions. Mutation of two redundant *A. thaliana* MAP65 genes leads to plant growth defects but not dramatically disorganized microtubule arrays (Lucas et al., 2011). These data suggest that these proteins have additional functions beyond array organization (Lucas et al., 2011). Additionally, there are MAPs specific to the plant lineage, such as the MAP70 family (Korolev, Buschmann, Doonan, & Lloyd, 2007). In plants, there is much microtubule biology yet to be discovered and live-cell imaging of fluorescent protein fusions will help researchers uncover this science.

This review focuses on techniques used to image microtubules and MAPs in living cells of *A. thaliana*. The first step is to build transgenic plants that express a fluorescent protein fused to a protein of interest in the appropriate genetic background. After transgenic lines are generated, live-cell image data are collected with confocal microscopy and subsequently analyzed using different software packages.

15.1 PROTOCOLS

15.1.1 Building transgenic *A. thaliana*

Transgenes can be introduced into the *A. thaliana* genome quickly and easily using the bacterium *Agrobacterium tumefaciens* (Clough & Bent, 1998). *A. tumefaciens* is a soil-dwelling plant pathogen that can stably transfer a segment of plasmid DNA into the nuclear genome of several dicotyledonous plant species (Lee & Gelvin, 2008; Pitzschke & Hirt, 2010). There are multiple ways to transform *A. thaliana* with

A. tumefaciens, and in this section, we will discuss the basics of a common method. We will provide guidelines for selecting appropriate transgenic plant lines that express GFP and mCherry fusions to tubulin and MAPs.

A. tumefaciens causes crown gall disease in plants by transferring a piece of DNA called the T-DNA (transfer DNA) from the tumor inducing, T_i, plasmid into the infected plant's genome (Lee & Gelvin, 2008; Pitzschke & Hirt, 2010). In nature, the T-DNA encodes proteins that alter the physiology of the infected plant in order to benefit the bacteria (Lee & Gelvin, 2008; Pitzschke & Hirt, 2010). The T-DNA also encodes the majority of proteins required for virulence and horizontal (Lee & Gelvin, 2008; Pitzschke & Hirt, 2010). The T-DNA is surrounded by short "border" sequences that are recognized by the *A. tumefaciens* machinery that catalyzes the genetic transfer (Lee & Gelvin, 2008). For routine use in the laboratory, the *A. tumefaciens* T_i plasmid has been split into two smaller pieces: the T-DNA binary vector and the helper plasmid (Komari et al., 2006; Lee & Gelvin, 2008). The binary vector contains the border sequences between which the fluorescent fusion sequence and plant selectable marker cassette are placed. When choosing a selectable antibiotic resistance gene, it is critical to pick a marker different than any the plant may be already expressing. Horizontal transfer of the T-DNA requires the proteins encoded by the helper plasmid (Hoekema, Hirsch, Hooykaas, & Schileroort, 1983; Lee & Gelvin, 2008). Because the doubling time of *A. tumefaciens* is twice that of *E. coli*, it is more efficient to perform the molecular cloning steps in *E. coli*. To facilitate this, the binary vector was engineered to contain an *E. coli* origin of replication, bacterial selectable markers, and multiple cloning sites (Komari et al., 2006; Lee & Gelvin, 2008). Once the cassette is completed in *E. coli*, the binary vector is moved into an *A. tumefaciens* strain containing a helper plasmid. Various binary vector systems are publically available from the Biological Resource Center, www.arabidopsis.org (Komari et al., 2006).

15.1.1.1 Transgene construction

Building transgenes for plants is similar to building transgenes in other organisms. In *A. thaliana*, scientists do not have the ability to use homologous recombination to control the site of transgene insertion into the genome. Because the expression of the transgene can be affected by its position in the genome, it is especially important in *A. thaliana* to follow best practices when building transgenic plants. The full genomic sequence including the native promoter, exons, introns, and UTRs should be used for transgene construction. In practice, the promoter is usually defined as ~2.5 kb of upstream sequence. The full genome sequence of *A. thaliana* is publically available at www.arabidopsis.org. Whenever possible, the gene fusion should be transformed into a null mutant background to verify that it is functional. The constitutively active 35S cauliflower mosaic virus or ubiquitin promoters can be when protein overexpression is desirable. In plants, the most commonly used fluorescent reporters currently are mCherry, green, yellow, and cyan fluorescent proteins (GFP, YFP, and CFP). Originally, the *Aequorea victoria* GFP-coding sequence needed to be altered for use in *A. thaliana* because plant cells recognized a cryptic splice site in the GFP mRNA resulting in a truncated, nonfluorescent protein (Haseloff, 1999). Fluorescent reporters attached the N-terminus of alpha and beta tubulin-6 (the most commonly used isoforms) are effective when imaging microtubules in living *Arabidopsis* cells. Transgenic plants expressing alpha tubulin fusions are prone to show abnormal growth phenotypes such as twisting (Abe & Hashimoto, 2005; Ueda, Matsuyama, & Hashimoto, 1999). When driven from the 35S promoter, microtubules are not visible in root tissues (Ueda et al., 1999). However, microtubules in root cells can be visualized when constructs are driven from the ubiquitin1 promoter (Ambrose, Allard, Cytrynbaum, & Wasteneys, 2011).

15.1.1.2 Plant transformation

A. thaliana transformation can be accomplished by simply coating young flower buds with a solution of *A. tumefaciens*, sugar, and a surfactant (Clough & Bent, 1998). This "floral dipping" protocol has been optimized and described in greater detail in several articles (Davis, Hall, Millar, Darrah, & Davis, 2009; Desfeux, Clough, & Bent, 2000; Logemann, Birkenbihl, Ulker, & Somssich, 2006; Zhang, Henriques, Lin, Niu, & Chua, 2006). Here, a common protocol that works well in our hands is outlined. One of the more important aspects of this procedure is to start with healthy, robust plants with ample developing flower buds because transformation occurs in the ovule of the flower (Desfeux et al., 2000; Ye et al., 1999). To start, sow nine seeds in each of ten 4 × 4 in. pots. Grow the plants till the inflorescence stem (the flower stalk) is 6–8 in. tall, which can take 3–5 weeks depending on the genotype and growth conditions (the light regime and temperature are the primary factors). Once 6–8 in. tall, the floral stem is removed by cutting it near the base with a clean razor blade. The plants are watered with a balanced fertilizer after cutting. The removal of the primary floral stem releases multiple axillary floral meristems from apical dominance and results in the outgrowth of the axillary floral buds. This increases the total number of flowers available for transformation several fold (at least 4 ×).

When the axillary floral stems reach the height of 5–7 in. and are actively producing fruit, they are gently submerged (dipped) in a suspension of *A. tumefaciens* in 5% sucrose and 0.02% of the surfactant Silwet L-77 (Clough & Bent, 1998). The Silwet concentration can be varied from 0.005% to 0.02%. The first step in producing the *A. tumefaciens* suspension is to grow a 500-ml culture in Luria broth (LB) at 28 °C for 48 h (*A. tumefaciens* grows slower than *E. coli*). This culture is then centrifuged at 4000 rpm for 20 min at room temperate. The resulting pinkish pellet is gently resuspended in the sucrose/Silwet solution by slowly pipetting up and down with a large bore pipette. The young pliable floral stems are then gently submerged in the bacterial suspension for 30–60 s. When dipping the plants, one should avoid wetting the rosette leaves and breaking the floral stems. Placing the plants in a humid low-light environment, like a covered bucket or tray, for 2–3 days after dipping the plants increases the transformation success (Clough & Bent, 1998). After 48–72 h, the plants are slowly transitioned to normal growth conditions over the period of a day. The plants will continue to grow and flower after dipping. *A. thaliana* readily self-pollinates, and generating large numbers of seeds to screen for antibiotic resistance is simple. To aid seed collection once the plants stop producing new flowers, the flowering stems are gently grouped together with string and tied to a bamboo skewer placed in the center of the pot. Progeny seeds are collected from the dipped plants after the fruit are dry and light brown in color. These collected seeds are usually known as the first transformed generation, the T1 seeds. After collection, all *A. thaliana* seeds should be dried in a desiccator for at least 1 week before sowing or the germination frequency will be low.

A. thaliana produces large volumes of seed, and so recovering many transformed lines from each pot of dipped plants is standard. In general, the transformation efficiency of floral dipping ranges from 1% to 3% (Clough & Bent, 1998). The efficiency of the transformation can be increased by treating the *A. tumefaciens* cultures with a virulence buffer containing acetosyringone for 2 h prior to plant transformation. Acetosyringone is a chemical released by dicotyledonous plants upon wounding that induces the virulence of the bacterium (Winans, 1991). Cotransformation with multiple transgenes can be accomplished by dipping the plants in an *A. tumefaciens* suspension containing different transgenic bacteria, as opposed to creating a bacterial strain harboring multiple binary vectors (Davis et al., 2009).

15.1.1.3 Transgenic plant recovery

Transgenic lines are recovered by growing T1 seeds on selective antibiotic media. The antibiotics kanamycin and hygromycin and the herbicide Basta are commonly used for selection. For selection, antibiotics are added to sterile media poured into plates; the herbicide is applied when watering young seedlings grown on soil. T1 seeds are surface sterilized and aseptically sown in Petri plates containing antibiotic media. After 10–18 days, antibiotic-resistant seedlings are visible because they have expanded green leaves and long roots. Resistant seedlings are removed from the plate and transplanted to soil and then grown to maturity for T2 seed collection (Fig. 15.2). In general, experiments are carried out with T2 seedlings or the subsequent T3 plants. At least three independent transgenic lines (lines derived from separate pots of dipped plants) should be used in every experiment because the T-DNA insertion is random.

There are multiple ways to surface sterilize seeds to prepare them for aseptic culture. Most methods involve bleach, ethanol, hydrogen peroxide, or chlorine gas. In our hands, the most efficient protocol is to place seed on a double layer of filter paper in a sterile laminar flow hood and then rinse the seed with an aqueous solution of 5% hydrogen peroxide and 75% ethanol. The hydrogen peroxide/ethanol solution is pipetting directly onto the seeds on the filter paper. About 2 ml of solution should be used for about 250 μl of seeds on 100-mm round filter paper. Depending on the flow rate of the laminar hood, the ethanol/peroxide solution will fully evaporate in 5–20 min. After the seeds and filter paper are dry, the seeds are sown at high density (10–20 seeds/cm^2) by sprinkling them from the dried filter paper onto the plates. In addition to the selective antibiotics, 100–500 μg/ml carbenicillin or 100 μg/ml vancomycin can be added to the media to halt the proliferation of any *A. tumefaciens* that escaped sterilization of the seed surface.

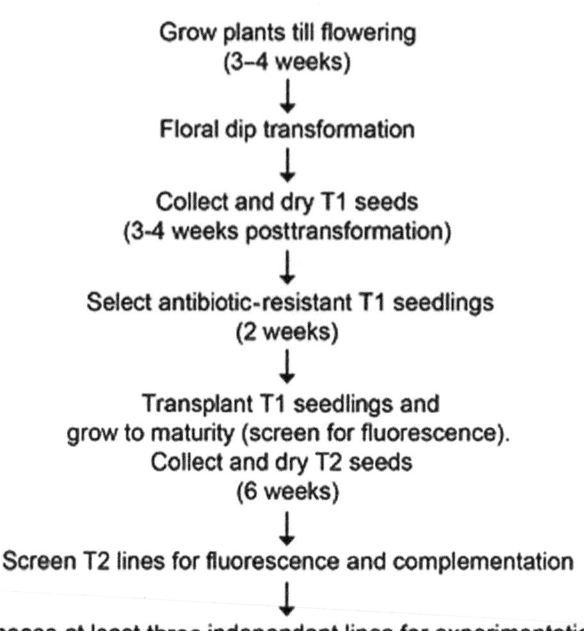

FIGURE 15.2

Timeline for *A. thaliana* transformation and subsequent recovery of fluorescent T2 seed lines for experimentation.

The outcome of the selection process is most obvious 10–18 days after sowing. The first set of true leaves will emerge on the antibiotic-resistant plants (botanically, the cotyledons are embryonic leaves and not "true" leaves). Resistant plants produce true leaves and long roots, while the nonresistant plants are stunted and do not develop true leaves or long roots. Resistant plants are darker green in color in comparison to the nonresistant seedlings. Resistant T1 seedlings are gently removed from the agar plates with forceps and carefully transplanted to soil. To acclimate the seedlings to life on soil, the plants should be covered with a clear plastic dome for 1 week after transplanting to maintain high humidity around the seedlings. The dome is progressively lifted away after a week. The plants are grown to maturity, allowed to set seed, and then T2 seeds are collected for experimentation. Multiple antibiotic-resistant T2 lines should be collected for the next screening step.

15.1.1.4 Screening T2 seed lines

At least three different transgenic lines should be used in every experiment to compensate for the random insertion of the T-DNA. The construct should be transformed into a null mutant background whenever possible, and then the T2 generation should be tested for complementation to demonstrate the proper functioning of a transgene.

Lines that do not rescue mutant phenotypes and/or show abnormal phenotypes should not be used in experiments. For microscopy experiments, fluorescent protein fusions need to accumulate at detectable levels. Ultimately, it is preferable to choose multiple transgenic lines that express the least amount of visible fluorescent protein and compliment the null mutant phenotype.

The location of T-DNA insertion into the genome is uncontrolled and therefore the T-DNA can act as a mutagen if inserted into a coding or regulatory sequence, or if the addition of the fluorescent tag changes protein function. When creating fluorescent protein fusions, a typical goal is to observe the protein in its native environment and so T-DNA-induced mutation is undesirable. Therefore, it is important to screen multiple transgenic lines to find those that do not display unexpected phenotypes. Mutant phenotypes can be observed in the T1 and T2 generation. We have observed organ twisting, swelling, dwarfing, and reduced fertility when GFP-tagged versions of tubulin and some MAPs are expressed. Observation of other nonwild-type phenotypes may indicate that the transgene has caused a genetic mutation.

The expression of the transgene will vary from line to line due to the number of transgenes inserted and the position of the transgene relative to transcriptional promoters, enhancers, and suppressors. This phenomenon is referred to as the "positional" effect. In some lines, no transgene expression will be observed. Screening for fluorescence can be accomplished with an epifluorescence or confocal microscope. When screening for mCherry transgenes, it is useful to remove the red chlorophyll autofluorescent signal with appropriate filters. Finally, some MAPs are very difficult to detect because they are naturally expressed at very low levels and/or have a high cytoplasmic localization.

15.1.2 Sample preparation

Many cellular and developmental processes that involve microtubules occur in the *A. thaliana* seedling thereby making the seedling an excellent system for experimentation. Upon germination, growth of the seedling stem (hypocotyl) is driven almost exclusively by cell expansion (Derbyshire, Findlay, McCann, & Roberts, 2007; Refregier, Pelletier, Jaillard, & Hofte, 2004). Cell divisions primarily occur at the apical meristems at the shoot and root tip. The root meristem is an accessible tissue and the patterns of cell division are predictable (Sedbrook & Kaloriti, 2008). The shoot apical meristem is shielded from view by young leaves that make live-cell imaging challenging, but possible (Hamant et al., 2008; Heisler et al., 2010). Stereotyped asymmetric divisions occur during the formation of stomata in the developing leaf epidermis (Lucas et al., 2006). Concurrent with stomatal formation, small box-like epidermal cells develop into large puzzle-piece-shaped cells (Fu, Gu, Zheng, Wasteneys, & Yang, 2005). All of these processes are available to researchers because the entire *A. thaliana* seedling fits between a standard slide and square cover glass. Next, we will discuss general sample preparation techniques for the *A. thaliana* seedling.

Plant shoots typically grow upward toward the light, and therefore it can be a fairly traumatic life event for a seedling to be sandwiched horizontally between a slide and a coverslip. The impact of this situation should be minimized to avoid

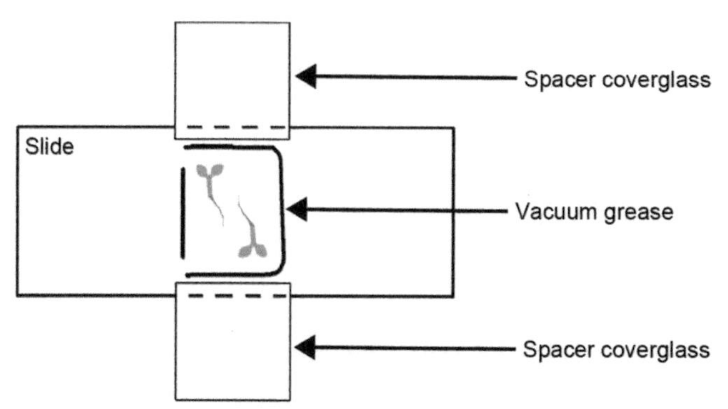

FIGURE 15.3

Diagram of a microscope slide before mounting the top cover glass to demonstrate the position of the spacers. Gaps in the vacuum grease allow for the displacement of some liquid mounting media while positioning the cover glass. (See color plate.)

damaging the seedling and altering its physiology. The distance between the slide and the coverslip is an important factor to consider and the width of the plant organ being observed dictates this distance. Cover glass spacers can be used to avoid squishing the plant while gently mounting the cover glass (Fig. 15.3). Spacers are made by gluing together multiple coverslips to achieve the proper distance between the slide and cover glass. We have found that the ideal spacer for 4- to 5-day-old light-grown Col-0 seedlings is two #2 coverslips glued together, for etiolated hypocotyls is a single #1 cover glass, and for root tips is a #0 slip. The coverslip is secured to the slide with vacuum grease to prevent evaporation of the mounting media during imaging, to maintain the space between the slide and cover glass, and to hold the cover glass in place when using an inverted microscope.

Before being moved to slides, seedlings are typically grown on sterile half-strength Murishage and Skoog (0.5 × MS) agar plates under the appropriate environmental conditions as determined by the experimental setup. To minimize osmotic disruptions, seedlings should be mounted in 0.5 × MS liquid media on the slide. Tap water should never be used as it can contain herbicides and fertilizers from agricultural runoff that vary seasonally. Seedlings should be allowed to acclimate on slides for at least 15 min before imaging. The media do not need to be exchanged during experiments lasting less than 2 h, and new media should be perfused onto the slide during longer-term experiments (see Buschmann, Sambade, Pesquet, Calder, & Lloyd, 2010; Shaw, 2006 for more information about perfusion chambers).

15.1.3 Imaging

Seedlings are thick specimens that produce many autofluorescent compounds that contribute extensive out-of-focus fluorescent information that substantially degrades image quality. Many plant biologists use confocal microscopes to capture thin optical sections, thereby excluding the out-of-focus information. TIRF microscopy is also an option for plant biologists needing optical sectioning. Chlorophyll is the main pigment that autofluoresces red from ~550 to 750 nm when excited with either the standard 488 or 543-nm laser lines. Band pass filters or acousto-optical tunable filters are required to remove the autofluorescence when imaging mCherry.

Each experiment requires different microscope and camera/PMT settings that must be determined prior to data collection. Many confocal microscope software packages have generic settings for each laser and/or probe that are helpful starting points. The magnification and imaging interval (single time point or time lapsed) should be determined first. When imaging, the usual goal is to collect as much signal over background noise as possible, while using the least amount of laser light possible in order to avoid bleaching the fluorescent protein or stressing the living specimen. When maximizing signal to noise on confocal microscopes, the PMT settings and laser intensity are linked. Increasing the PMT gain amplifies all signals, including the background noise. Frame and line averaging can be used to reduce the background noise. Increasing the laser intensity usually generates a higher signal-to-noise ratio, but bleaches the probe more quickly. The amount of acceptable bleaching is often dictated by the imaging interval. Probe bleaching is not a concern when imaging at a single time point and so laser intensity can be increased, rather than PMT gain. Photobleaching and plant health must be considered when collecting a time-lapsed image series. One must usually sacrifice some image contrast by decreasing the laser intensity and increasing the gain to image over time. It is difficult to image fluorescent reporter fusions that are not abundant. One strategy to image dim reporters is to sum or average images taken over time. The specimen is likely to drift while imaging, and this can be corrected after image acquisition with software such as the StackReg plugin for ImageJ (http://bigwww.epfl.ch/thevenaz/stackreg/). Deconvolution software can also be helpful to remove some of the background fluorescence from the data sets.

To measure the dynamics of microtubules and MAPs, we routinely used a 63 × water immersion objective and electronically zoomed 2–3 times so that a single light-grown hypocotyl cell filled the field of view. When imaging microtubules and MAPs, we have found it useful to measure the signal intensity using a line scan tool. To start, draw a line across the entire cell and quantify the fluorescence. Next, adjust the PMT gain so that almost no pixels are saturated, and ensure that the baseline is slightly above zero by adjusting the offset. Now measure the signal of a few single and bundled polymers by drawing a line across several microtubule structures. To ensure a wide and linear dynamic range, an incremental stepwise increase in signal should be observed due to the bundled microtubules (Eisinger, Kirik, Lewis, Ehrhardt, & Briggs, 2012). A linear increase in signal is especially important when quantifying fluorescent signal over time, such as in fluorescence recovery after photobleaching experiments. However, because some microtubule bundles are composed of numerous microtubules, we have often found it necessary to saturate the fluorescent signal of higher order bundles in order to distinguish single microtubules. To ensure that sufficient signal over noise is collected, it is important that the pixel value of single microtubules is at least 2–3 times the background value.

15.2 MATERIALS

15.2.1 Sterile media for sterile plant growth

Add 2.2 g MS powder, 500 μl Gamborg's micronutrients, and 0.5 g 2-(*N*-morpholino)ethanesulfonic acid (MES) to 1 l distilled water. Adjust the pH to 5.7 with potassium hydroxide and then add agar that has been tested for plant cell culture. For many experiments, 0.8% agar (w/v) is suitable. Less rigid (0.6%) plates are recommended when selecting T1 seedlings to ease the removal of the seedling roots embedded in the agar. Sterilize the media by autoclaving for at least 15 min and then cool the media to 50 °C in water bath. Supplements (50 μg/ml kanamycin, 25 μg/ml hygromycin B, and 500 μg/ml carbenicillin) can be added to the media after it has cooled. Pour the media into sterile Petri plates and allow to solidify.

15.2.2 *A. tumefaciens* suspension for plant transformation

Grow a 500-ml *A. tumefaciens* bacterial culture in LB supplemented with the appropriate antibiotic for selection determined by the T_i plasmid. The culture should be grown for 48 h at 28–30 °C. The common *A. tumefaciens* strain, GV3101, is resistant to both rifamycin (10 μg/ml) and gentamycin (25–50 μg/ml). Pellet the bacteria by centrifugation for 20 min at 4000 rpm at room temperature (the resulting pellet will be pinkish in color). Resuspend the bacterial pellet in 500 ml of virulence buffer (recipe below). If performing a cotransformation with two different transgenes, resuspend each pellet in 250 ml of buffer and then combine the resuspended pellets. To induce virulence, shake the cultures at 28–30 °C at 250 rpm for 2 h. Pellet the cultures by centrifugation as described above and then resuspend them in 500 ml of 5% sucrose (sugar from the grocery store is adequate) and 0.002% Silwet L-77.

15.2.2.1 *Virulence buffer*

85 ml of 20 × AT salts (recipe below)
17 ml of 50% Glucose (filter sterilized)
34 ml of 1 M MES buffer, pH 5.6
850 μl of 100 mM Phosphate
850 μl of 200 mM Acetosyringone
Add distilled water to 1.7 l.

20 × AT salts

80 g Ammonium sulfate
6.4 g Magnesium sulfate heptahydrate
0.4 g Calcium chloride dihydrate
0.48 g Manganese sulfate monohydrate
Add distilled water to 2 l and then autoclave.

200 mM Acetosyringone

780 mg Acetosyringone
20 ml of DMSO
Sterilize using a 0.2 μm filter and then store at −20 °C in 900 μl aliquots.

CONCLUDING COMMENTS

Plants are valuable members of our global ecosystem and many scientists are actively working to understand the biology of these important organisms. Microtubules and MAPs are the subject of much research because they play important roles in plant growth and development. The ability to observe proteins tagged with fluorescent proteins in intact living plants with confocal microscopes has helped the field progress tremendously. The guidelines here describe how to generate and image *A. thaliana* plants expressing fluorescent fusion proteins.

Acknowledgment

I am grateful for support from Drs Michelle Bezanson and Claire Walczak.

References

Abe, T., & Hashimoto, T. (2005). Altered microtubule dynamics by expression of modified alpha-tubulin protein causes right-handed helical growth in transgenic Arabidopsis plants. *The Plant Journal*, *43*, 191–204.

Ambrose, J. C., Allard, J., Cytrynbaum, E., & Wasteneys, G. O. (2011). A CLASP-modulated cell edge barrier mechanism drives cell-wide cortical microtubule organization in Arabidopsis. *Nature Communications*, *2*, 430.

Ambrose, J. C., Shoji, T., Kotzer, A. M., Pighin, J. A., & Wasteneys, G. O. (2007). The Arabidopsis CLASP gene encodes a microtubule-associated protein involved in cell expansion and division. *The Plant Cell*, *19*, 2763–2775.

Baskin, T. I., Wilson, J. E., Cork, A., & Williamson, R. E. (1994). Morphology and microtubule organization in Arabidopsis roots exposed to oryzalin or taxol. *Plant & Cell Physiology*, *35*, 935–942.

Burk, D. H., Liu, B., Zhong, R., Morrison, W. H., & Ye, Z. H. (2001). A katanin-like protein regulates normal cell wall biosynthesis and cell elongation. *The Plant Cell*, *13*, 807–827.

Buschmann, H., Sambade, A., Pesquet, E., Calder, G., & Lloyd, C. W. (2010). Microtubule dynamics in plant cells. *Methods in Cell Biology*, *97*, 373–400.

Chan, J. (2012). Microtubule and cellulose microfibril orientation during plant cell and organ growth. *Journal of Microscopy*, *247*, 23–32.

Chan, J., Calder, G. M., Doonan, J. H., & Lloyd, C. W. (2003). EB1 reveals mobile microtubule nucleation sites in Arabidopsis. *Nature Cell Biology*, *5*, 967–971.

Clough, S. J., Bent, A. F. (1998). Floral dip: a simplified method for Agrobacterium-mediated transformation of Arabidopsis thaliana. *The Plant Journal*, *16*, 735–743.

Corson, F., Hamant, O., Bohn, S., Traas, J., Boudaoud, A., & Couder, Y. (2009). Turning a plant tissue into a living cell froth through isotropic growth. *Proceedings of the National Academy of Sciences of the United States of America*, *106*, 8453–8458.

Crowell, E. F., Gonneau, M., Stierhof, Y. D., Hofte, H., & Vernhettes, S. (2010). Regulated trafficking of cellulose synthases. *Current Opinion in Plant Biology*, *13*, 700–705.

Davis, A. M., Hall, A., Millar, A. J., Darrah, C., & Davis, S. J. (2009). Protocol: Streamlined sub-protocols for floral-dip transformation and selection of transformants in Arabidopsis thaliana. *Plant Methods*, *5*, 3.

Derbyshire, P., Findlay, K., McCann, M. C., & Roberts, K. (2007). Cell elongation in Arabidopsis hypocotyls involves dynamic changes in cell wall thickness. *Journal of Experimental Botany*, *58*, 2079–2089.

Desfeux, C., Clough, S. J., & Bent, A. F. (2000). Female reproductive tissues are the primary target of Agrobacterium-mediated transformation by the Arabidopsis floral-dip method. *Plant Physiology*, *123*, 895–904.

De Smet, I., & Beeckman, T. (2011). Asymmetric cell division in land plants and algae: The driving force for differentiation. *Nature Reviews. Molecular Cell Biology*, *12*, 177–188.

Ehrhardt, D. W. (2008). Straighten up and fly right: Microtubule dynamics and organization of non-centrosomal arrays in higher plants. *Current Opinion in Cell Biology*, *20*, 107–116.

Ehrhardt, D. W., & Shaw, S. L. (2006). Microtubule dynamics and organization in the plant cortical array. *Annual Review of Plant Biology*, *57*, 859–875.

Eisinger, W. R., Kirik, V., Lewis, C., Ehrhardt, D. W., & Briggs, W. R. (2012). Quantitative changes in microtubule distribution correlate with guard cell function in Arabidopsis. *Molecular Plant*, *5*, 716–725.

Fu, Y., Gu, Y., Zheng, Z., Wasteneys, G., & Yang, Z. (2005). Arabidopsis interdigitating cell growth requires two antagonistic pathways with opposing action on cell morphogenesis. *Cell*, *120*, 687–700.

Hamada, T. (2007). Microtubule-associated proteins in higher plants. *Journal of Plant Research*, *120*, 79–98.

Hamant, O., Heisler, M. G., Jönsson, H., Krupinski, P., Uyttewaal, M., Bokov, P., et al. (2008). Developmental patterning by mechanical signals in Arabidopsis. *Science*, *322*, 1650–1655. http://dx.doi.org/10.1126/science.1165594.

Hardham, A. R., & Gunning, B. E. (1978). Structure of cortical microtubule arrays in plant cells. *The Journal of Cell Biology*, *77*, 14–34.

Haseloff, J. (1999). GFP variants for multispectral imaging of living cells. *Methods in Cell Biology*, *58*, 139–151.

Heisler, M. G., Hamant, O., Krupinski, P., Uyttewaal, M., Ohno, C., Jönsson, H., Traas, J., Meyerowitz, E. M. (2010). Alignment between PIN1 polarity and microtubule orientation in the shoot apical meristem reveals a tight coupling between morphogenesis and auxin transport. *PLoS Biology, 19*, 8(10).

Hoekema, A., Hirsch, P. R., Hooykaas, P. J. J., & Schileroort, R. A. (1983). A binary plant vector strategy based on separation of vir- and T-region of the Agrobacterium tumefaciens Ti plasmid. *Nature*, *303*, 179–180.

Jiang, C. J., & Sonobe, S. (1993). Identification and preliminary characterization of a 65kDa higher-plant microtubule-associated protein. *Journal of Cell Science*, *105*, 891–901.

Kirik, V., Herrmann, U., Parupalli, C., Sedbrook, J. C., Ehrhardt, D. W., & Hulskamp, M. (2007). CLASP localizes in two discrete patterns on cortical microtubules and is required for cell morphogenesis and cell division in Arabidopsis. *Journal of Cell Science*, *120*, 4416–4425.

Komari, T., Takakura, Y., Ueki, J., Kato, N., Ishida, Y., & Hiei, Y. (2006). Binary vectors and super-binary vectors. *Methods in Molecular Biology*, *343*, 15–41.

Korolev, A. V., Buschmann, H., Doonan, J. H., & Lloyd, C. W. (2007). AtMAP70-5, a divergent member of the MAP70 family of microtubule-associated proteins, is required for anisotropic cell growth in Arabidopsis. *Journal of Cell Science*, *120*, 2241–2247.

Lee, L. Y., Gelvin, S. B. (2008). T-DNA binary vectors and systems. *Plant Physiology*, *146*, 325–332.

Lloyd, C. (2011). Dynamic microtubules and the texture of plant cell walls. *International Review of Cell and Molecular Biology*, *287*, 287–329.

Logemann, E., Birkenbihl, R. P., Ulker, B., & Somssich, I. E. (2006). An improved method for preparing Agrobacterium cells that simplifies the Arabidopsis transformation protocol. *Plant Methods*, *2*, 16.

Lucas, J. R., Courtney, S., Hassfurder, M., Dhingra, S., Bryant, A., & Shaw, S. L. (2011). Microtubule-associated proteins MAP65-1 and MAP65-2 positively regulate axial cell growth in etiolated Arabidopsis hypocotyls. *The Plant Cell*, *23*, 1889–1903.

Lucas, J. R., Nadeau, J. A., & Sack, F. D. (2006). Microtubule arrays and Arabidopsis stomatal development. *Journal of Experimental Botany*, *57*, 71–79.

Lucas, J., & Shaw, S. L. (2008). Cortical microtubule arrays in the Arabidopsis seedling. *Current Opinion in Plant Biology*, *11*, 94–98.

Muller, S., Wright, A. J., & Smith, L. G. (2009). Division plane control in plants: New players in the band. *Trends in Cell Biology*, *19*, 180–188.

Murata, T., Sonobe, S., Baskin, T. I., Hyodo, S., Hasezawa, S., Nagata, T., et al. (2005). Microtubule-dependent microtubule nucleation based on recruitment of gamma-tubulin in higher plants. *Nature Cell Biology*, *7*, 961–968.

Nakamura, M., Ehrhardt, D. W., & Hashimoto, T. (2010). Microtubule and katanin-dependent dynamics of microtubule nucleation complexes in the acentrosomal Arabidopsis cortical array. *Nature Cell Biology*, *12*, 1064–1070.

Nick, P. (2012). Microtubules and the tax payer. *Protoplasma*, *249*(Suppl. 2), S81–S94.

Pastuglia, M., & Bouchez, D. (2007). Molecular encounters at microtubule ends in the plant cell cortex. *Current Opinion in Plant Biology*, *10*, 557–563.

Pitzschke, A., & Hirt, H. (2010). Mechanism of MAPK-targeted gene expression unraveled in plants. *Cell Cycle*, *9*, 18–19.

Rasmussen, C. G., Humphries, J. A., & Smith, L. G. (2011). Determination of symmetric and asymmetric division planes in plant cells. *Annual Review of Plant Biology*, *62*, 387–409.

Reddy, A. S., & Day, I. S. (2001). Kinesins in the Arabidopsis genome: A comparative analysis among eukaryotes. *BMC Genomics*, *2*, 2.

Refregier, G., Pelletier, S., Jaillard, D., & Hofte, H. (2004). Interaction between wall deposition and cell elongation in dark-grown hypocotyl cells in Arabidopsis. *Plant Physiology*, *135*, 959–968.

Sedbrook, J. C., & Kaloriti, D. (2008). Microtubules, MAPs and plant directional cell expansion. *Trends in Plant Science*, *13*, 303–310.

Shaw, S. L. (2006). Imaging the live plant cell. *The Plant Journal*, *45*, 573–598.

Smith, L. G. (2001). Plant cell division: Building walls in the right places. *Nature Reviews. Molecular Cell Biology*, *2*, 33–39.

Smith, L. G., & Oppenheimer, D. G. (2005). Spatial control of cell expansion by the plant cytoskeleton. *Annual Review of Cell and Developmental Biology*, *21*, 271–295.

Szymanski, D. B., & Cosgrove, D. J. (2009). Dynamic coordination of cytoskeletal and cell wall systems during plant cell morphogenesis. *Current Biology*, *19*, R800–R811.

Ueda, K., Matsuyama, T., & Hashimoto, T. (1999). Visualization of microtubules in living cells of transgenic Arabidopsis thaliana. *Protoplasma*, *206*, 201–206.

Van Damme, D., & Geelen, D. (2008). Demarcation of the cortical division zone in dividing plant cells. *Cell Biology International*, *32*, 178–187.

Van Damme, D., Vanstraelen, M., & Geelen, D. (2007). Cortical division zone establishment in plant cells. *Trends in Plant Science*, *12*, 458–464.

Wasteneys, G. O. (2004). Progress in understanding the role of microtubules in plant cells. *Current Opinion in Plant Biology*, *7*, 651–660.

Wasteneys, G. O., & Ambrose, J. C. (2009). Spatial organization of plant cortical microtubules: Close encounters of the 2D kind. *Trends in Cell Biology*, *19*, 62–71.

Wasteneys, G. O., & Fujita, M. (2006). Establishing and maintaining axial growth: Wall mechanical properties and the cytoskeleton. *Journal of Plant Research*, *119*, 5–10.

Whittington, A. T., Vugrek, O., Wei, K. J., Hasenbein, N. G., Sugimoto, K., Rashbrooke, M. C., et al. (2001). MOR1 is essential for organizing cortical microtubules in plants. *Nature*, *411*, 610–613.

Winans, S. C. (1991). An Agrobacterium two-component regulatory system for the detection of chemicals released from plant wounds. *Molecular Microbiology*, *5*, 2345–2350.

Ye, G. N., Stone, D., Pang, S. Z., Creely, W., Gonzalez, K., Hinchee, M. (1999). Arabidopsis ovule is the target for Agrobacterium in planta vacuum infiltration transformation. *The Plant Journal*, *19*, 249–257.

Zhang, X., Henriques, R., Lin, S. S., Niu, Q. W., & Chua, N. H. (2006). Agrobacterium-mediated transformation of Arabidopsis thaliana using the floral dip method. *Nature Protocols*, *1*, 641–646.

CHAPTER

16 Investigating Tubulin Posttranslational Modifications with Specific Antibodies

Maria M. Magiera and Carsten Janke
Institut Curie, CNRS UMR3306, INSERM U1005, Orsay, France

CHAPTER OUTLINE

Introduction **248**
16.1 Observations **249**
16.1.1 Impact of Cell Fixation on Antibody Labeling 249
16.1.2 The Importance of Antibody Concentration in Analyzing MT PTMs in Cells 252
16.1.3 Antibody and Sample Concentration in Immunoblot Analysis 253
16.2 Materials and Methods **255**
16.2.1 Antibodies Detecting Tubulin PTMs 255
16.2.1.1 Acetylation 255
16.2.1.2 Tyrosination/detyrosination/Δ2-tubulin 256
16.2.1.3 Polyglutamylation 258
16.2.1.4 Polyglycylation 258
16.2.2 Cell Culture and Fixation 259
16.2.2.1 Cell Culture and Transfection 259
16.2.2.2 Cold Methanol Fixation 259
16.2.2.3 PFA Fixation 259
16.2.2.4 DSP–PFA Fixation 259
16.2.3 Immunocytochemistry on Fixed Cells 260
16.2.4 Microscopy and Image Acquisition 260
16.2.5 Electrophoresis and Immunoblot 260
16.3 Buffer Composition **261**
16.3.1 Cell Fixation 261
16.3.1.1 PFA Method 261
16.3.1.2 DSP–PFA Method 261
16.3.2 Gel Electrophoresis and Immunoblotting 262
16.3.2.1 SDS-PAGE for α/β-tubulin Separation 262

Methods in Cell Biology, Volume 115

ISSN 0091-679X
http://dx.doi.org/10.1016/B978-0-12-407757-7.00016-5

Concluding Remarks 263
Acknowledgments 263
References 264

Abstract

Microtubules play highly diverse and essential roles in every eukaryotic cell. While built from conserved dimers of α- and β-tubulin, microtubules can be diversified by posttranslational modifications in order to fulfill specific functions in cells. The tubulin posttranslational modifications: acetylation, detyrosination, polyglutamylation, and polyglycylation play important roles in microtubule functions; however, only little functional and mechanistic insight has been gained so far. The modification state of microtubules can be visualized with specific antibodies. A drawback is that detailed information about the specificities and limitations of these antibodies are not easily accessible in the literature. We provide here a comprehensive description of the currently available set of antibodies specific to tubulin modifications. Focusing on glutamylation antibodies, we discuss specific protocols that allow using these antibodies to gain semi-quantitative information on the levels and distribution of tubulin modifications in immunocytochemistry and immunoblot.

INTRODUCTION

Microtubules (MTs) are core components of a complex filamentous system called the cytoskeleton. Present in every eukaryotic cell, they display a variety of functions including the regulation of cell shape, polarity and motility, cell division, as well as intracellular transport. Moreover, MTs are essential regulators of cell differentiation, as for instance in neurons. They are also the major building blocks of cilia, flagella, and centrosomes, where they assemble into highly complex structures called axonemes and centrioles. Considering the diversity of MT structures and associated functions, it is obvious that mechanisms are needed to create different MT identities.

In the past, great advances have been made in understanding the implication of MT-associated proteins (MAPs) and molecular motors in the functional specialization of MT populations. Various MAPs decorate discrete populations of MTs in a single cell, and motors selectively bind to MT subspecies (e.g., certain MT populations of the mitotic spindle, or axonal vs. dendritic MTs in neurons). However, in many of these well-studied systems, the signals that confer specificity to these specific MAP–MT interactions are not completely understood.

Two mechanisms have been considered as sources of MT diversity: the incorporation of different tubulin isoforms, also known as isotypes, and the posttranslational modifications (PTMs) of MTs. Of these two mechanisms, PTMs are the regulators that are specifically added to already assembled MTs and can be easily controlled in a spatial and temporal manner. It is not so for tubulin isoforms that can only slowly, by changing the gene expression profiles, get incorporated into newly formed MTs. Thus, tubulin PTMs, which include acetylation, detyrosination, polyglutamylation, and polyglycylation, are the prime candidates for fine-tuning MT functions (Box 16.1; reviews: Janke & Bulinski, 2011; Luduena, 2013).

Despite the variety of functions tubulin PTMs are expected to fulfill, current research in the MT field often fails to correctly assess the importance or even the presence of these modifications. Functional studies have been difficult because most of the enzymes catalyzing tubulin PTMs had not been discovered until recently (Akella et al., 2010; Janke et al., 2005; Kimura et al., 2010; Rogowski et al., 2009, 2010; Shida, Cueva, Xu, Goodman, & Nachury, 2010; van Dijk et al., 2007; Wloga et al., 2009). Moreover, tubulin PTMs are hard to quantify or even detect. For instance, glutamylated peptides are often not detected in classical mass spectrometry, a problem that has been overcome for purified tubulin, but that requires relatively large amounts of protein (Redeker, 2010). Furthermore, modifications do not significantly alter the migration properties of tubulin on a sodium dodecyl sulfate polyacrylamide gel electrophoresis (SDS-PAGE); hence, gel-shift cannot be used to assay these modifications.

Thus, the most reliable way for distinguishing differentially modified MTs and tubulin in tissues, cells, and by immunoblot is PTM-specific antibodies. While some of these antibodies have proven to be excellent tools, precise data about their specificities and limitations have not been comprehensively reported. Here, we present a collection of well-analyzed antibodies specific to tubulin PTMs, together with appropriate protocols that allow using them in a semi-quantitative fashion in immunofluorescence on cells, as well as in immunoblot. We illustrate the opportunities and limitations of the described methods based on the example of glutamylation-specific antibodies.

16.1 OBSERVATIONS

16.1.1 Impact of cell fixation on antibody labeling

We have found that the cell fixation method is crucial for the correct preservation of the MT cytoskeleton, as well as for the quality of detection with some of the modification-specific antibodies. Neither a paraformaldehyde (PFA)–sucrose fixation nor the fixation with cold methanol provides satisfying results, especially for the staining of MT modifications of certain structures, as for instance the midbody. Therefore, we use a method in which the cytoskeleton is prefixed with a bifunctional protein cross-linker, dithiobis succinimidyl propionate (DSP; Bell & Safiejko-Mroczka, 1995). To demonstrate the impact of these three methods on immunolabeling of cultured cells, we have fixed HeLa cells that were cultured on fibronectin-coated coverslips and stained the cells with polyE and 12G10 antibodies. Cells in anaphase, representing a midbody structure, were observed (Fig. 16.1).

Using the DSP fixation method, we have previously shown that midbody MTs carry long glutamate side chains (Lacroix et al., 2010); however, this specific labeling is only seen when cells are fixed with the DSP–PFA method. In contrast, fixing

Box 16.1 CHARACTERISTICS OF TUBULIN PTMs

Acetylation

The best-analyzed acetylation site of α-tubulin is lysine residue number 40 (LeDizet & Piperno, 1987). Recently, additional acetylation sites have been found in a proteomic study (Choudhary et al., 2009) but not been confirmed otherwise. Lysine-40 acetylation is catalyzed by the acetyl transferases Atat-1 (Mec-17) and Atat-2 (Akella et al., 2010; Shida et al., 2010); however, other acetyl transferases, such as Elp3 (Creppe et al., 2009), are also involved in tubulin acetylation. Tubulin acetylation is reversed by the deacetylases HDAC6 (Hubbert et al., 2002) and SIRT2 (North, Marshall, Borra, Denu, & Verdin, 2003). Because of its localization at the inside of MTs (Soppina, Herbstman, Skiniotis, & Verhey, 2012), lysine-40 acetylation does not directly regulate motor traffic on MTs (Walter, Beranek, Fischermeier, & Diez, 2012), but plays an important role in MT assembly (Cueva, Hsin, Huang, & Goodman, 2012; Topalidou et al., 2012).

Detyrosination/Tyrosination

Detyrosination removes the terminal, gene-encoded tyrosine from α-tubulin (Hallak, Rodriguez, Barra, & Caputto, 1977; Valenzuela et al., 1981). This modification is reversible (Arce, Rodriguez, Barra, & Caputo, 1975; Raybin & Flavin, 1975) and returosination is catalyzed by a well-characterized enzyme, tubulin tyrosine ligase (TTL; Ersfeld et al., 1993). Strikingly, the enzymes catalyzing detyrosination have remained unknown despite great efforts to identify them. Detyrosinated tubulin is sometimes referred to as Glu-tubulin, which often leads to confusion with polyglutamylated tubulin.

Following detyrosination, the C-terminal tail of α-tubulin can be further shortened by the removal of the penultimate glutamate residue generating Δ2-tubulin (Paturle-Lafanechere et al., 1991). This deglutamylation reaction is catalyzed by enzymes from the cytosolic carboxypeptidase (CCP) family, which are also involved in the removal of posttranslationally added polyglutamylation (see below; Rogowski et al., 2010). Further proteolysis of the C-terminal tubulin tail can occur to generate Δ3-tubulin (Berezniuk et al., 2012). Δ2-Tubulin is an irreversibly modified form of α-tubulin, as it cannot be retyrosinated (Paturle-Lafanechere et al., 1991; Prota et al., 2013). Detyrosination (and its derivate Δ2-tubulin) is accumulated on rather stable MT species (Gundersen, Kalnoski, & Bulinski, 1984; Paturle-Lafanechere et al., 1994; Schulze, Asai, Bulinski, & Kirschner, 1987). Detyrosination enhances kinesin-1-driven transport in neurons (Konishi & Setou, 2009), but inhibits the MT-plus-end tracking of CAP-Gly domain proteins (Bieling et al., 2008; Peris et al., 2006) and the MCAK-driven depolymerization of MTs (Peris et al., 2009).

Polyglutamylation

Polyglutamylation consists of the addition of glutamate side chains of different lengths on glutamate residues within the C-terminal tails of α- and β-tubulins (Eddé et al., 1990; Rüdiger et al., 1992). Polyglutamylation is catalyzed by enzymes of the tubulin tyrosine ligase-like (TTLL) family (Janke et al., 2005). Various members of this family have distinct reaction specificities (initiation vs. elongation of the side chains; α- vs. β-tubulin), which result in the generation of different polyglutamylation patterns on MTs in cells (van Dijk et al., 2007). Polyglutamylation is removed by deglutamylases from the CCP family that, equivalent to the TTLLs, specifically either shorten long side chains, or remove the short chains including the branching point (Rogowski et al., 2010).

Polyglutamylation is expected to regulate multiple interactions between MTs and their associated proteins, such as MAPs and molecular motors. The first regulatory events have been demonstrated with the MT-severing enzyme spastin (Lacroix et al., 2010), and several publications have shown an important role in ciliary beating (e.g., Lee et al., 2012; Suryavanshi et al., 2010).

Polyglycylation

Polyglycylation, analogous to polyglutamylation, generates glycine side chains on glutamate residue of C-terminal tails of α- and β-tubulins (Bré et al., 1996; Redeker et al., 1994). The modification sites for polyglutamylation and polyglycylation are overlapping. Polyglycylation is catalyzed by enzymes of the TTLL family (Rogowski et al., 2009; Wloga et al., 2009); however, reverse, deglycylating enzymes have not yet been described. In mammals, glycylases that initiate the side chains are strictly different from the elongating enzyme, whereas in Drosophila, both steps of the reaction are catalyzed by the same enzymes (Rogowski et al., 2009). Though mechanistic studies have not yet been performed, initial data suggest that the main function of glycylation is the stabilization of ciliary axonemes (Rogowski et al., 2009; Wloga et al., 2008).

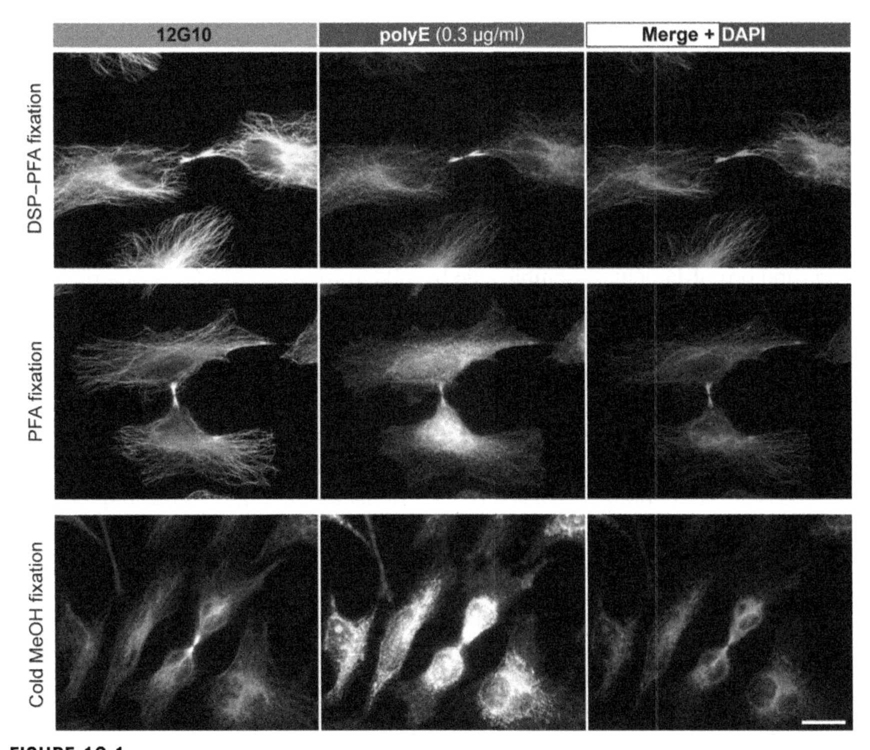

FIGURE 16.1

Impact of fixation method on the detection of posttranslational modifications of MTs. HeLa cells cultured on fibronectin-treated coverslips were fixed using the DSP–PFA, PFA, or the cold methanol fixation protocol. Cells were subjected to indirect immunofluorescence with 12G10 (α-tubulin) antibody to visualize the MT network (green), 0.3 μg/ml polyE antibody (polyglutamylation; red), and DAPI (blue). Note that only the DSP–PFA fixation method allows visualizing a specific polyglutamylation on the midbody MTs. Scale bar: 10 μm. (See color plate.)

the same cells with either the PFA or the cold methanol method prevents the detection of this polyglutamylated MT population (Fig. 16.1), indicating that polyE is sensitive to the fixation method. In contrast, in all three fixation methods, the tubulin antibody 12G10 shows a nice MT labeling (Fig. 16.1), which can be particularly misleading as it suggests that MTs have been fully preserved. A possible explanation for the divergence between polyE and 12G10 labeling on midbody MTs is the assumed low level of polyglutamylation on those MTs. Partial disassembly of MTs during PFA and methanol fixations would affect rare epitopes much more than the abundant (and anyway too dense—see Section 16.1.2) epitopes of tubulin. Though this effect has mostly been observed with polyE antibody, it is likely that other modification-specific antibodies are also sensitive to the cell fixation. We therefore strongly recommend testing different fixation conditions if specific PTMs of MTs are tested, even if the MTs are apparently intact if detected with a PTM-independent antibody, or with an antibody for a tubulin PTM that is highly enriched on the specific MTs investigated.

16.1.2 The importance of antibody concentration in analyzing MT PTMs in cells

A MT is a hollow tube of 25 nm of external diameter and is in most organisms composed of 13 protofilaments. Each protofilament is a linear assembly of α-/β-tubulin dimers, and protofilaments assemble side by side to form the MTs. Considering that the diameter of an α- or β-tubulin molecule is about 4–5 nm, the dense assembly of MTs generates a situation where similar epitopes are arranged in a distance of about 4 nm. An average antibody has an approximate diameter of 15 nm length and width, a size that is even further increased by the binding of secondary antibodies that can take the space of a cube, the side length of which could be around 30 nm or more. Thus, it is obvious that the spatial constraints on the MT will not allow for a stoichiometric decoration of epitopes on each tubulin molecule. In other words, only a fraction of the tubulin molecules present in the MT lattice can be labeled with antibodies due to space constrains. This creates a severe problem when antibodies directed against PTMs of tubulin are used to estimate the modification status and level of a given MT population using immunolabeling in fixed cells.

While it is easy to distinguish MTs carrying very low levels of modification from those with very high levels in cells, it is much harder to see subtle differences, which could still be highly relevant from a functional point of view. Moreover, even in situations where only a given percentage of the tubulin molecules in the MT lattice are posttranslationally modified, decoration with modification-specific antibodies could generate strong and virtually continuous signals that cannot be distinguished from those obtained from MTs with much higher modification levels. To avoid misinterpretations of the modification status of MTs based on antibody labeling, it is essential to test serial dilutions for each of the antibodies used. Moreover, it is important to perform such tests for each novel experimental setting.

To illustrate the importance of dilution in the correct assessment of changes in tubulin modifications, we have transfected the U2OS cell line with either the glutamylase TTLL4, known to generate glutamylation on MTs (van Dijk et al., 2007), or the deglutamylase CCP5, known to remove glutamate side chains (Rogowski et al., 2010). Strikingly, at dilutions commonly used for the antibody GT335 (0.5–0.1 μg/ml), cytoplasmic MTs are not clearly labeled in nontransfected cells (Fig. 16.2A, white contours), while all MTs are fully labeled in TTLL4-expressing cells (Fig. 16.2A, green contours). In contrast, no change in modification levels is seen in cells transfected with CCP5 with 0.5 μg/ml GT335 (Fig. 16.2B, lower panel, green contours). However, using a higher concentration of the antibody (5 μg/ml) reveals a clear MT staining in nontransfected U2OS cells (Fig. 16.2B, upper panel, white contours), while this staining is strongly decreased in cells expressing the deglutamylase CCP5 (Fig. 16.2B, upper panel, green contours).

This demonstrates that U2OS cells carry low levels of glutamylation on their interphase MTs (Regnard, Desbruyeres, Denoulet, & Eddé, 1999), which can only be revealed at higher concentrations of GT335 antibody. This antibody concentration is, of course, too elevated for the detection of MTs with higher levels of PTMs, where even much lower concentrations of the same antibody are sufficient for strong labeling (Fig. 16.2A, lower panel). While this conclusion might appear trivial in this overexpression experiments, it is important to point out that similar situations might be encountered in cells. For instance, single MTs in cells might acquire specific functions by slight changes in MT PTMs, which could be easily overseen using antibodies at high dilutions. Conversely, in cells containing MTs with very high modification levels, such as neurons, developmental changes in MT PTM patterns can only be observed if antibodies are sufficiently diluted. To illustrate this notion, we show here the staining of a 3-day-old rat cortical neuron, stained with 0.1 mg/ml of GT335. Despite the fact that MTs in young neurons are not fully glutamylated (Audebert et al., 1994; Fig. 16.2C), and despite the high dilution of the antibody, the MTs in these neurons appear already strongly labeled, suggesting that even higher dilutions of GT335 might be required to determine changes in MT glutamylation during neuronal development.

Finally, it should be noted that not only the density of the tubulin molecules within an MT but also the density of MTs within certain cellular structures, such as neuronal axons and dendrites (Fig. 16.2C), axonemes in cilia and flagella (Bressac et al., 1995), or other MT bundles in cells, such as midbodies in anaphase cells (Fig. 16.1), preclude the possibility of identifying specific PTM identities of single MTs within these MT arrays. Because there is as yet no easy approach to circumvent this problem (even super resolution light microscopy is not able to fully resolve single MTs in dense bundles), it should be taken into account when interpreting immunostaining results from fixed cells.

16.1.3 Antibody and sample concentration in immunoblot analysis

Strikingly, tubulin concentration is also a critical factor in immunoblot analyses, especially if semi-quantitative information about the levels of PTMs should be deduced from the detection levels with the different antibodies. Apart from the above-discussed importance of using a range of dilutions of the antibodies to determine

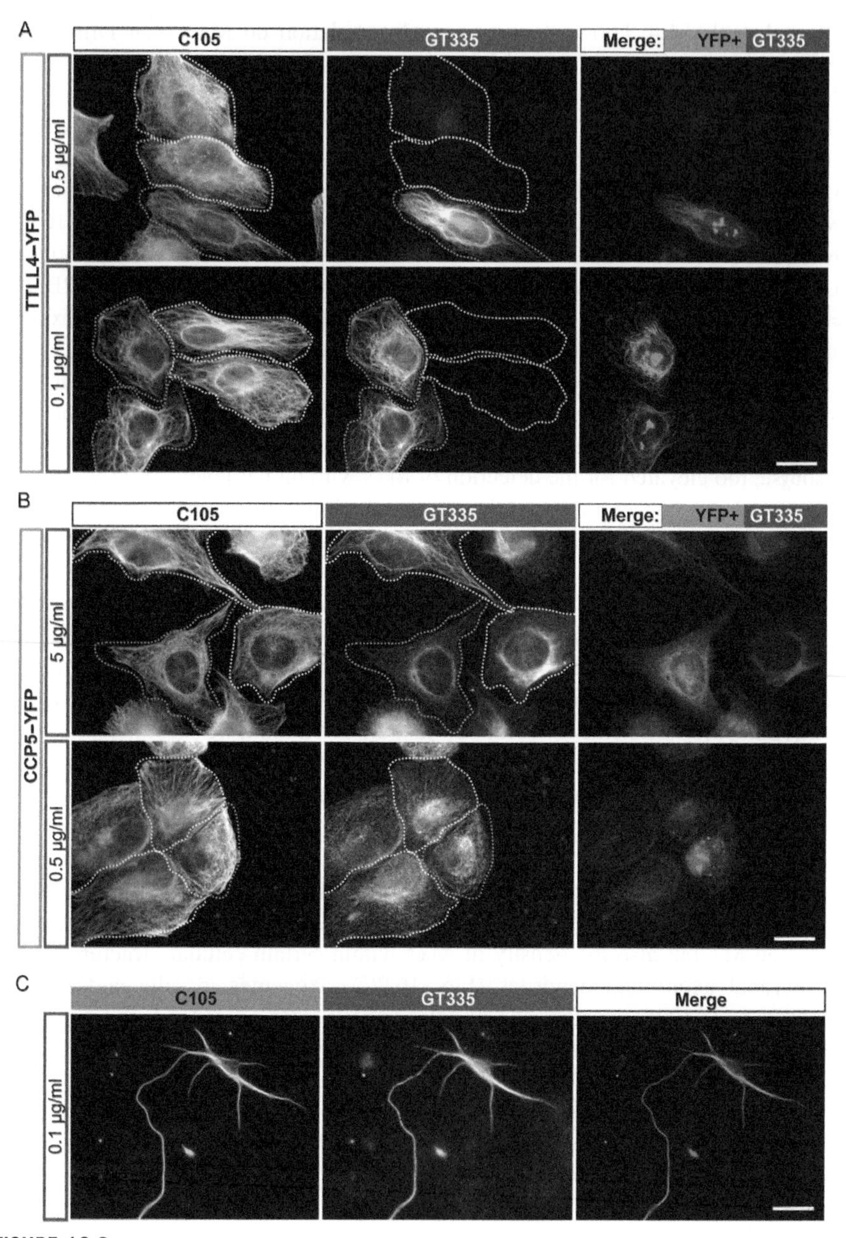

FIGURE 16.2

Varying antibody dilutions allows detecting slight changes in MT modification levels. U2OS cells were transfected with either TTLL4–YFP (A) or CCP5–YFP (B) constructs and fixed with the DSP–PFA method 20 h posttransfection (transfected, YFP-positive cells are labeled with green contours, nontransfected cells with white contours in A and B). Cells were

the optimal (or better minimal) concentration (Fig. 16.3A), another important factor is the dilution of the protein samples that are loaded on the SDS-PAGE. Because tubulin, even at relatively high concentrations, resolves into one (standard SDS-PAGE) or two (specific SDS-PAGE to resolve α- and β-tubulin; Lacroix & Janke, 2011 and described below) neat bands, overloading of the antigen is often not noticed. In contrast, differences in tubulin modification levels are easily overlooked if too much material has been loaded on the SDS-PAGE prior to immunoblot (Fig. 16.3B). As the presence of tubulin PTMs does not significantly alter the apparent molecular weight of tubulin separated on an SDS-PAGE, there is no additional hint, apart from the antibody detection levels, indicating the PTM state of tubulin.

An additional factor that can easily obscure the need to dilute the antibodies of tubulin PTMs is that most of these antibodies yield only very low background signals, even if used in high concentrations. This has been observed for both, immunocytochemistry and immunoblot (Fig. 16.3A), and could thus hamper the identification of differences in modification levels in these two experimental setups.

16.2 MATERIALS AND METHODS

16.2.1 Antibodies detecting tubulin PTMs (summarized in Table 16.1)

16.2.1.1 Acetylation

So far, only one site of tubulin acetylation has been studied in detail. Acetylation of lysine-40 of α-tubulin is specifically detected with the mouse monoclonal antibody 6-B11-1 (IgG2b isotype), which has been raised against the axonemal α-tubulin from sea urchin sperm flagella (Sigma #T6793, #T7451; LeDizet & Piperno, 1987, 1991;

subjected to indirect immunofluorescence using C105 (β-tubulin) to reveal MT network, and GT335 (glutamylation) at 0.1 μg/ml (A lower panel), 0.5 μg/ml (A upper panel and B lower panel), or 5 μg/ml (B upper panel) to reveal MT glutamylation. TTLL4-transfected cells are strongly stained with GT335 even at very low concentration (A lower panel), while nontransfected cells show no detectable MT staining. In contrast, higher concentrations of GT335 reveal a specific, but faint staining of the MT network in nontransfected cells (B upper panel). This staining is related to low levels of glutamylation, as it can be diminished by the deglutamylase CCP5 (B, green cells). These slight differences are barely visible at "standard" concentrations of GT335 (0.5 μg/ml; B lower panel). Strikingly, the presence of a strong antigen (A) depletes the antibody from the solution, as nontransfected (white contours) cells are weakly labeled in B, but not in A (though the same exposure conditions have been employed). (C) Embryonic rat cortical neurons were fixed after 3 days *in vitro* using the DSP–PFA protocol and stained with C105 and GT335 at 0.1 μg/ml. Scale bar in all panels: 10 μm. (See color plate.)

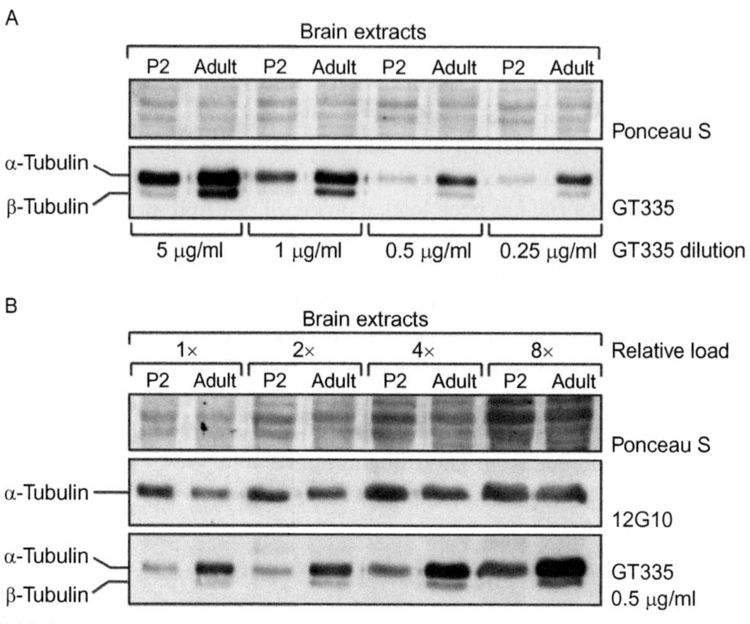

FIGURE 16.3

Impact of sample and antibody concentration on the assessment of tubulin modification levels by immunoblot. Comparative immunoblot analysis of protein extracts from 2 days old and adult mouse brain. The samples have been resolved on 10% gels allowing for separation of α- and β-tubulin. (A) Equal amounts of the two samples were separated on SDS-PAGE, and following the transfer, incubated with a different dilution of the GT335 antibody (glutamylation): 5, 1, 0.5, and 0.25 μg/ml. At higher concentrations of GT335, differences in modification levels of α-tubulin are not detected, while they are obvious at lower antibody concentrations. Differences in glutamylation in β-tubulin, which is less modified in brain (Audebert et al., 1994), are still detectable at higher GT335 concentrations. Ponceau S staining of the blot membranes indicates loading. (B) Increasing amounts of samples were loaded and revealed with highly diluted GT335 (0.5 μg/ml). Similarly to (A), differences in glutamylation levels are faint at higher sample load, though the tubulin bands still look very distinct and might thus not suggest an overload of the sample. Ponceau S and 12G10 (α-tubulin) indicate total load and α-tubulin levels.

Piperno & Fuller, 1985). The antibody detects acetylated tubulin in most species tested so far.

16.2.1.2 ***Tyrosination/detyrosination/Δ2-tubulin***

Tyrosinated α-tubulin is recognized by the YL1/2 rat monoclonal antibody (Millipore #MAB1964; Cumming et al., 1984; Kilmartin et al., 1982). This antibody was raised against purified yeast tubulin and thus recognizes the C-terminal -EEF and -EEY sequences on α-tubulin.

Table 16.1 Overview of antibodies detecting tubulin PTMs described in this work

Name	Citations	Type	Modification	Comment	Commercial availability
6-11B-1	LeDizet and Piperno (1991), Piperno and Fuller (1985)	Monoclonal mouse IgG2b	Acetylation K40 on α-tubulin	Raised against flagellar tubulin from sea urchin	#T6793, #T7451 (Sigma)
YL1/2	Cumming, Burgoyne, and Lytton (1984), Kilmartin, Wright, and Milstein (1982)	Monoclonal rat IgG2a	C-terminal -EEF (slightly preferred) and -EEY on α-tubulin	Initially raised against purified yeast tubulin (-EEF)	#MAB1864 (Millipore)
Anti-detyr-tubulin	Paturle-Lafanechere et al. (1994)	Polyclonal rabbit	C-terminal -GEE on α-tubulin	Raised against -CGEEEGEE	#AB3201 (Millipore)
Anti-Δ2-tubulin	Paturle-Lafanechere et al. (1994)	Polyclonal rabbit	C-terminal -GE on α-tubulin	Raised against -CEGEEEGE	#AB3203 (Millipore)
GT335	Wolff et al. (1992)	Monoclonal mouse IgG1κ	γ-Linked E_n side chain ($n=1, 2, 3...$) on modified E	Detects most glutamylated α- and β-tubulin	#AG-20B-0020 (Adipogen)
polyE	Rogowski et al. (2010), Shang, Li, and Gorovsky (2002)	Polyclonal rabbit	C-terminal -E_n ($n \geq 3$)	Detects polyglutamylation	Not available
1D5	Rüdiger et al. (1999), Warn, Harrison, Planques, Robert-Nicoud, and Wehland (1990)	Monoclonal mouse IgG1	C-terminal -E_n ($n \geq 2$)	Detects polyglutamylation and detyrosination on most α-tubulin isoforms (C-terminal sequence -EE)	#302 011 (Synaptic Systems)
TAP952	Bré et al. (1996), Callen et al. (1994), Levilliers, Fleury, and Hill (1995)	Monoclonal mouse IgG1	γ-Linked G side chain ($n=1$) on modified E	Detects strictly monoglycylation	Currently not available
AXO49	Bré et al. (1996), Callen et al. (1994), Levilliers et al. (1995)	Monoclonal mouse IgG1	γ-Linked G_n side chain ($n \geq 3$) on modified E	Detects polyglycylation	Currently not available
polyG	Rogowski et al. (2009), Shang et al. (2002)	Polyclonal rabbit	C-terminal polyG chains	Detects polyglycylation	Currently not available

Adapted from Janke and Bulinski (2011)

Anti-detyr-tubulin rabbit polyclonal antibody (Millipore #AB3201; Paturle-Lafanechere et al., 1994), raised against the -CGEEEGEE(-COOH) peptide, recognizes the C-terminal -GEE sequence of detyrosinated α-tubulin. Another antibody, the mouse monoclonal 1D5 (Synaptic Systems #302011; Rüdiger et al., 1999; Warn et al., 1990), also detects C-terminal stretches of two and more glutamate residues ($-E_n$; $n \geq 2$). However, while 1D5 detects detyrosinated α-tubulin (which ends with C-terminal -EE in most species), it cross-reacts with any polyglutamylated form of tubulin that contains glutamate side chains that contain two or more glutamate residues. Thus, the use of this antibody alone makes the interpretation of the data impossible.

Δ2-tubulin (see Box 16.1) can be detected by a rabbit polyclonal antibody (Millipore #AB3203; Paturle-Lafanechere et al., 1994) that has been raised against the peptide -CEGEEEGE(-COOH). Anti-Δ2-tubulin antibody detects specifically C-terminal -GE sequence of proteins (Rogowski et al., 2010).

16.2.1.3 ***Polyglutamylation***

Two antibodies allow assessing the extent of both α- and β-tubulin polyglutamylation: the mouse monoclonal GT335 (Adipogen #AG-20B-0020; Wolff et al., 1992), which recognizes the epitope formed at the branching point of the glutamate side chains (γ-linked glutamate on a main-chain glutamate), and thus stains both short and long side chains of tubulin polyglutamylation. It has been raised against the branched peptide -EGEGE*EEG(-$CONH_2$) with a bi-glutamate side chain (-EE(-COOH)) attached to the γ-carboxy group of the glutamate marked with the asterisk. GT335 also detects a range of nontubulin substrates of glutamylation (Regnard et al., 2000; van Dijk et al., 2008).

Polyglutamylation can be detected using the rabbit polyclonal antibody polyE (Rogowski et al., 2010; Shang et al., 2002; van Dijk et al., 2007). This antibody is still not commercially available. It detects stretches of at least three glutamate residues situated at the C-terminal extremity of proteins and has been raised against a -CEEEEEEEEE (-COOH) peptide, and is therefore specific to long side chains added by polyglutamylation to tubulin and other substrates, and it also detects proteins that have a genetically encoded C-terminal tail with three or more glutamate residues (Rogowski et al., 2010).

The use of these two antibodies, GT335 and polyE, allows distinguishing between long (GT335- and polyE-positive) and short (GT335-positive, polyE-negative) side chains generated by polyglutamylation.

16.2.1.4 ***Polyglycylation***

The mouse monoclonal antibody TAP952 detects monoglycylated tubulin (Callen et al., 1994), but unlike GT335 (detecting both short and long glutamate chains), it does not cross-react with elongated glycine side chains. TAP952 has been raised against Paramecium axonemal tubulin and been shown to detect γ-linked glycine residues on glutamic acids 437, 438, 439, and 441 of peptides corresponding to C-terminal tails of paramecium β-tubulin (Bré, Redeker, Vinh, Rossier, & Levilliers, 1998). Though the antibody has been shown to detect strongly glycylated tubulin in cilia and flagella of many species, it is not yet clear whether all possible modification sites are detected, especially because of the high-sequence variability of tubulin within the C-terminal tails, where the modification takes place.

AXO49 is a mouse monoclonal antibody raised against Paramecium axonemes that recognize γ-linked glycine side chains of at least three residues (Callen et al., 1994). It has been mapped to detect side chains of three and more glycine residues attached to the Glu 437 of the paramecium β-tubulin C-terminal peptide (Bré et al., 1998).

The rabbit polyclonal polyG antibody has been raised against the -CGGGGGGGGG(-COOH) peptide and detects, similar to the polyE antibody, C-terminal polyglycine chains (Rogowski et al., 2009; Wloga et al., 2009).

None of the antibodies against glycylation are commercially available.

16.2.2 Cell culture and fixation

16.2.2.1 *Cell culture and transfection*

HeLa cells (ATCC® #CCL-2™) were grown on glass coverslips (Marienfeld #01 115 20) in DMEM (Life technologies #41965-039) supplemented with 10% fetal bovine serum (Sigma #F7524) and 2 mM L-glutamine (Life technologies #25030-024) at 37 °C and in 5% CO_2. Before use, coverslips were incubated in 60-μl drops of a 5 μg/ml aqueous fibronectin (Sigma #F1141) solution for 1 h at 37 °C in a humid chamber. Coverslips were then transferred to a 24-well plate and washed gently three times with phosphate-buffered saline (PBS) and left in PBS until the seeding of cells.

The U2OS cell line (ATCC® #40342™) was grown as described above. Cells were transfected with expression vectors encoding the murine TTLL4 (van Dijk et al., 2007) or CCP5 (Rogowski et al., 2010) genes using jetPei® (Polyplus #101-10) according to manufacturers' instructions. Cells were fixed 20 h posttransfection.

Rat cortical neurons were cultured as previously described (Saudou, Finkbeiner, Devys, & Greenberg, 1998) and let to grow until 3 days *in vitro* in neurobasal-B27 medium.

16.2.2.2 *Cold methanol fixation*

Coverslips were washed once with PBS, then rapidly immersed in −20 °C methanol, and incubated at −20 °C for 10 min. Methanol was removed, and the coverslips were washed three times with PBS and stored in PBS at 4 °C if required.

16.2.2.3 *PFA fixation*

Coverslips were rinsed once with PBS and incubated in 3.7% PFA, 2% sucrose solution for 15 min at room temperature (RT). Coverslips were rinsed three times with PBS and could be stored in PBS at 4 °C if required.

16.2.2.4 *DSP–PFA fixation*

This method has been described previously (Bell & Safiejko-Mroczka, 1995) and involves several incubation steps that are all performed at RT:

- 10 min incubation in 1 mM DSP in Hank's balanced salt solution (HBSS)
- 10 min incubation in 1 mM DSP in microtubule stabilizing buffer (MTSB)

- 5 min incubation in 0.5% Triton X in MTSB
- 15 min incubation in 4% PFA in MTSB
- 5 min wash in PBS
- 5 min incubation in 100 mM glycine in PBS (in order to quench remaining aldehyde groups)
- 5 min wash in PBS at RT, followed by an optional storage at 4 °C in PBS.

All the solutions should be at RT in order to preserve intact MTs inside the cells (in all steps prior to the PFA fixation, MTs depolymerize if cold solutions are used).

16.2.3 Immunocytochemistry on fixed cells

Coverslips with fixed cells were subjected to indirect immunochemistry. Antibodies have been diluted to the desired concentration in PBS, 3% bovine serum albumin, 0.1% Triton X-100 (this solution can be filtered, aliquoted, and stored at −20 °C). In order to use a minimum amount of antibodies, we incubate coverslips with attached cells facing downward on a drop of 30 μl of antibody solution in a humid chamber for 1 h at RT. 12G10 is a monoclonal antibody detecting α-tubulin (Developmental Studies Hybridoma Bank, University of Iowa). C105 is a rabbit polyclonal, anti-β-tubulin antibody (gift of Jose M. Andreu, Madrid; Arevalo, Nieto, Andreu, & Andreu, 1990). The coverslips were then transferred to a 24-well plate and rinsed three times with PBS, 0.1% Triton X-100, and subsequently incubated with the fluorescent secondary antibodies as previously described (Life technologies #A11001 Alexa 488 anti-mouse IgG, #A11019 Alexa 568 anti-mouse IgG, #A11036 Alexa 568 anti-rabbit IgG, all used at 2 μg/ml, and #A21068 for Alexa 350 anti-rabbit IgG used at 10 μg/ml). The coverslips were again transferred to a 24-well plate and rinsed four times with PBS, 0.1% Triton X-100 (if DNA staining was required, the first wash was performed in PBS, 0.1% Triton X-100, 0.1 μg/ml DAPI for 3 min at RT). Coverslips were then mounted carefully on microscopy slides using 6 μl of Mowiol solution. Mowiol is left to polymerize by leaving the slides in horizontal position at RT and in the dark for at least 4 h, and most commonly overnight.

16.2.4 Microscopy and image acquisition

Images were acquired on a Leica DMRXA upright microscope using Metamorph (Molecular Devices) software. Images were treated using ImageJ (http://imagej.nih.gov/ij/) or Photoshop (Adobe Systems).

16.2.5 Electrophoresis and immunoblot

We use a variation of SDS-PAGE allowing for the separation of α- and β-tubulins as two distinct bands. The method has been described in detail in Lacroix and Janke (2011). It differs from the standard SDS-PAGE protocol by a different acrylamide to bis-acrylamide ratio, the pH of the Tris buffer, and the type of SDS (see "buffer composition," Section 3.2.1).

Brains of either 2 days old or adult mice were rapidly homogenized in Laemmli sample buffer and using an Ultra-Turrax® homogenizer (IKA®) and heated to 95 °C for 5 min. Subsequently, two short pulses of sonication were applied to fragment genomic DNA.

The samples were loaded onto the SDS-PAGE gels (for tubulin, 10% acryl amide gels are used) and run until the bromophenol blue front reaches the bottom of the gel. The gels were then shortly incubated in transfer buffer, and proteins were transferred onto a nitrocellulose membrane (Protran BA 85, Whatman, GE Healthcare #10 401 196) for 1 h at 4 °C at 100 V and approximately 400 mA using a tank blot system (Bio-Rad #170-4070).

After transfer, the nitrocellulose membranes were incubated for 1 min in Ponceau S staining solution and destained twice for 2 min in 1% acidic acid. The membranes were scanned and transferred to Tris-buffered saline (TBS) containing 0.1% Tween 20 (TBS-T). Subsequently, the membrane was incubated in 5% fat-free dry milk in TBS-T at RT for 1 h. The primary antibodies were diluted in 2.5% milk–TBS and were incubated with the membrane at RT for 1 h. After three washes of 5 min with TBS-T, the membranes were transferred into a dilution of secondary, horseradish peroxidase (HRP)-labeled antibody (anti-mouse- or anti-rabbit-HRP-conjugated IgG; GE Healthcare #NA931V or #NA934V, respectively) in TBS-T and incubated at RT for 45 min. After five washes of 3 min with TBS-T, membranes were developed by incubation within the ECL Western Blotting Detection Reagents (GE Healthcare #RPN2209) for 1 min. The membrane was then exposed to photographic films (Amersham Hyperfilm™ MP, GE Healthcare #28906844) for 1 min. The films were developed using a photographic developing machine (Curix60, Agfa) and scanned with trough light using a linear gray scale gradient.

16.3 BUFFER COMPOSITION

16.3.1 Cell fixation

16.3.1.1 *PFA method*

PFA–sucrose in PBS

Prepare 4% PFA (Sigma #P6148) in PBS (first dissolve in water by adding NaOH, bring to pH 7.0, and adjust the concentration with 10 × PBS) and then use this solution to prepare 3.7% PFA, 2% sucrose solution in PBS (can be aliquoted and stored at −20 °C).

16.3.1.2 *DSP–PFA method*

DSP stock solution

DSP 20 mg/ml in DMSO (50 mM; Perbio Pierce #22585) can be aliquoted and stored at −20 °C.

HBSS: Hank's balanced salt solution (can be prepared as 5 × buffer and stored at 4 °C); 1 × solution can be obtained from Gibco (#14025-050)

- 1.26 mM $CaCl_2$
- 5.33 mM KCl
- 0.44 mM KH_2PO_4
- 0.5 mM $MgCl_2$
- 0.41 mM $MgSO_4$
- 138 mM NaCl
- 4 mM $NaHCO_3$
- 0.3 mM Na_2HPO_4
- 5.6 mM glucose

MTSB: Microtubule stabilizing buffer (store at 4 °C)

- 1 mM EGTA
- 4% PEG 8000
- 100 mM PIPES, pH 6.9

TSB

- 0.5% Triton X-100 in MTSB

PFA in MTSB

- 4% PFA in MTSB: PFA powder (Sigma #P6148) is heated in MTSB to approximately 50 °C and dissolved by adding a drop of 10 M NaOH; the pH is then brought to pH 7.0 with 6 M HCl. The solution is left to cool down to RT.

Mowiol mounting medium

- 10% Mowiol (polyvinyl alcohol) 4-88 (Sigma #81381)
- 100 mM Tris/HCl, pH 8.5
- 25% glycerol

16.3.2 Gel electrophoresis and immunoblotting

16.3.2.1 SDS-PAGE for α/β-tubulin separation

5 × Laemmli sample buffer

- 450 mM DTT
- 10% SDS (BDH #442444H)
- 400 mM Tris/HCl, pH 6.8
- 50% glycerol
- bromophenol blue

Electrophoresis cell: Mini-PROTEAN® cell with a PowerPac™ power supply (Bio-Rad #165-8001EDU and #164-5052MP, respectively).

Separating gel buffer (4 ×)

- 1.5 M Tris/HCl, pH 9.0
- 0.4% SDS (specific type required: Sigma #L5750)

Stacking gel buffer (4×)
  0.5 M Tris/HCl, pH 6.8
  0.4% SDS (Sigma #L5750)
Acrylamide/bis-acrylamide stock solution (74:1)
  40% acrylamide solution (Bio-Rad #161-0140) supplemented with 0.54% bis-acryl amide (w/v) powder (Bio-Rad #161-0210).
Running buffer
  50 mM Tris/HCl
  384 mM glycine
  0.1% SDS (Sigma #L5750) in deionized water
Transfer buffer
  50 mM Tris/HCl
  40 mM glycine
  0.4 mM SDS
  in deionized water
  20% (v/v) ethanol
Ponceau S staining solution
  0.2% (w/v) Ponceau S in 1% acetic acid

CONCLUDING REMARKS

Here, we have summarized our experiences in the detection of tubulin PTMs using immunocytochemistry and immunoblot. We have described appropriate methods to detect changes in MT PTMs using a battery of modification-specific antibodies and suggest several experimental precautions to take in order to avoid misleading interpretations. Because many functional aspects of MT PTMs might be related to gradual changes rather than to abrupt changes, careful evaluation of changes in MT modification patterns could be essential for the understanding of their roles in cells.

Acknowledgments

This work was supported by the Institut Curie, the CNRS, the INSERM, the Labex CelTisPhyBio 11-LBX-0038, the French National Research Agency (ANR) awards 08-JCJC-0007, the Human Frontier Science (HFSP) grants RGP0023/2008, the Institut National du Cancer (INCA) grant 2009-1-PL BIO-12-IC-1, and an EMBO Young Investigator Program grant to C. J.

We thank C. Benstaali, P. Marques (Institut Curie, Orsay, France), A. Fleury-Aubusson (CBM, Gif-sur-Yvette, France) for technical assistance, and R. Basto, G. Raposo, A. Wehenkel (Institut Curie), C. Antony (EMBL Heidelberg), and N. Spassky (ENS, Paris, France) for instructive discussions. We are grateful to J.M. Andreu (CSIC, Madrid, Spain) for the kind gift of the antibody C105. The monoclonal antibody 12G10 developed by J. Frankel and M. Nelson was obtained from the Developmental Studies Hybridoma Bank developed under the auspices of the NICHD and maintained by the University of Iowa.

The authors declare no competing financial interests.

References

Akella, J. S., Wloga, D., Kim, J., Starostina, N. G., Lyons-Abbott, S., Morrissette, N. S., et al. (2010). MEC-17 is an alpha-tubulin acetyltransferase. *Nature*, *467*, 218–222.

Arce, C. A., Rodriguez, J. A., Barra, H. S., & Caputto, R. (1975). Incorporation of L-tyrosine, L-phenylalanine and L-3,4-dihydroxyphenylalanine as single units into rat brain tubulin. *European Journal of Biochemistry*, *59*, 145–149.

Arevalo, M. A., Nieto, J. M., Andreu, D., & Andreu, J. M. (1990). Tubulin assembly probed with antibodies to synthetic peptides. *Journal of Molecular Biology*, *214*, 105–120.

Audebert, S., Koulakoff, A., Berwald-Netter, Y., Gros, F., Denoulet, P., & Eddé, B. (1994). Developmental regulation of polyglutamylated alpha- and beta-tubulin in mouse brain neurons. *Journal of Cell Science*, *107*, 2313–2322.

Bell, P. B., Jr., & Safiejko-Mroczka, B. (1995). Improved methods for preserving macromolecular structures and visualizing them by fluorescence and scanning electron microscopy. *Scanning Microscopy*, *9*, 843–857, discussion 858–860.

Berezniuk, I., Vu, H. T., Lyons, P. J., Sironi, J. J., Xiao, H., Burd, B., et al. (2012). Cytosolic carboxypeptidase 1 is involved in processing alpha- and beta-tubulin. *The Journal of Biological Chemistry*, *287*, 6503–6517.

Bieling, P., Kandels-Lewis, S., Telley, I. A., van Dijk, J., Janke, C., & Surrey, T. (2008). CLIP-170 tracks growing microtubule ends by dynamically recognizing composite EB1/tubulin-binding sites. *The Journal of Cell Biology*, *183*, 1223–1233.

Bré, M. H., Redeker, V., Quibell, M., Darmanaden-Delorme, J., Bressac, C., Cosson, J., et al. (1996). Axonemal tubulin polyglycylation probed with two monoclonal antibodies: Widespread evolutionary distribution, appearance during spermatozoan maturation and possible function in motility. *Journal of Cell Science*, *109*, 727–738.

Bré, M. H., Redeker, V., Vinh, J., Rossier, J., & Levilliers, N. (1998). Tubulin polyglycylation: Differential posttranslational modification of dynamic cytoplasmic and stable axonemal microtubules in Paramecium. *Molecular Biology of the Cell*, *9*, 2655–2665.

Bressac, C., Bré, M. H., Darmanaden-Delorme, J., Laurent, M., Levilliers, N., & Fleury, A. (1995). A massive new posttranslational modification occurs on axonemal tubulin at the final step of spermatogenesis in Drosophila. *European Journal of Cell Biology*, *67*, 346–355.

Callen, A. M., Adoutte, A., Andreu, J. M., Baroin-Tourancheau, A., Bré, M. H., Ruiz, P. C., et al. (1994). Isolation and characterization of libraries of monoclonal antibodies directed against various forms of tubulin in Paramecium. *Biology of the Cell*, *81*, 95–119.

Choudhary, C., Kumar, C., Gnad, F., Nielsen, M. L., Rehman, M., Walther, T. C., et al. (2009). Lysine acetylation targets protein complexes and co-regulates major cellular functions. *Science*, *325*, 834–840.

Creppe, C., Malinouskaya, L., Volvert, M.-L., Gillard, M., Close, P., Malaise, O., et al. (2009). Elongator controls the migration and differentiation of cortical neurons through acetylation of alpha-tubulin. *Cell*, *136*, 551–564.

Cueva, J. G., Hsin, J., Huang, K. C., & Goodman, M. B. (2012). Posttranslational acetylation of alpha-tubulin constrains protofilament number in native microtubules. *Current Biology*, *22*, 1066–1074.

Cumming, R., Burgoyne, R. D., & Lytton, N. A. (1984). Immunocytochemical demonstration of alpha-tubulin modification during axonal maturation in the cerebellar cortex. *The Journal of Cell Biology*, *98*, 347–351.

Eddé, B., Rossier, J., Le Caer, J. P., Desbruyeres, E., Gros, F., & Denoulet, P. (1990). Post-translational glutamylation of alpha-tubulin. *Science*, *247*, 83–85.

Ersfeld, K., Wehland, J., Plessmann, U., Dodemont, H., Gerke, V., & Weber, K. (1993). Characterization of the tubulin-tyrosine ligase. *The Journal of Cell Biology*, *120*, 725–732.

Gundersen, G. G., Kalnoski, M. H., & Bulinski, J. C. (1984). Distinct populations of microtubules: Tyrosinated and nontyrosinated alpha tubulin are distributed differently in vivo. *Cell*, *38*, 779–789.

Hallak, M. E., Rodriguez, J. A., Barra, H. S., & Caputto, R. (1977). Release of tyrosine from tyrosinated tubulin. Some common factors that affect this process and the assembly of tubulin. *FEBS Letters*, *73*, 147–150.

Hubbert, C., Guardiola, A., Shao, R., Kawaguchi, Y., Ito, A., Nixon, A., et al. (2002). HDAC6 is a microtubule-associated deacetylase. *Nature*, *417*, 455–458.

Janke, C., & Bulinski, J. C. (2011). Post-translational regulation of the microtubule cytoskeleton: Mechanisms and functions. *Nature Reviews. Molecular Cell Biology*, *12*, 773–786.

Janke, C., Rogowski, K., Wloga, D., Regnard, C., Kajava, A. V., Strub, J.-M., et al. (2005). Tubulin polyglutamylase enzymes are members of the TTL domain protein family. *Science*, *308*, 1758–1762.

Kilmartin, J. V., Wright, B., & Milstein, C. (1982). Rat monoclonal antitubulin antibodies derived by using a new nonsecreting rat cell line. *The Journal of Cell Biology*, *93*, 576–582.

Kimura, Y., Kurabe, N., Ikegami, K., Tsutsumi, K., Konishi, Y., Kaplan, O. I., et al. (2010). Identification of tubulin deglutamylase among Caenorhabditis elegans and mammalian cytosolic carboxypeptidases (CCPs). *The Journal of Biological Chemistry*, *285*, 22936–22941.

Konishi, Y., & Setou, M. (2009). Tubulin tyrosination navigates the kinesin-1 motor domain to axons. *Nature Neuroscience*, *12*, 559–567.

Lacroix, B., & Janke, C. (2011). Generation of differentially polyglutamylated microtubules. *Methods in Molecular Biology*, *777*, 57–69.

Lacroix, B., van Dijk, J., Gold, N. D., Guizetti, J., Aldrian-Herrada, G., Rogowski, K., et al. (2010). Tubulin polyglutamylation stimulates spastin-mediated microtubule severing. *The Journal of Cell Biology*, *189*, 945–954.

LeDizet, M., & Piperno, G. (1987). Identification of an acetylation site of Chlamydomonas alpha-tubulin. *Proceedings of the National Academy of Sciences of the United States of America*, *84*, 5720–5724.

LeDizet, M., & Piperno, G. (1991). Detection of acetylated alpha-tubulin by specific antibodies. *Methods in Enzymology*, *196*, 264–274.

Lee, J. E., Silhavy, J. L., Zaki, M. S., Schroth, J., Bielas, S. L., Marsh, S. E., et al. (2012). CEP41 is mutated in Joubert syndrome and is required for tubulin glutamylation at the cilium. *Nature Genetics*, *44*, 193–199.

Levilliers, N., Fleury, A., & Hill, A. M. (1995). Monoclonal and polyclonal antibodies detect a new type of post-translational modification of axonemal tubulin. *Journal of Cell Science*, *108*, 3013–3028.

Luduena, R. F. (2013). A hypothesis on the origin and evolution of tubulin. *International Review of Cell and Molecular Biology*, *302*, 41–185.

North, B. J., Marshall, B. L., Borra, M. T., Denu, J. M., & Verdin, E. (2003). The human Sir2 ortholog, SIRT2, is an NAD^+-dependent tubulin deacetylase. *Molecular Cell*, *11*, 437–444.

Paturle-Lafanechere, L., Eddé, B., Denoulet, P., Van Dorsselaer, A., Mazarguil, H., Le Caer, J. P., et al. (1991). Characterization of a major brain tubulin variant which cannot be tyrosinated. *Biochemistry*, *30*, 10523–10528.

Paturle-Lafanechere, L., Manier, M., Trigault, N., Pirollet, F., Mazarguil, H., & Job, D. (1994). Accumulation of delta 2-tubulin, a major tubulin variant that cannot be tyrosinated, in neuronal tissues and in stable microtubule assemblies. *Journal of Cell Science*, *107*, 1529–1543.

Peris, L., Thery, M., Faure, J., Saoudi, Y., Lafanechere, L., Chilton, J. K., et al. (2006). Tubulin tyrosination is a major factor affecting the recruitment of CAP-Gly proteins at microtubule plus ends. *The Journal of Cell Biology*, *174*, 839–849.

Peris, L., Wagenbach, M., Lafanechere, L., Brocard, J., Moore, A. T., Kozielski, F., et al. (2009). Motor-dependent microtubule disassembly driven by tubulin tyrosination. *The Journal of Cell Biology*, *185*, 1159–1166.

Piperno, G., & Fuller, M. T. (1985). Monoclonal antibodies specific for an acetylated form of alpha-tubulin recognize the antigen in cilia and flagella from a variety of organisms. *The Journal of Cell Biology*, *101*, 2085–2094.

Prota, A. E., Magiera, M. M., Kuijpers, M., Bargsten, K., Frey, D., Wieser, M., et al. (2013). Structural basis of tubulin tyrosination by tubulin tyrosine ligase. *The Journal of Cell Biology*, *200*, 259–270.

Raybin, D., & Flavin, M. (1975). An enzyme tyrosylating alpha-tubulin and its role in microtubule assembly. *Biochemical and Biophysical Research Communications*, *65*, 1088–1095.

Redeker, V. (2010). Mass spectrometry analysis of C-terminal posttranslational modifications of tubulins. *Methods in Cell Biology*, *95*, 77–103.

Redeker, V., Levilliers, N., Schmitter, J. M., Le Caer, J. P., Rossier, J., Adoutte, A., et al. (1994). Polyglycylation of tubulin: A posttranslational modification in axonemal microtubules. *Science*, *266*, 1688–1691.

Regnard, C., Desbruyeres, E., Denoulet, P., & Eddé, B. (1999). Tubulin polyglutamylase: Isozymic variants and regulation during the cell cycle in HeLa cells. *Journal of Cell Science*, *112*, 4281–4289.

Regnard, C., Desbruyeres, E., Huet, J. C., Beauvallet, C., Pernollet, J. C., & Eddé, B. (2000). Polyglutamylation of nucleosome assembly proteins. *The Journal of Biological Chemistry*, *275*, 15969–15976.

Rogowski, K., Juge, F., van Dijk, J., Wloga, D., Strub, J.-M., Levilliers, N., et al. (2009). Evolutionary divergence of enzymatic mechanisms for posttranslational polyglycylation. *Cell*, *137*, 1076–1087.

Rogowski, K., van Dijk, J., Magiera, M. M., Bosc, C., Deloulme, J.-C., Bosson, A., et al. (2010). A family of protein-deglutamylating enzymes associated with neurodegeneration. *Cell*, *143*, 564–578.

Rüdiger, M., Plessman, U., Kloppel, K. D., Wehland, J., & Weber, K. (1992). Class II tubulin, the major brain beta tubulin isotype is polyglutamylated on glutamic acid residue 435. *FEBS Letters*, *308*, 101–105.

Rüdiger, A. H., Rudiger, M., Wehland, J., & Weber, K. (1999). Monoclonal antibody ID5: Epitope characterization and minimal requirements for the recognition of polyglutamylated alpha- and beta-tubulin. *European Journal of Cell Biology*, *78*, 15–20.

Saudou, F., Finkbeiner, S., Devys, D., & Greenberg, M. E. (1998). Huntingtin acts in the nucleus to induce apoptosis but death does not correlate with the formation of intranuclear inclusions. *Cell*, *95*, 55–66.

Schulze, E., Asai, D. J., Bulinski, J. C., & Kirschner, M. (1987). Posttranslational modification and microtubule stability. *The Journal of Cell Biology*, *105*, 2167–2177.

Shang, Y., Li, B., & Gorovsky, M. A. (2002). Tetrahymena thermophila contains a conventional gamma-tubulin that is differentially required for the maintenance of different microtubule-organizing centers. *The Journal of Cell Biology*, *158*, 1195–1206.

Shida, T., Cueva, J. G., Xu, Z., Goodman, M. B., & Nachury, M. V. (2010). The major alpha-tubulin K40 acetyltransferase alphaTAT1 promotes rapid ciliogenesis and efficient mechanosensation. *Proceedings of the National Academy of Sciences of the United States of America*, *107*, 21517–21522.

Soppina, V., Herbstman, J. F., Skiniotis, G., & Verhey, K. J. (2012). Luminal localization of alpha-tubulin K40 acetylation by Cryo-EM analysis of Fab-labeled microtubules. *PLoS One*, *7*, e48204.

Suryavanshi, S., Edde, B., Fox, L. A., Guerrero, S., Hard, R., Hennessey, T., et al. (2010). Tubulin glutamylation regulates ciliary motility by altering inner dynein arm activity. *Current Biology*, *20*, 435–440.

Topalidou, I., Keller, C., Kalebic, N., Nguyen, K. C. Q., Somhegyi, H., Politi, K. A., et al. (2012). Genetically separable functions of the MEC-17 tubulin acetyltransferase affect microtubule organization. *Current Biology*, *22*, 1057–1065.

Valenzuela, P., Quiroga, M., Zaldivar, J., Rutter, W. J., Kirschner, M. W., & Cleveland, D. W. (1981). Nucleotide and corresponding amino acid sequences encoded by alpha and beta tubulin mRNAs. *Nature*, *289*, 650–655.

van Dijk, J., Miro, J., Strub, J.-M., Lacroix, B., van Dorsselaer, A., Eddé, B., et al. (2008). Polyglutamylation is a post-translational modification with a broad range of substrates. *The Journal of Biological Chemistry*, *283*, 3915–3922.

van Dijk, J., Rogowski, K., Miro, J., Lacroix, B., Eddé, B., & Janke, C. (2007). A targeted multienzyme mechanism for selective microtubule polyglutamylation. *Molecular Cell*, *26*, 437–448.

Walter, W. J., Beranek, V., Fischermeier, E., & Diez, S. (2012). Tubulin acetylation alone does not affect kinesin-1 velocity and run length in vitro. *PLoS One*, *7*, e42218.

Warn, R. M., Harrison, A., Planques, V., Robert-Nicoud, N., & Wehland, J. (1990). Distribution of microtubules containing post-translationally modified alpha-tubulin during Drosophila embryogenesis. *Cell Motility and the Cytoskeleton*, *17*, 34–45.

Wloga, D., Rogowski, K., Sharma, N., Van Dijk, J., Janke, C., Edde, B., et al. (2008). Glutamylation on alpha-tubulin is not essential but affects the assembly and functions of a subset of microtubules in Tetrahymena thermophila. *Eukaryotic Cell*, *7*, 1362–1372.

Wloga, D., Webster, D. M., Rogowski, K., Bre, M.-H., Levilliers, N., Jerka-Dziadosz, M., et al. (2009). TTLL3 Is a tubulin glycine ligase that regulates the assembly of cilia. *Developmental Cell*, *16*, 867–876.

Wolff, A., de Nechaud, B., Chillet, D., Mazarguil, H., Desbruyeres, E., Audebert, S., et al. (1992). Distribution of glutamylated alpha and beta-tubulin in mouse tissues using a specific monoclonal antibody, GT335. *European Journal of Cell Biology*, *59*, 425–432.

CHAPTER

Purification and Assembly of Bacterial Tubulin BtubA/B and Constructs Bearing Eukaryotic Tubulin Sequences

17

José M. Andreu and María A. Oliva

Centro de Investigaciones Biológicas, CSIC, Madrid, Spain

CHAPTER OUTLINE

Introduction and Rationale ... **270**

17.1 Materials: Genes, Constructs, and Expressed Proteins ... **272**

17.2 Methods ... **273**

17.2.1 BtubA/B Purification ... 273

17.2.2 BtubA and BtubB Purification ... 275

17.2.3 BtubA/B Chimera Containing Eukaryotic Sequences ... 276

17.2.4 Methods to Study Assembly of Purified BtubA/B ... 276

Conclusions ... **280**

Acknowledgments ... **280**

References ... **280**

Abstract

Bacterial tubulin BtubA/B is a close structural homolog of eukaryotic $\alpha\beta$-tubulin, thought to have originated by transfer of ancestral tubulin genes from a primitive eukaryotic cell to a bacterium, followed by divergent evolution. BtubA and BtubB are easily expressed homogeneous polypeptides that fold spontaneously without eukaryotic chaperone requirements, associate into weak BtubA/B heterodimers and assemble forming tubulin-like protofilaments. These protofilaments coalesce into pairs and bundles, or form five-protofilament tubules proposed to share the architecture of microtubules.

Bacterial tubulin is an attractive framework for tubulin engineering. Potential applications include humanizing different sections of bacterial tubulin with the aims of

Methods in Cell Biology, Volume 115

ISSN 0091-679X
http://dx.doi.org/10.1016/B978-0-12-407757-7.00017-7

creating recombinant binding sites for antitumor drugs, obtaining well-defined substrates for the enzymes responsible for tubulin posttranslational modification, or bacterial microtubule-like polymeric trails for motor proteins. Several divergent sequences from the surface loops of bacterial tubulin have already been replaced by the corresponding eukaryotic sequences, yielding soluble folded chimeras. We describe the purification protocol of untagged bacterial tubulin BtubA/B by means of ion exchange, size exclusion chromatography, and an assembly–disassembly cycle. This is followed by methods and examples to characterize its assembly, employing light scattering, sedimentation, and electron microscopy.

INTRODUCTION AND RATIONALE

Tubulin-like proteins share the structural fold of eukaryotic tubulin and are widespread in the prokaryotic world. These include essential cell division protein FtsZ (widely distributed through bacteria and archaea; Erickson, Anderson, & Osawa, 2010), bacterial tubulin BtubA/B (found only in several *Prosthecobacter* species; Jenkins et al., 2002), and the plasmid partition protein TubZ (present in large toxin encoding plasmids from *Bacillus* species; Larsen et al., 2007). A TubZ has also been found in a partition system encoded by phage cs-t of *Clostridium botulinum* (Oliva, Martin-Galiano, Sakaguchi, & Andreu, 2012), and a related protein PhuZ is encoded by phage 201phi2.1 of *Pseudomonas chlororaphis* (Kraemer et al., 2012).

Bacterial tubulin is the closest known prokaryotic homolog of eukaryotic tubulin. The *btubA* and *btubB* genes were discovered during the genome sequencing of the bacterium *Prosthecobacter dejongeii* (Jenkins et al., 2002). They perform an unclear cytoskeletal function in several *Prosthecobacter* species, whose genomes also encode the *ftsZ* essential for bacterial cell division (Pilhofer, Rosati, Ludwig, Schleifer, & Petroni, 2007). BtubA/B share higher sequence identity with eukaryotic tubulin ($\sim$35%) than the other tubulin structural homologs FtsZ ($\sim$17%), TubZ and PhuZ ($<$10%). Both proteins show a surprisingly high structural similarity with an root-mean square deviation (RMSD) of only of 1.3–1.5 Å with α- or β-tubulin, whereas when compared with FtsZ from *Methanococcus jannaschii*, the RMSD is 2.7 Å (Schlieper, Oliva, Andreu, & Lowe, 2005). A structural alignment between the bacterial and eukaryotic tubulin sequences is shown in Fig. 17.1. Bacterial tubulin includes the typical C-terminal helices H11 and H12 present in eukaryotic tubulin, which is the more divergent region between tubulin-like proteins. Though BtubA and BtubB lack the highly acidic 15–20 residue tails of eukaryotic tubulin that participate in interactions with microtubule-associated proteins and motors (Nogales, 2000), only BtubA has a charged C-terminal extension. Bacterial tubulin is homogeneous, lacking the isotype diversity and the posttranslational modifications of eukaryotic $\alpha\beta$-tubulin. The ease of expression and better stability of bacterial tubulin compared to eukaryotic tubulin make it an attractive framework for tubulin engineering, in addition to yeast tubulin (Ayaz, Ye, Huddleston, Brautigam, & Rice, 2012).

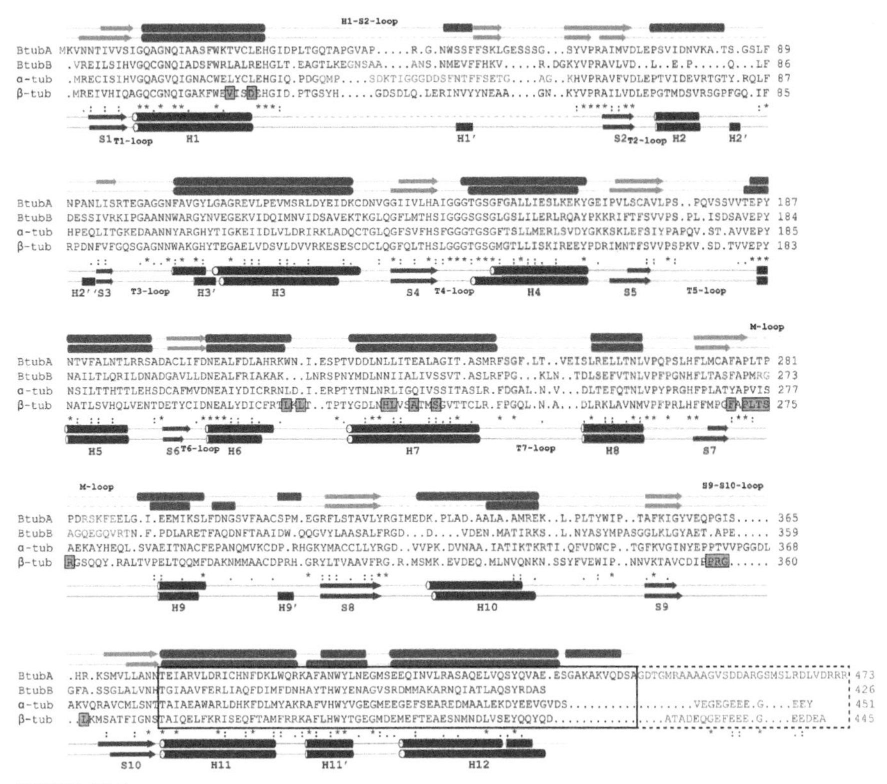

FIGURE 17.1

A structure-based alignment of bacterial BtubA and BtubB and eukaryotic α- and β-tubulin. The secondary structures of bacterial tubulin (PDB entry 2btq) and pig brain tubulin (PDB 1jff) were aligned with PDBeFold (http://www.ebi.ac.uk/msd-srv/ssm/; Krissinel & Henrick, 2004). The secondary structure element names of tubulin are as described before (Löwe, Downing, and Nogales, 2001) and the β-tubulin residues in contact with the anticancer drug taxol are in gray boxes. The amino acid residues which are missing from each crystal structure were added (dash box marked in blue) solely to show the complete protein sequences, whose loops should be separately aligned. The C-terminal domains are boxed. (See color plate.)

The BtubA/B structure showed a heterodimer and both proteins copolymerize in the presence of GTP and magnesium into tubulin-like protofilaments (Martin-Galiano et al., 2011; Schlieper et al., 2005; Sontag, Sage, & Erickson, 2009; Sontag, Staley, & Erickson, 2005). But in the BtubA/B heterodimer crystal structure, it is not possible to assign BtubA or BtubB to α- or β-tubulin, because it contains a continuous . . .ABABAB. . . filament, and both BtubA and BtubB have an activating T7 loop sequence and a short S9-S10 loop in the taxol binding pocket (Schlieper et al., 2005) as in α-tubulin. BtubA and BtubB have indeed mosaic sequences with intertwining features from both α- and β-tubulins (Martin-Galiano et al., 2011). An interesting difference between bacterial and eukaryotic tubulin, considering the structural similarity, is that BtubA/B is expressed soluble in *Escherichia coli* and can fold and form a weak dimer *in vitro* without chaperone requirements (Schlieper et al., 2005). In fact, the more divergent zones of bacterial tubulin correspond to the tubulin loops involved in binding to eukaryotic cytosolic chaperonin and in microtubule assembly (Martin-Galiano et al., 2011). Digging deeper into bacterial tubulin biochemical properties has showed that BtubA/B has more primitive assembly features, including a wider range of buffer conditions for polymerization. The distinct loop sequences and primitive assembly properties of bacterial tubulin support its origin from a spontaneously folding α- and β-tubulin ancestor shortly after heterodimer duplication, possibly by horizontal gene transfer from a primitive eukaryotic cell, followed by divergent heteropolymer evolution (Martin-Galiano et al., 2011).

Purified bacterial tubulin protofilaments further associate apparently forming filament pairs and bundles, which were observed in negative stain electron microscopy (Martin-Galiano et al., 2011; Schlieper et al., 2005; Sontag et al., 2005). However, using cryo-electron tomography, bacterial tubulin has been more recently found to form five-protofilament tubules in *Prosthecobacter* and in *E. coli* cells and BtubA/B purified with C-terminal His tags can also form five-protofilament tubules. Therefore, these structures have been suggested to be a primitive tubular architecture that later evolved into the typical 13-protofilaments eukaryotic microtubules (Pilhofer, Ladinsky, McDowall, Petroni, & Jensen, 2011). This opens the possibility of using recombinant bacterial tubules for microtubule research, although it would be necessary to quantify first the abundance of tubules and other polymers. The tubule formation should be confirmed as well with untagged bacterial tubulin, because the His tags interfere with BtubA/B assembly (Schlieper et al., 2005; Sontag et al., 2009).

The purpose of this chapter is to describe the necessary methods for purifying untagged bacterial tubulin and characterizing its assembly, with examples of BtubA/B constructs bearing sequences from eukaryotic tubulin studied in the authors' laboratory.

17.1 MATERIALS: GENES, CONSTRUCTS, AND EXPRESSED PROTEINS

The *btubA* and *btubB* genes from *P. dejongeii* DSM 1225 were cloned for biscistronic expression in a pHis17 vector, leaving the intergenic region intact and not adding any extra nucleotides to the protein genes (Schlieper et al., 2005). From this construct, untagged BtubA and BtubB are simultaneously expressed in *E. coli* C41 (DE3) cells, induced in an exponentially growing culture at OD 0.4 (600 nm), with 1 mM isopropyl β-D-thiogalactoside for 3 h at 37 °C. Bacterial cell pellets are resuspended to 15–20 mL of 50 mM Tris/HCl, pH 8.0, aliquots checked for BtubA and BtubB expression by SDS-PAGE and stored frozen at −75 °C. The plasmid carrying the *btubA* and *btubB* genes was employed as template to generate chimeras containing sequences from the bovine α-1 and β-2 tubulin isotypes, using internal oligonucleotides carrying the desired mutations or insertions by the QuikChange site-directed mutagenesis kit (Stratagene). Several BtubA/B chimeras were constructed in which the more divergent parts of bacterial loops were replaced by the corresponding residues from the eukaryotic H1-S1 loop (loop 1), the S7-H9 loop (M loop), and the S9-S10 loop (loop S) (Martin-Galiano et al., 2011). These sections from α- and β-tubulin were, respectively, exchanged into BtubA, BtubB, or both, and all constructions were checked by sequencing both *btubA* and *btubB* genes. These BtubA/B chimeras were expressed as the wild-type protein, but some of them were partially soluble and required reducing the induction temperature to 30 or 25 °C.

Standard centrifugation, chromatographic and electrophoresis equipment, and a spectrophotometer are employed for proteins purification and quantification. A fluorometer, a table-top ultracentrifuge, and access to electron microscopy are additionally required for the characterization of protein assemblies.

17.2 METHODS

17.2.1 BtubA/B purification

Bacterial tubulin BtubA/B (without affinity tags) was initially purified by anion exchange chromatography and a size exclusion gel filtration (Schlieper et al., 2005). Later, we added a polymerization–depolymerization cycle (Martin-Galiano et al., 2011), inspired by classical tubulin preparation procedures (Miller & Wilson, 2010). This additional step significantly improves the protein quality by removing the assembly incompetent BtubA and BtubB monomers. The following method starts from a 2-L cell culture pellet.

1. *Cells Breaking*: Once the bacterial cells pellet is completely thawed on ice, add 2 mg/mL lysozyme (Sigma) and 10 μg/mL DNAse (Roche) and keep the suspension on ice for five more minutes. Then break the cells by two passes in a cold French press and remove the cell debris by centrifugation of the lysate for 1 h at 100,000 × *g*, 4 °C.
2. *Anion Exchange Chromatography*: The supernatant should be filtered through a 0.22-μm filter to avoid clogging the chromatography column, and taken to 50 mL with Lysis buffer (50 mM Tris/HCl, pH 8) to facilitate protein binding to the anion exchange column. This is loaded on a 10-mL home-packed refrigerated Q Sepharose HP column preequilibrated with buffer A (20 mM Tris/HCl, 1 mM sodium azide, pH 8.5). BtubA and BtubB are eluted around 150 mM of NaCl concentration in a 0–50% gradient of buffer B (A plus 1 M NaCl) in 20 column volumes at 4 °C (flow, 5 ml/min) and collected in 10 mL fractions. All fractions containing BtubA or BtubB (easily distinguished in 12% acrylamide

SDS-PAGE) are pooled and concentrated at 4 °C to 2 ml with Centriprep 30 K (Amicon).

3. *Size Exclusion Chromatography*: Next, the sample is loaded onto a refrigerated home-packed Sephacryl S300 HR column (1.6 × 60 cm) that has been preequilibrated in buffer C (20 mM Tris, 1 mM EDTA, 1 mM sodium azide, pH 7.5) at 4 °C. Proteins are eluted at 1 mL/min, due to column pressure requirements and to allow maximum peak resolution, collecting 4 mL fractions. The main fractions containing BtubA and/or BtubB are then pooled and concentrated to about 0.5 mL at 4 °C with Centriprep 30 K. Note that BtubA and BtubB partially separate during the two chromatographic steps (2 and 3) and that the fractions should be chosen in this step to approach a 1:1 ratio of BtubA to BtubB. At this stage, purified BtubA/B is partially active in quantitative assembly experiments.
4. *Assembly/Disassembly Cycle*: To induce BtubA/B polymerization, 1 mM EGTA, 5 mM $MgCl_2$, 2 mM GTP, and 300 mM potassium glutamate (from a 3 M stock solution) are added to BtubA/B (~4 mg/mL final concentration; see measurement below). All components are mixed well avoiding foam formation and incubated for 10 min at 25 °C without agitation. The mixture is then centrifuged for 30 min at 100,000 × *g* 25 °C (in a prewarmed rotor). The semitransparent assembled protein pellet (P1) (about 40% of the purified BtubA/B) can be easily distinguished. After the supernatant is separated, the pellet is carefully redissolved in one volume of cold 20 mM Tris/HCl, 1 mM EGTA, pH 7.5, buffer, avoiding foam formation, during 20 min on ice to allow for the complete depolymerization of the BtubAB-assembled polymers. The solution is then centrifuged another 30 min, at 100,000 × *g* (at 4 °C in a second, precooled rotor). BtubA and BtubB are now recovered in the supernatant (S2), which must be clear and colorless, indicative of no protein aggregation or assembly. Finally, the protein is concentrated to approximately 0.5 mL and stored at −75 °C.
5. *Protein Concentration Measurement*: The BtubA/B concentration is accurately measured following a spectrophotometric method which takes into account the contribution of the bound guanine nucleotide. First, we determine the nucleotide concentration after extraction from small protein aliquots, diluted 25–50 times into ice-cold 0.5 N $HClO_4$ (final concentration). The acid solution is incubated on ice for 10 min and the denatured protein is removed by centrifugation (10 min, 10,000 × *g*, 4 °C). The guanine nucleotide remains in the supernatant and its concentration is then measured employing an extinction coefficient 12,400 M^{-1} cm^{-1} at 254 nm (Correia, Baty, & Williams, 1987). Second, the BtubA/B concentration is measured spectrophotometrically from its absorption spectrum in 6 M guanidinium chloride, after subtraction of the absorption due to the guanine nucleotide, employing a protein extinction coefficient 86,550 M^{-1} cm^{-1} at 280 nm (calculated from the protein sequence) and a nucleotide extinction coefficient 8100 M^{-1} cm^{-1} at 280 nm (Diaz et al., 2001). Note that these BtubA/B preparations could also be approximately measured employing a practical value of extinction coefficient of 100,000 ± 3000 M^{-1} cm^{-1} at 280 nm in aqueous buffer, including the bound

nucleotide contribution (0.68 ± 0.08 GTP and 1.01 ± 0.11 GDP bound per BtubA/B; Martin-Galiano et al., 2011).

BtubA/B prepared in this manner contains BtubA and BtubB in equimolar ratio and is ~100% active in polymerization above a critical concentration (Cr). Wild-type BtubA/B can also be purified for routine purposes performing the polymerization cycle on the high-speed supernatant of the cell lysate (following step 1) instead of doing it on the purified protein (following step 3). These preparations gave a somewhat higher yield of a similar BtubA/B, ~90% active in polymerization above Cr. However, we found that this procedure is not suited for BtubA/B constructs (below) with modified polymerization activity due to mutations or insertions.

17.2.2 BtubA and BtubB purification

Because *btubA* and *btubB* genes are cloned for bicistronic expression, both proteins are purified together. Nevertheless, for some experiments, it is necessary to use isolated BtubA and BtubB. For this purpose, we perform the size exclusion chromatography (step A3 above, using nonpolymerized BtubA/B) with a prepacked Sephacryl S200 HR 26/60 HP column (GE Healthcare) at 4 °C (run at 0.5 mL/min, taking 2 mL fractions). The BtubA and BtubB subunits separate despite their similar size, possibly because of BtubA dimerization or BtubB interaction with the gel matrix (Fig. 17.2). The nucleotide content measurements of BtubA and BtubB samples analyzed by HPLC after protein precipitation with ice-cold 0.5N $HClO_4$ showed that

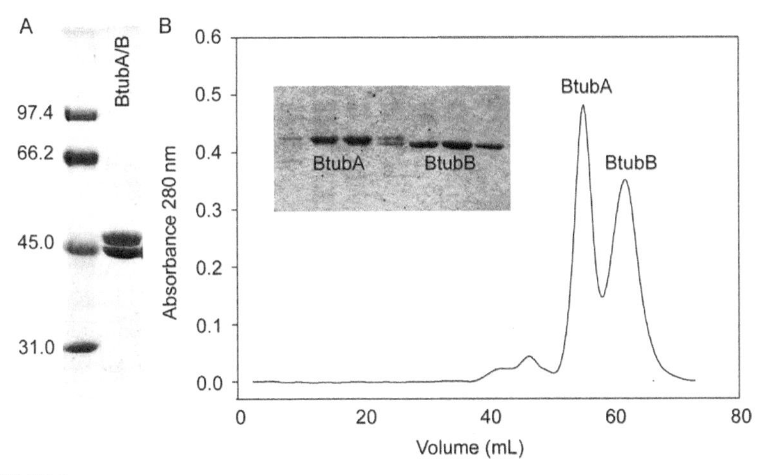

FIGURE 17.2

(A) Purified bacterial tubulin BtubA/B in gel electrophoresis. Molecular weight markers (kDa) are on the left lane. (B) Separation of BtubA and BtubB by gel chromatography in a Sephacryl S200 column (Methods, B).

isolated BtubA contained 0.13 $\pm$ 0.06 GTP and 0.53 $\pm$ 0.11 GDP and isolated BtubB contained 0.08 $\pm$ 0.07 GTP and 0.47 $\pm$ 0.09 GDP. Protein concentration can be determined spectrophotometrically in 6 M guanidinium chloride (as above), employing the extinction coefficients BtubA 47,700 M^{-1} cm^{-1} and BtubB 38,780 M^{-1} cm^{-1} at 280 nm, after subtracting the contribution of bound guanine nucleotide (as described in purification section 5 above). Separated untagged BtubA and BtubB were not able to assemble by themselves (in homo-polymers), and titration experiments showed that the maximum level of heteropolymer recovery by centrifugation is when BtubA and BtubB are mixed in an equimolar ratio.

17.2.3 BtubA/B chimera containing eukaryotic sequences

In a previous work, we constructed bacterial–eukaryotic tubulin chimeras and analyzed their biochemical behavior (Martin-Galiano et al., 2011). These chimeras, containing eukaryotic loop sequences 1, M, and S (see Section 1) from α- or β-tubulin, were purified as described above, as well as the wild-type protein for comparison purposes. We showed that every construct was folded and shared similar average secondary structure compared with wild-type BtubA/B when analyzed by circular dichroism. Furthermore, these chimeras assembled into polymers similar to the wild-type protein. This included introducing the M loop of α-tubulin into BtubA and that of β-tubulin into BtubB, or vice versa, the M loop of α-tubulin into BtubB and that of β-tubulin into BtubA. However, upon successive introduction of the eukaryotic sequences, the overall expression and purification yield decreased. A special case were the chimeras containing the α-tubulin S-loop, because only one-third of the expressed protein was soluble and only a small proportion of that was able to assemble during the polymerization step (section 4 in purification protocol) (Fig. 17.3). Therefore, when working with BtubA/B constructs with suspected inhibited polymerization activity, this step should be omitted.

17.2.4 Methods to study assembly of purified BtubA/B

There are several methods for determining the ability of BtubA/B and their constructs to polymerize into large head-to-tail ordered structures. The more frequent methods include right-angle light scattering, sedimentation, and electron microscopy.

The time course of BtubA/B assembly can be conveniently monitored by right-angle light scattering in a fluorometer with excitation and emission wavelengths set at 350 nm, employing a temperature controller. As other tubulin-like proteins, BtubA/B assembly requirements include Mg^{2+}, K^{+}, and a guanosine nucleotide triphosphate. It is important to consider that untagged BtubA/B hardly polymerizes in typical microtubule assembly buffers. Instead, we use the polymerization buffer (20 mM Tris, 500 mM KCl, 1 mM EGTA, pH 7.5) and add 5 mM $MgCl_2$ followed by GTP (0.1 or 1 mM), or the slowly hydrolysable analogue GMPCPP (0.1 mM), to start the assembly reaction at 25 °C (Martin-Galiano et al., 2011). To ensure the

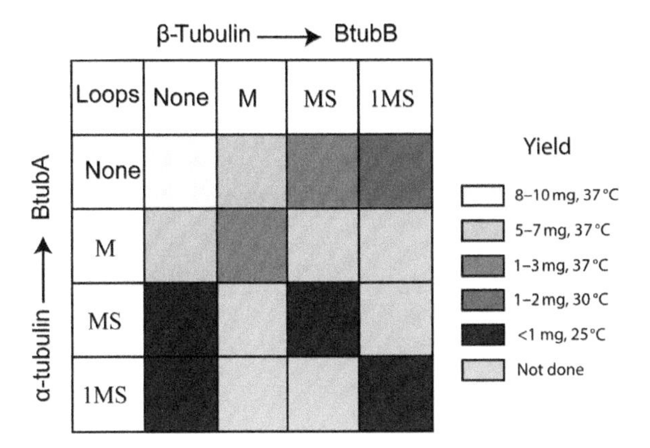

FIGURE 17.3

Recovery of BtubA/B chimera containing eukaryotic tubulin loop sequences 1, M, and S in one or both subunits, following expression and purification as for wild-type BtubA/B (Martin-Galiano et al., 2011). (See color plate.)

This research was originally published in the Journal of Biological Chemistry. (2011) 286, 19789-19803 © the American Society for Biochemistry and Molecular Biology.

correct protein behavior, it is important to examine the BtubA/B polymer reversibility, monitoring disassembly by GTP (0.1 mM) exhaustion or upon excess GDP addition. Notice that this buffer does not support eukaryotic tubulin assembly, but we have observed that both proteins polymerize with GTP in 20 mM Tris, 300 mM potassium glutamate, 5 mM $MgCl_2$, 1 mM EGTA, pH 7.5, at 30 °C. Our results also showed that BtubA/B polymerization is more weakly dependent on pH, temperature, and divalent cations than eukaryotic tubulin (Martin-Galiano et al., 2011). This method requires a relatively high amount of protein but directly monitors the filaments assembly and disassembly dynamicity.

Sedimentation experiments allow measuring the amount of polymer formation by centrifuging assembled reaction mixtures and by carefully quantifying pellets and supernatants in stained SDS-PAGE gels. We usually prepare 0.1 mL samples and after an incubation period, centrifuge them at 250,000 × g 10 min in a Beckman Optima table-top ultracentrifuge. The advantage of this method is that it requires less amount of protein and makes monitoring a wide range of conditions in a single experiment possible. Figure 17.4 shows an example of BtubA/B polymerization analyzed by sedimentation (top) and the comparison using both sedimentation and the maximal light scattering signal for critical concentration measurement (Cr = 0.9 μM).

Notice the importance of confirming the morphology of the BtubA/B polymers by negative stain electron microscopy of unfixed samples. This is necessary to truly relate sedimentation and light-scattering profiles to the formation of ordered

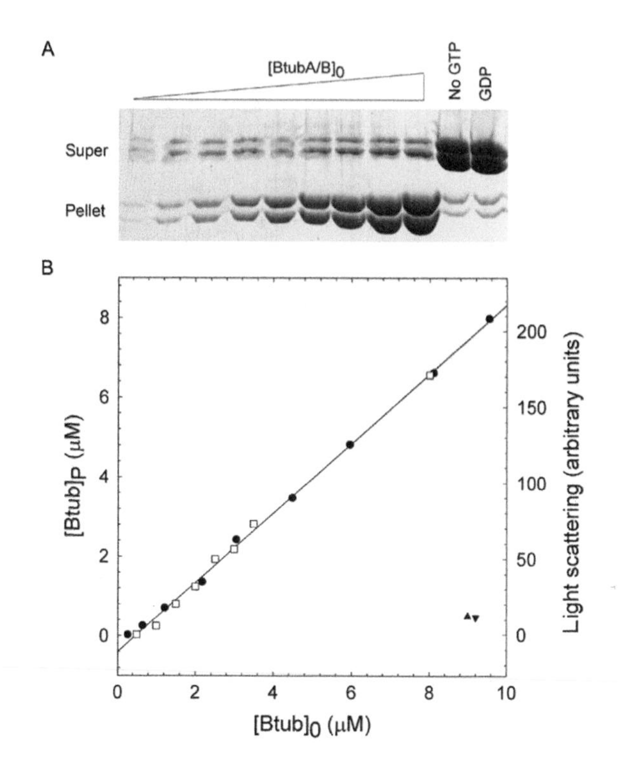

FIGURE 17.4

Polymerization of BtubA/B in 20 mM Tris, 300 mM potassium glutamate, 1 mM EGTA, 5 mM $MgCl_2$, 1 mM GTP, pH 7.5, at 25 °C, measured by sedimentation. (A) SDS-PAGE of the pellets and supernatants, sequentially loaded with an electrophoretic shift in the same lanes. (B) Protein sedimentation measurements (left axis, filled circles, pellet protein concentrations; the triangles are controls without nucleotide and with GDP), compared with light scattering values (right axis, empty squares, plateau scattering). The solid line is the least squares linear fit to the centrifugation data. This BtubA/B was prepared with the polymerization–depolymerization cycle right after the high-speed centrifugation of the cell lysate (purification step 1).

polymers and rule out the possible formation of aggregates. As an example, Fig. 17.5 shows the assembly a BtubA/B chimera in which the M loop of β-tubulin was exchanged into BtubB, in comparison with wild-type BtubA/B and microtubules, in Pipes microtubule assembly buffer with the slowly hydrolysable nucleotide GMPCPP, examined with light scattering, sedimentation, and electron microscopy. The repeat distances between monomers in the BtubA/B polymer lattices can be analyzed with diffraction patterns computed from their micrographs (Martin-Galiano et al., 2011; Pilhofer et al., 2011; Schlieper et al., 2005).

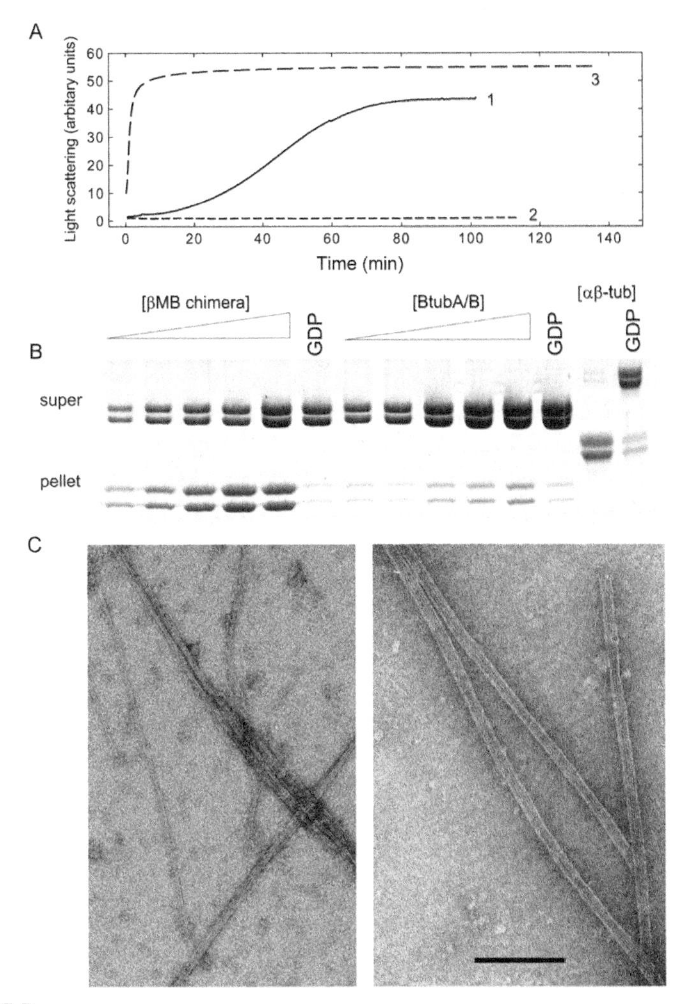

FIGURE 17.5

(A) Light scattering time courses of assembly of 10 μM (1) chimera BtubA/βMB containing the M loop from β-tubulin in BtubB, (2) wild-type BtubA/B, and (3) calf brain αβ-tubulin in 80 mM Pipes/NaOH, 1 mM EGTA, 6 mM $MgCl_2$, 0.1 mM GMPCPP, pH 6.8 at 30 °C. (B) Corresponding SDS-PAGE electrophoresis, following 2 h assembly and centrifugation, of BtubA/βMB (5, 10, 15, 20, 25 μM), BtubA/B (10, 15, 20, 25, 30 μM), and αβ-tubulin (10 μM) and controls with 1 mM GDP instead of GMPCPP. (C) Representative electron micrographs of negatively stained polymers of chimera BtubA/β MB (left, similar to 20μM wild-type BtubA/B) and microtubules assembled from αβ-tubulin (right). The bar indicates 200 nm.

CONCLUSIONS

In this chapter, we have described methods for purification of bacterial tubulin BtubA/B and chimera containing eukaryotic tubulin sequences, as well as for characterization of their assembly, aiming to facilitate tubulin engineering for microtubule research, based on easily expressed bacterial constructs. In principle, any different eukaryotic tubulin sequences can be replaced into bacterial tubulin (see Section on "Introduction"; Fig. 17.1), although this may come at the expense of reduced solubility. For example, we have observed how an accumulation of several eukaryotic tubulin loops, and in particular the S9-S10 loop from α-tubulin, results in a decrease in the yield of soluble purified chimera (Fig. 17.3; Martin-Galiano et al., 2011).

Given their close structural similarity, it is conceivable that different sections of bacterial tubulin may be humanized with the aims of creating recombinant binding sites for antitumor drugs (Jordan & Wilson, 2004), obtaining well-defined substrates for the tubulin posttranslational modification enzymes (Janke & Bulinski, 2011), or bacterially produced microtubule-like polymers that may be employed as trails for microtubule motor proteins (Vale, 2003). The latter two potential applications require replacing C-terminal segments of bacterial tubulin by eukaryotic sequences. This appears especially feasible, because the α(404–451) and β(394–445) C-terminal sequences from the α-1 and β-2 vertebrate tubulin isotypes are well soluble following expression in *E. coli*, and they form in trifluoroethanol helical segments similar to native tubulin (Jimenez et al., 1999). This suggests that it may be feasible to replace these sections of eukaryotic tubulin, including helix H12 and the C-terminal tails, or perhaps the whole C-terminal domain, into bacterial tubulin.

Acknowledgments

We thank our collaborators A.J. Martin-Galiano and the laboratories of J. Löwe and J.M. Valpuesta. Work in our laboratory was supported by Grant BFU 2011–23416 and CAM S2010/2353 (J. M. A.) and a Ramon y Cajal contract (M. A. O.).

References

Ayaz, P., Ye, X. C., Huddleston, P., Brautigam, C. A., & Rice, L. M. (2012). A TOG:alpha beta-tubulin complex structure reveals conformation-based mechanisms for a microtubule polymerase. *Science*, *337*, 857–860.

Correia, J. J., Baty, L. T., & Williams, R. C. (1987). Mg-2+ dependence of guanine-nucleotide binding to tubulin. *The Journal of Biological Chemistry*, *262*, 17278–17284.

Diaz, J. F., Kralicek, A., Mingorance, J., Palacios, J. M., Vicente, M., & Andreu, J. M. (2001). Activation of cell division protein FtsZ—Control of switch loop T3 conformation by the nucleotide gamma-phosphate. *The Journal of Biological Chemistry*, *276*, 17307–17315.

Erickson, H. P., Anderson, D. E., & Osawa, M. (2010). FtsZ in bacterial cytokinesis: Cytoskeleton and force generator All in One. *Microbiology and Molecular Biology Reviews*, *74*, 504–528.

Janke, C., & Bulinski, J. C. (2011). Post-translational regulation of the microtubule cytoskeleton: Mechanisms and functions. *Nature Reviews. Molecular Cell Biology*, *12*, 773–786.

Jenkins, C., Samudrala, R., Anderson, I., Hedlund, B. P., Petroni, G., Michailova, N., et al. (2002). Genes for the cytoskeletal protein tubulin in the bacterial genus Prosthecobacter. *Proceedings of the National Academy of Sciences of the United States of America*, *99*, 17049–17054.

Jimenez, M. A., Evangelio, J. A., Aranda, C., Lopez-Brauet, A., Andreu, D., Rico, M., et al. (1999). Helicity of alpha(404-451) and beta(394-445) tubulin C-terminal recombinant peptides. *Protein Science*, *8*, 788–799.

Jordan, M. A., & Wilson, L. (2004). Microtubules as a target for anticancer drugs. *Nature Reviews. Cancer*, *4*, 253–265.

Kraemer, J. A., Erb, M. L., Waddling, C. A., Montabana, E. A., Zehr, E. A., Wang, H. N., et al. (2012). A phage tubulin assembles dynamic filaments by an atypical mechanism to center viral DNA within the host cell. *Cell*, *149*, 1488–1499.

Krissinel, E., & Henrick, K. H. (2004). Secondary-structure matching (PDBeFold), a new tool for fast protein structure alignment in three dimensions. *Acta Crystallographica. Section D, Biological Crystallography*, *60*, 2256–2268.

Larsen, R. A., Cusumano, C., Fujioka, A., Lim-Fong, G., Patterson, P., & Pogliano, J. (2007). Treadmilling of a prokaryotic tubulin-like protein, TubZ, required for plasmid stability in Bacillus thuringiensis. *Genes & Development*, *21*, 1340–1352.

Löwe, J., Li, H., Downing, K. H., & Nogales, E. (2001). Refined structure of alpha beta-tubulin at 3.5 Å resolution. *Journal of Molecular Biology*, *313*, 1045–1057.

Martin-Galiano, A. J., Oliva, M. A., Sanz, L., Bhattacharyya, A., Serna, M., Yebenes, H., et al. (2011). Bacterial tubulin distinct loop sequences and primitive assembly properties support its origin from a eukaryotic tubulin ancestor. *The Journal of Biological Chemistry*, *286*, 19789–19803.

Miller, H. P., & Wilson, L. (2010). Preparation of microtubule protein and purified tubulin from bovine brain by cycles of assembly and disassembly and phosphocellulose chromatography. *Methods in Cell Biology*, *95*, 3–15.

Nogales, E. (2000). Structural insights into microtubule function. *Annual Review of Biochemistry*, *69*, 277–302.

Oliva, M. A., Martin-Galiano, A. J., Sakaguchi, Y., & Andreu, J. M. (2012). Tubulin homolog TubZ in a phage-encoded partition system. *Proceedings of the National Academy of Sciences of the United States of America*, *109*, 7711–7716.

Pilhofer, M., Ladinsky, M. S., McDowall, A. W., Petroni, G., & Jensen, G. J. (2011). Microtubules in bacteria: Ancient tubulins build a five-protofilament homolog of the eukaryotic cytoskeleton. *PLoS Biology*, *9*, e1001213.

Pilhofer, M., Rosati, G., Ludwig, W., Schleifer, K. H., & Petroni, G. (2007). Coexistence of tubulins and ftsZ in different Prosthecobacter species. *Molecular Biology and Evolution*, *24*, 1439–1442.

Schlieper, D., Oliva, M. A., Andreu, J. M., & Lowe, J. (2005). Structure of bacterial tubulin BtubA/B: Evidence for horizontal gene transfer. *Proceedings of the National Academy of Sciences of the United States of America*, *102*, 9170–9175.

Sontag, C. A., Sage, H., & Erickson, H. P. (2009). BtubA-BtubB heterodimer is an essential intermediate in protofilament assembly. *PloS One*, *4*, e7253.

Sontag, C. A., Staley, J. T., & Erickson, H. P. (2005). In vitro assembly and GTP hydrolysis by bacterial tubulins BtubA and BtubB. *The Journal of Cell Biology*, *169*, 233–238.

Vale, R. D. (2003). The molecular motor toolbox for intracellular transport. *Cell*, *112*, 467–480.

CHAPTER

Microtubule-Associated Proteins and Tubulin Interaction by Isothermal Titration Calorimetry

18

P.O. Tsvetkov*, P. Barbier[†], G. Breuzard[†], V. Peyrot[†] and F. Devred[†]

*Engelhardt Institute of Molecular Biology, Russian Academy of Sciences, Moscow, Russia

[†]Aix-Marseille University, Inserm, CRO2 UMR_S 911, Faculté de Pharmacie, 13385 Marseille, France

CHAPTER OUTLINE

Introduction **284**
18.1 Isothermal Titration Calorimetry **286**
18.1.1 Principles 286
18.1.2 Experimental Procedure 286
18.1.3 Binding Models 287
18.1.4 Thermodynamic Profile of Binding (Meaning of Enthalpy and Entropy Signs) 289
18.1.5 Temperature Dependence of ΔH 289
18.2 Tubulin and MAPs SAMPLE PREPARATION **290**
18.2.1 Equilibration 290
18.2.2 Determination of Protein Concentrations 291
18.2.3 Temperature 292
18.2.4 Buffer Conditions 293
18.2.5 Reaction Volume and Duration 293
18.3 Results: Tubulin/MAPs by ITC **294**
18.3.1 Stathmin–Tubulin Interaction 295
18.3.2 Tau–Tubulin 297
Conclusion **298**
Acknowledgment **299**
References **299**

Methods in Cell Biology, Volume 115

ISSN 0091-679X
http://dx.doi.org/10.1016/B978-0-12-407757-7.00018-9

Abstract

Microtubules play an important role in a number of vital cell processes such as cell division, intracellular transport, and cell architecture. The highly dynamic structure of microtubules is tightly regulated by a number of stabilizing and destabilizing microtubule-associated proteins (MAPs), such as tau and stathmin. Because of their importance, tubulin–MAPs interactions have been extensively studied using various methods that provide researchers with complementary but sometimes contradictory thermodynamic data. Isothermal titration calorimetry (ITC) is the only direct thermodynamic method that enables a full thermodynamic characterization (stoichiometry, enthalpy, entropy of binding, and association constant) of the interaction after a single titration experiment. This method has been recently applied to study tubulin–MAPs interactions in order to bring new insights into molecular mechanisms of tubulin regulation. In this chapter, we review the technical specificity of this method and then focus on the use of ITC in the investigation of tubulin–MAPs binding. We describe technical issues which could arise during planning and carrying out the ITC experiments, in particular with fragile proteins such as tubulin. Using examples of stathmin and tau, we demonstrate how ITC can be used to gain major insights into tubulin–MAP interaction.

INTRODUCTION

Microtubules, which consist of polymerized tubulin heterodimers, play key roles in numerous cell processes, including mitosis, active intracellular transport, and neuronal plasticity. Microtubules are highly dynamic structures that switch from elongation to shrinking phase and *vice versa,* depending on different regulatory factors (Mitchison & Kirschner, 1984). Under physiological conditions, microtubule dynamics is tightly controlled by stabilizing microtubule-associated proteins (MAPs), such as tau, and destabilizing factors or proteins, such as stathmin (Fig. 18.1). These proteins are in turn regulated by posttransitional modifications, the most studied one being phosphorylation. Targeting microtubules and their dynamics is also a well-established area of anticancer research. Indeed, altering microtubule dynamics by drugs named microtubule targeting agents (MTAs) leads to mitotic block and cell death. MTAs that can also be divided into two classes, molecules that stabilize microtubules, like taxanes, and molecules that induce microtubule depolymerization, like vinca-alkaloids, have been studied using a broad spectrum of techniques (for review, see Calligaris et al., 2010). Deciphering the molecular mechanisms of microtubule regulation by MAPs is fundamental to understanding the biology of the cell. This is of particular interest in oncology as it was proposed that resistance to anticancer drugs could be linked to the level of expression of MAPs (Alli, Yang, Ford, & Hait, 2007; Cucchiarelli et al., 2008; see Chapter 5). A better knowledge of MAP–MTA interplay would help orient treatments and possibly lead to the discovery of new therapeutic strategies. Since the discovery of the tau family

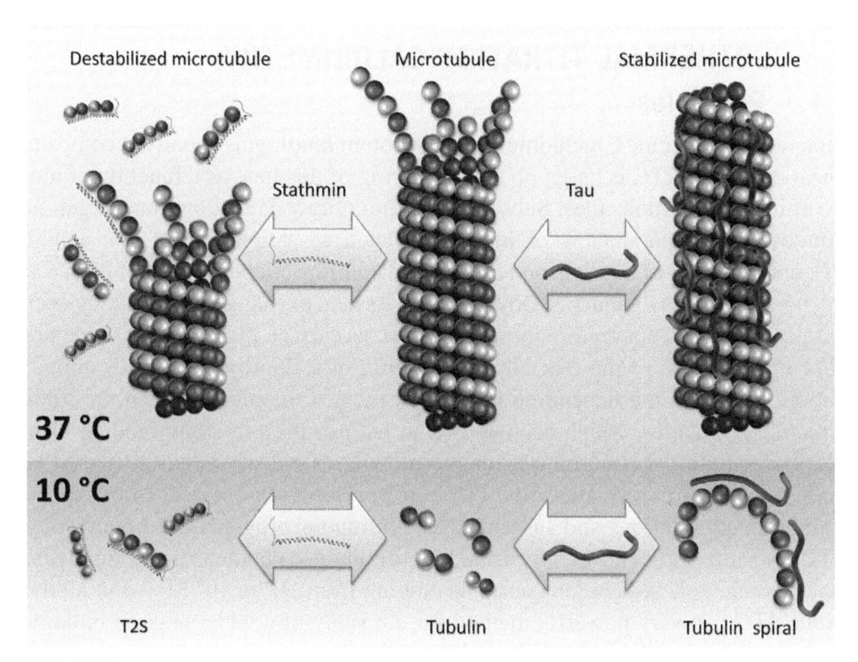

FIGURE 18.1

MAPs–tubulin interactions and their consequences on tubulin self-assembly at 10 and 37 °C. This scheme represents tubulin self-assembly in a classical polymerization buffer (in which tubulin spontaneously polymerizes into MT at 37 °C). At 10 °C, free tubulin forms T2S complex in the presence of stathmin and circular protofilaments in the presence of tau. At 37 °C, MTs depolymerize in the presence of stathmin and tubulin and stathmin form T2S complex, whereas MTs are stabilized in the presence of tau. (See color plate.)

(Weingarten, Lockwood, Hwo, & Kirschner, 1975) and until the identification of stathmin (Belmont & Mitchison, 1996), many MAPs had been discovered and investigated. Their interaction with tubulin has been and is still being studied, using numerous biochemical and biophysical methods both *in vitro* (Devred et al., 2010; Kiris, Ventimiglia, & Feinstein, 2010; Ross & Dixit, 2010; Wilson & Correia, 2010) and *in vivo* (Drubin & Kirschner, 1986; Konzack, Thies, Marx, Mandelkow, & Mandelkow, 2007; Samsonov, Yu, Rasenick, & Popov, 2004; Weissmann et al., 2009). In this chapter, we focus on the use of isothermal titration calorimetry (ITC). ITC is a very useful technique that has been extensively used in other fields, but that is still rarely utilized to study the tubulin cytoskeleton. After describing briefly the principle of ITC, we focus on the important steps (from sample preparation to data analysis) when applying ITC to the tubulin cytoskeleton and MAPs. Taking as examples the case of a destabilizing MAP (stathmin) and stabilizing MAP (tau), we illustrate the pitfalls that must be avoided and the precautions that must be taken in order to use this technique efficiently to gain new insights into the molecular mechanism of action of MAPs on MTs.

18.1 ISOTHERMAL TITRATION CALORIMETRY

18.1.1 Principles

Interaction of molecules, including protein–protein binding, is usually accompanied by heat exchange. ITC is based on the measuring of this heat as a function of molar ratio of interacting molecules. Subsequent fitting of raw data allows one to get most thermodynamical parameters of interaction such as stoichiometry (N), enthalpy (ΔH), entropy (ΔS) of binding, and association equilibrium constant (K_a) from a single 1-h-experiment (Ladbury & Doyle, 2004). As heat exchange upon binding occurs naturally, ITC does not require immobilization, as surface plasmon resonance does, and/or modification of the reactants by addition of a fluorophore, for example. In addition, ITC does not depend on the size or mass difference between the studied interacting molecules, which enables ITC to be also used to study binding of low molecular weight drugs and metal ions to proteins (Tsvetkov et al., 2010). At last, contrary to spectroscopic methods, ITC can be used with colored, turbid or even not transparent solutions and suspensions. Continuous progress and improvements in ITC instrumentation now allow using ITC routinely to characterize thermodynamics of binding with association constants ranging from 10^3 to 10^8 M^{-1}. For all these reasons, ITC is a very powerful method for the study of a wide range of biological systems under near physiological conditions (Ladbury & Doyle, 2004).

18.1.2 Experimental procedure

An ITC apparatus consists of a calorimetric cell in an isothermal jacket and a syringe that is inserted in the cell (Ladbury & Doyle, 2004; Pierce, Raman, & Nall, 1999). To perform ITC experiments, the calorimetric cell is filled with protein solution at a concentration close to the expected dissociation constant, and the titration syringe is filled with about a 10-fold more concentrated solution of the second interactant (usually called ligand) in an identical buffer. During each successive injection of small aliquots of ligand into the cell, the microcalorimeter registers heat exchange. The released heat is proportional to the amount of complex formed after each injection and decreases as the protein gets saturated by the ligand (Freyer & Lewis, 2008). During final injections, as there is no more binding of the ligand to the protein, the measured heat corresponds to the heat of mixing of the two solutions, which is often referred to as heat of dilution. The signal due to dilution can be significant in the case of imbalance in compositions of buffers between the calorimetric cell and the syringe, and/or a high intrinsic dilution heat of the ligand. This is why the sample preparation is a crucial step in the experimental procedure (see Section 18.2.1). Areas under each peak corresponding to heat exchanges are then plotted against molar ratio of ligand over protein in order to obtain a thermogram or binding isotherm. To determine the thermodynamic parameters, this binding isotherm should be fitted with a theoretical binding isotherm curve. All ITC instruments are supplied with the software that offers several standard models of interaction.

18.1.3 Binding models

Generally, the list of models includes one-set-of-sites, two-sets-of-sites, and sequential-binding-sites models. To choose the appropriate binding model, reasonable assumptions should be made, based on preliminary knowledge about the investigated system and/or the shape of the binding isotherm. The first model, one-set-of-sites, is the most simple one and can be used for the systems where the ligand has one or more equal independent (noninteracting) sites on the target molecule. The binding isotherm for such a system represents a monotonic curve which has no or one inflection point (Fig. 18.2). When concentration of investigated molecule $[M]$ in the calorimetric cell is close to optimal value ($10/NK_a < [M] < 100/NK_a$), the binding isotherm has a sigmoid shape with an inflection point close to the molar ratio corresponding to the stoichiometry of the interaction (N) (Ladbury & Doyle, 2004). This value should be

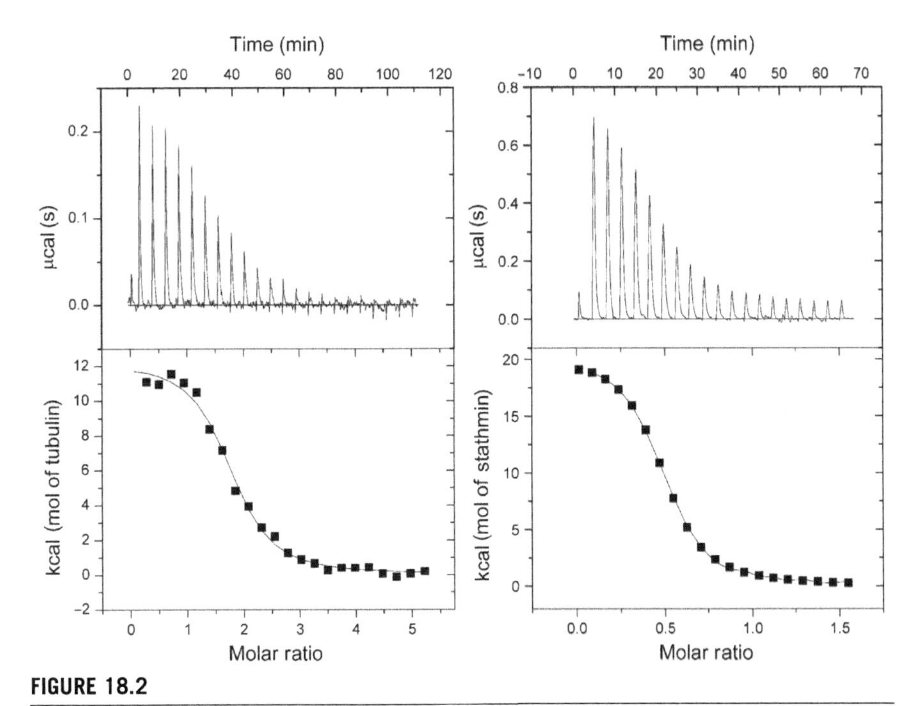

FIGURE 18.2

Direct and reverse titration by ITC. Raw ITC curves (top panels) and binding isotherms with fitting curves (bottom panels) of stathmin titration by tubulin (left panels) and tubulin titration by stathmin (right panels). Both thermograms are monophasic curves with an inflection point at 2 for direct titration and 0.5 for reverse titration. Thermodynamic parameters of both titrations are in good agreement. Fitting binding isotherms with model of one-set-of-sites yielded the same constants for both titrations and the enthalpy of binding around 14 kcal/(mole of tubulin) for direct binding and 27 kcal/(mole of stathmin) for reverse titration. The last value corresponds to overall enthalpy of formation of the T2S complex (ΔH_{T2S}).

equal to n or $1/n$, where n is an integer (Fig. 18.2). When stoichiometry of the interaction is slightly different from these "allowed" values, it indicates that there is an error in concentration of either interacting molecules or that a portion of one of the interactants is not active (mis-folding, aggregation, etc.). In both cases, concentrations should be corrected to the closest allowed value. Substoichiometry ($1/n$) can be observed in two cases: when the titration syringe is filled with the solution of a studied molecule that has several binding sites and the ligand is placed in the calorimetric cell (usually called "reverse titration"), and/or when titrated molecules are oligomers that bind only one molecule of ligand. For stathmin–tubulin binding, direct titration leads to a stoichiometry of 2, whereas reverse titration leads to a stoichiometry of 0.5 (Fig. 18.2). It should always be kept in mind that ΔH result is given per mole of molecule in the syringe. For example, in the case of the formation of T2S complex between stathmin (S) and tubulin (T), direct titration is stathmin titrated by tubulin, and reverse titration is tubulin titrated by stathmin and $\Delta H_{T2S} = \Delta H_{revers} = 2\Delta H_{direct}$. In case of substoichiometry, it is strongly recommended to also perform the opposite titration to obtain stoichiometry more precisely. For example, the difference between a stoichiometry of three and a stoichiometry of four would correspond to a 4–3 = 1 shift in direct titration, whereas it would correspond only to a ¼ − ⅓ = 0.08 shift of inflection points in reverse titration. It should be noted that even if experimental data can be fitted with the model of one-set-of-sites, it does not necessarily mean that the interaction occurs through identical or noninteracting sites. For example, if there are two identical (with the same ΔH and K_a) strongly interacting sites and if the binding with the first site modifies only the constant of binding to the second site (but not the enthalpy), then the resulting binding isotherm obtained by ITC will be indistinguishable from that for "equal noninteracting sites" model with higher constants of binding. If a binding isotherm has several inflection points or extremes, it indicates that there is more than one nonequal or interacting binding site. In order to fit such binding isotherms, more complex models should be used. The standard software gives a choice between two-sets-of-sites and sequential-binding-sites models. The first one implies the existence of two sets of noninteracting sites and allows the determination of stoichiometry, binding constant, enthalpy, and entropy of binding for each set of sites ($[n_1, K_{a1}, \Delta H_1, \Delta S_1]$, $[n_2, K_{a2}, \Delta H_2, \Delta S_2]$). The second one assumes the preliminary knowledge of the stoichiometry of binding, which should be set before fitting. If any reasonable assumption about stoichiometry cannot be made, then fitting with a model of interacting binding sites should be started from two sites. In the case of an unsatisfactory fitting result, the number of binding sites used in the model should be sequentially increased until a further rise in the number of sites will not give a significant drop in fitting error. Unfortunately, in some cases, the experimental binding isotherm could be well fitted using both models and additional experiments are necessary to determine the precise model of binding. If none of the models described above can be used to fit the experimental data, researchers should implement their own model of binding. The choice between models of binding could be a rather tricky problem that goes beyond the scope of this chapter. Nevertheless, authors should always take into account that in biology, as in any other field, "all models are wrong, but some are useful" (Box & Draper, 1987).

18.1.4 Thermodynamic profile of binding (meaning of enthalpy and entropy signs)

Interactions can occur only if the variation of free Gibbs energy of the process is negative ($\Delta G < 0$). This is a fundamental thermodynamic law, which is valid for all interacting systems. Gibbs energy has two components, enthalpic and entropic: $\Delta G = \Delta H - T\Delta S$. Fitting a binding isotherm allows us to determine ΔH of binding and association constant K_a. Using the above equation and standard thermodynamic relationship ($\Delta G = -\mathrm{RT} \ln K_a$), entropy of binding could be easily calculated (Ladbury & Doyle, 2004). Thus, contrary to other methods, ITC allows one to determine both components of Gibbs energy after one single experiment, providing us with information about the nature of the interaction. The values of enthalpy and entropy can be either positive or negative. They constitute the energetic signature of the interaction, also referred to as the thermodynamic profile of binding. Interactions are going to be favored by negative ΔH and/or positive ΔS. If ΔH is negative (exothermic reaction), the entropic component of free Gibbs energy could be either favorable ($\Delta S > 0$) or unfavorable ($\Delta S < 0$), as long as ΔG stays negative. Otherwise, if ΔH is positive (endothermic reaction), then entropy of binding should be favorable ($\Delta S > 0$). In the last case, it can be typically concluded that binding is driven by hydrophobic interactions. For example, during binding, there is a burying of hydrophobic areas in the interface of interaction or conformational changes in one interacting molecule that lead to hiding of hydrophobic surfaces. Otherwise, a highly favorable enthalpy and an unfavorable entropy of binding are usually associated with a high degree of hydrogen bonding formed upon interaction, in addition to conformational changes (Ladbury & Doyle, 2004; Ross & Subramanian, 1981). In addition to providing information about driving forces, the thermodynamic parameters of interaction of tubulin with different ligands can sometimes be correlated with the differences in biological activity between these ligands (Buey et al., 2004, 2005).

18.1.5 Temperature dependence of ΔH

It should be noted that ΔH of binding depends on temperature. In the temperature range where interacting molecules are not denatured, this dependence is linear and usually has a negative slope. This slope corresponds to heat capacity change of binding (ΔC_p) which is generally correlated with the surface of the area buried upon complex formation (Ladbury & Doyle, 2004). A consequence of this temperature dependence is that at certain temperatures, the enthalpy of binding could be equal to zero, making such binding undetectable by ITC. In other words, the absence of signal during an ITC experiment does not necessarily mean that there is no interaction between molecules, but could signify that ΔH of binding is equal to zero at the chosen experimental temperature. In this case, entropy is the driving force of the interaction. Fortunately, modern microcalorimeters allow carrying out titration experiments at a wide range of temperatures.

18.2 TUBULIN AND MAPs SAMPLE PREPARATION

Because of certain peculiarities of tubulin, its interaction with regulatory proteins has been studied by ITC only occasionally, despite the growing popularity of this method and its obvious advantages. The difficulties that could arise during such a study necessitate a deep knowledge of both method details and tubulin cytoskeleton regulatory mechanisms. In this section, we want to draw attention to some important points about tubulin and certain MAPs sample preparations for ITC experiments.

18.2.1 Equilibration

To minimize the heat signal due to the dilution of the samples during injections, a balance between the composition of buffers in the calorimetric cell and the syringe needs to be established. Due to the high sensitivity of microcalorimeters, the two solutions must be matched with regard to composition, pH, buffer, and salt concentrations. A slight mismatch between the two solutions may lead to heat of dilution that could overwhelm the heat of the binding reaction. Usually, to achieve the perfect match between buffers in the cell and syringe, dialysis of both interactant solutions against the same buffer is used. Unfortunately, due to tubulin instability, buffer specificity, and the necessity of keeping a high concentration of ligand, this option is not appropriate. After purification, when tubulin is stored, 1 M sucrose buffer to stabilize its conformation upon freezing (Frigon & Lee, 1972), the sucrose should be completely removed from buffer before ITC experiment, since it significantly contributes to the dilution effect. However, extensive dialysis cannot be used because of low tubulin stability over an extended period of time in the absence of a stabilizer. Previously, we described a tubulin equilibration procedure using two custom-made columns filled with Sephadex G25 (Andreu & Timasheff, 1982; Barbier, Peyrot, Leynadier, & Andreu, 1998; Devred et al., 2010; Na & Timasheff, 1982; Peyrot et al., 1992). Later, we optimized the protocol by replacing these two columns by a single desalting Hitrap column (GE Healthcare) on an AKta Purifier FPLC system. This allowed us to reduce the time and to increase the yield of tubulin preparation. Tubulin can also be commercially bought as a powder, which contains stabilizers of tubulin that should be removed by running the tubulin preparation on a desalting column. MAPs, such as stathmin and tau, can be dry-lyophilized and then stored as powders (Devred et al., 2004, 2008). Direct dilution of lyophilized proteins in experimental buffer often results in an increase in the dilution signal, even if MAPs were dialyzed against water to eliminate salt before dry-lyophilization. Thus, when used for ITC experiments, dry-lyophylized MAPs should be resuspended in the buffer of interest, centrifuged to remove aggregated proteins, and ran on the desalting column identical to the one used for tubulin.

18.2.2 Determination of protein concentrations

ITC is based on measuring the heat exchange during the interaction as a function of the ratio of interacting molecules. This is why, just like any other quantitative analysis of interaction, knowing the concentration of tubulin and tau or stathmin at the beginning of the experiment is also a critical point that should not be overlooked. Tubulin concentration is usually determined spectrophotometrically at 275 nm using an extinction coefficient of 109,000 M^{-1} cm^{-1} in 6 M guanidine hydrochloride (Andreu & Timasheff, 1982; Na & Timasheff, 1982). This determination should be done after tubulin full equilibration and as late as possible (just before the ITC experiment). Indeed, in a buffer without glycerol or other tubulin stabilizer, tubulin rapidly degrades, leading to significant errors in determination of thermodynamic parameters of interaction (Fig. 18.3). Thus, prior to each subsequent ITC titration, aggregated tubulin should be eliminated by centrifugation and concentration should be measured again. As tau protein is unstructured and elongated, it induces some scattering of light. Thus, to measure tau concentration, it is necessary to do a full UV-spectrum of the sample and then correct it for light scattering to avoid overestimation of tau concentration (Winder & Gent, 1971). It should also be noted that GTP, which has to be present in the final buffer in all binding experiments with tubulin, strongly absorbs in the range used for measurement of tau concentration. Thus, it is recommended to equilibrate tau in the absence of GTP, which should be added just prior to the ITC titration.

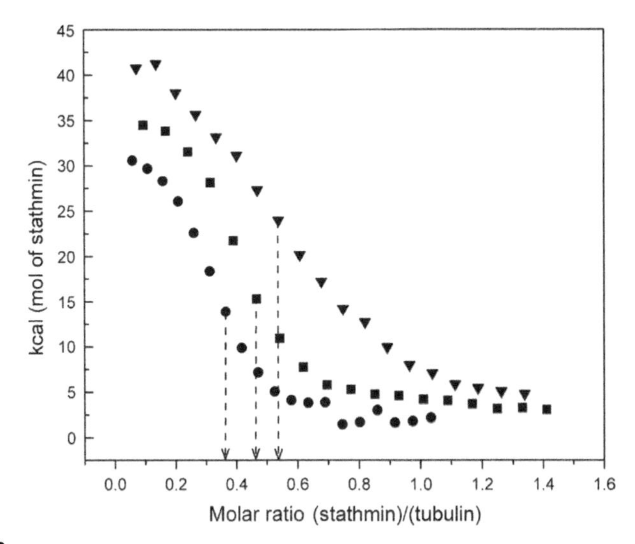

FIGURE 18.3

Effect of tubulin degradation with time on binding isotherms. Three sequential ITC experiments made with the same protein samples with a period of 1 h. The arrows indicate the stoichiometry which decreases as time goes by, which indicates that less active tubulin is available due to degradation over time.

Determination of stathmin concentration is even more challenging, as stathmin bears no tyrosine or tryptophane residues. We tested several approaches, including colorimetric methods (DC Protein Assay, Biorad) with BSA as standards. As none of these techniques proved to be satisfactory enough for ITC, we often had to adjust stathmin concentration after ITC experiments in order to reach the expected stathmin:tubulin stoichiometry of 0.5. Ideally, the most precise method would be to constitute a stock of stathmin of known concentration (e.g., previously determined by amino acid composition) and aliquot them to use as standards for colorimetric methods (instead of BSA) every time stathmin concentration is measured.

18.2.3 Temperature

As described above, the enthalpy of binding (ΔH) depends on experimental temperature. At a certain temperature, when $\Delta H=0$, it is impossible to carry out ITC experiments (Fig. 18.4). This is why it is necessary to collect ITC titration at least at two different temperatures before considering that an interaction cannot be measured using ITC. In case of stathmin–tubulin interaction, the absolute value of ΔH is minimal at a temperature close to 25 °C, which is traditionally used as a standard temperature for ITC experiments. This means that for temperatures lower than 25 °C, the stathmin–tubulin interaction can be monitored by the endothermic signal ($\Delta H>0$), whereas above 25 °C the interaction will be monitored by the exothermic signal ($\Delta H<0$).

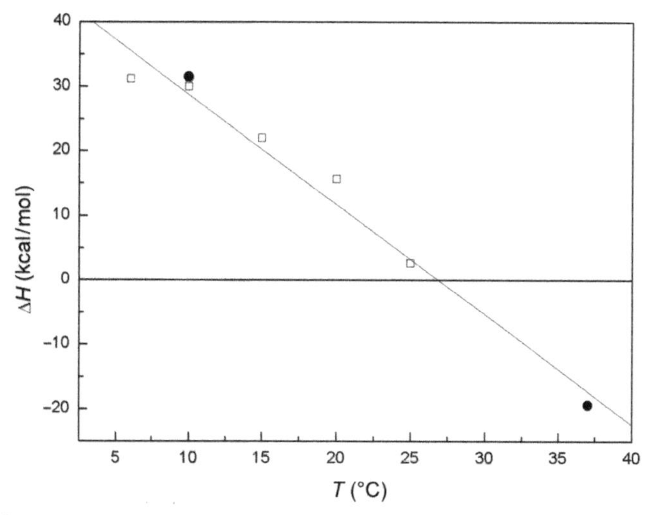

FIGURE 18.4

Temperature dependence of enthalpy of stathmin binding to tubulin. Plot is based on data obtained by Honnappa with coauthors (Honnappa, Cutting, Jahnke, Seelig, & Steinmetz, 2003) (open squares) and our data (circles). The slope of temperature dependence of enthalpy corresponds to molar heat capacity change of interaction (ΔC_p).

For the study of the interaction of tubulin with MAPs, the choice of temperature is also critical, due to temperature dependence of the self-assembly properties of tubulin (see Fig. 18.1). Indeed, in a minimum buffer (Devred et al., 2004) in the presence of tau, tubulin forms rings at 10 °C and microtubules at 37 °C. Stability of the proteins should also be taken into account when one chooses the experimental temperature. For tubulin, whenever it is possible, lower temperatures should be chosen in order to guarantee as little degradation as possible. In summary, to study the binding of depolymerizing MAPs, such as stathmin, or the binding of ligands to free tubulin, a temperature of 10 °C presents the advantage of guarantying better/longer stability for tubulin. When studying a stabilizing MAP, such as tau, or the binding of ligands to tubulin in microtubules, the temperature should be as close as possible to 37 °C.

18.2.4 Buffer conditions

As equilibrium between free tubulin dimers and microtubules can be easily perturbed by buffer components, the investigation of tubulin binding with proteins that regulate its assembly/disassembly should be performed in the minimum buffer to avoid contribution of cofactors to the thermodynamics of interaction. In our studies of MAPs binding to tubulin, we use a minimum buffer containing only phosphate (NaPi), which does not favor tubulin self-assembly, and GTP, which is necessary for tubulin structural integrity and stability, especially in the absence of any stabilizer. To study the binding of monomeric tau with tubulin, reducing agents such as DTT or TCEP should be present in the buffer to prevent formation of tau intra- or inter-molecular S–S bridges. Using buffers that stabilize microtubules, such as PIPES or MES, or molecules known to favor tubulin self-assembly, such as Mg^{2+} or glycerol, will increase the stability of tubulin compared to the so-called nonpolymerizing buffer (Devred et al., 2004). However, these molecules might also completely change the mode of binding.

18.2.5 Reaction volume and duration

One of the downsides of this technique has always been the amount of protein necessary for the measure of the interaction. For the longest time (for older generation MCS ITC and VP ITC machines), the volume of reaction chamber had to be 1.4 mL with tubulin in the 10 μM scale and syringe volume of 500 μL in the 100 μM range for ligand. Therefore, experiments could not be performed for a number of proteins only available in small quantities. For our studies, we used tubulin concentration in the 5–20 μM range in the cell, whereas stathmin or tau was in the 15–100 μM range (Devred et al., 2008; Tsvetkov, Makarov, Malesinski, Peyrot, & Devred, 2012). Practically, 2.5 mL of tubulin sample needed to be prepared to fill the 1.4 mL cell, as well as a minimum of 750 μL of ligand (tau or stathmin) to fill the injection syringe. Another limiting factor was the duration of the titration (around 60 min), which rendered the work with tubulin difficult to reproduce during the same day. The new generation model iTC200 has drastically improved both time and quantity requirements. With this new apparatus, experiments require 300 μL to fill the sample cell and 70 μL to fill the injection syringe. With fast equilibration times, up to two runs per hour can be accomplished. Nevertheless, in some cases, a large reaction volume, as in VP ITC, is still necessary. For example, it might be difficult to obtain or to work with high concentrations of the ligand due to the possibility of aggregation. Also, larger volumes of reaction might be needed for a low binding constant, or when the ΔH is small and requires the sum of many interactions to be detected. In these latter cases, the ITC200 can still be used, but it will imply conducting several consecutive experiments with the resultant curves concatenated (for a comparison, see Fig. 18.5).

18.3 RESULTS: TUBULIN/MAPs BY ITC

All the requirements described above may explain why ITC has not been used more often to study such complex systems as the cytoskeleton network. Nevertheless, ITC has been used to study the mechanism of bacterial tubulin homologue FtsZ assembly (Caplan & Erickson, 2003; Huecas et al., 2007) or to characterize the binding of several modulators of FtsZ assembly in order to use them in new anti-bacterial treatments (Chen, Milam, & Erickson, 2012; Domadia, Bhunia, Sivaraman, Swarup,

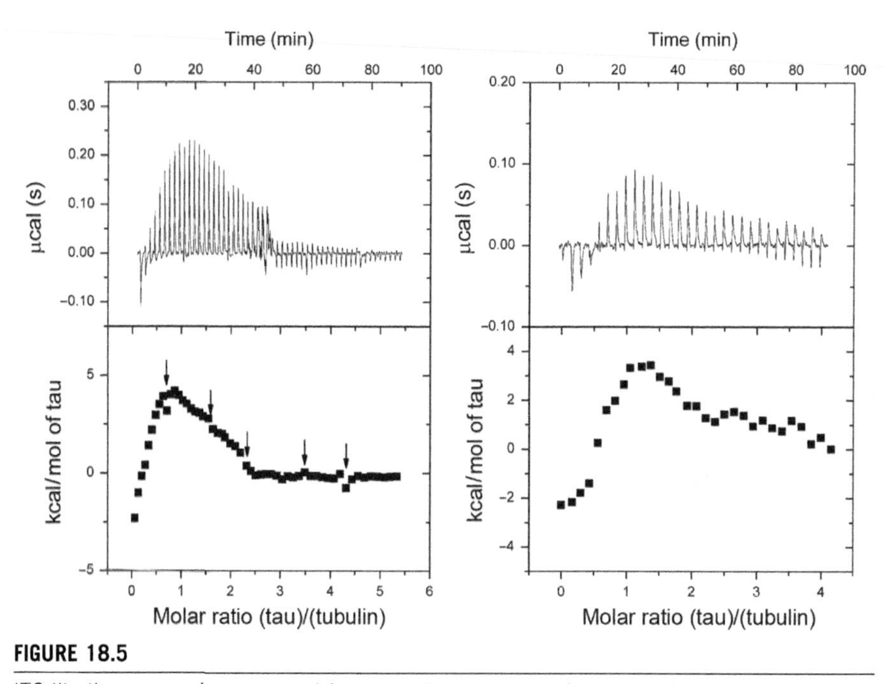

FIGURE 18.5

ITC titration curves (upper panels) and binding isotherms (low panels) of tau–tubulin interactions registered on ITC200 (left panels) and on VP ITC (right panels) at 10 °C in 20 mM NaPi, 0.1 mM GTP, 1 mM TCEP, buffer at pH 6.5. The arrows show syringe refilling with the same tau solution.

& Dasgupta, 2008; Domadia, Swarup, Bhunia, Sivaraman, & Dasgupta, 2007). It has also been used to characterize the binding of modulators of microtubule assembly on tubulin (Banerjee et al., 2005; Das et al., 2009; Gupta et al., 2003; Menendez, Laynez, Medrano, & Andreu, 1989; Rappl et al., 2006; Tsvetkov et al., 2011), to study stathmin–tubulin binding (Honnappa et al., 2003) and more recently tau–tubulin binding (Tsvetkov et al., 2012).

18.3.1 Stathmin–tubulin interaction

In 2003, Steinmetz and coauthors from Paul Scherrer Institute published an extensive characterization of the thermodynamics of the stathmin–tubulin interaction (Honnappa et al., 2003). They determined the stoichiometry, binding constant, variation of enthalpy and of entropy under different conditions of pH, temperature, and nucleotide presence (GTP/GDP). Under all investigated conditions, they obtained simple sigmoid binding isotherms, which can be well fitted with a simple one-set-of-sites binding model, described by following equations:

$$\begin{gathered}\mathrm{T}+\mathrm{S} \rightleftarrows \mathrm{TS}(K_0, \Delta H_0) \\ \mathrm{T}+\mathrm{TS} \rightleftarrows \mathrm{T2S}(K_0, \Delta H_0)\end{gathered} \tag{18.1}$$

They reported two binding sites of equal affinity with an equilibrium binding constant of $K_0 = 6.0 \times 10^6\ \mathrm{M}^{-1}$ and large negative molar heat capacity change ($\Delta C_\mathrm{p} = -860\ \mathrm{cal\ mol^{-1}\ K^{-1}}$), which suggest that the major driving force of the binding reaction was hydrophobic interactions (Fig. 18.4). Nevertheless, earlier studies using several techniques, including pull-down assays (Holmfeldt et al., 2001; Larsson et al., 1999) and analytical ultracentrifugation (Amayed, Carlier, & Pantaloni, 2000; Jourdain, Curmi, Sobel, Pantaloni, & Carlier, 1997), suggested the existence of two highly cooperative binding sites. These findings led Honnappa and coauthors to conclude that ITC data contrasted with earlier studies proposing that the second tubulin subunit is bound distinctly tighter than the first one. Nevertheless, several models can fit the same curve. Indeed, the fact that ITC titration results in a simple thermogram does not guarantee that the simplest model is the real one. In other words, in this case, the principle of Occam's razor could be summarized as "other things being equal, a simpler explanation is better than a more complex one." As mentioned above, the sigmoid form of binding isotherm could also be observed for more complex models in the case of degenerate parameters. For example, for a model of nonequal interacting sites (Fig. 18.6) described by the following equations

$$\begin{gathered}\mathrm{T}+\mathrm{S} \rightleftarrows \mathrm{TS}(K_{A1}, \Delta H_{A1}) \\ \mathrm{T}+\mathrm{TS} \rightleftarrows \mathrm{T2S}(K_{A2}, \Delta H_{A2}) \\ \mathrm{S}+\mathrm{T} \rightleftarrows \mathrm{ST}(K_{B1}, \Delta H_{B1}) \\ \mathrm{ST}+\mathrm{T} \rightleftarrows \mathrm{T2S}(K_{B2}, \Delta H_{B2})\end{gathered} \tag{18.2}$$

if there is strong cooperativity ($K_{A2} \gg K_{A1}$ and $K_{B2} \gg K_{B1}$) between two equal sites ($\Delta H_{A1} = \Delta H_{B1}, K_{A1} = K_{B1}$) and if binding of first ligand molecule does not change the

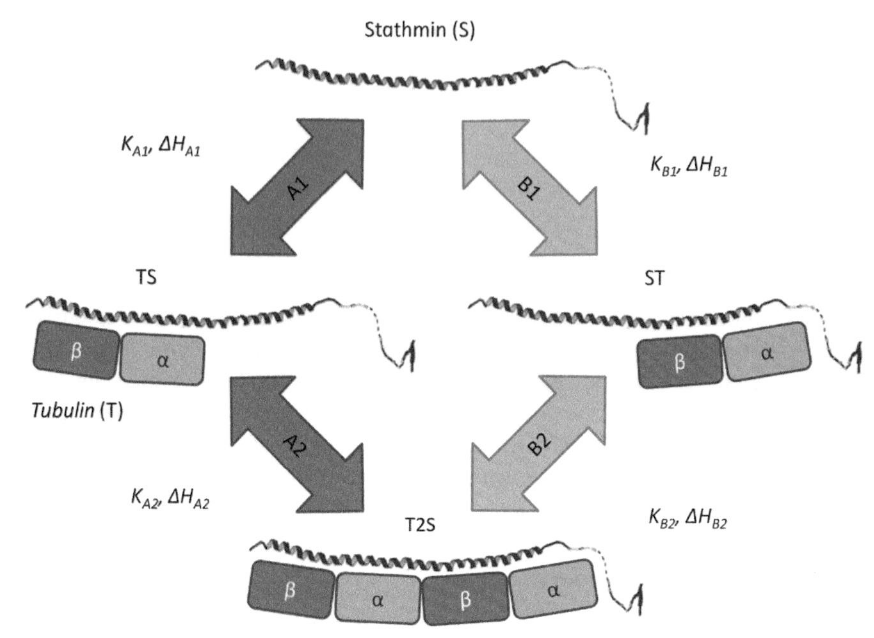

FIGURE 18.6

Schema of tubulin binding to stathmin. The general model of tubulin interaction with stathmin supposes or assumes the existence of two nonequal interacting sites described by six independent parameters K_{A1}, K_{B1}, K_{A2}, ΔH_{A1}, ΔH_{B1}, and ΔH_{A2} (although in the full thermodynamic cycle K_{B2}, ΔH_{B2} are not independent). (See color plate.)

enthalpy of binding for the second site ($\Delta H_{A1} = \Delta H_{A2} = \Delta H_{B1} = \Delta H_{B2}$). In this case, treating a binding isotherm with a model of one-set-of-sites results in the determination of wrong binding constants (K_0) for each site. However, in both cases, the overall reaction can be written as follows:

$$2\mathrm{T} + \mathrm{S} \rightleftarrows \mathrm{T2S}(K_{\mathrm{T2S}}, \Delta H_{\mathrm{T2S}}) \tag{18.3}$$

with an overall constant of formation of the T2S complex $K_{\mathrm{T2S}} = K_0 * K_0 = K_{A1} * K_{A2} = K_{B1} * K_{B2}$.

Considering all the earlier evidence about the cooperativity, the current structural view of the asymmetric T2S complex (Gigant et al., 2000), and the fact that ITC cannot distinguish between two binding sites that would have the same ΔH, it is likely that ΔH of binding of individual tubulin to stathmin changes neither with the position on stathmin, nor with the presence of another tubulin on stathmin. A possible way to explain how ΔH can be similar despite the asymmetry of stathmin would be to consider that the heat exchanged during tubulin–stathmin interaction is mostly due to the lateral interaction of tubulin with the long alpha helix of stathmin, with very little (negligible) heat exchanged at the interface between β-tubulin and the consecutive α-tubulin or between α-tubulin and the N-terminal cap of stathmin. In this context, whichever model is hypothesized, only the overall constant of formation of T2S complex (K_{T2S}) can be determined via K_0, as K_{A1}, K_{A2}, K_{B1}, or K_{B2} cannot be determined from the model of nonequal interacting sites if the binding isotherm is a degenerate sigmoid curve. Measurement of K_0 or further extrapolation to K_{T2S} enables only the characterization of T2S complex formation (described by Eq. 18.3) with values of stoichiometry, entropy, enthalpy, and free energy. The knowledge of these thermodynamic parameters allows one to characterize the nature of the forces involved in the interaction. In the example presented in Fig. 18.4, below 28 °C, since $\Delta H > 0$ (Fig. 18.4), the only driving force of T2S complex formation is hydrophobic interactions ($\Delta S > 0$), whereas above 28 °C, the reaction is enthalpy ($\Delta H < 0$) and entropy driven ($\Delta S > 0$). Rather than the intrinsic values of these parameters, which can vary greatly depending on the buffer conditions and temperature, it is the comparison of the parameters obtained in different conditions that will bring new information about the interaction. And despite the open question about the true nature (cooperative vs. noncooperative) of stathmin–tubulin binding, ITC enabled investigators to quantify the impact of each one of the four stathmin phosphorylations and different combinations of them, on its affinity for tubulin (Honnappa, Jahnke, Seelig, & Steinmetz, 2006). It enabled the authors to provide *in vitro* the biophysical basis for understanding the mechanism by which stathmin activity gradients will regulate local microtubule growth. This approach has also been used to determine the consequences of the presence of anticancer agents, such as vinblastine, on the activity of stathmin (Devred et al., 2008). Comparison of stathmin–tubulin binding in the presence or absence of vinblastine revealed an increase in the stathmin affinity for tubulin in the presence of vinblastine, setting the molecular basis of a new or revised mechanism of action of this MTA.

18.3.2 Tau–tubulin

ITC has also been used to study the interaction of tau with tubulin. We can expect that the binding of a stabilizing MAP such as tau, whose individual repeat domains can bind and stabilize microtubules (Aizawa et al., 1989; Butner & Kirschner, 1991; Devred, Douillard, Briand, & Peyrot, 2002; Ennulat, Liem, Hashim, & Shelanski, 1989; Goedert, Wischik, Crowther, Walker, & Klug, 1988; Gustke, Trinczek, Biernat, Mandelkow, & Mandelkow, 1994), is more complex to study than the binding of a destablizer, such as stathmin. The presence of any factor that favors or inhibits tubulin polymerization may have an impact on the extent of the tau-induced self-assembly and potentially on the thermodynamic parameters determined. This is why the recent ITC study of tau–tubulin interaction was conducted in a minimum phosphate-GTP buffer in the absence of Mg^{2+} (Tsvetkov et al., 2012). Even though tau has been studied for more than 40 years, very little is known about its structure (Harbison, Bhattacharya, & Eliezer, 2012). In addition, there are several discrepancies regarding its mode and parameters of binding to tubulin, probably in part due to the fact that microtubules can induce the formation of tau filaments (Duan & Goodson, 2012). Nevertheless, several studies have suggested the existence of two binding sites, one that may overlap the paclitaxel binding site and that would be located in the lumen, and another one on the outside wall of MT (Ackmann, Wiech, & Mandelkow, 2000; Kar, Fan, Smith, Goedert, & Amos, 2003; Makrides, Massie, Feinstein, & Lew, 2004). These two sites would not be equally accessible depending on the nature of experimental study, such as tau-induced MT self-assembly versus tau binding to stabilized MT. ITC titration of tubulin by tau results in a complex two-phase binding isotherm that could be well fitted using two-sets-of-sites model, compatible with the two types of tau–tubulin binding modes described in the literature: one corresponding to a high affinity binding site with a tau:tubulin stoichiometry of 0.2 and the other one to a low affinity binding site with a stoichiometry of 0.8. Nevertheless, it cannot be excluded that tau–tubulin binding follows a more complex model. To assign the real model, many complementary experiments will need to be performed. Like in the case of stathmin–tubulin interaction, even the simplest binding model which resulted in the determination of only apparent thermodynamic parameters helped us to gain new insights into the mechanism of tau binding to tubulin. Indeed, even though tau induces the formation of curved tubulin protofilament at 10 °C, and the formation of microtubules at 37 °C (Devred et al., 2004; Fig. 18.1), tau–tubulin binding isotherm obtained at 10 and 37 °C were both biphasic with a maximum at a tau:tubulin molar ratio of one, indicating a similar binding model (Fig. 18.5). The fact that tau would bind similarly on an MT and on a circular protofilament indicates that on MT the interaction is longitudinal (along the same protofilament) and not transversal (bridging several parallel protofilaments). This allowed us to rule out the models which hypothesized that tau stabilizes MT by binding across several protofilaments on the MT lattice.

CONCLUSION

ITC is one of the latest and most powerful techniques to be used in characterizing the binding affinity of ligands for proteins or proteins for proteins. But like most techniques, it would be useless without other methods. ITC measures the heat exchange and thus often relies on complementary studies to hint at or confirm what reaction is really happening in the calorimetric cell. For example, analytical ultracentrifugation is a technique of choice to determine stoichiometry, changes in conformation or assembly state of the molecules studied (Correia & Stafford, 2009; Demeler, Brookes, & Nagel-Steger, 2009; Lebowitz, Lewis, & Schuck, 2002; Schuck, 2003). Through the examples detailed in this chapter, we have shown that if a certain number of precautions, due mostly to the nature of tubulin, are taken, ITC can be used to thermodynamically characterize molecular interactions between tubulin and MAPs. If stathmin binding to tubulin is now well characterized by ITC, there is still a lot to understand about tau binding to tubulin. In summary, even though the tubulin cytoskeleton is a challenging system to work on, ITC is a powerful technique able to provide significant advances in our understanding of tubulin interaction with its partners.

Acknowledgment

This work was supported by the Molecular and Cellular Biology Program of the Russian Academy of Sciences.

References

Ackmann, M., Wiech, H., & Mandelkow, E. (2000). Nonsaturable binding indicates clustering of tau on the microtubule surface in a paired helical filament-like conformation. *Journal of Biological Chemistry*, *275*, 30335–30343.

Aizawa, H., Kawasaki, H., Murofushi, H., Kotani, S., Suzuki, K., & Sakai, H. (1989). A common amino acid sequence in 190-kDa microtubule-associated protein and tau for the promotion of microtubule assembly. *Journal of Biological Chemistry*, *264*(10), 5885–5890.

Alli, E., Yang, J. M., Ford, J. M., & Hait, W. N. (2007). Reversal of stathmin-mediated resistance to paclitaxel and vinblastine in human breast carcinoma cells. *Molecular Pharmacology*, *71*, 1233–1240.

Amayed, P., Carlier, M. F., & Pantaloni, D. (2000). Stathmin slows down guanosine diphosphate dissociation from tubulin in a phosphorylation-controlled fashion. *Biochemistry*, *39*, 12295–12302.

Andreu, J. M., & Timasheff, S. N. (1982). Interaction of tubulin with single ring analogues of colchicine. *Biochemistry*, *21*, 534–543.

Banerjee, M., Poddar, A., Mitra, G., Surolia, A., Owa, T., & Bhattacharyya, B. (2005). Sulfonamide drugs binding to the colchicine site of tubulin: Thermodynamic analysis of the drug-tubulin interactions by isothermal titration calorimetry. *Journal of Medicinal Chemistry*, *48*, 547–555.

Barbier, P., Peyrot, V., Leynadier, D., & Andreu, J. M. (1998). The active GTP- and ground GDP-liganded states of tubulin are distinguished by the binding of chiral isomers of ethyl 5-amino-2-methyl-1,2-dihydro-3-phenylpyrido[3,4-b]pyrazin-7-yl carbamate. *Biochemistry*, *37*, 758–768.

Belmont, L. D., & Mitchison, T. J. (1996). Identification of a protein that interacts with tubulin dimers and increases the catastrophe rate of microtubules. *Cell*, *84*, 623–631.

Box, G. E. P., & Draper, N. R. (1987). *Empirical model-building and response surfaces*. New York, NY, USA: Wiley.

Buey, R. M., Barasoain, I., Jackson, E., Meyer, A., Giannakakou, P., Paterson, I., et al. (2005). Microtubule interactions with chemically diverse stabilizing agents: Thermodynamics of binding to the paclitaxel site predicts cytotoxicity. *Chemistry and Biology*, *12*(12), 1269–1279.

Buey, R. M., Díaz, J. F., Andreu, J. M., O'Brate, A., Giannakakou, P., Nicolaou, K. C., et al. (2004). Interaction of epothilone analogs with the paclitaxel binding site: Relationship between binding affinity, microtubule stabilization, and cytotoxicity. *Chemistry and Biology*, *11*(2), 225–236.

Butner, K. A., & Kirschner, M. W. (1991). Tau protein binds to microtubules through a flexible array of distributed weak sites. *The Journal of Cell Biology*, *115*(3), 717–730.

Calligaris, D., Verdier-Pinard, P., Devred, F., Villard, C., Braguer, D., & Lafitte, D. (2010). Microtubule targeting agents: From biophysics to proteomics. *Cellular and Molecular Life Sciences*, *67*, 1089–1104.

Caplan, M. R., & Erickson, H. P. (2003). Apparent cooperative assembly of the bacterial cell division protein FtsZ demonstrated by isothermal titration calorimetry. *Journal of Biological Chemistry*, *278*, 13784–13788.

Chen, Y., Milam, S. L., & Erickson, H. P. (2012). SulA inhibits assembly of FtsZ by a simple sequestration mechanism. *Biochemistry*, *51*, 3100–3109.

Correia, J. J., & Stafford, W. F. (2009). Extracting equilibrium constants from kinetically limited reacting systems. *Methods in Enzymology*, *455*, 419–446.

Cucchiarelli, V., Hiser, L., Smith, H., Frankfurter, A., Spano, A., Correia, J. J., et al. (2008). Beta-tubulin isotype classes II and V expression patterns in nonsmall cell lung carcinomas. *Cell Motility and the Cytoskeleton*, *65*, 675–685.

Das, L., Gupta, S., Dasgupta, D., Poddar, A., Janik, M. E., & Bhattacharyya, B. (2009). Binding of indanocine to the colchicine site on tubulin promotes fluorescence, and its binding parameters resemble those of the colchicine analogue AC. *Biochemistry*, *48*, 1628–1635.

Demeler, B., Brookes, E., & Nagel-Steger, L. (2009). Analysis of heterogeneity in molecular weight and shape by analytical ultracentrifugation using parallel distributed computing. *Methods in Enzymology*, *454*, 87–113.

Devred, F., Barbier, P., Douillard, S., Monasterio, O., Andreu, J. M., & Peyrot, V. (2004). Tau induces ring and microtubule formation from alphabeta-tubulin dimers under nonassembly conditions. *Biochemistry*, *43*, 10520–10531.

Devred, F., Barbier, P., Lafitte, D., Landrieu, I., Lippens, G., & Peyrot, V. (2010). Microtubule and MAPs: Thermodynamics of complex formation by AUC, ITC, fluorescence, and NMR. *Methods in Cell Biology*, *95*, 449–480.

Devred, F., Douillard, S., Briand, C., & Peyrot, V. (2002). First tau repeat domain binding to growing and taxol-stabilized microtubules, and serine 262 residue phosphorylation. *FEBS Letters*, *523*(1–3), 247–251.

Devred, F., Tsvetkov, P. O., Barbier, P., Allegro, D., Horwitz, S. B., Makarov, A. A., et al. (2008). Stathmin/Op18 is a novel mediator of vinblastine activity. *FEBS Letters*, *582*, 2484–2488.

Domadia, P. N., Bhunia, A., Sivaraman, J., Swarup, S., & Dasgupta, D. (2008). Berberine targets assembly of Escherichia coli cell division protein FtsZ. *Biochemistry*, *47*, 3225–3234.

Domadia, P., Swarup, S., Bhunia, A., Sivaraman, J., & Dasgupta, D. (2007). Inhibition of bacterial cell division protein FtsZ by cinnamaldehyde. *Biochemical Pharmacology*, *74*, 831–840.

Drubin, D. G., & Kirschner, M. W. (1986). Tau protein function in living cells. *The Journal of Cell Biology*, *103*, 2739–2746.

Duan, A. R., & Goodson, H. V. (2012). Taxol-stabilized microtubules promote the formation of filaments from unmodified full-length tau in vitro. *Molecular Biology of the Cell*, *23*, 4796–4806.

Ennulat, D. J., Liem, R. K., Hashim, G. A., & Shelanski, M. L. (1989). Two separate 18-amino acid domains of tau promote the polymerization of tubulin. *Journal of Biological Chemistry*, *264*(10), 5327–5330.

Freyer, M. W., & Lewis, E. A. (2008). Isothermal titration calorimetry: Experimental design, data analysis, and probing macromolecule/ligand binding and kinetic interactions. *Methods in Cell Biology*, *84*, 79–113.

Frigon, R. P., & Lee, J. C. (1972). The stabilization of calf-brain microtubule protein by sucrose. *Archives of Biochemistry and Biophysics*, *153*, 587–589.

Gigant, B., Curmi, P. A., Martin-Barbey, C., Charbaut, E., Lachkar, S., Lebeau, L., et al. (2000). The 4 A X-ray structure of a tubulin:stathmin-like domain complex. *Cell*, *102*, 809–816.

Goedert, M., Wischik, C. M., Crowther, R. A., Walker, J. E., & Klug, A. (1988). Cloning and sequencing of the cDNA encoding a core protein of the paired helical filament of Alzheimer disease: Identification as the microtubule-associated protein tau. *Proceedings of the National Academy of Sciences of the United States of America*, *85*, 4051–4055.

Gupta, S., Chakraborty, S., Poddar, A., Sarkar, N., Das, K. P., & Bhattacharyya, B. (2003). BisANS binding to tubulin: Isothermal titration calorimetry and the site-specific proteolysis reveal the GTP-induced structural stability of tubulin. *Proteins*, *50*, 283–289.

Gustke, N., Trinczek, B., Biernat, J., Mandelkow, E. M., & Mandelkow, E. (1994). Domains of tau protein and interactions with microtubules. *Biochemistry*, *33*(32), 9511–9522.

Harbison, N. W., Bhattacharya, S., & Eliezer, D. (2012). Assigning backbone NMR resonances for full length tau isoforms: Efficient compromise between manual assignments and reduced dimensionality. *PLoS One*, *7*, e34679.

Holmfeldt, P., Larsson, N., Segerman, B., Howell, B., Morabito, J., Cassimeris, L., et al. (2001). The catastrophe-promoting activity of ectopic Op18/stathmin is required for disruption of mitotic spindles but not interphase microtubules. *Molecular Biology of the Cell*, *12*, 73–83.

Honnappa, S., Cutting, B., Jahnke, W., Seelig, J., & Steinmetz, M. O. (2003). Thermodynamics of the Op18/stathmin-tubulin interaction. *Journal of Biological Chemistry*, *278*, 38926–38934.

Honnappa, S., Jahnke, W., Seelig, J., & Steinmetz, M. O. (2006). Control of intrinsically disordered stathmin by multisite phosphorylation. *Journal of Biological Chemistry*, *281*, 16078–16083.

Huecas, S., Schaffner-Barbero, C., Garcia, W., Yebenes, H., Palacios, J. M., Diaz, J. F., et al. (2007). The interactions of cell division protein FtsZ with guanine nucleotides. *Journal of Biological Chemistry*, *282*, 37515–37528.

Jourdain, L., Curmi, P., Sobel, A., Pantaloni, D., & Carlier, M. F. (1997). Stathmin: A tubulin-sequestering protein which forms a ternary T2S complex with two tubulin molecules. *Biochemistry*, *36*, 10817–10821.

Kar, S., Fan, J., Smith, M. J., Goedert, M., & Amos, L. A. (2003). Repeat motifs of tau bind to the insides of microtubules in the absence of taxol. *EMBO Journal*, *22*, 70–77.

Kiris, E., Ventimiglia, D., & Feinstein, S. C. (2010). Quantitative analysis of MAP-mediated regulation of microtubule dynamic instability in vitro focus on tau. *Methods in Cell Biology*, *95*, 481–503.

Konzack, S., Thies, E., Marx, A., Mandelkow, E. M., & Mandelkow, E. (2007). Swimming against the tide: Mobility of the microtubule-associated protein tau in neurons. *Journal of Neuroscience*, *27*, 9916–9927.

Ladbury, J. E., & Doyle, M. L. (2004). *Biocalorimetry 2: Applications of calorimetry in the biological sciences*. Chichester, West Sussex, England: John Wiley & Sons.

Larsson, N., Segerman, B., Gradin, H. M., Wandzioch, E., Cassimeris, L., & Gullberg, M. (1999). Mutations of oncoprotein 18/stathmin identify tubulin-directed regulatory activities distinct from tubulin association. *Molecular and Cellular Biology*, *19*, 2242–2250.

Lebowitz, J., Lewis, M. S., & Schuck, P. (2002). Modern analytical ultracentrifugation in protein science: A tutorial review. *Protein Science*, *11*(9), 2067–2079, Review.

Makrides, V., Massie, M. R., Feinstein, S. C., & Lew, J. (2004). Evidence for two distinct binding sites for tau on microtubules. *Proceedings of the National Academy of Sciences of the United States of America*, *101*, 6746–6751.

Menendez, M., Laynez, J., Medrano, F. J., & Andreu, J. M. (1989). A thermodynamic study of the interaction of tubulin with colchicine site ligands. *Journal of Biological Chemistry*, *264*, 16367–16371.

Mitchison, T., & Kirschner, M. (1984). Dynamic instability of microtubule growth. *Nature*, *312*, 237–242.

Na, G. C., & Timasheff, S. N. (1982). Physical properties of purified calf brain tubulin. *Methods in Enzymology*, *85 Pt. B*, 393–408.

Peyrot, V., Leynadier, D., Sarrazin, M., Briand, C., Menendez, M., Laynez, J., et al. (1992). Mechanism of binding of the new antimitotic drug MDL 27048 to the colchicine site of tubulin: Equilibrium studies. *Biochemistry*, *31*, 11125–11132.

Pierce, M. M., Raman, C. S., & Nall, B. T. (1999). Isothermal titration calorimetry of protein-protein interactions. *Methods*, *19*, 213–221.

Rappl, C., Barbier, P., Bourgarel-Rey, V., Gregoire, C., Gilli, R., Carre, M., et al. (2006). Interaction of 4-arylcoumarin analogues of combretastatins with microtubule network of HBL100 cells and binding to tubulin. *Biochemistry*, *45*, 9210–9218.

Ross, J. L., & Dixit, R. (2010). Multiple color single molecule TIRF imaging and tracking of MAPs and motors. *Methods in Cell Biology*, *95*, 521–542.

Ross, P. D., & Subramanian, S. (1981). Thermodynamics of proteine association reactions: Forces contributing to stability. *Biochemistry*, *20*, 3096–3102.

Samsonov, A., Yu, J. Z., Rasenick, M., & Popov, S. V. (2004). Tau interaction with microtubules in vivo. *Journal of Cell Science*, *117*, 6129–6141.

Schuck, P. (2003). On the analysis of protein self-association by sedimentation velocity analytical ultracentrifugation. *Analytical Biochemistry*, *320*(1), 104–124.

Tsvetkov, F. O., Kulikova, A. A., Devred, F., Zernii, E., Lafitte, D., & Makarov, A. A. (2011). Thermodynamics of calmodulin and tubulin binding to the vinca-alkaloid vinorelbine. *Molecular Biology (Mosk)*, *45*, 697–702.

Tsvetkov, P. O., Kulikova, A. A., Golovin, A. V., Tkachev, Y. V., Archakov, A. I., Kozin, S. A., et al. (2010). Minimal Zn(2+) binding site of amyloid-beta. *Biophysical Journal*, *99*, L84–L86.

Tsvetkov, P. O., Makarov, A. A., Malesinski, S., Peyrot, V., & Devred, F. (2012). New insights into tau-microtubules interaction revealed by isothermal titration calorimetry. *Biochimie*, *94*, 916–919.

Weingarten, M. D., Lockwood, A. H., Hwo, S. Y., & Kirschner, M. W. (1975). A protein factor essential for microtubule assembly. *Proceedings of the National Academy of Sciences of the United States of America*, *72*, 1858–1862.

Weissmann, C., Reyher, H. J., Gauthier, A., Steinhoff, H. J., Junge, W., & Brandt, R. (2009). Microtubule binding and trapping at the tip of neurites regulate tau motion in living neurons. *Traffic*, *10*, 1655–1668.

Wilson, L., & Correia, J. J. (2010). *Microtubules, in vivo*. Oxford, England: Academic Press.

Winder, A. F., & Gent, W. L. (1971). Correction of light-scattering errors in spectrophotometric protein determinations. *Biopolymers*, *10*, 1243–1251.

CHAPTER

19 Methods for Studying Microtubule Binding Site Interactions: Zampanolide as a Covalent Binding Agent

Jessica J. Field*, Enrique Calvo†, Peter T. Northcote‡, John H. Miller*, Karl-Heinz Altmann§ and José Fernando Díaz¶

**School of Biological Sciences, Victoria University of Wellington, Wellington, New Zealand*

†Unidad de Proteómica, Centro Nacional de Investigaciones Cardiovasculares, Madrid, Spain

‡School of Chemical and Physical Sciences, Victoria University of Wellington, Wellington, New Zealand

§Department of Chemistry and Applied Biosciences, Institute of Pharmaceutical Sciences, Swiss Federal Institute of Technology, Zürich, Switzerland

¶Centro de Investigaciones Biológicas, CSIC, Madrid, Spain

CHAPTER OUTLINE

Introduction 304

19.1 Materials 305

19.1.1 Tubulin and Stabilized Microtubules 305

19.1.2 Buffers 306

19.1.3 Software for Data Analysis 307

19.1.4 Ultracentrifugation 307

19.2 Methods and Results 307

19.2.1 Induction of Microtubule Assembly 307

19.2.1.1 Microtubule Assembly in Glycerol Buffer 308

19.2.1.2 Microtubule Assembly in PEDTA Buffer 309

19.2.1.3 Transmission Electron Microscopy 310

19.2.2 Flutax Displacement Assay 310

19.2.3 Laulimalide/peloruside Competition Assay 312

19.2.3.1 Notes 314

19.2.4 Stoichiometry 314

19.2.5 Binding Kinetics 316

19.2.5.1 Flutax-2 Method 316

19.2.5.2 High-performance Liquid Chromatography Analysis 317

Methods in Cell Biology, Volume 115

ISSN 0091-679X
http://dx.doi.org/10.1016/B978-0-12-407757-7.00019-0

19.2.6 Mass Spectrometry 318
19.2.6.1 Protein Preparation and Digestion for MS Analysis 319
19.2.7 Visualization of Covalent Binding in Cells 320
19.2.7.1 Adherent Cell Lines 322
19.2.7.2 Cells in Suspension 322
19.2.8 Further Experiments for Characterization 324
Summary **324**
Acknowledgments **324**
References **324**

Abstract

In this chapter, we describe the methods used to determine the binding site and binding profile of zampanolide, a novel microtubule-stabilizing agent (MSA) that binds covalently to tubulin. These methods can be applied to other novel MSAs in which the binding site and mechanism of binding are unknown. Using the described methods, we have shown that zampanolide binds to the taxoid site on β-tubulin, but unlike most other MSAs is able to covalently modify this site. The purpose of this chapter is to provide a step-by-step protocol for determining the binding site of a novel MSA.

INTRODUCTION

Microtubules (MTs) are one of the most successful targets for anticancer therapy because of their essential role in mitosis. There are various MT-targeting drugs used in the clinic today. However, mechanisms of resistance toward these drugs have appeared, most importantly, the upregulation of drug efflux pumps, such as P-glycoprotein (P-gp) (Gottesman, Fojo, & Bates, 2002). Because tumor cells can become resistant to drugs with long-term chemotherapy, there is a need to find new compounds with a similar mechanism of action but which avoid these mechanisms of resistance. One way of circumventing resistance is to use an MSA that binds covalently. Covalent binding agents are effective in P-gp overexpressing cells because they cannot be pumped out of the cell once they covalently bind. It has also been suggested that covalent binding agents would not be affected by upregulation of different tubulin isotypes or mutations in tubulin (Singh, Petter, Baillie, & Whitty, 2011).

One of the first steps in characterizing a novel MSA is to determine the site where it binds. Currently, there are two known MSA binding sites, the well-characterized taxoid site and the laulimalide/peloruside site. The taxoid site is located in the lumen of the β-tubulin subunit in a pocket where α-tubulin has eight extra amino acids (Nogales, Whittaker, Milligan, & Downing, 1999). There is also evidence that taxoid site compounds interact with the pore type I of MTs on their way to the taxoid site (Buey et al., 2007). The laulimalide/peloruside site is biochemically distinct from the taxoid site; however, its exact location is unknown, with evidence suggesting it could be present in both subunits. A number of different assays can be performed to determine the binding site of a novel MSA, and competition methods appear to be the most popular. The mechanism by which an MSA settles into its binding site is an important factor in its characterization. It is well known that MSAs may bind in a reversible manner, like the taxanes for example, or in an irreversible manner, like cyclostreptin (Buey et al., 2007). Covalent adduct formation causes permanent disruption of the target's biological function, increasing the compound's potency, and prolonging its effect. This can lead to a more desirable clinical profile and therapeutic window as this gives potential for less-frequent dosing and lower drug concentrations (Singh et al., 2011).

Zampanolide (ZMP), isolated from a marine sponge, is a potent MSA but is not susceptible to the P-gp efflux pump (Field et al., 2009), thus it evades the main mechanism of resistance presented toward these drugs in the clinic. Its enantiomer, dactylolide, also stabilizes MTs, but with much lower potency (Zurwerra, Gertsch, & Altmann, 2010). Both compounds bind in a covalent manner, but the binding kinetics of dactylolide is much slower. Interestingly, ZMP is unique to both cyclostreptin and dactylolide as it binds covalently to tubulin with high affinity but, unlike the other two drugs, is a potent inducer of MT polymerization (Field et al., 2012).

In this chapter, we describe how the stabilizing activity of a "test" compound can be confirmed and the location of its binding site determined. It is possible that a ligand can bind to one of the known MSA binding sites or to a novel site. We then probe the mechanism of binding to ascertain whether the compound binds in a reversible or irreversible manner. The methods described have been used to establish the stabilizing activity, the location of the binding site, and the mechanism of binding of ZMP (Field et al., 2012).

19.1 MATERIALS

19.1.1 Tubulin and stabilized microtubules

Prepare purified calf brain tubulin for *in vitro* experiments as described by Andreu (2007) using a modified Weisenberg procedure. Freeze the tubulin in 5–20 mg aliquots in liquid nitrogen and thaw when required. Alternatively, purified bovine brain tubulin can be purchased (fer@cib.csic.es).

1. Take an aliquot and remove sucrose using a Sephadex G-25 medium-size column (25 × 0.9 cm) (GE Healthcare Bioscience) equilibrated with two volumes of cold buffer (Section 1.2).
2. Collect the tubulin and using the buffer as the blank, measure the absorbance at 295 nm, pooling all fractions with an absorbance greater than 1.0.

3. Remove aggregates by ultracentrifugation (90,000 × g, 4 °C, 10 min), collect the supernatant and keep on ice.
4. To measure the protein concentration, dilute a small aliquot 20-fold in 10 mM phosphate buffer (1% SDS), pH 7.0, and determine the tubulin concentration from the absorption at 275 nm versus a 20-fold diluted buffer blank. The extinction coefficient at 275 nm for tubulin is 107,000 M^{-1} cm^{-1} (Andreu, Gorbunoff, Lee, & Timasheff, 1984). Dilute the protein to the required concentration and alter the buffer if assembly conditions are required.

Tubulin can be assembled into MTs and stabilized by mild glutaraldehyde cross-linking, and the concentration of taxoid binding sites measured. This protocol is described in Díaz and Buey (2007). Stabilized MTs are snap-frozen and stored in liquid nitrogen. Glutaraldehyde-stabilized MTs are an extremely useful tool for studying MT–ligand interactions and have been essential for identifying the binding site and binding affinity for a number of important MSAs. Because of the glutaraldehyde stabilization, these MTs do not assemble or disassemble in response to changes in ligand concentration, nor are they affected by changes in temperature.

19.1.2 Buffers

Two buffers are required depending on the experimental conditions.

Glycerol–EGTA buffer (GAB)—Assembly buffer
- 3.4 M glycerol
- 10 mM sodium phosphate (NaPi)
- 1 mM EGTA
- 6 mM $MgCl_2$
- 1 mM GTP
- pH 6.5

Prepare on the day of use without $MgCl_2$ and with 0.1 mM GTP, pH 6.8, keep on ice. After column equilibration, add 6 mM $MgCl_2$, and increase the GTP to 1 mM for assays requiring assembly conditions (pH shifts to 6.5). In experiments that use stabilized-MTs, the GAB buffer left over from the dialysis is used to dilute the sites and act as the experimental buffer.

Phosphate–EDTA buffer (PEDTA), 0.1 mM GTP—Buffer unfavorable for assembly
- 10 mM NaPi
- 1 mM EDTA
- 0.1 mM GTP
- pH 7.0

Prepare on the day of use, keep cold on ice. When required, increase the GTP concentration to 1 mM and add the required amount of $MgCl_2$.

19.1.3 Software for data analysis

Equigra v5.0 (F. Díaz, unpublished) for Flutax competition assay analysis (available from the author). This program fits the experimental data by least-squares to the equilibrium binding constant of the ligand under investigation, using the known values for a reference ligand. Instructions for program use can be found in Díaz and Buey (2007).

The binding constant of the test ligand, $K(1)$, can be determined from the known values of the binding constant of Flutax, $K(r)$, and the concentrations of binding sites, Flutax and test ligand by solving the simultaneous mass action equations (Díaz & Buey, 2007):

$$K(1) = [\text{Ligand}]_{\text{bound}}/[\text{Sites}]_{\text{free}} * [\text{Ligand}]_{\text{free}}$$
$$K(r) = [\text{Flutax}]_{\text{bound}}/[\text{Sites}]_{\text{free}} * [\text{Flutax}]_{\text{free}}$$
$$[\text{Flutax}]_{\text{free}} = [\text{Flutax}]_{\text{total}} - [\text{Flutax}]_{\text{bound}}$$
$$[\text{Ligand}]_{\text{free}} = [\text{Ligand}]_{\text{total}} - [\text{Ligand}]_{\text{bound}}$$
$$[\text{Sites}]_{\text{free}} = [\text{Sites}]_{\text{total}} - [\text{Ligand}]_{\text{bound}} - [\text{Flutax}]_{\text{bound}}$$

19.1.4 Ultracentrifugation

In all experiments, ultracentrifugation is performed using a Beckman Optima TLX ultracentrifuge employing a TLA 120.2 (2-mL tubes) or TLA 100 (200-μL tubes) rotor at $90{,}000 \times g$ (temperature and running time is experiment-dependent). Alternatively, other ultracentrifuges and rotors can be used providing the parameters are kept the same as described in these experiments. For example, we have used a Beckman Coulter Optima™ L-100 XP ultracentrifuge, employing a type 70.1 Ti rotor, using Delrin adaptors for thick-walled 4.0-mL polycarbonate tubes (Beckman Coulter). If only large rotors and tubes are available, mineral oil can be used to make up the final volume to the minimum required for centrifugation, and the aqueous sample can then be removed from under the oil layer after centrifugation.

19.2 METHODS AND RESULTS

The first task to characterize a suspected MSA is to confirm that it is a true MSA and determine how potent it is *in vitro*. This provides information on the assay conditions needed for further experiments. Strong MSAs can induce tubulin assembly in conditions that tubulin on its own is not able to assemble or assembles very slowly. A weaker MSA should at least be able to enhance tubulin assembly in conditions where tubulin is already able to assemble. Ligand-induced MTs also need to be confirmed visually by electron microscopy.

19.2.1 Induction of microtubule assembly

Tubulin polymerization can be easily monitored by quantifying polymer amounts in a pellet following centrifugation. Tubulin assembles into MTs from a critical concentration (C_r), the concentration of tubulin at which polymerization rate overcomes depolymerization rate. Calf-brain tubulin self-polymerizes at a C_r of 3.3 μM when in the described GAB buffer, so-called favorable conditions (Buey et al., 2005). The C_r will change in different conditions and is decreased upon addition of an MSA. The decrease can be used to measure the assembly induction power of an MSA. Strong MSAs promote MT assembly in conditions that are "hostile" to tubulin assembly, such as in PEDTA buffer that lacks glycerol. In this buffer, tubulin is unable to assemble unless a strong MSA is present, or the tubulin is at high concentrations (>200 μM) (Díaz, Menendez, & Andreu, 1993). Determination of the potency of an MSA involves using a fixed concentration of tubulin and altering the concentration of the MSA. As the MSA concentration increases, the supernatant tubulin concentration decreases as MTs form.

19.2.1.1 Microtubule assembly in glycerol buffer

In glycerol buffer, strong and weak MSAs can decrease the C_r of tubulin for self-assembly.

1. Prepare 20 mg of tubulin in GAB buffer as in Section 1.1, add 6 mM $MgCl_2$, and then increase the concentration of GTP to 1 mM to create assembly conditions.
2. Incubate samples of 15 and 25 μM tubulin with 20 and 30 μM ligand to ensure site saturation (or equivalent volume of DMSO) at 37 °C for 30–40 min. For a blank, use buffer with 15 μM MSA. Use a known MSA as a positive control (decrease Cr) and a destabilizing agent as a negative control (increase C_r).
3. Separate the MTs by ultracentrifugation (20 min, 37 °C). Collect the supernatant by pipet and dilute 50 μL in 450 μL of 10 mM NaPi, 1% SDS buffer, pH 7.0.
4. To the pellet, add 200 μL of 10 mM NaPi, 1% SDS buffer, pH 7.0 and then dilute 50 μL in 400 μL buffer (10 mM NaPi, 1% SDS pH 7.0) and 50 μL GAB. (This ensures that all samples are in the same sample buffer for the standard curve).
5. Prepare a tubulin standard curve in GAB buffer with tubulin concentrations ranging from 0 to 20 μM. Dilute 50 μL of each sample in 450 μL of 10 mM NaPi, 1% SDS buffer, pH 7.0.
6. Determine the tubulin concentration of the samples fluorometrically ($\lambda_{exc} = 280$ nm and $\lambda_{em} = 323$ nm) using a spectrofluorometer.
7. Subtract the blank and determine the concentration of tubulin in the supernatants and pellets from the standard curve. The supernatant and pellet concentrations should add up to approximately give the total tubulin concentration in the sample. The C_r of tubulin required for assembly changes between different batches of protein. In order to be able to compare data obtained from different tubulin batches, the C_r data obtained need to be normalized. To do so, determine the C_r data in the presence of DMSO, divide all data by this value, and multiply by 3.3×10^6. This makes the C_r value in the presence of the vehicle equal to the reference value of 3.3 μM. The C_r of the test ligand is then equal to the tubulin concentration in the supernatant of that sample.

Using this method, tubulin in the presence of ZMP was found to have a C_r of 0.81 μM, similar to that found in the presence of docetaxel; whereas in the presence of dactylolide tubulin had a C_r of 2.10 μM (Field et al., 2012; Fig. 19.1A).

19.2.1.2 Microtubule assembly in PEDTA buffer

To determine if a compound is a strong MSA, its ability to induce tubulin assembly in conditions that are unfavorable for assembly is determined. In these conditions, weak MSAs such as cyclostreptin cannot induce MT assembly.

1. Prepare 20 mg tubulin in PEDTA buffer as described in Section 1.1. Add 6 mM $MgCl_2$ to the buffer and increase the concentration of GTP to 1 mM.
2. Prepare samples of 20 μM tubulin with increasing concentrations of MSA and incubate at 37 °C for 30–40 min. Use a sample with 0 μM tubulin and 15 μM MSA as a blank.
3. Separate the supernatant from the pellet by ultracentrifugation (20 min, 37 °C).
4. Collect the supernatant and resuspend the pellet in 10 mM NaPi, 1% SDS buffer, pH 7.0.
5. Dilute the samples, prepare a standard curve, and measure tubulin polymerization fluorometrically as in step 6 above.
6. Analyze the data as in step 7 above but do not normalize. A decreasing amount of tubulin in the supernatant will be seen as MSA concentration increases. When this value stabilizes, it represents the C_r.

The C_r for tubulin in the presence of ZMP in these conditions is 4.1 μM (Fig. 19.1B).

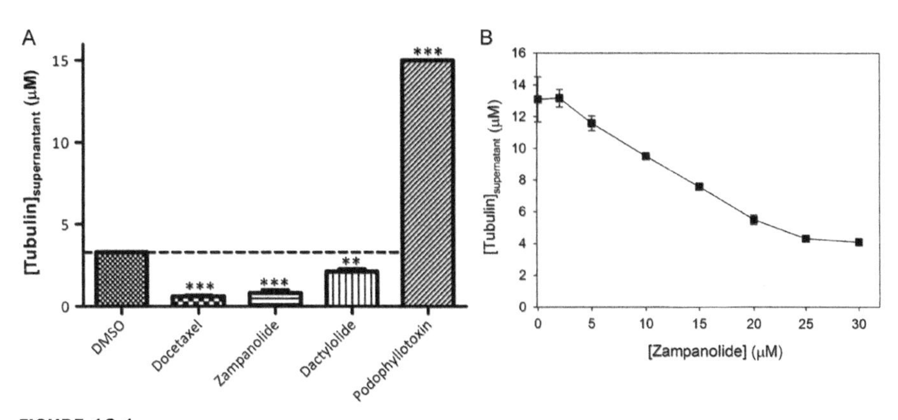

FIGURE 19.1

Induction of microtubule assembly by ZMP. (A) Induction of MT assembly by ZMP in conditions favorable for assembly. In these conditions, the C_r for self-induced assembly is 3.3 μM (DMSO bar and dashed line). Both docetaxel and ZMP significantly decrease the C_r to 0.65 μM and 0.81 μM, respectively. Dactylolide, less potent than ZMP, is able to decrease the C_r to 2.10. (B) Induction of assembly by ZMP in conditions unfavorable to tubulin assembly. In these conditions, ZMP has a C_r of 4.1 μM. Normally in these conditions, the tubulin concentration required for assembly is >200 μM.

This figure was modified from Field et al. (2012) with permission from the publisher.

19.2.1.3 Transmission electron microscopy

Once the MT-stabilizing activity of a test compound has been confirmed, it is important to visually confirm that the formed polymers are MTs. The reason for this is that some compounds induce tubulin aggregation rather than MT assembly, yet the presence of aggregates increases light scattering and fluorescence similar to MTs. To check for MT morphology, pipet the polymerized tubules onto 400-mesh carbon- and Formvar-coated copper grids (or equivalent), stain with 2% phosphotungstate, and observe in a transmission electron microscope at 20,000× magnification.

If the above assays are positive for MSA activity, the next step in its characterization is to determine its binding site. In principle, the best strategy to probe the binding site is to perform various competition assays against compounds that bind to the two known binding sites-the taxoid site, and the laulimalide/peloruside site.

19.2.2 Flutax displacement assay

A binding competition assay using stabilized taxoid binding sites and a fluorescent derivative of paclitaxel that binds the taxoid site (Flutax-1 from Calbiochem® or Flutax-2 purchased from us) can be performed to determine whether an MSA binds to the taxoid site. This assay measures the binding affinity and kinetic binding constant of a ligand in reference to its ability to displace the reference ligand from its binding site.

The equilibrium binding constants of test ligands can be determined from anisotropy titration measurements carried out at different temperatures by measuring fluorescence intensities. The displacement isotherm of each ligand is determined in a black 96-well plate (Nunc® FluoroNunc™, Sigma–Aldrich) in a fluorescence polarization microplate reader. This protocol is described in Díaz and Buey (2007) or is available from the author. Equilibrium constants for Flutax-2 are presented in Table 19.1. Taxoid site ligands can be used as positive controls to compete with Flutax-2. The displacement isotherm for the test ligand needs to be measured a number of times at different temperatures in different plates.

A typical experiment is presented in Fig. 19.2A. A decrease in the anisotropy indicates displacement of Flutax from the binding site. Docetaxel, ZMP, and

Table 19.1 Equilibrium constants of Flutax-2 binding to microtubules

	26 °C	27 °C	30 °C	32 °C	35 °C	37 °C	40 °C	42 °C
K_a (10^7 M^{-1})	6.5	5.9	4.6	4.2	3.0	2.2	2.0	1.8
v_o (Flutax-2_{bound}/Flutax-2_{total})	0.378	0.563	0.522	0.508	0.451	0.398	0.382	0.364

Table 19.2 Apparent binding constants of ligands to the taxoid site ($K_b \times 10^7\ M^{-1}$)

	26 °C	27 °C	30 °C	32 °C	35 °C	37 °C	40 °C	42 °C
Paclitaxel	2.64±0.17	2.19±0.05	1.83±0.09	1.81±0.21	1.43±0.17	1.07±0.11	0.96±0.14	0.94±0.23
Cyclostreptin	1.02±1.4	1.43±0.9	1.59±0.7	2.05±1.7	2.06±1.1	2.12±2.1	2.67±2.4	2.84±2.4
Zampanolide	13.7±2.9	15.0±0.4	14.3±4.8	18.7±7.6	21.4±9.3	25.8±11.0	41.6±17.3	42.4±17.5

Paclitaxel data from Buey et al. (2004); Cyclostreptin data from Edler et al. (2005); and Zampanolide data from Field et al. (2012).

epothilone A all compete with Flutax-2; whereas peloruside acts as a negative control as it binds to a different site, showing no change in anisotropy as its concentration is increased. The majority of taxoid site ligands bind exothermically to preformed glutaraldehyde-stabilized MTs, and the drug binding reaction is enthalpy driven, with the ligands binding reversibly to the taxoid binding site. The binding constants therefore decrease with temperature (as stabilized MTs are being used, the true binding constant is being measured not an apparent one). However, for ZMP, the apparent binding constants measured increased with temperature (Table 19.2), the opposite of what is expected for an enthalpy-driven reaction. This indicates that either there is a strong entropic effect or that the reaction observed is not reversible but an irreversible covalent reaction. In this case, the binding constants measured are apparent, not true, given the nature of the binding reaction. With covalent binding, the observed reaction is only kinetically controlled, as with any other irreversible reaction. Thus, the extension of the reaction (apparent displacement) increases as temperature increases and is limited by the rate at which ZMP binds to the MTs. The same relationship was seen previously with cyclostreptin, which also binds irreversibly to the taxoid site (Calvo et al., 2012).

19.2.3 Laulimalide/peloruside competition assay

If the compound of interest does not bind to the taxoid site, it is possible that it may bind to the laulimalide/peloruside site. Like the taxoid site, this site can be probed using a number of competition techniques. HPLC techniques are useful to quantify the bound ligand.

1. Treat stabilized cross-linked MTs with equimolar concentrations of test ligand and either DMSO or a 5- to 10-fold excess of competitor ligand in GAB-0.1 mM GTP buffer for 45 min at 37 °C. Alternatively, purified tubulin can be assembled into MTs in GAB buffer (1 mM GTP, 6 mM $MgCl_2$), and the competition experiments carried out.
2. Separate the supernatant and pellet by ultracentrifugation (10 min, 37 °C), resuspend the pellet in 10 mM NaPi, and determine the presence of ligands in both pellets and supernatants by HPLC.
3. Add 10 μM internal standard to the samples (e.g., 10 μM docetaxel, although any other ligand with a detectable HPLC trace can be used).
4. Extract the ligands from the samples with three volumes of dichloromethane, dry in a vacuum, and resuspend in solvent for analysis.
5. Analyze the samples by HPLC. We use an Agilent-1100 Series instrument employing a Supercosil, LC18 DB, 250 × 4.6 mm, 5 mm bead diameter column at a flow rate of 1 mL/min. Follow the absorbance of the compounds at their respective wavelengths.
6. Run solvent blanks at the start of each experiment and quantify the concentration of ligands in each of the samples by comparison of the integrated areas of the HPLC peak with that of the internal standard.

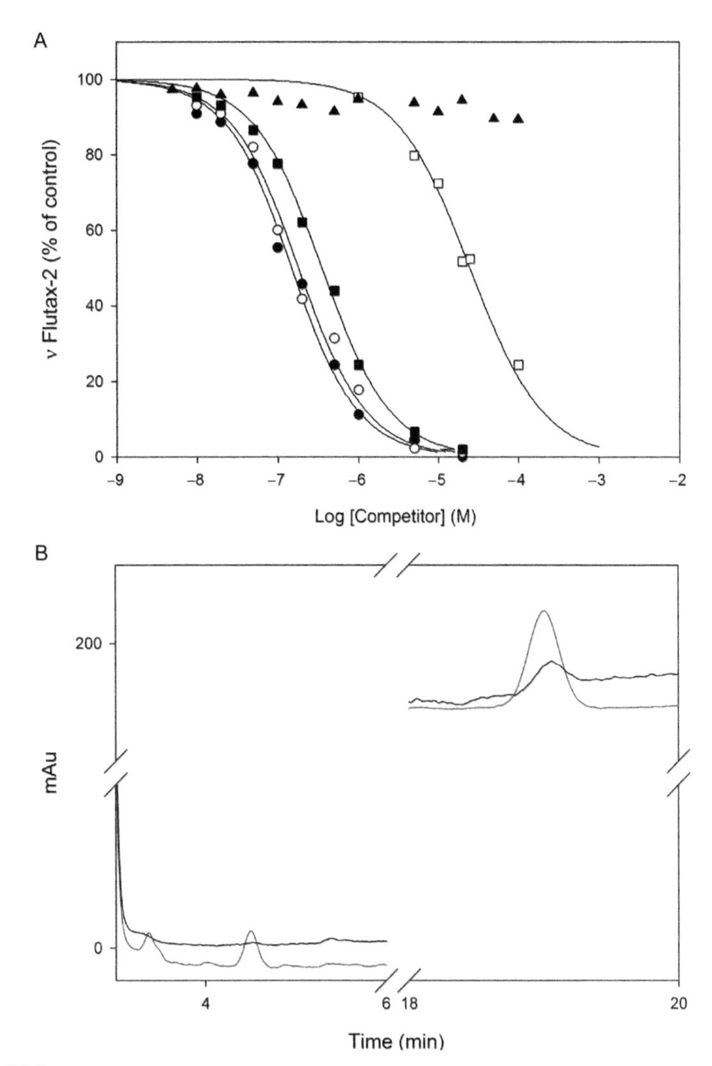

FIGURE 19.2

Determination of the binding site. (A) Competition experiments between Flutax-2 and MSAs for binding to MTs. Displacement of Flutax-2 (50 nM) from MT binding sites (50 nM) by ZMP (black circles), docetaxel (white circles), epothilone A (black squares), and dactylolide (white squares). Peloruside A (black triangles) was unable to displace Flutax-2 from the binding site. See Table 19.2 for the binding constants. The peloruside data were taken from Gaitanos et al. (2004). The figure was modified from Field et al. (2012) with permission from the publisher. (B) Competition experiments between 1 μM peloruside and 4 μM laulimalide for MTs containing 1 μM taxoid binding sites. First section of the graph shows elution of peloruside and the second elution of laulimalide (mAu = milli absorbance units). Peak at 4.5 min shows peloruside in the supernatant (gray), with none bound to the MTs in the pellet (black). Peak at 18.8 min shows laulimalide saturation of the pellet with excess laulimalide in the supernatant, indicating that these two compounds bind the same site on MTs since excess laulimalide binding prevents peloruside from binding.

19.2.3.1 Notes

The solvent used to develop the column depends on which compounds are being analyzed and their polarity. Preliminary test runs with different percentages of methanol/water need to be carried out to work out the optimal solvent and elution time of each compound. For example, when running ZMP and docetaxel, 70% (v/v) methanol/water works well, and the column is developed in the same buffer with no gradient. When ZMP, docetaxel, and peloruside are run, the samples are dissolved in (v/v) 55% methanol/water and the column developed with three isocratic steps of 13 min 55% methanol/water, 10 min 70% methanol/water, and 10 min 55% methanol/water. The absorbance of each compound is followed at its respective wavelength.

Ligand in the supernatant is unbound, whereas that in the pellet is bound. If the test ligand is in the pellet fraction, it is occupying its binding site. In the case of a covalently reacting ligand, the compound will disappear from the supernatant but will not be seen in the pellet as the covalently bound ligand cannot be extracted. For a test ligand that binds reversibly, displacement of the ligand from its site with excess control ligand results in the test ligand only being present in the supernatant, and this indicates the test ligand and the competitor ligand share the same site (Fig. 19.2B).

If no competition is detected by this second experiment and both test ligand and laulimalide/peloruside are detected simultaneously bound to the MT, it means that a new binding site is likely to be present.

19.2.4 Stoichiometry

Once it is known that a compound binds to MTs, it is important to determine the stoichiometry in which it binds to its binding site. Currently, all known taxoid site and laulimalide/peloruside site compounds bind to tubulin in a 1:1 stoichiometry, with one ligand per heterodimer (Díaz & Andreu, 1993; Pera et al., 2010). To confirm the test ligand is reacting in the same way, the stoichiometry needs to be investigated.

1. Measure the stoichiometry of MSA binding to cross-linked MTs by incubating 10 μM taxoid binding sites in stabilized MTs with increasing amounts of ligand (0–50 μM) in GAB-0.1 mM GTP buffer. Incubate a sample with no tubulin in the same way with 10 μM ligand.
2. Separate MTs with their bound MSA from unbound MSA by ultracentrifugation (20 min, 25 °C). Collect the supernatant and resuspend the pellet in 10 mM NaPi buffer pH 7.0.
3. Add an internal standard to the samples. Extract the ligands from the samples with three volumes of dichloromethane, dry, and resuspend the sample in solvent (Section 2.3.1). Analyze the samples by HPLC as described in Section 2.3, steps 5 and 6.

If the stoichiometry is 1:1, then given that there are 10 μM binding sites, the amount of ligand in the pellet should increase until it plateaus at 10 μM, at which concentration the binding sites are saturated. The ligand concentration in the supernatant should then begin to increase as more ligand is added (Fig. 19.3A).

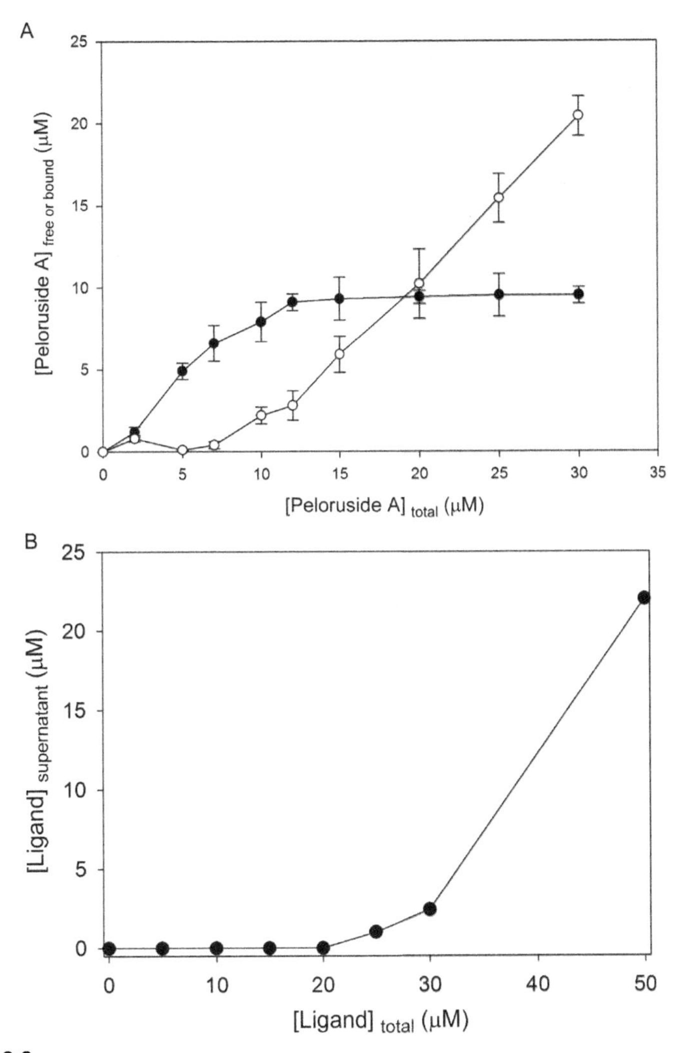

FIGURE 19.3

Stoichiometry of MSAs binding to microtubules. (A) Increasing concentrations of peloruside bound to stabilized taxoid binding sites (10 μM). Black circles represent peloruside found in the pellet; white circles represent peloruside in the supernatant. The maximum sites available are 10 μM. Once peloruside has reached this concentration, its concentration in the pellet remains constant, whereas its concentration in the supernatant increases relative to the increasing peloruside concentration, indicating 1:1 stoichiometry. (B) Stoichiometric reaction of increasing ZMP concentrations with stabilized taxoid binding sites (25 μM). ZMP in the supernatant represents compound that has not reacted with the taxoid site. ZMP does not appear in the supernatant until at least 25 μM is present in the reaction (same concentration as the binding sites). This indicates a 1:1 stoichiometry.

Graph modified from Field et al. (2012) with permission from the publisher.

With compounds that bind irreversibly, it is not possible to extract the compound from the pellets due to the covalent bond (reversible MSAs are easily recovered from the pellets). The stoichiometry of ZMP or any other irreversible ligand can be monitored by its presence in the supernatant, which will increase as extra ligand is added once saturation of the binding site has occurred (which is the same as for a reversible binder). Only the supernatant samples need to be analyzed in this case. If 1:1 stoichiometry is occurring, the test compound will not appear in the supernatant samples until its concentration has increased to greater than the concentration of the total number of binding sites in the reaction (concentration in which the binding sites are saturated) (Fig. 19.3B).

19.2.5 Binding kinetics

If the test ligand is an irreversible MSA, there are no thermodynamic equilibrium parameters to characterize. The kinetic parameters of the binding can, however, be studied by two different methods using either Flutax-2 or HPLC.

19.2.5.1 *Flutax-2 method*

The kinetics of the reaction of compounds with cross-linked MTs is measured by determining the inhibition of Flutax-2 binding.

1. Incubate 5 μM taxoid binding sites with 6 μM test compound (or equivalent volume of DMSO) at 25 °C in GAB buffer over a series of time points (30 min–overnight). The time points can be staggered to all finish incubating at the same time.
2. After the desired incubation, add 10 μM Flutax-2, incubate for 5 min, and centrifuge the samples (20 min, 37 °C).
3. Remove the supernatant and resuspend the pellet in 10 mM NaPi buffer with 1% SDS, pH 7.0 buffer.
4. Dilute 50 μL of supernatant sample in 200 μL of 10 mM NaPi buffer with 1% SDS, pH 7.0, and 50 μL of the pellet in 50 μL GAB 0.1 mM GTP and 150 μL of 10 mM NaPi buffer with 1% SDS, pH 7.0.
5. Prepare a standard curve (0–10 μM Flutax-2) in 1:4 GAB 0.1 mM GTP:10 mM NaPi buffer with 1% SDS, pH 7.0.
6. Measure concentration of Flutax-2 in the pellet and supernatant spectrofluorometrically (λ_{exe} 495 nm and λ_{ems} 520 nm) in a black 96-well plate. Calculate the free (supernatant) and bound (pellet) Flutax-2 concentrations from the standard curve.

Fast covalent binders such as ZMP will immediately bind to the MT in an irreversible manner, and no Flutax-2 will be associated with the pellet after 30 min. Compounds that do not bind irreversibly will have 100% of the Flutax-2 associated with the pellet, given that it is in excess concentration. Slow covalent binders such as dactylolide bind covalently over time, will show a decrease in Flutax associated with the pellet over time (Fig. 19.4).

19.2.5.2 High-performance liquid chromatography analysis

The kinetics of the reaction of compounds with cross-linked stabilized MTs and with unassembled tubulin can be measured by HPLC.

1. Prepare 20 mg tubulin in PEDTA buffer as described in Section 1.1. Add 1.5 mM $MgCl_2$ to the buffer and increase the concentration of GTP to 1 mM.
2. Incubate 20 μM compounds for different times with either 25 μM tubulin in PEDTA buffer or 25 μM cross-linked MTs in GAB buffer.
3. After the desired incubation, centrifuge the samples (37 °C, 20 min), separate the pellet, and resuspend the supernatant in 10 mM NaPi buffer.
4. Add an internal standard and extract compounds from the supernatant three times with one reaction volume of dichloromethane and vacuum dry.
5. Resuspend the sample in solvent and analyze by HPLC as described in Section 2.3, steps 5 and 6, and Section 2.3.1.

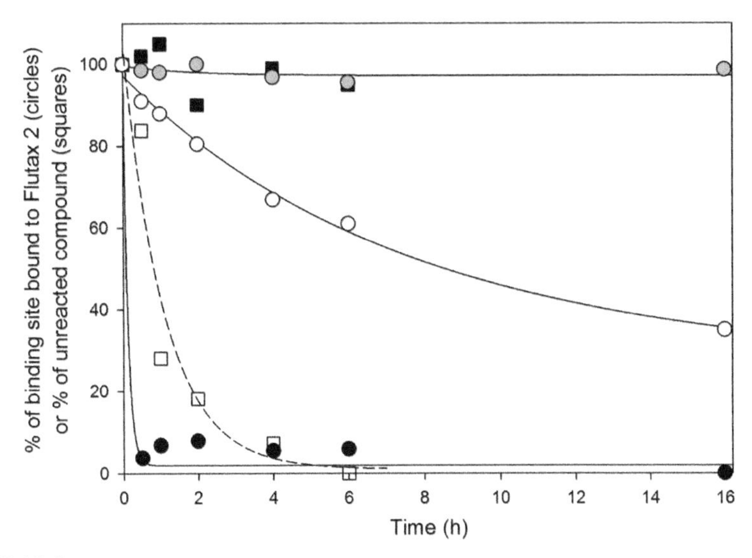

FIGURE 19.4

Binding kinetics of ZMP and dactylolide to the microtubule. The percentage of binding sites bound by Flutax-2 (circles) and the percentage of unreacted compound detected by HPLC (squares): ZMP (black circles), dactylolide (white circles), DMSO control (gray circles), dactylolide (white squares), and dactylolide control with no MTs (black squares). Solid lines represent the loss of available binding sites, and the dashed line represents the fit for decay of unbound dactylolide. The difference in the two measurements for dactylolide indicates that the covalent reaction is slow. All FTX-2 was displaced by ZMP, and all ZMP was found bound to the pellet in 30 min, indicating covalent binding immediately follows the noncovalent reaction.

Graph modified from Field et al. (2012) with permission from the publisher.

Using these kinetic experiments, two different processes can be monitored to provide an estimate of the kinetics of the covalent reaction once the compound is noncovalently bound to the binding site. The first experiment measures the number of binding sites that have not yet reacted and therefore are available for Flutax-2 binding. This measures the kinetics of the covalent reaction. In the second kinetics measurement, MTs are pelleted and the concentration of unreacted compound measured, such that, both covalently and noncovalently reacted compounds are pelleted. This measurement accounts for the kinetics of the noncovalent binding reaction. If the covalent reaction between the compound and the sites is immediate, such as with ZMP, no difference will be observed between the two measurements. In this case, the covalent reaction immediately follows noncovalent binding. However, if the covalent reaction is slow, as with dactylolide, and the noncovalently bound compound can exchange with unbound compound, a significant difference between measurements will be obtained. This indicates a significant delay between the noncovalent and the covalent reaction of dactylolide (Fig. 19.4; Field et al., 2012).

19.2.6 Mass spectrometry

To determine where on the protein covalent binding is occurring, a detection method for the ligand bound to the protein is used. Mass spectrometry (MS) is the most suitable method to characterize ligand–protein covalent binding. Specialized instruments based on hybrid configurations allow the detailed analysis of covalently modified peptides. MS is also used to determine the modified amino acid residue (s) at which the ligand is binding by the comprehensive analysis of the corresponding fragmentation spectra, giving direct evidence of the binding site (Buey et al., 2007; Calvo et al., 2012; Field et al., 2012).

To determine the modified peptides, first characterize the fragmentation spectrum of the MSA by MS. This information is crucial to determine the putative diagnostic ion(s) for subsequent ion-filtering experiments, mainly performed in triple quadrupole systems. Precursor ion scanning (PIS) and selective reaction monitoring (SRM) allow the selection of a given subset of peptides according to their masses and the presence of specific ions in the fragmentation spectrum when collision dissociation is induced. Thus, only those ions that satisfy both conditions are analyzed, whereas the rest of the ions are deflected. In the hybrid systems, it is possible to combine the benefits from a conventional triple quadrupole system and a linear ion trap. Ions filtered by PIS or SRM are detected with improved resolution to determine mass–charge of the precursor ion and the corresponding fragmentation spectrum. The comprehensive analysis of the fragmentation spectra can determine not only the masses of the MSA-attached peptides but also the modified residue(s) within the sequence.

In a normal experiment to characterize the peptides/residues modified by an MSA, the ligand is first bound to the protein and the samples are subjected to digestion using a protease. The tryptic peptide mixtures for ion-filtering experiments are then subjected to liquid chromatography coupled to hybrid MS (LC–MS). Peptides are separated by HPLC in a reversed phase C18 nanocolumn. A gradient consisting of 5–35% acetonitrile in 80 min was performed to elute peptides from the column to a nanospray emitter for ionization and fragmentation in the mass spectrometer. Depending on the target protein sequence, a variety of endopeptidases (such as chymotrypsin and endopeptidase V8) can be used to yield peptide species amenable for MS analysis. It is also possible to use multiple proteolytic enzyme combinations to produce different fragment profiles. For specific protocols, see Buey et al. (2007) and Field et al. (2012).

It is important to note that all ion-filtering experiments are performed in low resolution, and it is therefore difficult to characterize the exact modification site(s), especially when the ions of interest are larger than 2000 Da. In these cases, after the first ion filtering experiments, a targeted or semitargeted high-resolution MS experiment can be performed. High-resolution fragmentation spectra are much more easily interpreted, as the mass and the charge of all fragments are properly determined. For these experiments, an LC–MS analysis is performed using hybrid mass spectrometers based on orbital analyzers (Orbitrap) (Field et al., 2012). The combination of low-resolution ion filtering and high-resolution targeted experiments represents a powerful tool to study and determine the MSA modification of MTs.

19.2.6.1 Protein preparation and digestion for MS analysis

Depending on the conditions, a ligand may modify tubulin at different residues. Tubulin can exist in a number of different aggregation states which can result in differential binding of MSAs. ZMP is able to covalently modify MTs, tubulin oligomers, and dimeric tubulin at the same two residues, indicating binding to the taxoid site in all tubulin aggregation states (Field et al., 2012). In contrast, cyclostreptin labels two different residues in MTs, but only one of these amino acids is labeled in dimeric tubulin (Buey et al., 2007). It is therefore important to determine if the test compound labels the same or different residues in different preparations of tubulin.

19.2.6.1.1 Assembled microtubules

MSAs have high affinity for MTs and much lower affinity for unassembled tubulin. Use the following method to assemble MTs if the MSA is a strong inducer of assembly (determined in Section 2.1).

1. Prepare 20 mg tubulin in GAB buffer as described in Section 1.1. Add 6 mM $MgCl_2$ to the buffer and increase the concentration of GTP to 1 mM.
2. To 20 μM tubulin, add 25 μM ligand or an equivalent volume of DMSO and incubate for 1 h, 37 °C. Different concentrations can be used but ensure the ligand is in 10% excess over the number of binding sites to ensure site saturation.
3. Pellet the MTs by ultracentrifugation (20 min, 25 °C), discard the supernatant, and resuspend the pellet in 200 μL of 50 mM NH_4HCO_3. Add 20 μL of the pellet solution to 1 μg/μL trypsin (sequencing grade) and add another 20 μL of NH_4HCO_3.

4. Incubate the samples (2 h, 37 °C) and vacuum dry. Store the samples (−20 °C) until analysis.

19.2.6.1.2 Stabilized cross-linked microtubules

If the MSA is a weak inducer of MT assembly, use cross-linked stabilized MTs. This allows the ligand to bind straight to preformed MTs.

1. Incubate 20 μM stabilized taxoid sites with 25 μM ligand or an equivalent volume of DMSO in GAB-0.1 mM GTP buffer for 4 h at 25 °C (4 h should be sufficient time for weaker compounds to react).
2. Pellet the MTs by centrifugation (20 min, 25 °C) and discard the supernatant. Resuspend the pellet in 200 μL of 50 mM NH_4HCO_3 and process the samples as per steps 3–4 above.

19.2.6.1.3 Unassembled tubulin

It is important to check if the ligand interacts with a different residue in oligomeric or unassembled tubulin compared to MTs.

1. Prepare 20 mg tubulin in PEDTA buffer as described in Section 1.1.
2. Add 1.5 mM $MgCl_2$ for oligomer formation. For dimeric tubulin, omit this step.
3. Incubate 20 μM tubulin with 25 μM ligand in this buffer at 25 °C for 4 h or overnight, depending on how strong a binding agent the test MSA is.
4. Collect the tubulin supernatant by centrifugation (20 min, 25 °C), add 20 μL of the supernatant to 1 μg/μL trypsin, and then add 20 μL of NH_4HCO_3.
5. Incubate the samples at 37 °C for 2 h and vacuum dry overnight. Store the samples at −20 °C until analysis.

19.2.6.1.4 Compounds

To determine the best ion tracer fragments, prepare compounds without protein (1–20 mM). Keep at −80 °C for 1 h before lyophilizing overnight (Fig. 19.5).

19.2.7 Visualization of covalent binding in cells

Covalent binding *in vitro* needs to be confirmed in a cellular setting to ensure that the same mechanism of binding is occurring. For taxoid binding site ligands, this can be carried out using Flutax and conventional fluorescence or confocal microscopy. Alternatively, MS of cellular tubulin can be used to confirm the covalent binding in cells, as described by Buey et al. (2007).

Using Flutax, covalent binding can be visualized in cells using simple competition methods. Known taxoid site MSAs can be used as positive controls and those that bind the laulimalide/peloruside site as negative controls. In the case of a reversible taxoid site ligand, when added in excess over Flutax, there will be no Flutax staining of the MTs. When the Flutax is given in excess, however, the MTs will be stained as only one of the two ligands can bind the site at any one time, and Flutax in excess outcompetes the test MSA. Using an MSA that binds the laulimalide/

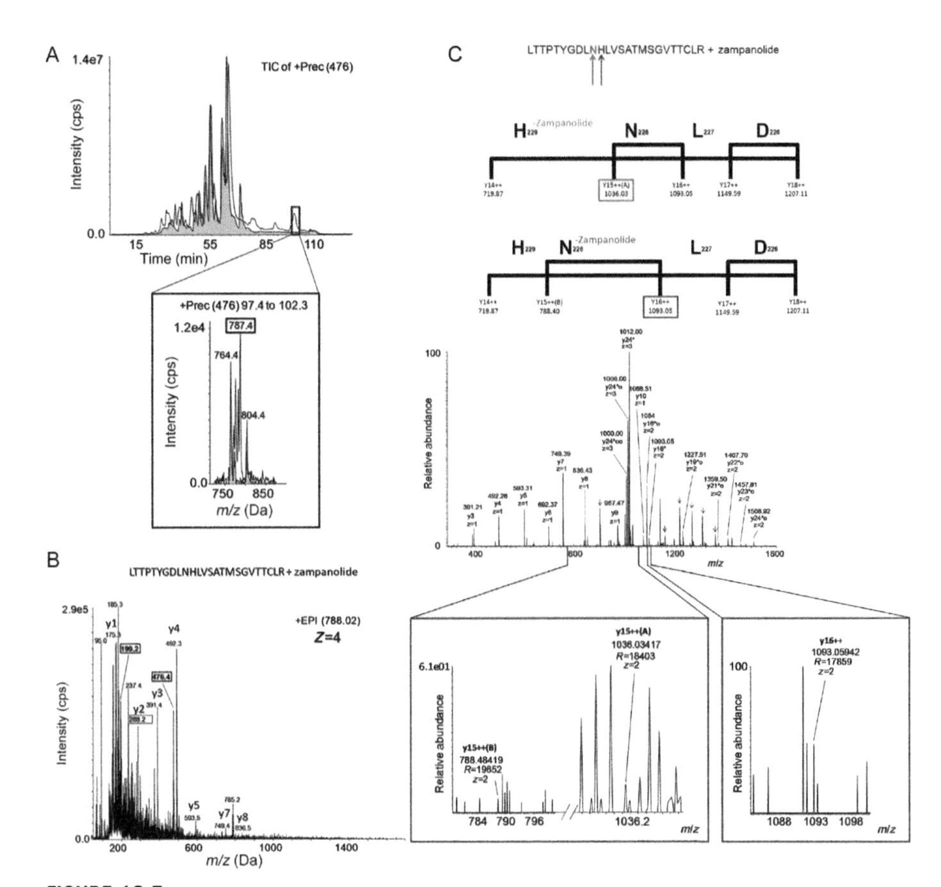

FIGURE 19.5

Mass spectrometry. (A) Total ion chromatogram of a PIS experiment at selected *m/z* values for control MTs (gray tracing) or ZMP MTs (white tracing). (B) Fragmentation spectrum for the tubulin-derived tryptic peptides bound to ZMP obtained in a low-resolution triple quadrupole system. Signals correspond to peptides with four charges. Boxed numbers correspond to ZMP fragments. The ZMP interaction domain is on the tryptic β-peptide 219-LTTPTYDGLNHLVSATMSGVTTCLR-243. (C) High-resolution MS/MS spectrum ($R=30{,}000$) for the triply charged tryptic peptides bound to ZMP acquired on an Orbitrap Elite mass spectrometer. The asterisks mark the y ions bearing the bound ZMP. Water losses are labeled with the symbol "o." Some ions corresponding to the accompanying "a" ion series are marked with an arrow. The value of *z* indicates the charge state on each fragment. Detailed information about the key fragment ions (in the y series) containing the modified residue (N228 or H229) is shown. (See color plate.)

Figures modified from Field et al. (2012) with permission from the publisher.

peloruside site, regardless of concentration, the MTs stain as each ligand binds a different site on the MT, and there is no competition between them. If the test ligand is binding covalently to the taxoid site, no staining of MTs will be seen, whether Flutax is in excess or not. Barasoain, Díaz, and Andreu (2010) describe a number of different methods for staining and visualizing MTs with fluorescent taxoids, and variations of these methods may be used to confirm covalent binding.

19.2.7.1 *Adherent cell lines*

1. Plate cells onto glass coverslips in culture dishes. Allow the cells to attach overnight. Cell number can be varied depending on the cell line used. For human 1A9 ovarian carcinoma cells, 1×10^5 cells/dish works well.
2. Replace the medium with MSA- and/or Flutax-2-containing medium (25–200 nM) or drug-free (control) medium, and treat cells for 12–16 h.
3. Incubate with DAPI (10 μg/mL) for 30 min at 37 °C to stain the nuclei.
4. Wash twice in PBS and fix the coverslips onto clean glass slides with a small volume (~10 μL) of glycerol buffer (0.13 M glycine/NaOH, 0.2 M NaCl, 70% glycerol, pH 8.6). Other methods of mounting the slides and fixing the cells do not work as the glycerol is needed to permeabilize the cells and preserve the taxoid binding sites.
5. Fluorescent staining of the MTs can then be examined using a confocal or an epifluorescence microscope.

19.2.7.2 *Cells in suspension*

1. Seed cells into a 24-well plate (cell number, volume, and plate size can be varied depending on the cell line being used). For HL-60 promyelocytic leukemic cells, 2×10^5 cells/well are seeded in 500 μL.
2. Treat with MSA and Flutax for 12–16 h, varying the concentrations. For HL-60 cells, a concentration range of 25–200 μM works well with an incubation time of 16 h.
3. Add 10 μg/mL DAPI and incubate for 30 min, 37 °C.
4. Cytospin 100 μL from each well onto slides (1265 rpm for 5 min) and mount coverslips onto slides using glycerol buffer as described earlier.
5. Examine the fluorescent staining.

Alternatively, the cells can be seeded onto plates with coverslips in the wells, and the plate then centrifuged ($400 \times g$, 5 min) in a cytospin centrifuge to attach the cells to the coverslips. The coverslips can then be mounted onto slides using a small volume of glycerol buffer.

To examine staining in the plate directly, cells in suspension can be centrifuged onto the bottom of the plate. Centrifuge at $400 \times g$ for 5 min, remove the supernatant, and carefully wash the cells once with PBS and then add glycerol buffer to cover the surface. Cells can then be examined in an epifluorescence microscope. For higher-resolution images, examine the cells under a confocal microscope using glass-bottom plates.

Covalent binding in cells can also be examined using flow cytometery. This is done by monitoring cells after MSA treatment and comparing the number of cells in the G_2/M phase of the cell cycle after drug treatment and after drug washout. If the MSA is covalently bound, then the cells will remain in G_2/M even after washout of the unbound drug. This method is described in Buey et al. (2007) (Fig. 19.6).

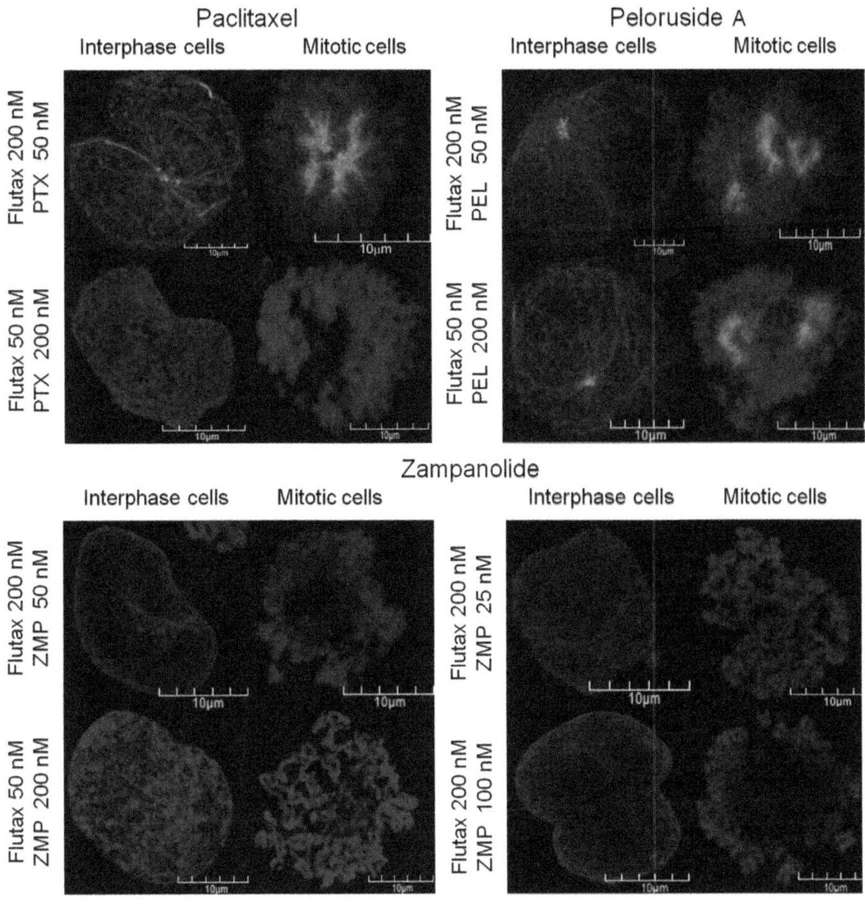

FIGURE 19.6

Flutax-2 staining of HL-60 cells showing covalent binding of ZMP. Flutax-2 staining is shown in light gray (or green for color images). DAPI staining of the nucleus is in dark gray (or blue). When in excess, Flutax-2 stains the MTs in both paclitaxel (PTX)- and peloruside (PEL)-treated cells. When paclitaxel is in excess, Flutax cannot stain the MTs due to competition for the taxoid binding site; however, when peloruside is in excess, Flutax-2 can still stain the MT network as both ligands can bind simultaneously. Lack of MT staining by Flutax-2 in ZMP-treated cells indicates the irreversible, covalent binding of ZMP, as even in excess, Flutax-2 cannot compete with ZMP to stain the MTs. (See color plate.)

19.2.8 Further experiments for characterization

If the above experiments test positive for covalent binding, a number of other experiments can be carried out to further characterize the MSA. For example, by covalently binding an MSA to tubulin, an activated adduct is essentially formed, and this may have profound effects on the MT structure and properties. For example, the ligand/MT interactions at other ligand binding sites may be altered, including those of the MT-destabilizing agents: the vinca domain and the colchicine binding site, or even the exchangeable nucleotide binding site. In addition, the aggregation properties of tubulin may differ if dimeric tubulin is structurally altered by the ligand.

SUMMARY

In this chapter, we have provided a number of different protocols for determining the binding site and mechanism of binding of novel MSAs, using ZMP as an example. Given the increasing occurrence of resistance toward MSAs in the clinic, the characterization and development of novel MSAs are important for the clinical success of this class of compounds. Covalently binding MSAs may offer excellent value in this field because covalent binding produces prolonged duration of action on the target, even after the residual unbound drug is cleared by excretion or metabolism. This would potentially enable a dosing regimen that supports less-frequent dosing and lower drug doses.

Acknowledgments

We thank our colleagues who have contributed to the development and application of these techniques, including but not limited to, J. M. Andreu, B. Pera, R. Buey, J. Antonio López, D. Zurwerra, R. Matesanz, A. Kanakkanthara, E. Hamel, and I. Barasoain. W. S. Fan for Flutax-2, Rhône Poulenc Rorer Aventis for docetaxel, and Matadero Municipal Vicente de Lucas de Segovia for the calf brains for tubulin purification. This work was supported in part by Grants BIO2010-16351 (Ministerio de Economia y Competitividad) and S2010/BMD-2457 BIPEDD2 (Comunidad Autónoma de Madrid) (J. F. D.), Cancer Society NZ, and WMRF (JM). The CNIC is supported by the Ministerio de Ciencia e Innovación and the Fundación Pro CNIC.

References

Andreu, J. M. (2007). Large scale purification of brain tubulin with the modified Weisenberg procedure. In J. Zhou (Ed.), *Microtubule protocols*: *Vol. 137*. (pp. 17–28). Totowa, NJ: Humana Press.

Andreu, J. M., Gorbunoff, M. J., Lee, J. C., & Timasheff, S. N. (1984). Interaction of tubulin with bifunctional colchicine analogs—An equilibrium study. *Biochemistry*, *23*, 1742–1752.

Barasoain, I., Díaz, J. F., & Andreu, J. M. (2010). Fluorescent taxoid probes for microtubule research. In L. Wilson & J. J. Correia (Eds.), *Methods in cell biology, microtubules, in vitro*: *Vol. 95*. (pp. 354–370). USA: Elsevier.

Buey, R. M., Barasoain, I., Jackson, E., Meyer, A., Giannakakou, P., Paterson, I., et al. (2005). Microtubule interactions with chemically diverse stabilizing agents: Thermodynamics of binding to the paclitaxel site predicts cytotoxicity. *Chemistry & Biology*, *12*, 1269–1279.

Buey, R. M., Calvo, E., Barasoain, I., Pineda, O., Edler, M. C., Matesanz, R., et al. (2007). Cyclostreptin binds covalently to microtubule pores and luminal taxoid binding sites. *Nature Chemical Biology*, *3*, 117–125.

Buey, R. M., Díaz, J. F., Andreu, J. M., O'Brate, A., Giannakakou, P., Nicolaou, K. C., et al. (2004). Interaction of epothilone analogs with the paclitaxel binding site: Relationship between binding affinity, microtubule stabilization, and cytotoxicity. *Chemistry & Biology*, *11*, 225–236.

Calvo, E., Barasoain, I., Matesanz, R., Pera, B., Camafeita, E., Pineda, O., et al. (2012). Cyclostreptin derivatives specifically target cellular tubulin and further map the paclitaxel site. *Biochemistry*, *51*, 329–341.

Díaz, J. F., & Andreu, J. M. (1993). Assembly of purified GDP tubulin into microtubules induced by taxol and taxotere—Reversibility, ligand stoichiometry, and competition. *Biochemistry*, *32*, 2747–2755.

Díaz, J. F., & Buey, R. M. (2007). Characterizing ligand-microtubule binding by competition methods. In J. Zhou (Ed.), *Microtubule protocols*: *Vol. 137*. (pp. 245–260). Totowa, NJ: Humana Press.

Díaz, J. F., Menendez, M., & Andreu, J. M. (1993). Thermodynamics of ligand-induced assembly of tubulin. *Biochemistry*, *32*, 10067–10077.

Edler, M. C., Buey, R. M., Gussio, R., Marcus, A. I., Vanderwal, C. D., Sorensen, E. J., et al. (2005). Cyclostreptin (FR182877), an antitumor tubulin-polymerizing agent deficient in enhancing tubulin assembly despite its high affinity for the taxoid site. *Biochemistry*, *44*, 11525–11538.

Field, J. J., Pera, B., Calvo, E., Canales, A., Zurwerra, D., Trigili, C., et al. (2012). Zampanolide, a potent new microtubule-stabilizing agent, covalently reacts with the taxane luminal site in tubulin α, β-heterodimers and microtubules. *Chemistry & Biology*, *19*, 686–698.

Field, J. J., Singh, A. J., Kanakkanthara, A., Halafihi, T., Northcote, P. T., & Miller, J. H. (2009). Microtubule-stabilizing activity of zampanolide, a potent macrolide isolated from the Tongan marine sponge *Cacospongia mycofijiensis*. *Journal of Medicinal Chemistry*, *52*, 7328–7332.

Gaitanos, T. N., Buey, R. M., Díaz, J. F., Northcote, P. T., Teesdale-Spittle, P., Andreu, J. M., et al. (2004). Peloruside A does not bind to the taxoid site on beta-tubulin and retains its activity in multidrug-resistant cell lines. *Cancer Research*, *64*, 5063–5067.

Gottesman, M. M., Fojo, T., & Bates, S. E. (2002). Multidrug resistance in cancer: Role of ATP-dependent transporters. *Nature Reviews. Cancer*, *2*, 48–58.

Nogales, E., Whittaker, M., Milligan, R. A., & Downing, K. H. (1999). High-resolution model of the microtubule. *Cell*, *96*, 79–88.

Pera, B., Razzak, M., Trigili, C., Pineda, O., Canales, A., Buey, R. M., et al. (2010). Molecular recognition of peloruside A by microtubules. The C24 primary alcohol is essential for biological activity. *ChemBioChem*, *11*, 1669–1678.

Singh, J., Petter, R. C., Baillie, T. A., & Whitty, A. (2011). The resurgence of covalent drugs. *Nature Reviews. Drug Discovery*, *10*, 307–317.

Zurwerra, D., Gertsch, J., & Altmann, K.-H. (2010). Synthesis of (−)-dactylolide and 13-desmethylene-(−)-dactylolide and their effects on tubulin. *Organic Letters*, *12*, 2302–2305.

CHAPTER

Studying Kinetochore-Fiber Ultrastructure Using Correlative Light-Electron Microscopy

20

Daniel G. Booth[*,†,1], **Liam P. Cheeseman**[*,1], **Ian A. Prior**[*] and **Stephen J. Royle**[*,‡]

[*]*Department of Cellular & Molecular Physiology, Institute of Translational Medicine, University of Liverpool, Liverpool, United Kingdom*

[†]*Wellcome Trust Centre for Cell Biology, University of Edinburgh, Edinburgh, United Kingdom*

[‡]*Centre for Mechanochemical Cell Biology, Division of Biomedical Cell Biology, Warwick Medical School, University of Warwick, Coventry, United Kingdom*

CHAPTER OUTLINE

Introduction 328
20.1 Materials 329
20.2 Methods 331
20.2.1 Cell Transfection and Observation 331
20.2.2 Fixation and Sample Preparation 331
20.2.2.1 Fixative Solution Osmolarity 331
20.2.2.2 Light Microscopy 332
20.2.2.3 Resin Embedding 333
20.2.3 Longitudinal Sectioning 333
20.2.4 Orthogonal Sectioning 336
20.2.5 Imaging and Sample Tilting 336
20.3 Discussion 338
Acknowledgments 341
References 341

[1]These authors contributed equally to this work.

Methods in Cell Biology, Volume 115

ISSN 0091-679X
http://dx.doi.org/10.1016/B978-0-12-407757-7.00020-7

Abstract

Electron microscopy (EM) has dominated high-resolution cellular imaging for over 50 years, thanks to its ability to resolve on nanometer-scale intracellular structures such as the microtubules of the mitotic spindle. It is advantageous to view the cell of interest prior to processing the sample for EM. Correlative light-electron microscopy (CLEM) is a technique that allows one to visualize cells of interest by light microscopy (LM) before being transferred to EM for ultrastructural examination. Here, we describe how CLEM can be applied as an effective tool to study the spindle apparatus of mitotic cells. This approach allows transfected cells of interest, in desirable stages of mitosis, to be followed from LM to EM. CLEM has often been considered as a technically challenging and laborious technique. In this chapter, we provide step-by-step pictorial guides that allow successful CLEM to be achieved. In addition, we explain how it is possible to vary the sectioning plane, allowing spindles and microtubules to be analyzed from different angles, and the outputs that can be obtained from these methods when applied to the study of kinetochore fiber ultrastructure.

INTRODUCTION

The mitotic spindle is a complex machine consisting of microtubules, motor proteins, and nonmotor proteins which, together, generate the forces needed to separate the sister chromatids between the two daughter cells (Scholey, Brust-Mascher, & Mogilner, 2003). A better visualization of its ultrastructure is necessary to understand the mechanisms underlying its functions.

Light microscopy (LM) and the discovery of the green fluorescent protein (GFP) led to many important discoveries due to the possibility of tracking protein dynamics in live cells. However, LM has a relatively low resolution, which does not allow one to visualize structures smaller than ~200 nm. This diffraction limit has been a major imaging weakness, and electron microscopy (EM) has been one of the few techniques to overcome it. Another disadvantage of LM is the restricted number of separate wavelength channels which can be used on a single sample without overlap, while the rest of the cell remains unobservable.

EM also possesses its share of drawbacks, other than the tricky and time-consuming nature of sample preparation. Only static samples can be observed, making the analysis of dynamic changes impossible. Also, routine EM does not allow one to easily locate cells of interest, such as cells expressing a fluorescent protein or in a particular stage of the cell cycle. It is possible to overcome these limitations by combining the ease and dynamic nature of LM with the subnanometer resolving power of EM in the form of correlative light-electron microscopy (CLEM).

CLEM techniques are useful for studying the mitotic spindle. The complexity of spindle microtubules means that they cannot be viewed individually by LM. Also, mitosis is a very dynamic process; each of its stages lasts less than

30 min, so pinpointing the exact stage of the cell cycle for a particular cell is critical before engaging in time-costly EM sample preparation. This is why the ability to observe and select cells of interest using LM prior to EM sample processing is a great advantage; allowing both the stages of mitosis to be chosen carefully, and to ensure that the cell is adequately expressing a fluorescent protein of interest.

Studies using EM to research mitotic spindles have yielded outstanding data, such as the quantification of microtubule polarity by Euteneuer and McIntosh (1981), the study of microtubule spacing, position, displacement, and length (McDonald, O'Toole, Mastronarde, & McIntosh, 1992), or the more recent whole-cell reconstruction by electron tomography to study cytoskeletal elements (Hoog et al., 2007).

Here, we describe our own application of CLEM to study the ultrastructure of the mitotic spindle, particularly kinetochore fibers (K-fibers) (Booth, Hood, Prior, & Royle, 2011; Cheeseman, Booth, Hood, Prior, & Royle, 2011). We describe both longitudinal and orthogonal sectioning relative to the spindle axis (Fig. 20.1), which reveal different information about spindle architecture (Fig. 20.1D), and how we can quantify such results. Longitudinal sectioning has allowed us to quantify microtubule cross-linkers between K-fiber microtubules, whereas sample-tilting of orthogonally sectioned K-fibers allowed the quantification of the number of microtubules forming the fiber. Subsequent analysis of the spacing of these microtubules allows us to measure their density and distribution.

20.1 MATERIALS

1. 35-mm glass-bottomed dishes with etched coordinates (MatTek Corporation, P35G-2-14-C-GRD)—referred to here as CLEM dishes
2. 0.1 M phosphate buffer (PB): mix 0.2 M Na_2HPO_4 with 0.2 M NaH_2PO_4 and dilute to 0.1 M. Solution should be at pH 7.4
3. Fix solution (EM grade fixatives: 3%, w/v, glutaraldehyde (Agar Scientific R1020), 0.5%, w/v, paraformaldehyde (Agar Scientific R1026) in 0.05 M PB)
4. Wash solution (0.05 M PB, 0.1 M sucrose)
5. DNA stain solution (0.1%, w/v, Hoechst-33342 in wash solution, or other similar DNA dye)
6. 1% osmium tetroxide (Agar Scientific R1015) in water
7. 0.5% (w/v) uranyl acetate (Agar Scientific R1260A) in 30% ethanol
8. Molecular grade 100% ethanol (Sigma-Aldrich 270741-1L)
9. EPON resin (Agar Scientific R1031, made up using the supplier's specifications for a "medium" block. Make sure resin mix is fully homogenized, and containing as few bubbles as possible. 200 ml of resin can be made up at a time, aliquoted into small glass vials, and kept frozen at −20 °C)

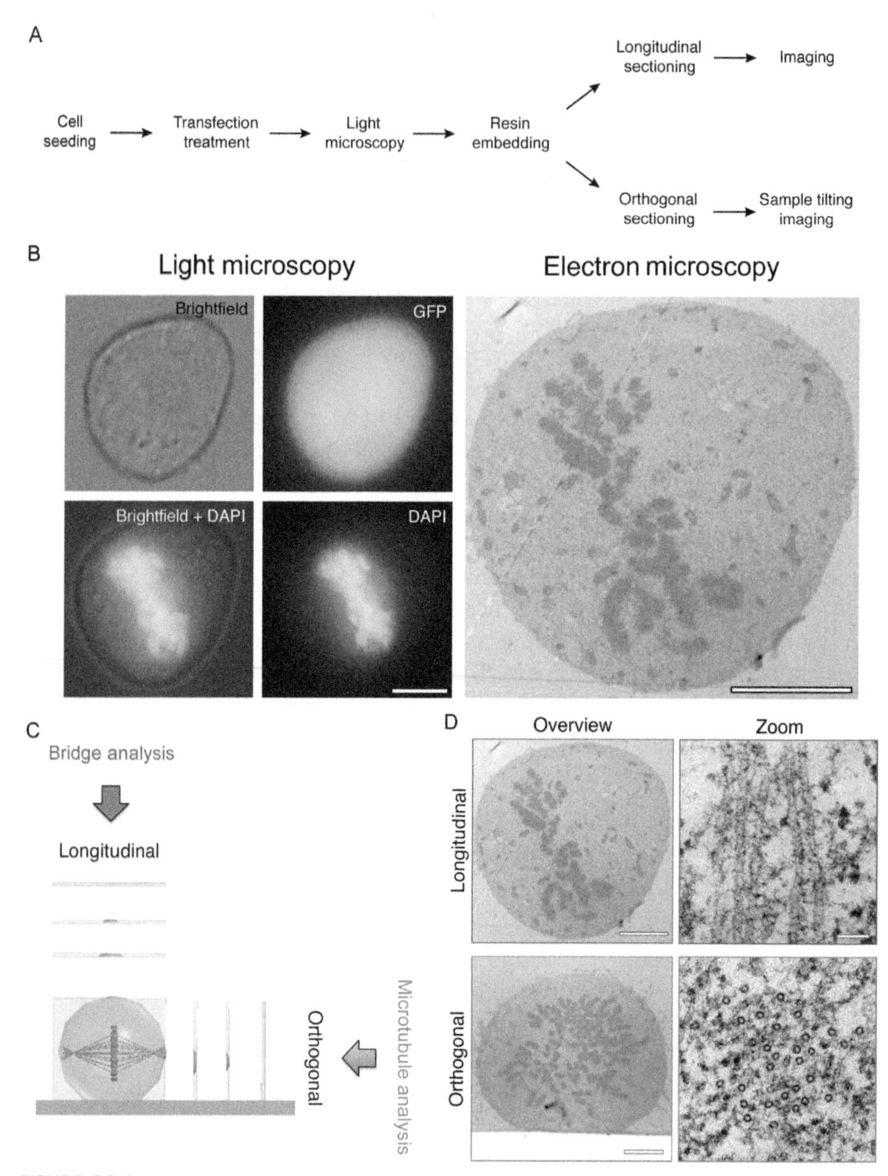

FIGURE 20.1

CLEM performed on mitotic cells. (A) A workflow to achieve CLEM using longitudinal or orthogonal sectioning. (B) A transfected mitotic HeLa observed by LM (Brightfield, GFP, and DAPI) and by electron microscopy. Scale bar 5 μm. (C) Schematic of longitudinal and orthogonal EM sectioning, and examples of output analysis. (D) Representative electron micrographs of cells sectioned longitudinally (above) and orthogonally (below) with high magnification of microtubules (right). Scale bar 4 μm (overview) and 50 nm (zoom). (See color plate.)

10. Gelatin capsules (Size 0, Agar Scientific AGG29210)
11. Copper mesh sample grids, coated with formvar (Agar Scientific R1202). We routinely use 200 hexagonal mesh grids (TAAB GG017/C), but 100 mesh or slot grids can also be used. Beware, as the larger the gaps between the copper bars, the more easily the sample will distort and can tear
12. High precision tweezers. We prefer self-closing tweezers as they facilitate the handling of sample grids
13. 5% (w/v) uranyl acetate (Agar Scientific R1260A) in 50% ethanol
14. Reynold's lead citrate solution (see Reynolds 1963)

20.2 METHODS

20.2.1 Cell transfection and observation

Cells are seeded into CLEM dishes that contain a coordinate-engraved glass coverslip, providing a pattern to be left in the base of the resin, once embedded. The coordinates are essential for the LM to EM transfer as they allow cells of interest to be tracked throughout the entire CLEM process.

Seeding the appropriate amount of cells into the dishes is important: too many will make locating the cell of interest among many other unwanted cells difficult once the sample is embedded in resin; it will also make reading the coordinates under the light microscope difficult. However, seeding too few cells reduces the chances of finding a suitable cell of interest. We, therefore, seed cells at 5% density, or 40,000 HeLa cells per 35-mm dish in preparation for imaging and resin-embedding the sample the following day. If the cells require transfection for over 24 h, we usually transfect in separate plates (such as 6-well plates) and reseed them into the CLEM dishes at the appropriate time to attain the required density. Aim for a cell density of 10–15% on the day of processing for CLEM.

20.2.2 Fixation and sample preparation

20.2.2.1 Fixative solution osmolarity

The physiological osmolality of mammalian tissue is ~290 mOsm, depending on species, tissue type, and hydration status (Loqman, Bush, Farquharson, & Hall, 2010; Mathieu, Claassen, & Weibel, 1978). Fixative solutions should mimic physiological osmolality, providing an iso-osmotic equilibrium between intracellular and extracellular fluids. Figure 20.2 shows the examples of orthogonally sectioned cells fixed with solutions of varying osmotic strengths. At 440 and 1100 mOsm, a large amount of cell shrinkage can be observed, with poor spindle apparatus preservation and unusually dense cytosol. Therefore, we routinely use a fixing solution of ~280 mOsm, consisting of 3% glutaraldehyde, 0.5% paraformaldehyde in 0.05 M PB.

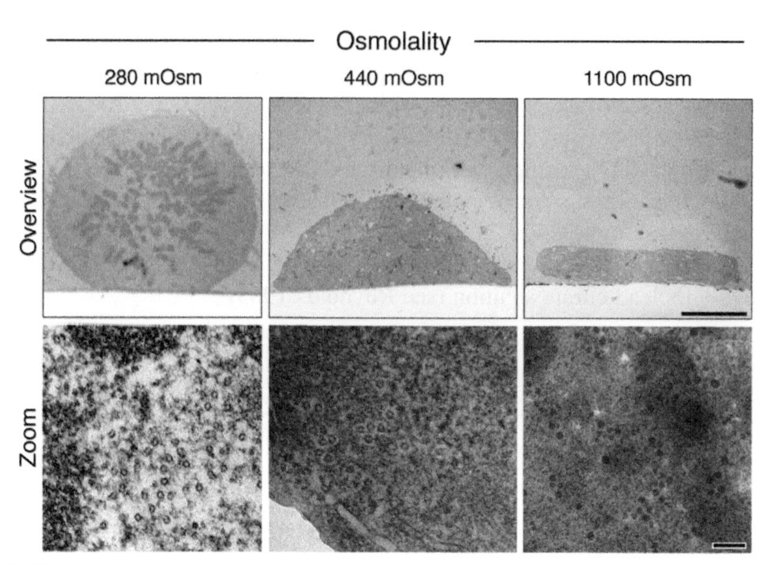

FIGURE 20.2

Optimization of mitotic spindle and cell structure preservation. Orthogonal sections of cells fixed with 280, 440, or 1100 mOsm solutions. Representative high-magnification electron micrographs of the cytosol in each condition are shown below. Scale bars 5 μm (overview) and 100 nm (bottom).

20.2.2.2 *Light microscopy*

On the day of sample processing, cells of interest can be identified under the light microscope with a 20× air objective. The low magnification allows images to be acquired containing a large field of view, useful for cell relocation during later processing. Some cells expressing fluorescent proteins require prefixation imaging as their fluorescence is obscured by the autofluorescence created by glutaraldehyde. Once the cell of interest has been located and imaged, add fixative solution for 1 h. It should be noted that microtubules are sensitive to temperature changes (Engelborghs, Heremans, Demaeyer, & Hoebeke, 1976); we, therefore, recommend that imaging of unfixed cells is carried out using an appropriate live imaging chamber at 37 °C.

After fixation, replace the fixative solution with 1–2 ml wash solution with 0.1% Hoechst-33342 (or similar DNA dye), incubate for ~20 min, rinse three times with wash solution (leaving on 1 ml of the final wash), and return the dish to the microscope. This second round of imaging is an opportunity to acquire high-magnification images of cells of interest, using 60× or 100× oil-immersion objectives (Fig. 20.1B, left). Take fluorescent and white light images of the cell, and also images of the same field of view focused on the coordinates. These will serve as references later to pinpoint the cell in the resin block and determine the orientation of the spindle axis. It helps at this stage to carefully wipe off any immersion oil with ethanol, and mark the approximate location of the cell with a fine marker pen on the underside of the dish.

20.2.2.3 *Resin embedding*

Cells become round during mitosis and are, therefore, less adherent to their substrate. This means that during all steps up to resin embedding, dishes must be handled with extreme care so as to not detach or change the orientation of the cell.

Next, replace the wash solution with a few drops of 1% osmium tetroxide on the coverslip for 1 h. Remove the osmium and gently rinse the cells twice for 30 min with double-distilled water. Remove the water and replace with 30% ethanol for 30 min, then replace with a small amount of 0.5% uranyl acetate in 30% ethanol for 1 h. Next, the cells need to be dehydrated using a gradient of sequential solutions containing increasing amounts of ethanol. Replace stepwise with each of the following solutions: 30%, 50%, 60%, 70%, 80%, and 90% ethanol, then twice with 100% ethanol, incubating at each step for 10 min.

The cells can now be infiltrated with resin. Mix a 1:2 ratio of resin:ethanol solution, making sure that it is fully homogenized using a 3-ml plastic Pasteur pipette. Remove the ethanol from the dish and lightly cover the bottom of the dish with the resin infiltration mix for 20 min. Remove and replace with a 1:1 ratio of resin:ethanol solution for 20 min. Remove the mix and replace with a ~2-mm layer of 100% resin covering the bottom of the dish. If the sample is to be sectioned orthogonally, the dish can be placed in a 60 °C oven for 48–72 h. For longitudinal sectioning, fill either half of an embedding capsule with 100% resin and gently place it open-side down onto the cell (Fig. 20.3A), which you should be able to locate using the pen mark placed earlier. The dish can then be placed in the oven.

We recommend the use of EPON resin as other resin types (e.g., Agar Scientific Low Viscosity Resin) that we have tested react and bind the CLEM dish, making the separation steps (below) much more difficult.

20.2.3 Longitudinal sectioning

Longitudinal sectioning is the conventional EM method for viewing cells. Sections parallel to the plane of the coverslip are taken from the base of the cell moving progressively upward (see Fig. 20.1C). This plane of sectioning allows extended lengths of microtubules to be observed and is, therefore, particularly useful for analyzing microtubule attachment to the kinetochore, or quantifying microtubule cross-linkers (Fig. 20.1D).

Once the resin has fully polymerized, the dishes can be removed from the oven. Figure 20.3 contains a pictorial guide of the steps required to separate the resin and dish up to the sectioning. Using pliers, start by cutting-off the edges of the dish entirely (Fig. 20.3A–C) so that the seam between resin and plastic is accessible all the way around the dish. Very carefully insert a razor blade between the plastic and resin (Fig. 20.3D), slowly forcing the razor toward the center of the dish all the way around

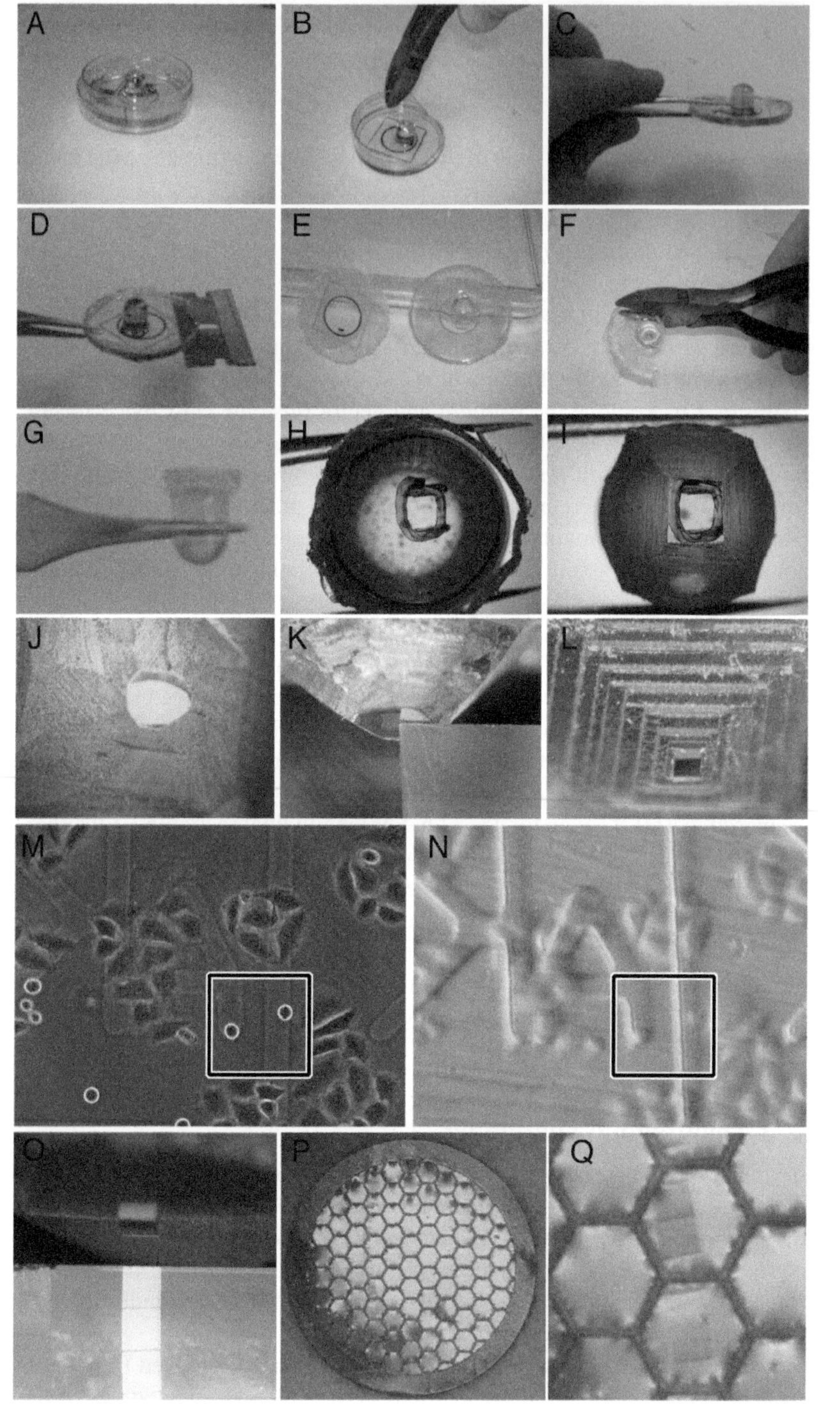

FIGURE 20.3

Pictorial guide to CLEM processing for sectioning longitudinally to the spindle axis. Following polymerization, resin was separated from the CLEM dish. Unwanted plastic was removed from the edges of the dish using pliers (A–C) allowing a razor to be inserted between the resin

the resin, to separate them (Fig. 20.3E). This must be performed with extreme care, as too much leverage by the razor will shatter the glass. This shattering usually renders the sample unusable, as removing all the glass fragments from the resin is very difficult without damaging the sample, and any microscopic shards of glass remaining will damage the diamond knife during sectioning. Dipping the resin and dish into liquid nitrogen for 1–2 s can help separate them, as the difference in thermal expansion between resin and plastic will eventually detach them. Other protocols use 40% hydrofluoric acid to dissolve the glass, bypassing this tricky step (Polishchuk, Polishchuk, & Luini, 2012).

Once detached, excess resin can be trimmed away until only the capsule remains (Fig. 20.3F and G). The coordinates imprinted on the underside of the capsule can be observed using a tissue dissection microscope; draw around the coordinate containing your cell of interest using a thin marker pen, the LM images serving as reference (Fig. 20.3H). Using a microtome chuck and bench-top vice to firmly hold the resin block in place, remove excess resin around the coordinate using a junior hacksaw (Fig. 20.3I), making sure that it never scratches the coordinate surface. This risk can be minimized by trimming away the resin using razor blades (Fig. 20.3J).

The remainder of the resin trimming and sectioning is performed using an ultramicrotome with glass knives (Fig. 20.3K) and a diamond knife (Fig. 20.3O), respectively. It is possible to make out the cell of interest and the etched coordinate using microtome binoculars (Fig. 20.3M and N). We routinely trim a square block face, up to ~50 μm from the cell edge (Fig. 20.3L), but a wider space can be left according to one's experience. The larger the block face created during sectioning, the more difficult it will be to locate the cell in the sections under the EM. A square block face is optimal, as this helps acquire serial sections during sectioning. Sections 80 nm in thickness are taken using the diamond knife and collected using the copper grids coated with formvar (Fig. 20.3P and Q). To handle the grids, high-precision tweezers should be used at all times, carefully gripping the grid by its outer edge only, so as to not tear or damage the formvar or sample sections.

To attain optimal contrast under the microscope, we post-stain the sections by placing each grid section-side-down onto a drop of 5% uranyl acetate in 50% ethanol for 7–8 min, gently rinse in distilled water for 1 min, and place face-down on a drip of Reynold's lead citrate solution for 7–8 min. The grid is then rinsed in water again for 1 min. Both solutions should be centrifuged in 1.5-ml Eppendorf tubes at

---

and the dish base (D). Following the separation of resin and dish (E), excess resin was removed using pliers (F) until just the capsule remained (G). The cell of interest was marked (H) with the aid of LM images (M) and resin coordinates (N). Unwanted resin was removed using a junior hacksaw (I) and a razor (J). Resin was trimmed using a microtome and a glass knife (K) until a neat block was generated at the top of a pyramid (L). Blocks were sectioned using a diamond knife (O) and ribbons collected using 100 mesh copper grids (P and Q), coated with formvar. (See color plate.)

$8000 \times g$ for 5 min before use to remove unwanted precipitate. Grids should be dried face-up for at least 2 h on clean filter paper before imaging, and kept in a clean, dust-free environment (such as a Petri dish or grid storage box).

20.2.4 Orthogonal sectioning

Orthogonal sectioning involves taking sections that are perpendicular to the spindle axis. In mitotic cells, this is useful to view and quantify most K-fiber microtubules within a single section. Quantifying K-fiber microtubules is possible using longitudinal serial sections (McEwen, Heagle, Cassels, Buttle, & Rieder, 1997); however, we avoided this method because (1) serial sections are particularly difficult to acquire and (2) spatial distribution analysis cannot be performed, as the compression forces exerted on each section of a serial reconstruction by the knife will likely deform the sample more than a single orthogonal cross-section through a K-fiber.

Figure 20.4 contains a pictorial guide of the steps required to prepare the sample for orthogonal sectioning. The samples to be sectioned orthogonally should consist of a flat layer of resin (without the resin capsule; Fig. 20.4A). Once removed from the oven, separate the dish from the resin using the same method as for the longitudinal samples (see above) and with the same amount of care (Fig. 20.4B–D). Using a tissue dissection microscope, find the coordinate containing your cell of interest and circle it using a marker pen (Fig. 20.4H). Using the coordinate grid and the LM images taken previously as references (Fig. 20.4E–G), determine the position and direction of the spindle axis and draw an elongated rectangle around the cell of interest (Fig. 20.4H–J). Cut out the rectangle using the hacksaw (Fig. 20.4K and L), paying attention not to touch or damage the surface of the marked coordinate. You should end up with a strip of resin (Fig. 20.4M). Carefully remove excess off one end, so the cell is near the tip, and insert it into a microtome chuck (Fig. 20.4N and O).

Using an ultramicrotome, you should be able to see the coordinates and can trim excess resin from the tip using glass knives until you approach the cell of interest (Fig. 20.4P–R). Trim away resin from either side of the cell to a depth of ~100 μm, leaving a 50- to 100-μm buffer zone around the cell. Finally, trim away excess resin from the "upper" side of the strip, which is the block face positioned reverse-parallel to the one imprinted with coordinates. The thickness should be similar to the width on either side of the cell edge so that the block face is square shaped. The cell can now be sectioned using a diamond knife; we routinely take 80–100 nm slices. Sections should be collected and treated as described for longitudinal sectioning (see above), along with the same poststaining method.

20.2.5 Imaging and sample tilting

In longitudinal sections, K-fibers can be identified as bundles of microtubules in parallel conformation terminating at the kinetochore. During image capture, we typically take 4 μm by 3 μm images at 60,000× magnification. This allows us to distinguish adjacent microtubules and the material that cross-links them with enough

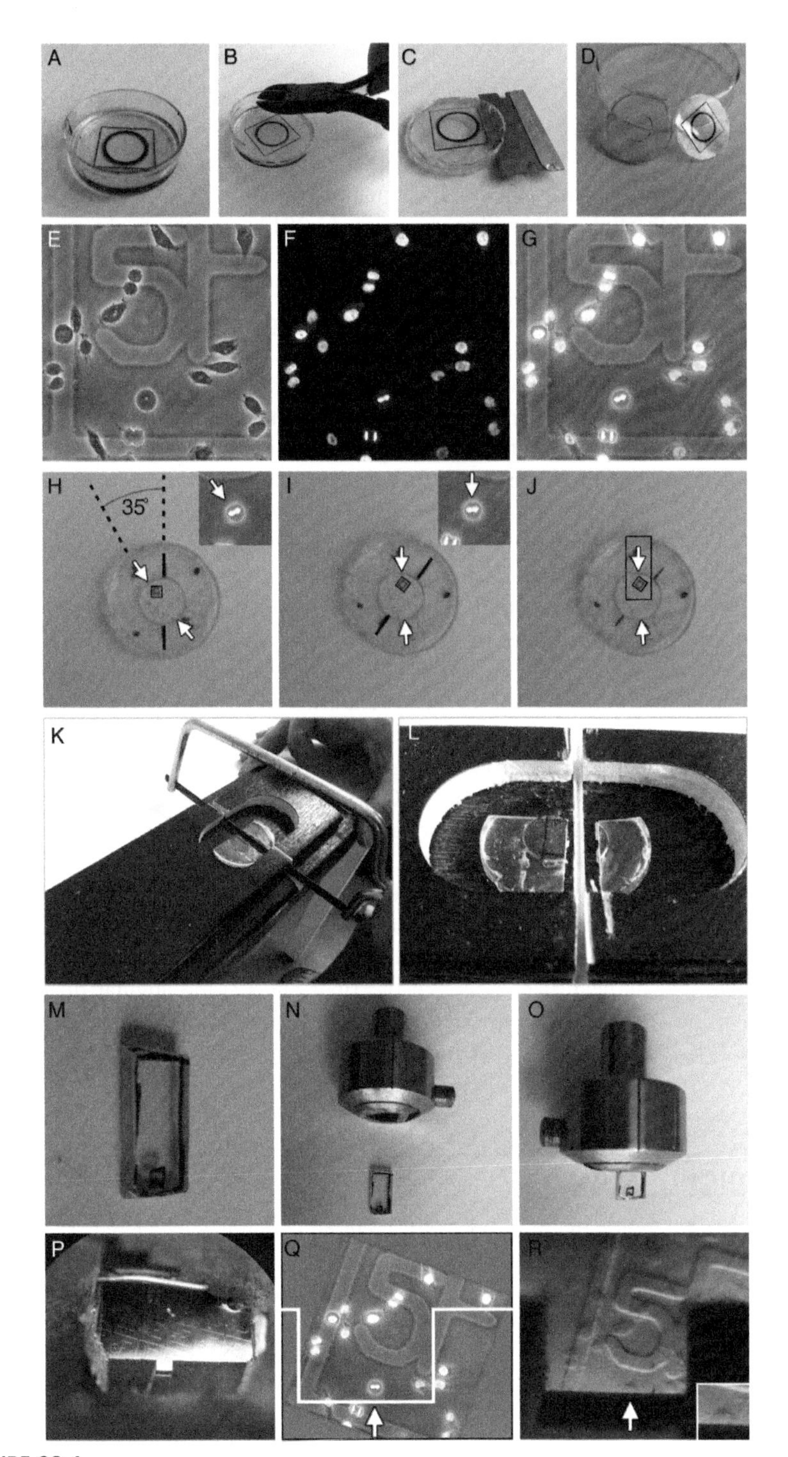

FIGURE 20.4

A pictorial guide to orthogonal CLEM processing. Following polymerization, resin was separated from the CLEM dish (A–D). Unwanted plastic was removed from the edges of the CLEM dish using pliers (B) allowing a razor to be inserted between the resin and the dish

Continued

resolution to measure the length of each element. One particular type of analysis that we have performed on such images is the quantification of microtubule cross-linker frequency (Booth et al., 2011; Cheeseman et al., 2011), but the qualitative assessment of microtubule attachment to kinetochores and of the overall organization of the fiber is also possible.

To image orthogonal sections under the microscope, full analysis of the K-fiber requires sample tilting. This is because microtubules are most easily recognizable when they are perpendicular to the imaging plane, as they appear as characteristic electron-dense rings. Not all microtubules will be at the correct angle, which increases the risk of quantification error. We can minimize this error by imaging the sample at various tilt angles (Fig. 20.5A). Our optimization shows that a single axis tilt of $\pm 45°$ ($90°$ in total) is sufficient to reveal 80% of microtubules in a K-fiber (Fig. 20.5B and C). A dual tilt along perpendicular axes is necessary to obtain 100% coverage. We perform image acquisition for a typical mammalian K-fiber at 60,000–90,000 × magnification.

Additionally, once the image tilt series (.raw file) is acquired, it can be assembled into a tomogram using IMOD software's Etomo package (Boulder Laboratory for Three-Dimensional Electron Microscopy). The final tomogram is a stack of images detailing the sample section in three dimensions and also removes some background compared to an unaltered electron micrograph.

We use ImageJ/Fiji software and IMOD's Neighbor Density Analysis (NDA) package to analyze the spatial distribution of microtubules within a fiber (McDonald et al., 1992). The output is a probability distribution graph, indicating the distance from any given microtubule at which one is most likely to find another microtubule. Other examples of types of spatial analyses which can be performed are (1) nearest neighbor analysis, which calculates the average distance between a microtubule and its nearest neighbor in the fiber and (2) the angular distribution of neighboring microtubules surrounding any given microtubule (also performed using the IMOD NDA package) (Ding, McDonald, & McIntosh, 1993).

20.3 DISCUSSION

CLEM remains among the most powerful imaging techniques available. The ability to view a live cell, in any state, or undergoing any particular or rare event, and to take an EM snapshot to be viewed at ~100,000 × magnification remains an outstanding

FIGURE 20.4—Cont'd base (C). Following separation (D), the position of the spindle was estimated using the reference LM images (E–G). These images allowed the reorientation of the resin (H and I) so that an appropriate block could be marked (J) before excision using a junior hacksaw and a miter block (K and L). The excised block (M) was inserted into a microtome chuck (N and O) and fine trimmed using a glass knife (P) before serial sections were taken of the cell, in the desired orientation (Q and R). (See color plate.)

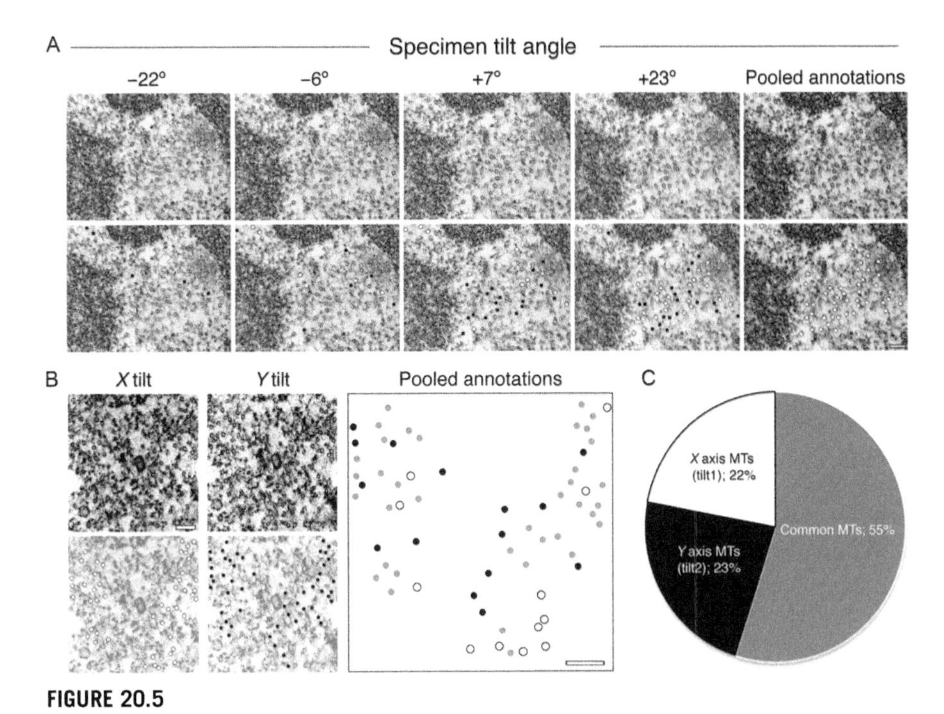

FIGURE 20.5

Sample tilting is necessary to achieve full coverage of microtubules in orthogonal sections. (A) Example electron micrographs taken from a −45° to +45° tilt series of a K-fiber. Observable microtubules are marked with dots (bottom row). Black dots represent microtubules that are unique to that tilt frame. White dots represent the accumulating microtubules identified in previous tilt frames. The total number of microtubule annotations were pooled together onto one frame (far right) giving a fair overview of the whole K-fiber. Scale bar 100 nm. (B and C) A dual tilt series of one K-fiber was carried out. (B) Representative electron micrographs taken from the central region of both *X* and *Y* tilts (A-top). All microtubules observed in each tilt series were annotated (bottom; *X* axis in white, *Y* axis in black). The sum of microtubules from both tilts were pooled together on to a single blank image, any microtubules that were common to both tilts were marked gray. Scale bar 100 nm. (C) A pie chart showing the percentage of total microtubules that were unique to each tilt and also the common ones.

tool to study cellular processes. However, this technique is often overlooked as it is considered too time-consuming and technically challenging.

Our CLEM protocol has been optimized for the study of mitotic spindles, where a great amount of attention has been placed on the preservation of the ultrastructure of microtubules and other spindle components. Our osmolality tests have shown that fixative solution osmolality must be as close to physiological conditions as possible. But further improvements could potentially be achieved during sample dehydration steps, at which point cell shrinkage can occur, as well as partial cytosolic washout.

Our protocol uses chemical fixation which is suboptimal for microtubule preservation. High-pressure freezing is an alternative fixation method, which uses the combination of ultra-low temperature to snap freeze the cell while applying pressure to inhibit the formation of ice crystals which would rupture and damage cellular structures. This fixation has been shown to substantially improve preservation of cellular architecture and organelle appearance (Wolf, Stockem, Wohlfarthbottermann, & Moor, 1981). Some studies have used CLEM with high-pressure freezing to study mitotic or meiotic events with remarkable success (Pelletier, O'Toole, Schwager, Hyman, & Mueller-Reichert, 2006). However, the implementation of this method with CLEM considerably increases the difficulty of the overall protocol, particularly when studying mitotic spindles. The size of the sample that can be frozen is very small, and as microtubules are particularly sensitive to temperature variation, a fast transfer from the light microscope/incubator to the high-pressure freezer is needed.

A useful addition to our CLEM protocol would be the ability to readily view proteins of interest under both light and electron microscopes. There has been recent focus on developing hybrid genetic tags that are both fluorescent and can be converted into an electron-dense signal to serve this purpose such as MiniSOG (Shu et al., 2011) and GFP-APEX (Martell et al., 2012). However, these tags have yet to be used to study the mitotic spindle.

Although our experimental purposes have only required standard epifluorescence micrographs before switching to EM, confocal microscopy could easily be implemented instead. This would allow the above protocol to be expanded, by combining confocal Z stacks and serial EM section imaging to create correlated 3D reconstructions in both light (confocal) and electron micrographs or electron tomograms. However, obtaining serial sections remains a challenge even for experienced electron microscopists. Nonetheless, there are currently several labs attempting the EM reconstruction of entire mitotic spindles, and whole-cell tomographic reconstruction has been achieved to study cytoskeletal structures (Hoog & Antony, 2007; Hoog et al., 2007), indicating the feasibility of this approach. Moreover, the recent effort by EM equipment suppliers to develop dual-beam EM microscopes which are able to both section and image the sample in an automated fashion could revolutionize this field. These machines, which are able to repeatedly remove 5 nm layers of sample and image the back-scattering of electrons using high-resolution scanning EM, bypass all the major difficulties involved with EM. So far, the resolution of this equipment is sufficient to comfortably reconstruct synaptic vesicles and other organelles (Knott, Rosset, & Cantoni, 2011), but it is not yet enough to view cytoskeletal elements such as microtubules in high detail.

Overall, CLEM is a powerful imaging method able to give unrivaled cellular structural detail, which we have applied to the study of K-fiber ultrastructure. We believe that the further integration of such tools as hybrid tags and dual-beam microscopes with CLEM will unlock a vast potential for the field of EM, which will maintain a firm place in research, regardless of the development of other super-resolution imaging systems.

Acknowledgments

We would like to thank Alison Beckett for technical help and discussions. D. G. B. and L. P. C. were supported by Prize Studentships from The Wellcome Trust. I. A. P. is a Royal Society University Research Fellow. S. J. R. is a Senior Cancer Research Fellow for Cancer Research UK.

References

Booth, D. G., Hood, F. E., Prior, I. A., & Royle, S. J. (2011). A TACC3/ch-TOG/clathrin complex stabilises kinetochore fibres by inter-microtubule bridging. *EMBO Journal*, *30*, 906–919. http://dx.doi.org/10.1038/emboj.2011.15.

Cheeseman, L. P., Booth, D. G., Hood, F. E., Prior, I. A., & Royle, S. J. (2011). Aurora A kinase activity is required for localization of TACC3/ch-TOG/clathrin inter-microtubule bridges. *Communicative & Integrative Biology*, *4*, 409–412.

Ding, R., McDonald, K. L., & McIntosh, J. R. (1993). Three-dimensional reconstruction and analysis of mitotic spindles from the yeast, Schizosaccharomyces pombe. *The Journal of Cell Biology*, *120*, 141–151.

Engelborghs, Y., Heremans, K. A. H., Demaeyer, L. C. M., & Hoebeke, J. (1976). Effect of temperature and pressure on polymerization equilibrium of neuronal microtubules. *Nature*, *259*, 686–689. http://dx.doi.org/10.1038/259686a0.

Euteneuer, U., & McIntosh, J. R. (1981). Structural polarity of kinetochore microtubules in Ptk1-cells. *The Journal of Cell Biology*, *89*, 338–345. http://dx.doi.org/10.1083/jcb.89.2.338.

Hoog, J. L., & Antony, C. (2007). Whole-cell investigation of microtubule cytoskeleton architecture by electron tomography. *Methods Cell Biol*, *79*, 145–167.

Hoog, J. L., Schwartz, C., Noon, A. T., O'Toole, E. T., Mastronarde, D. N., McIntosh, J. R., et al. (2007). Organization of interphase microtubules in fission yeast analyzed by electron tomography. *Developmental Cell*, *12*, 349–361. http://dx.doi.org/10.1016/j.devcel.2007.01.020.

Knott, G., Rosset, S., & Cantoni, M. (2011). Focussed ion beam milling and scanning electron microscopy of brain tissue. *Journal of Visualized Experiments*, *53*, e2588. http://dx.doi.org/10.3791/2588.

Loqman, M. Y., Bush, P. G., Farquharson, C., & Hall, A. C. (2010). A cell shrinkage artefact in growth plate chondrocytes with common fixative solutions: Importance of fixative osmolarity for maintaining morphology. *European Cells & Materials*, *19*, 214–227.

Martell, J. D., Deerinck, T. J., Sancak, Y., Poulos, T. L., Mootha, V. K., Sosinsky, G. E., et al. (2012). Engineered ascorbate peroxidase as a genetically encoded reporter for electron microscopy. *Nature Biotechnology*, *30*, 1143–1148. http://dx.doi.org/10.1038/nbt.2375 nbt.2375 [pii].

Mathieu, O., Claassen, H., & Weibel, E. R. (1978). Differential effect of glutaraldehyde and buffer osmolarity on cell dimensions—Study on lung-tissue. *Journal of Ultrastructure Research*, *63*, 20–34. http://dx.doi.org/10.1016/s0022-5320(78)80041-0.

McDonald, K. L., O'Toole, E. T., Mastronarde, D. N., & McIntosh, J. R. (1992). Kinetochore microtubules in Ptk cells. *The Journal of Cell Biology*, *118*, 369–383. http://dx.doi.org/10.1083/jcb.118.2.369.

McEwen, B. F., Heagle, A. B., Cassels, G. O., Buttle, K. F., & Rieder, C. L. (1997). Kinetochore fiber maturation in PtK1 cells and its implications for the mechanisms of chromosome congression and anaphase onset. *The Journal of Cell Biology, 137*, 1567–1580. http://dx.doi.org/10.1083/jcb.137.7.1567.

Pelletier, L., O'Toole, E., Schwager, A., Hyman, A. A., & Mueller-Reichert, T. (2006). Centriole assembly in Caenorhabditis elegans. *Nature*, *444*, 619–623. http://dx.doi.org/10.1038/nature05318.

Polishchuk, R. S., Polishchuk, E. V., & Luini, A. (2012). Visualizing live dynamics and ultrastructure of intracellular organelles with preembedding correlative light-electron microscopy. *Methods in Cell Biology*, *111*, 21–35.

Reynolds, E. S. (1963). The use of lead citrate at high pH as an electron- opaque stain in electron microscopy. *The Journal of Cell Biology*, *17*, 208–212.

Scholey, J. M., Brust-Mascher, I., & Mogilner, A. (2003). Cell division. *Nature*, *422*, 746–752.

Shu, X., Lev-Ram, V., Deerinck, T. J., Qi, Y., Ramko, E. B., Davidson, M. W., et al. (2011). A genetically encoded tag for correlated light and electron microscopy of intact cells, tissues, and organisms. *PLoS Biology*, *9*, e1001041. http://dx.doi.org/10.1371/journal.pbio.1001041.

Wolf, K. V., Stockem, W., Wohlfarthbottermann, K. E., & Moor, H. (1981). Cytoplasmic actomyosin fibrils after preservation with high-pressure freezing. *Cell and Tissue Research*, *217*, 479–495.

CHAPTER

Fluorescence-Based Assays for Microtubule Architecture

21

Susanne Bechstedt and Gary J. Brouhard

Department of Biology, McGill University, Montreal, Quebec, Canada

CHAPTER OUTLINE

Introduction 344

21.1 Materials 347

21.2 Methods 348

21.2.1 Tubulin and Microtubule Preparations 348

21.2.2 Setup for the Single-Molecule Fluorescence Assay 348

21.2.3 Comparison of Controlled 13-pf with Controlled 14-pf Microtubules 348

21.2.4 Comparison of Mixed Populations of Microtubules with an Internal 14-pf Control 349

21.2.5 Purification of DCX-GFP for Labeling 13-pf Microtubules 351

21.3 Discussion 352

Acknowledgments 352

References 352

Abstract

In vitro fluorescence-based assays have enabled the direct observation of single microtubule-associated proteins (MAPs) alongside the measurement of microtubule growth and shrinkage. Fluorescence-based assays have not, however, been able to address questions of "microtubule architecture." Tubulin can form diverse polymer structures *in vitro*. Importantly, microtubules nucleated spontaneously have different numbers of protofilaments (pfs), ranging from 11-pf to 16-pf, as well as sheet-like structures, indicating flexibility in tubulin–tubulin bonds. This structural diversity influences microtubule dynamics and the binding of MAPs to microtubules. Observation of microtubule architecture has required the imaging of microtubules by electron microscopy (EM). Because EM requires chemical fixation or freezing, it has not been possible to observe, in real time, how microtubule dynamics might influence structure and vice versa; it also remains technically challenging to directly observe

Methods in Cell Biology, Volume 115

ISSN 0091-679X
http://dx.doi.org/10.1016/B978-0-12-407757-7.00021-9

some MAPs, especially small ones, by EM. It is therefore imperative to develop fluorescence-based assays that enable the direct, real-time observation of microtubule architecture alongside growth, shrinkage, and MAP binding. In this chapter, we describe our efforts to control microtubule architecture for fluorescence-based assays. We also describe how microtubule structure can be probed with the help of GFP-tagged doublecortin, a MAP that binds preferentially to 13-pf microtubules.

INTRODUCTION

In vitro assays using total internal reflection fluorescence (TIRF) microscopy have become essential tools to observe the interaction of microtubule-associated proteins (MAPs) with microtubules. In these assays, microtubules are polymerized *in vitro* from purified tubulin. Single microtubules are adhered to a cover glass surface and fluorescently labeled MAPs are introduced and visualized. While simple in concept, TIRF assays have provided insights into the complex mechanisms of microtubule polymerases (Brouhard et al., 2008), depolymerases (Helenius, Brouhard, Kalaidzidis, Diez, & Howard, 2006), severing enzymes (Diaz-Valencia et al., 2011), end-binding proteins (Bieling et al., 2007), kinesins (Vale et al., 1996), and dynein (Reck-Peterson et al., 2006).

Fluorescence-based assays have not, however, been able to address questions of microtubule "architecture." The canonical microtubule is a polymer built from 13 protofilaments (pfs, see Fig. 21.1A). Microtubules are "plastic" polymers, however, meaning that tubulin forms polymers that differ in their longitudinal and lateral curvature (Kueh & Mitchison, 2009). These polymers include microtubules with different numbers of pfs (Fig. 21.1B) (Sui & Downing, 2010), as well as open sheets (Erickson, 1974), pf fragments (Elie-Caille et al., 2007), pf rings (Howard & Timasheff, 1986), pf spirals (Erickson, 1975), inside–out helical ribbons (Wang, Long, Finley, & Nogales, 2005), and zinc-induced "macrotubes" (Wolf, Mosser, & Downing, 1993). The analysis of these diverse polymers by electron crystallography and electron microscopy (EM) has characterized the tubulin–tubulin bonds that drive polymerization (Nogales, Wolf, & Downing, 1998; Wang & Nogales, 2005). Microtubule growth and shrinkage depend on the ability of tubulin to transition between different structural states; indeed, MAPs that regulate the microtubule cytoskeleton may recognize different structural states as part of their core mechanism. The microtubule polymerase Stu2, for example, may bind preferentially to longitudinally curved polymers (Ayaz, Ye, Huddleston, Brautigam, & Rice, 2012).

The basis of these diverse structures, especially the distribution of pf numbers, is the flexibility of interprotofilament bonds in the microtubule lattice. Interprotofilament interactions are mediated by the M-loops and H1′-S2 and H2-S3 loops of α- and β-tubulin. These loops are able to accommodate lateral deformations associated with changes in pf number and may form lateral bonds with different intrinsic curvatures (Wang & Nogales, 2005). Indeed, the ability to deform laterally without breaking

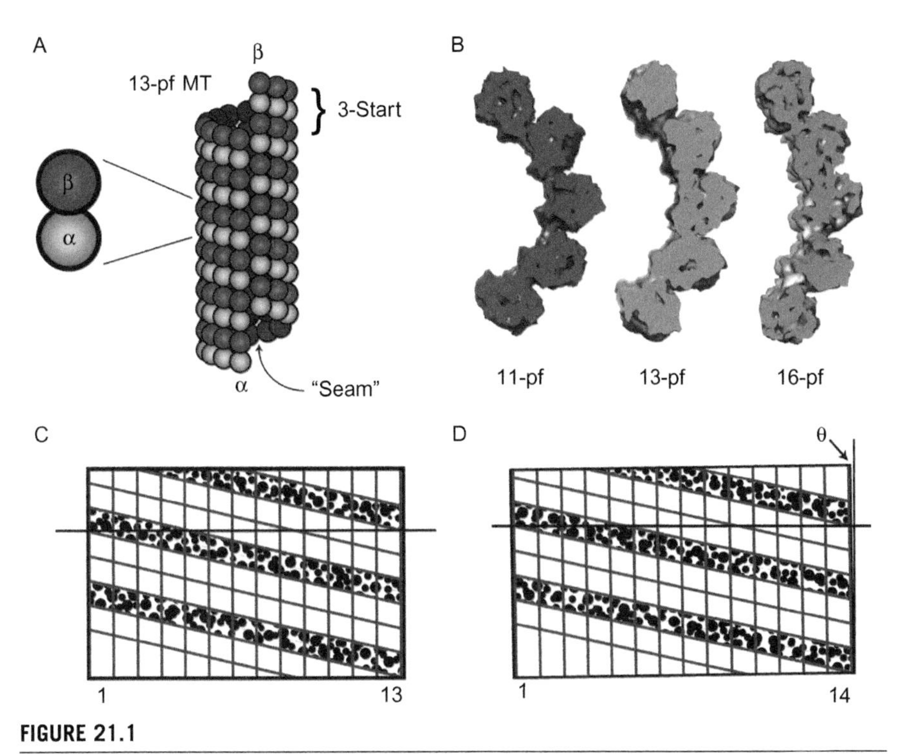

FIGURE 21.1

Microtubule architecture. (A) Schematic of a 13-pf microtubule showing the αβ-tubulin dimer, the 3-start helix, and the "seam." (B) Cross-sections of electron density maps for three different pf numbers (11-pf, EMD-5191; 13-pf, EMD-5193; and 16-pf, EMD-5196). (C) Schematic representation of the 13-pf lattice, viewed from the outside of the microtubule, indicating the left-handed helix. Dotted rows are used to indicate the 3-start helix. Note the horizontal line shows that one dotted row ends where another begins.(D) Schematic representation of a 14-pf lattice. The lattice was rotated by a small angle (θ) such that the dotted rows retained their alignment. This angle is the supertwist. (See color plate.)

Panels C and D adapted from Chretien and Wade (1991)

bonds may account for the unique mechanical properties of microtubules (Sui & Downing, 2010).

A consequence of changing the pf number is a change in the pitch and supertwist of the microtubule lattice. Tubulin dimers associate laterally in a "B-lattice" configuration, in which β-tubulins bind laterally to β- and α-tubulins bind to α-tubulins (Fig. 21.1A). The B-lattice bond has a small longitudinal offset, however, which gives the microtubule lattice a helical pitch. The helix is described by its monomer repeat; a 13-pf microtubule, for example, has a "3-start" helix based on a 3-monomer repeat (Fig. 21.1A, labeled). The 3-start helix creates a discontinuity in the B-lattice, in which one β-tubulin binds laterally to an α-tubulin, a feature known as the "seam"

(Fig. 21.1A, labeled). Changing the pf number can change the helical pitch of the microtubule. 15-pf microtubules, for example, have a 4-start helix and thus no seam (Sui & Downing, 2010).

For 13-pf microtubules, the pfs run parallel to the long axis of the microtubule (see Fig. 21.1A and C). Adding or removing pfs, however, introduces a supertwist (Fig. 21.1D). Chretien and Wade (1991) developed a theoretical explanation of the supertwist, the "surface lattice accommodation model," shown in Fig. 21.1C and D, which was confirmed by EM. This supertwist creates a moiré pattern specific to each microtubule type in EM (Langford, 1980), which is the basis for most EM measurements of pf number. Although the pf number and supertwist of a microtubule is presumably established during the nucleation process, the pf number may change along the length of a microtubule during growth (Chretien, Metoz, Verde, Karsenti, & Wade, 1992), thereby introducing a "lattice defect." Collectively, "microtubule architecture" is the combination of lateral curvature, longitudinal curvature, pitch, supertwist, and defects that describe an individual microtubule lattice.

Importantly, MAPs are sensitive to microtubule architecture, and thus different microtubule types can be used to reveal the intrinsic properties of MAPs. For example, the supertwist of non-13-pf microtubules was used to demonstrate that kinesin-1 travels on the path of single pfs, due to the fact that surface-immobilized kinesins caused non-13-pf microtubules to rotate (Ray, Meyhofer, Milligan, & Howard, 1993). The rotation of kinesin-1 around a supertwisted microtubule lattice allows for an indirect measurement of pf number by TIRF microscopy in which the 3D path of a quantum dot-labeled kinesin is tracked (Nitzsche, Ruhnow, & Diez, 2008). It was suggested that end-binding proteins bind specifically to the microtubule seam (Sandblad et al., 2006), although further experiments have cast doubt on this model (Maurer, Fourniol, Bohner, Moores, & Surrey, 2012) (see the chapter 23 in this volume). Finally, lattice defects are thought to be the preferred site of interaction for severing enzymes with microtubules (Davis, Odde, Block, & Gross, 2002; Diaz-Valencia et al., 2011). It has not been possible, however, to visualize pf transitions and severing enzymes concurrently.

What is needed is a suite of assays that visualize microtubule architecture, microtubule dynamics, and MAP activity concurrently by fluorescence. Investigating microtubule architecture by fluorescence will also allow us to gain insight into the nature of spontaneous nucleation and microtubule polymerization. After all, why do microtubules not form a more perfect lattice? We do not understand, for example, the mechanistic basis for pf number distributions. The pf number distributions measured in the literature vary widely; in our hands, day-to-day variability is commonplace. What is needed is an assay through which the basis of pf number variability could be investigated.

In this chapter, we describe methods we have developed for assaying microtubule architecture by fluorescence microscopy. First, we describe how to use controlled growth conditions to generate microtubules of predetermined architecture. These "control" microtubules can be used to determine whether a MAP binds preferentially to one type or another. Second, we describe how to use GFP-labeled doublecortin

(DCX), a protein that binds preferentially to 13-pf microtubules, as a readout for microtubule architecture in mixed populations.

21.1 MATERIALS

1. Unlabeled tubulin
2. Alexa Fluor 546-labeled tubulin, Alexa Fluor 546 (Succinimidyl Ester, Life Technologies A-20002)
3. TAMRA-labeled tubulin, TAMRA (Succinimidyl Ester, Life Technologies C1171)
4. Axonemes (Supporting protocol 2 in Waterman-Storer, 2001)
5. PIPES (Sigma, P6757)
6. Paclitaxel (Sigma, T7191)
7. GTP (Jena Biosciences #NU-1012)
8. GMPCPP (Jena Biosciences #NU-405S)
9. BRB80 buffer: 80 mm PIPES pH 6.9 (KOH), 1 mM EGTA, 1 mM $MgCl_2$ filtered (0.22 μm), degassed, and stored at −20 °C
10. Silanized cover glass (Gell et al., 2010)
11. Cover glass holders or silanized microscope slides (Gell et al., 2010)
12. Double-sided scotch tape
13. Whatman filter paper (#1001-090)
14. Antitetramethylrhodamine antibody (Life Technologies A6397)
15. Pluronic F-127 (Sigma P2443), stock solution of 1% in BRB80, filtered (0.22 μm filter), and stored at 4 °C
16. β-Mercaptoethanol (Sigma, M3148), 10 μl aliquots stored at 4 °C
17. BSA (Sigma, A3059), 10 mg/ml in BRB80, filtered and stored at 4 °C
18. Glucose (Sigma, G7528), 2 mg/ml in water and stored at −20 °C
19. Glucose oxidase (Sigma, G2133), 2 mg/ml in BRB80, flash-frozen and stored at −20 °C
20. Catalase (Sigma, C9322), 0.8 mg/ml in BRB80, flash-frozen and stored at −20 °C
21. Imaging buffer: BRB80 + 10 μM paclitaxel + 0.1 mg/ml β-Mercaptoethanol + 0.1 mg/ml BSA + 1:100 dilution of antifade reagents (glucose, glucose oxidase, catalase)
22. Beckman Airfuge
23. TIRF microscope with an EMCCD camera (e.g., Andor iXon +). Commercial implementations of TIRF (e.g., Zeiss Laser TIRF III) are largely sufficient for these experiments
24. Objective heater with controller 37-2 (Pecon GmbH, Germany #0280.004)
25. DCX-GFP expression vector (Bechstedt & Brouhard, 2012)
26. BL21(DE3) bacteria
27. LB medium
28. IPTG (GoldBio, I2481C)

29. His60 Ni Resin (Clontech, #635660)
30. Strep-tactin sepharose resin (IBA, Berlin, #2-1201-010)
31. Buffer A (50 mM Na_2HPO_4, 300 mM NaCl, pH 7.8)
32. Buffer B (50 mM Na_2HPO_4, 300 mM NaCl, 250 mM imidazole, pH 7.8)
33. Strep-tactin wash buffer (100 mM Tris/HCl, 1 mM EDTA, 150 mM NaCl, pH 8.0)
34. Strep-tactin elution buffer 100 mM Tris/HCl, 1 mM EDTA, 150 mM NaCl, 2.5 mM D-Desthiobiotin (DTB, IBA Berlin, #2-1000-002, 10% glycerol, pH 8.0)
35. Emulsiflex C3 (Avestin)
36. Disposable plastic columns (Thermo Scientific, # 29922)

21.2 METHODS

21.2.1 Tubulin and microtubule preparations

Tubulin was purified from juvenile bovine brain homogenates by three cycles of polymerization/depolymerization, followed by passage through a phosphocellulose column, as described in Gell et al. (2011). Labeling of cycled tubulin with Alexa Fluor 546 or TAMRA should be performed as described in Hyman et al. (1991). Other fluorophores can be chosen to suit the microscope equipment available.

21.2.2 Setup for the single-molecule fluorescence assay

The basics of the single-molecule fluorescence assay are well described in other chapters in this series, especially in Gell et al. (2011). In brief, we construct flow chambers using two silanized cover glass separated by double-stick tape. Microtubules are adhered to the cover glass surfaces by antibodies. The flow chambers allow us to visualize the surface-immobilized microtubules and exchange solutions in the chamber during microscopy.

21.2.3 Comparison of controlled 13-pf with controlled 14-pf microtubules

Two techniques can be used to generate microtubules of controlled thickness: (1) 13-pf microtubules are nucleated from sea urchin sperm axonemes and (2) 14-pf microtubules are nucleated using the slowly hydrolyzable GTP analog guanylyl 5′-α,β-methylenediphosphonate (GMPCPP). Microtubules nucleated from axonemes are templated by the A-tubule and central pair of the axoneme and are >90% 13-pf (Ray et al., 1993). Microtubules nucleated with GMPCPP are >98% 14-pf (Meurer-Grob, Kasparian, & Wade, 2001).

Sea urchin sperm axonemes can be purified according to existing protocols based on sucrose density gradients (Waterman-Storer, 2001). Purified centrosomes can be used as an alternative to axonemes, as the γ-tubulin ring complexes also provide a stable, 13-pf nucleation template (Evans, Mitchison, & Kirschner, 1985).

Dilute the purified axoneme stock (e.g., 1:4000) in BRB80 such that only a few axonemes appear in each field of view. Introduce diluted axonemes into a microscope chamber together with an anti-TAMRA antibody (1:100 in BRB80) and incubate for 5 min. Both the axonemes and the antibody adhere nonspecifically to the cover glass surfaces. Rinse the chamber with BRB80. Incubate the chamber with BRB80 + 1% Pluronic F-127 for 5–60 min to passivate the surfaces. Rinse four times with BRB80. Place the chamber onto the microscope with the objective heater set to 35 °C. Introduce polymerization buffer and incubate for ~10 min to elongate microtubules from the axoneme templates. Progress can be monitored by taking images in the tubulin color channel. When the microtubules are 10–20 μm in length (or as desired), rinse the chamber with BRB80 + 10 μM paclitaxel to stabilize the newly formed microtubules. Take an image to record the location of the 13-pf axonemal microtubules.

While preparing the 13-pf axonemal microtubules, also prepare 14-pf GMPCPP microtubules. To polymerize GMPCPP microtubules, add 2 μM TAMRA-labeled tubulin, 1 mM GMPCPP, and 1 mM $MgCl_2$ and adjust with BRB80 to 50 μl in a microcentrifuge tube. Incubate the mixture on ice for ~10 min. Polymerize the microtubules by incubating at 37 °C for 2 h. Add 200 ml BRB80, prewarmed to 37 °C. Centrifuge at 150,000 × *g* for 5 min (e.g., in a Beckman Airfuge or tabletop ultracentrifuge). Discard the supernatant and resuspend the pellet by gentle pipetting in 200 ml BRB80 + paclitaxel. GMPCPP-stabilized microtubules depolymerize very slowly and should be used on the same day.

After recording an image of the 13-pf axonemal microtubules, the TAMRA-labeled GMPCPP microtubules diluted into BRB80 + paclitaxel can be introduced into the flow chamber. The GMPCPP microtubules bind to the chamber surfaces via the anti-TAMRA antibodies introduced prior to the axonemes, creating a microscope chamber containing both axonemal microtubules and GMPCPP microtubules (Fig. 21.2). At this point, fluorescent MAPs of interest can be introduced, and preferential binding to either the axonemal microtubules or the GMPCPP microtubules can be detected (Bechstedt & Brouhard, 2012).

21.2.4 Comparison of mixed populations of microtubules with an internal 14-pf control

Microtubules nucleated in the presence of GTP will range from 11-pf to 16-pf (Sui & Downing, 2010). The ratio of the different pf species varies within the literature, but the consensus is that the fraction of 13-pf microtubules is between 35% (Moores et al., 2004; Ray et al., 1993) and 75% (Vitre et al., 2008), with the remainder consisting primarily of 14-pf. The larger and smaller species are a small minority. Mixed populations are a useful tool, as variations due to the nucleation conditions, stabilizing agent, tubulin preparation, fluorescent labeling, etc., are minimized. In our hands, however, day-to-day variability in the ratio of 13-pf microtubules is commonplace.

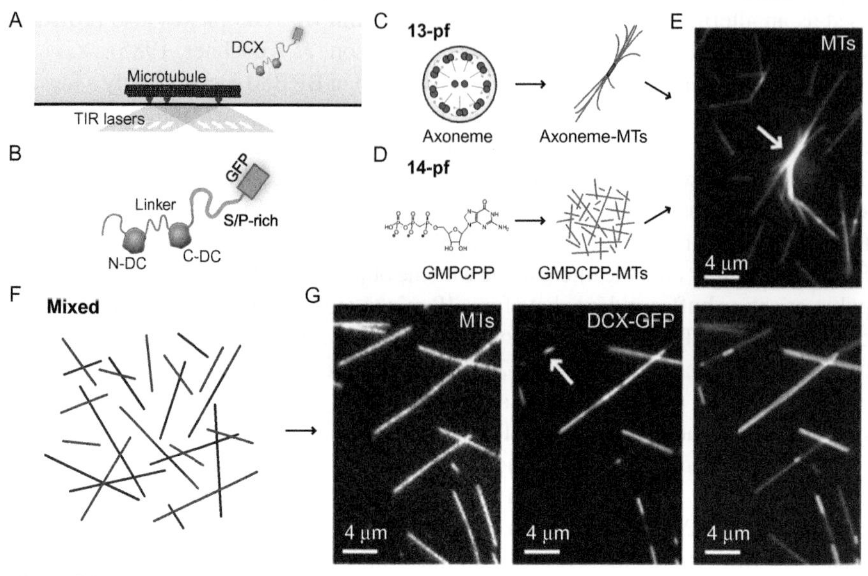

FIGURE 21.2

Fluorescence-based assays for microtubule architecture. (A) Schematic of the single-molecule assay. (B) Schematic of DCX-GFP. The two DC-domains (labeled) are joined by a linker (labeled) and flanked by polypeptides. The C-terminal polypeptide is enriched in S/P residues (labeled). The GFP-tag is C-terminal. (C) Schematic drawing of an axoneme (labeled), which nucleates axoneme microtubules (MTs) with >90% 13-pf. (D) Chemical drawing of GMPCPP, which nucleates GMPCPP-MTs with >96% 14-pf. (E) Image of axoneme-nucleated microtubules (white arrow) and GMPCPP microtubules in the same microscope chamber. (F) Schematic of a mixed population of MTs nucleated from purified tubulin. (G) Image of the mixed population of rhodamine-labeled microtubules (MTs); image of DCX-GFP exposed to this mixed population (DCX-GFP); color-combined image of MTs and DCX-GFP. Note that DCX-GFP binds preferentially to a subset of the mixed population, corresponding to the 13-pf subset, and to segments within individual microtubules (white arrow). (See color plate.)

Adapted from Bechstedt and Brouhard (2012)

Therefore, adding 14-pf GMPCPP microtubules into the flow chamber provides an important internal control for the experiment.

To polymerize the mixed microtubules, add 32 μM tubulin, 1 mM GTP, 4 mM $MgCl_2 + 5\%$ DMSO and adjust with BRB80 to 12 μl in a microcentrifuge tube. Incubate this mix at 37 °C for 30 min. Add 200 ml BRB80 + 10 μM paclitaxel, prewarmed to 37 °C, and centrifuge at 150,000 × *g* for 5 min (e.g., in a Beckman Airfuge or tabletop ultracentrifuge such as a Beckman Optima MAX). Discard the supernatant and resuspend the pellet in 200 ml BRB80 + 10 μM paclitaxel.

At this point, the paclitaxel-microtubules might be too concentrated for direct use and should be diluted up to fivefold before introduction into the flow chamber.

GMPCPP-microtubules are prepared as described earlier, with the modification that the GMPCPP-microtubules should be prepared with a higher ratio of fluorescently labeled tubulin, creating bright GMPCPP-MTs and dim paclitaxel-microtubules. Combine both microtubules into a single tube and introduce the mixture to the flow chamber. The mixing ratio that gives a desirable number of paclitaxel- and GMPCPP-microtubules in each field of view should be determined experimentally each day. At this point, fluorescent MAPs of interest can be introduced, and preferential binding to subsets of the mixed, paclitaxel-microtubules can be observed (Bechstedt & Brouhard, 2012). If such preferential binding occurs, the bright and dim subsets can be compared to the 14-pf GMPCPP-microtubules as a reference.

21.2.5 Purification of DCX-GFP for labeling 13-pf microtubules

DCX is a MAP that binds preferentially to 13-pf microtubules within certain concentration ranges. The following purification describes how to obtain human DCX (Acc Nr. NP_835365) containing an N-terminal 6xHis tag and C-terminal coding sequence for EGFP and StrepTagII (Bechstedt & Brouhard, 2012).

Inoculate 2 ml LB medium with a colony from a plate of freshly transformed BL21(DE3) bacteria and incubate with agitation (250 rpm) for ~14 h at 25 °C. Use this overnight culture to inoculate 1 L LB medium at 37 °C and let cells grow to OD 0.4–0.6. Cool cells down and induce at 18 °C with 1 mM IPTG. Incubate overnight (~14 h) with agitation (250 rpm) for protein expression. Spin down expression cultures at 5000 × *g* for 15 min and reconstitute pellets in 4 ml/g buffer A. Pellets can be stored at −80 °C when not directly processed. For DCX purification, lyse the pellets using, for example, an Emulsiflex C3 (Avestin). Clear the lysate by centrifugation at 45,000 × *g* for 1 h. Incubate the supernatant with mild agitation with 1 ml His60 Ni resin for 30 min at 4 °C. Wash the resin twice with buffer A and load the resin on a disposable plastic column. Elute the bound protein, including DCX-GFP, with four column volumes of buffer B. Load the green-fluorescent elution fractions on a 1-ml Strep-tactin Superflow column. Wash the column twice with two column volumes Strep-tactin wash buffer. Elute the purified DCX-GFP in 0.5 ml fractions with Strep-tactin elution buffer. Peak fractions can be combined and either used fresh or flash-frozen in 20 μl aliquots and stored at −80 °C.

DCX-GFP should be introduced to a mixed population of paclitaxel-microtubules with 14-pf GMPCPP-microtubules present as an internal control. Serial dilutions of DCX-GFP should be tested, as DCX-GFP binds equivalently to all microtubule types at very low concentrations (<1 nM) and saturates all microtubule types at very high concentrations (>1 μM). For example, prepare a solution of 0.1 μM DCX-GFP in imaging buffer: BRB80 + 10 μM paclitaxel + 0.1 mg/ml BSA + antifade reagents; this solution is introduced into the microscope chamber and images are recorded. The serial dilutions will identify a concentration range at which a subset of the paclitaxel-microtubules is brighter than the 14-pf GMPCPP-microtubules and the dim subset of the paclitaxel-microtubules. This concentration of DCX-GFP can then be used in future experiments to label the 13-pf fraction.

21.3 DISCUSSION

This chapter describes basic tools for controlling pf number *in vitro* and distinguishing 13-pf microtubules from others using DCX-GFP. It is our hope, of course, that fluorescence-based assays for microtubule architecture and microtubule structure will continue to evolve. In the future, we may be able to nucleate microtubules of any given type using controlled conditions or specialized reagents. Likewise, we may be able to define or control the longitudinal curvature of the lattice. The development of these tools will enable the next generation of single-molecule assays to ask deeper questions about how microtubule architecture influences the behavior of MAPs.

EM will retain pride of place in the study of microtubule architecture. Advances in helical reconstruction algorithms (Li et al., 2002) have enabled the generation of exceptional density maps for 11-pf through 16-pf microtubules (Sui & Downing, 2010). These density maps have described tubulin–tubulin bonds in atomic detail. By coupling this knowledge with dynamic assays, we can explore the "structural plasticity" of tubulin polymers.

Acknowledgments

We would like to thank Sami Chaaban for critical reading of the chapter. This work is supported by grants from the Natural Sciences and Engineering Research Council of Canada (#372593-09) and the Canadian Institutes of Health Research (MOP-111265). G. J. B. is supported by a Canadian Institutes of Health Research New Investigator Award.

References

Ayaz, P., Ye, X., Huddleston, P., Brautigam, C. A., & Rice, L. M. (2012). A TOG:alphabeta-tubulin complex structure reveals conformation-based mechanisms for a microtubule polymerase. *Science*, *337*, 857–860.

Bechstedt, S., & Brouhard, G. J. (2012). Doublecortin recognizes the 13-protofilament microtubule cooperatively and tracks microtubule ends. *Developmental Cell*, *23*, 181–192.

Bieling, P., Laan, L., Schek, H., Munteanu, E. L., Sandblad, L., Dogterom, M., et al. (2007). Reconstitution of a microtubule plus-end tracking system in vitro. *Nature*, *450*, 1100–1105.

Brouhard, G. J., Stear, J. H., Noetzel, T. L., Al-Bassam, J., Kinoshita, K., Harrison, S. C., et al. (2008). XMAP215 is a processive microtubule polymerase. *Cell*, *132*, 79–88.

Chretien, D., Metoz, F., Verde, F., Karsenti, E., & Wade, R. H. (1992). Lattice defects in microtubules: Protofilament numbers vary within individual microtubules. *The Journal of Cell Biology*, *117*, 1031–1040.

Chretien, D., & Wade, R. H. (1991). New data on the microtubule surface lattice. *Biology of the Cell*, *71*, 161–174.

Davis, L. J., Odde, D. J., Block, S. M., & Gross, S. P. (2002). The importance of lattice defects in katanin-mediated microtubule severing in vitro. *Biophysical Journal*, *82*, 2916–2927.

Diaz-Valencia, J. D., Morelli, M. M., Bailey, M., Zhang, D., Sharp, D. J., & Ross, J. L. (2011). Drosophila katanin-60 depolymerizes and severs at microtubule defects. *Biophysical Journal, 100*, 2440–2449.

Elie-Caille, C., Severin, F., Helenius, J., Howard, J., Muller, D. J., & Hyman, A. A. (2007). Straight GDP-tubulin protofilaments form in the presence of taxol. *Current Biology, 17*, 1765–1770.

Erickson, H. P. (1974). Microtubule surface lattice and subunit structure and observations on reassembly. *The Journal of Cell Biology, 60*, 153–167.

Erickson, H. P. (1975). Negatively stained vinblastine aggregates. *Annals of the New York Academy of Sciences, 253*, 51–52.

Evans, L., Mitchison, T., & Kirschner, M. (1985). Influence of the centrosome on the structure of nucleated microtubules. *The Journal of Cell Biology, 100*, 1185–1191.

Gell, C., Bormuth, V., Brouhard, G. J., Cohen, D. N., Diez, S., Friel, C. T., et al. (2010). Microtubule dynamics reconstituted in vitro and imaged by single-molecule fluorescence microscopy. *Methods in Cell Biology, 95*, 221–245.

Gell, C., Friel, C. T., Borgonovo, B., Drechsel, D. N., Hyman, A. A., & Howard, J. (2011). Purification of tubulin from porcine brain. *Methods in Molecular Biology, 777*, 15–28.

Helenius, J., Brouhard, G., Kalaidzidis, Y., Diez, S., & Howard, J. (2006). The depolymerizing kinesin MCAK uses lattice diffusion to rapidly target microtubule ends. *Nature, 441*, 115–119.

Howard, W. D., & Timasheff, S. N. (1986). GDP state of tubulin: Stabilization of double rings. *Biochemistry, 25*, 8292–8300.

Hyman, A., Drechsel, D., Kellogg, D., Salser, S., Sawin, K., Steffen, P., et al. (1991). Preparation of modified tubulins. *Methods in Enzymology, 196*, 478–485.

Kueh, H. Y., & Mitchison, T. J. (2009). Structural plasticity in actin and tubulin polymer dynamics. *Science, 325*, 960–963.

Langford, G. M. (1980). Arrangement of subunits in microtubules with 14 profilaments. *The Journal of Cell Biology, 87*, 521–526.

Li, H., DeRosier, D. J., Nicholson, W. V., Nogales, E., & Downing, K. H. (2002). Microtubule structure at 8 Å resolution. *Structure, 10*, 1317–1328.

Maurer, S. P., Fourniol, F. J., Bohner, G., Moores, C. A., & Surrey, T. (2012). EBs recognize a nucleotide-dependent structural cap at growing microtubule ends. *Cell, 149*, 371–382.

Meurer-Grob, P., Kasparian, J., & Wade, R. H. (2001). Microtubule structure at improved resolution. *Biochemistry, 40*, 8000–8008.

Moores, C. A., Perderiset, M., Francis, F., Chelly, J., Houdusse, A., & Milligan, R. A. (2004). Mechanism of microtubule stabilization by doublecortin. *Molecular Cell, 14*, 833–839.

Nitzsche, B., Ruhnow, F., & Diez, S. (2008). Quantum-dot-assisted characterization of microtubule rotations during cargo transport. *Nature Nanotechnology, 3*, 552–556.

Nogales, E., Wolf, S. G., & Downing, K. H. (1998). Structure of the alpha beta tubulin dimer by electron crystallography. *Nature, 391*, 199–203.

Ray, S., Meyhofer, E., Milligan, R. A., & Howard, J. (1993). Kinesin follows the microtubule's protofilament axis. *The Journal of Cell Biology, 121*, 1083–1093.

Reck-Peterson, S. L., Yildiz, A., Carter, A. P., Gennerich, A., Zhang, N., & Vale, R. D. (2006). Single-molecule analysis of dynein processivity and stepping behavior. *Cell, 126*, 335–348.

Sandblad, L., Busch, K. E., Tittmann, P., Gross, H., Brunner, D., & Hoenger, A. (2006). The Schizosaccharomyces pombe EB1 homolog Mal3p binds and stabilizes the microtubule lattice seam. *Cell, 127*, 1415–1424.

Sui, H., & Downing, K. H. (2010). Structural basis of interprotofilament interaction and lateral deformation of microtubules. *Structure*, *18*, 1022–1031.

Vale, R. D., Funatsu, T., Pierce, D. W., Romberg, L., Harada, Y., & Yanagida, T. (1996). Direct observation of single kinesin molecules moving along microtubules. *Nature*, *380*, 451–453.

Vitre, B., Coquelle, F. M., Heichette, C., Garnier, C., Chretien, D., & Arnal, I. (2008). EB1 regulates microtubule dynamics and tubulin sheet closure in vitro. *Nature Cell Biology*, *10*, 415–421.

Wang, H. W., Long, S., Finley, K. R., & Nogales, E. (2005). Assembly of GMPCPP-bound tubulin into helical ribbons and tubes and effect of colchicine. *Cell Cycle*, *4*, 1157–1160.

Wang, H. W., & Nogales, E. (2005). Nucleotide-dependent bending flexibility of tubulin regulates microtubule assembly. *Nature*, *435*, 911–915.

Waterman-Storer, C. M. (2001). Microtubule/organelle motility assays. *Current Protocols in Cell Biology*, *1998*, 13.1.1–13.1.21. Chapter 13.

Wolf, S. G., Mosser, G., & Downing, K. H. (1993). Tubulin conformation in zinc-induced sheets and macrotubes. *Journal of Structural Biology*, *111*, 190–199.

CHAPTER

Structure–Function Analysis of Yeast Tubulin

22

Anna Luchniak*,[1], Yusuke Fukuda†,[1] and Mohan L. Gupta, Jr.†

*Department of Biochemistry and Molecular Biology, University of Chicago, Chicago, Illinois, USA

†Department of Molecular Genetics and Cell Biology, University of Chicago, Chicago, Illinois, USA

CHAPTER OUTLINE

Introduction **356**

22.1 Reagents and Equipment **358**

22.2 Introducing Tubulin Mutations Into Yeast **360**

22.2.1 Preparation of Modified Tubulin DNA 360

22.2.2 Transformation of Modified Tubulin DNA Into Yeast 360

22.2.3 Verification of Tubulin Mutations in Yeast 362

22.3 Analysis of Tetrad Viability **363**

22.3.1 Sporulation 364

22.3.2 Microdissection of Haploid Spores 364

22.3.3 Genotype Analysis of Haploid Spores 365

22.3.4 Mating-type Analysis 365

22.4 Assessing Microtubule Stability By Drug Sensitivity **366**

22.4.1 Benomyl Plate Assay 366

22.5 Direct Analysis of Microtubule Dynamics *In Vivo* **367**

22.5.1 Yeast Strains for *in Vivo* Microtubule Dynamics Analysis 367

22.5.2 Maintaining Comparable *GFP–TUB1* Expression Levels 367

22.5.3 Preparation of Samples for Microscopy 368

22.5.4 Microscopy and Analysis of Microtubule Dynamics 368

22.6 Localization of Microtubule-Associated Proteins *In Vivo* **370**

22.6.1 Yeast Strains for Localization of Microtubule-Associated Proteins ... 370

22.6.2 Maintaining Equal Levels of YFP-fusion Proteins 371

22.6.3 Microscopy and Analysis of Microtubule-Associated Proteins 371

Acknowledgments **372**

References **373**

[1]Equally contributing authors.

Methods in Cell Biology, Volume 115

ISSN 0091-679X
http://dx.doi.org/10.1016/B978-0-12-407757-7.00022-0

Abstract

Microtubules play essential roles in a wide variety of cellular processes including cell division, motility, and vesicular transport. Microtubule function depends on the polymerization dynamics of tubulin and specific interactions between tubulin and diverse microtubule-associated proteins. To date, investigation of the structural and functional properties of tubulin and tubulin mutants has been limited by the inability to obtain functional protein from overexpression systems, and by the heterogeneous mixture of tubulin isotypes typically isolated from higher eukaryotes. The budding yeast, *Saccharomyces cerevisiae*, has emerged as a leading system for tubulin structure–function analysis. Yeast cells encode a single beta-tubulin gene and can be engineered to express just one of two alpha isotypes. Moreover, yeast allows site-directed modification of tubulin genes at the endogenous loci expressed under the native promoter and regulatory elements. These advantageous features provide a homogeneous and controlled environment for analysis of the functional consequences of specific mutations. Here, we present the techniques to generate site-specific tubulin mutations in diploid and haploid cells, assess the ability of the mutated protein to support cell viability, measure overall microtubule stability, and define changes in the specific parameters of microtubule dynamic instability. We also outline strategies to determine whether mutations disrupt interactions with microtubule-associated proteins. Microtubule-based functions in yeast are well defined, which allows the observed changes in microtubule properties to be related to the role of microtubules in specific cellular processes.

INTRODUCTION

Microtubules are cytoskeletal polymers composed of αβ-tubulin heterodimers that exhibit highly dynamic growth behavior, which allows for rapid rearrangement of microtubule networks to support diverse cellular tasks (Desai & Mitchison, 1997). Specific mutations in tubulin genes result in a spectrum of congenital neurological disorders in humans (Cederquist et al., 2012; Jaglin et al., 2009; Tischfield, Cederquist, Gupta, & Engle, 2011). Moreover, the pathology of Alzheimer's and Parkinson's disease has been at least in part attributed to microtubule dysfunction (Brandt, Hundelt, & Shahani, 2005; Lei et al., 2010). The organization and dynamics of microtubule polymer are also essential for proper function of the mitotic spindle and fidelity of chromosome segregation. Indeed, small molecules that interfere with microtubule dynamics are a powerful class of antimitotic agents and represent a proven strategy in the treatment of various cancers (Jordan, 2002). Therefore, understanding the structural and functional properties underlying the regulation of microtubule dynamics is of great interest for human health.

Studies directed toward elucidating the structure–function relationship of tubulin and microtubules have been challenging. The folding of tubulin polypeptides into native heterodimers requires interactions with chaperones and protein cofactors (Lewis, Tian,

Vainberg, & Cowan, 1996), which has precluded expression in bacterial systems. Moreover, unbalanced levels of α- and β-subunits produce toxic effects in eukaryotes (Katz, Weinstein, & Solomon, 1990), while overexpression hinders folding and produces nonfunctional aggregates (M. L. Gupta, unpublished results). Consequently, the large-scale expression of exogenous tubulin has not been successful to date.

Conventionally, *in vitro* assays addressing microtubule function and regulation by microtubule-associated proteins have largely utilized tubulin purified from mammalian brain tissue. This method yields large amounts of protein; however, the material obtained is not homogeneous as the majority of higher eukaryotes express multiple α- and β-tubulin isotypes. Additionally, site-specific mutagenesis of tubulin expressed from the native locus is not possible or feasible in most higher eukaryotes. Thus, purified brain tubulin is a heterogeneous mixture that confounds the analysis of particular isotypes or mutations.

The budding yeast, *Saccharomyces cerevisiae*, is a powerful tool for systematic analysis of the functional consequences of tubulin mutations. Budding yeast encodes a single β-tubulin, *TUB2* (Neff, Thomas, Grisafi, & Botstein, 1983), and two α-tubulin isotypes, *TUB1* and *TUB3* (Schatz, Pillus, Grisafi, Solomon, & Botstein, 1986), which all share ~75% sequence conservation with human homologues. Furthermore, yeast can tolerate deletion of *TUB3*, which can clarify *in vivo* readouts by eliminating influence of multiple isotypes and confer a source of homogeneous tubulin for *in vitro* analyses (Bode, Gupta, Suprenant, & Himes, 2003). In budding yeast, site-specific mutations can be inserted into the endogenous locus. Mutant proteins are expressed under the control of the native promoter and regulatory elements, overcoming challenges and complexity of mammalian systems. This approach eliminates difficulties associated with exogenous expression and overexpression that may lead to nonphysiological results.

Budding yeast has been utilized for over 2 decades to study structural and functional properties of tubulin and microtubules. Alanine-scanning mutagenesis of α- and β-tubulin uncovered clusters of charged amino acids crucial for structural and functional integrity of tubulin and microtubules (Reijo, Cooper, Beagle, & Huffaker, 1994; Richards et al., 2000). Directed mutagenesis demonstrated the relationship between GTP hydrolysis and dynamic instability *in vivo* and *in vitro* (Anders & Botstein, 2001; Davis, Sage, Dougherty, & Farrell, 1994) and revealed the contributions of microtubule dynamics to assembly and positioning of the mitotic spindle (Gupta et al., 2002; Huang & Huffaker, 2006). Moreover, studies of yeast tubulin have been important in defining binding sites of anticancer therapeutics (Gupta, Bode, Georg, & Himes, 2003) and for discriminating among proposed models for their interactions with microtubules (Entwistle et al., 2012). Recently, structure–function analysis of yeast tubulin has been key for elucidating the molecular etiology of a class of human neurological disorders that result from missense mutations in distinct tubulin isotypes (Cederquist et al., 2012; Jaglin et al., 2009; Tischfield et al., 2010).

Here, we present techniques to study the structural and functional consequences of tubulin mutations using the budding yeast *S. cerevisiae*. This genetically tractable system provides a well-controlled environment to relate the consequences of tubulin mutations on microtubule dynamics and interactions with specific regulatory proteins, to distinct microtubule-dependent processes.

22.1 REAGENTS AND EQUIPMENT

General equipment

- Temperature-controlled waterbath
- 30 °C incubator and shaker
- Thermocycler
- Tabletop centrifuge
- Tabletop microfuge
- Spectrophotometer
- Dissection microscope with micromanipulator
- Replica plate apparatus with sterile velvet towels
- Fluorescence microscope equipped with a 63 × high N.A. objective, CCD camera
- Microscope slides and coverglass
- SDS-PAGE/western blot apparatus and supplies
- Multichannel pipette or "frogger"
- Sterile wooden toothpicks
- 15-ml Conical tubes
- Culture tubes and flasks
- PCR tubes
- 96-Well plates

General reagents

- Yeast-extract peptone dextrose (YPD) media (1% yeast extract, 2% peptone, 2% glucose)
- YPD plates (YPD, 2% agar)
- Synthetic complete (SC) media (0.67% yeast nitrogen base without amino acids, 2% glucose, 0.2% appropriate amino acid mix (Sunshine Science Products, CA))
- SC plates (SC, 2% agar)
- Sporulation plates (1% potassium acetate, 0.1% Bacto-yeast extract, 0.05% glucose, 2% agar)
- YNB plates (0.67% yeast nitrogen base without amino acids, 2% glucose, 2% agar)
- For more details on preparing yeast media, see Treco and Lundblad (2001)
- Single-stranded salmon sperm carrier DNA; for preparation protocol, see Becker and Lundblad (2001)
- Adenine, NaOH, KCl, lithium acetate (LiAc), glycerol, polyethylene glycol MW 3350, benomyl, DMSO (Sigma, St. Louis, MO)
- Restriction enzymes (see Table 22.1)
- TE buffer (10 mM Tris, 1 mM EDTA, pH 8)

Table 22.1 Plasmids used for genetic manipulation of tubulin in yeast

Plasmid name	Tubulin gene	Linked prototrophic marker	Antibiotic selection	Digest using	Digestion products (~bp)	Desired digestion product (bp)	Reference
pCS6	*TUB1*	*TRP1*	*Ampicillin*	*Kpn*I, *Sal*I	3500, 4500	4500	Gift from C. Sage and K. Farrell
pCS3	*TUB2*	*URA3*	*Ampicillin*	*Sac*I, *Sph*I, *Pvu*II	2300, 2500, 5000	5000	Sage, Davis, Dougherty, Sullivan, and Farrell (1995)
pMG1	*TUB2–6 × His*	*URA3*	*Ampicillin*	*Sac*I, *Sph*I, *Pvu*II	2300, 2500, 5000	5000	Gupta, Bode, Dougherty, Marquez, and Himes (2001)
pAFS125	*GFP–TUB1*	*URA3*	*Ampicillin*	*Stu*I	8500	8500	Straight, Marshall, Sedat, and Murray (1997)
pMG3	*GFP–TUB1*	*LEU2*	*Ampicillin*	*Kas*I	8500	8500	Gupta et al. (2002)
pMG130	*CFP–TUB1*	*URA3*	*Ampicillin*	*Stu*I	8500	8500	Cederquist et al. (2012)

- Agarose
- 10 mM dNTP mix
- Taq polymerase and buffer
- Zymolyase (Zymoresearch, CA)
- Valap (equal weights of Vaseline (petrolatum), lanolin, and paraffin wax)
- Polyclonal anti-GFP antibody (Applied Biological Materials, Canada)

22.2 INTRODUCING TUBULIN MUTATIONS INTO YEAST

The most common method used for structure–function studies of yeast tubulin leverages homologous recombination-mediated gene replacement to introduce desired mutations into tubulin genes (Fig. 22.1). Directed mutagenesis is performed *in vitro* using standard molecular biology techniques. Following gene replacement, the modified tubulin is expressed as a single copy under control of the native promoter and regulatory elements.

Another useful approach is the "plasmid shuffle" in which a plasmid containing mutated tubulin is exchanged in the cell for a plasmid carrying a wild-type copy: described in Sikorski and Boeke (1991). This approach offers increased throughput relative to gene replacement. However, the number of gene copies cannot be tightly controlled using plasmids, which may confound interpretation of cell-based assays. Thus, we focus here on the strategy of direct gene replacement.

22.2.1 Preparation of modified tubulin DNA

Specific mutations are introduced into a plasmid containing the tubulin open reading frame (ORF) flanked by upstream and downstream genomic sequences and a prototrophic marker cassette. The prototrophy gene is inserted several hundred bases downstream from the stop codon and positioned not to interfere with neighboring genes. Site-directed mutagenesis can be accomplished with commercially available systems such as QuikChange (Agilent Technologies). Three commonly used starting plasmids are described in Table 22.1. In pMG1, *TUB2* has been tagged with 6 × His to aid in purification of mutated tubulin (Gupta et al., 2002; Johnson, Ayaz, Huddleston, & Rice, 2011).

> *Note:* It is essential to sequence the entire tubulin gene and promoter region after mutagenesis to ensure additional, unwanted mutations did not occur during *in vitro* manipulation.

To generate linear DNA for yeast transformation, digest >5 μg plasmid as described in Table 22.1, verify complete digestion by agarose gel electrophoresis, and gel-purify the appropriate fragment (Table 22.1).

22.2.2 Transformation of modified tubulin DNA into yeast

Tubulin is an essential gene. Thus, introducing a nonfunctional mutant into haploid cells may select for unwanted suppressor mutations. Thus, unless a mutation has been previously verified as nonlethal, the mutated gene should be transformed into a diploid strain and then sporulated to generate a haploid.

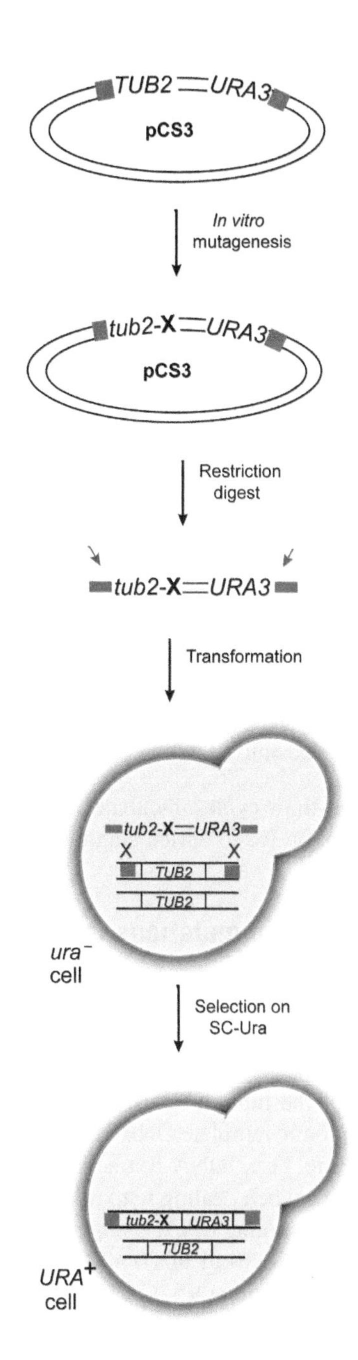

FIGURE 22.1

Schematic of direct gene replacement of yeast β-tubulin. Green shading and red arrows indicate homologous regions flanking β-tubulin ORF exposed by restriction digest. Purple "X" denotes regions of homologous recombination. See text for details. (See color plate.)

To transform yeast with the linear DNA fragment containing modified tubulin, grow a 5-ml yeast culture in YPD media overnight at 30 °C with shaking/rotating. The following day, dilute the near-saturated culture 1:50 into fresh media and continue shaking until cells enter log phase; 4 h is usually sufficient with typical growth rates. Use 10–20 ml of culture per transformation. Harvest cells by centrifuging 5 min at ~1000 × *g* and wash by resuspending in 1/10 original volume of sterile 0.1 M LiAc in TE buffer. Harvest cells again and resuspend in 1/100 original volume of 0.1 M LiAc in TE. For each transformation, combine in a sterile 15-ml tube the purified linear DNA fragment containing the mutated tubulin gene (≥0.5 μg in <30 μl) with 10 μl single-stranded salmon sperm carrier DNA. Subsequently, add 100 μl LiAc-treated yeast cells and mix gently. Add 0.7 ml sterile 40% PEG, 0.1 M LiAc in TE and mix well. Incubate for 30 min at 30 °C with shaking. Heat-shock the cells for 20 min in a 42 °C waterbath. Add 10 ml YPD and allow cells to recover for 2 h at 30 °C with shaking. Harvest cells, resuspend in ~0.5 ml YPD, spread onto appropriate selection plates, and incubate at least 2 days at 30 °C to allow colony growth.

When colonies reach sufficient size, potential transformants must be taken through two rounds of single-colony isolation on appropriate selection media to generate clonal strains. To isolate single colonies, streak cells across a fresh plate with a sterile loop or wooden toothpick to obtain sparse colony growth. For long-term storage, isolated clones should be grown in YPD or appropriate selection media, resuspended in sterile 15% glycerol, and stored at −80 °C in cryovials. To revive, transfer a small scraping from the still frozen tube onto a fresh plate using a sterile wooden applicator.

Note: After several freeze-thaw cycles, incubate salmon sperm DNA aliquots at 95 °C for 10 min and cool on ice to enrich single-stranded fragments.

22.2.3 Verification of tubulin mutations in yeast

Although fragment-mediated gene replacement is efficient, it is possible for the prototrophic marker to integrate into the genome independently from the tubulin ORF. Thus, it is essential to verify that the isolated clonal strain carries the desired mutation before proceeding with analyses. Verification can be accomplished by PCR amplification and sequencing of the tubulin locus. Primers should anneal ~50 bases outside the tubulin ORF. Genomic template DNA is readily obtained using commercial products, for example, the Yeast DNA Extraction Kit (Pierce). Colony-based PCR can be a useful alternative when dealing with many samples, although it is less robust than PCR from purified genomic DNA. Using a sterile pipette tip, resuspend a small portion of a fresh colony in 10 μl of 20 mM sterile NaOH in a PCR tube. Incubate in a thermocycler for 10 min at 95 °C. Use 3 μl of the NaOH-treated cells as a template in 25-μl reactions and perform PCR using standard conditions. Sequence the PCR product to verify presence of the mutation. If the mutation is in a diploid strain, the sequencing chromatogram should reflect the presence of both the wild-type and mutated alleles as a double peak at the position corresponding to the mutation (Fig. 22.2).

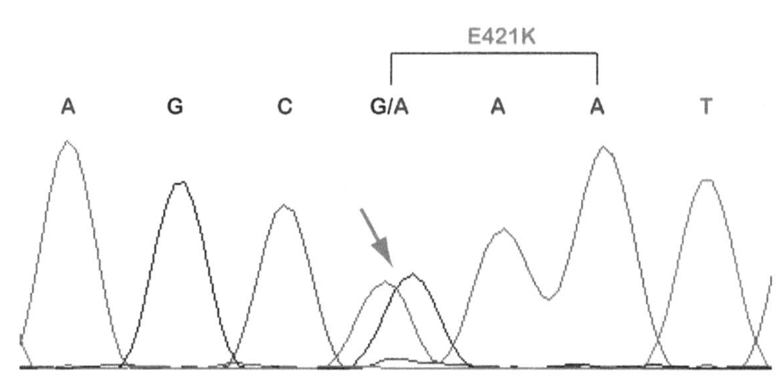

FIGURE 22.2

Sequencing chromatogram from a diploid strain heterozygous for *tub2–E421K* mutant tubulin. Double peak (arrow) indicates the presence of both wild-type and mutant alleles. (See color plate.)

Note: We recommend sequencing the full gene of interest to exclude the possibility of additional mutations in the final strain.

22.3 ANALYSIS OF TETRAD VIABILITY

Linking genetic alterations to phenotypic changes through tetrad dissection and subsequent analysis of spores is fundamental for determining the functional consequences of specific mutations. In response to nitrogen starvation and in the presence of a nonfermentable carbon source, diploid yeast cells sporulate. During this process, meiotic segregation gives rise to four haploid daughter cells or spores. Once spores have formed, the intact anucleate mother cell collapses around them forming a mature tetrahedral ascus (Fig. 22.3A). Microdissection and germination of the spores from a single tetrad yields four haploid strains where a given genetic locus originating from the pair of homologous parent chromosomes is represented in two of the four strains.

Mutations that disrupt the function of essential genes, such as *TUB1* and *TUB2*, typically exhibit recessive lethality and produce a characteristic pattern of two viable and two inviable spores (2+/2−). The surviving spores contain the wild-type version of the gene, which is reflected by the phenotype. Alternatively, viability of all four spores is indicative of less harmful mutations and allows analysis of the resultant haploids that contain only the mutated tubulin. In certain cases, spore viability may digress from the classic "all or nothing" 2+/2− hypothesis. With regard to tubulin mutations, low viability can result from severe disruption of meiotic chromosome segregation, which is a microtubule-dependent process. However, the survival of none, one, or three spores can also indicate the presence of independent genomic variations that influence spore viability, for example, suppressor mutations.

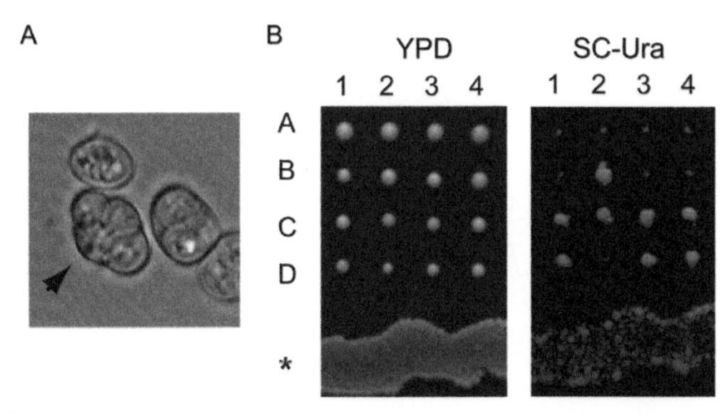

FIGURE 22.3

Yeast tetrad dissection and spore viability analysis. (A) Diamond-shaped tetrad (arrow) consisting of four haploid spores encapsulated by the ascus. (B) Cells heterozygous for a *TUB2* mutation linked to uracil prototrophy (*tub2–URA3*) were sporulated, Zymolyase-treated, and spread across a dissection plate (asterisk). Four tetrads (1–4) were dissected and the individual spores placed in positions A–D. All spores grew on YPD demonstrating the mutation is nonlethal. Replica-plating onto SC-Ura revealed two spores contained the mutated *tub2–URA3* allele (growth) and two contained the wild-type allele (no growth).

Assessing spore viability in independently created diploid mutants or elevated chromosome missegregation rates in the surviving mutant spores can help discriminate between these possibilities.

22.3.1 Sporulation

Grow yeast strains overnight on YPD agar plates and then transfer onto sporulation plates as a patch using a sterile toothpick. Verify sporulation progress after 3–4 days by picking a small amount of cells off the plate and scoring for the presence of asci. The appearance of dyads/triads is usually a positive indicator of sporulation progress. Formation of mature tetrads can take approximately a week depending on the genetic background of the yeast strain. Proceed with dissection when sufficient tetrads have formed.

> *Note I*: Dissection is facilitated when sporulation efficiency is >20%. Dissection is possible, albeit more challenging, when tetrad frequency is lower.
> *Note II*: Tetrads can be stored at 4 °C for an extended period of time.

22.3.2 Microdissection of haploid spores

Each complete ascus encapsulates four spores in what typically appears as a pyramidal or diamond-shaped structure (Fig. 22.3A). To isolate spores, the ascus is digested with Zymolyase. Pick up a small amount of sporulated culture using a toothpick or pipette tip and resuspend in 40 μl of filtered 0.8 M KCl, 25 mM Tris, pH 7.5 buffer, containing 0.4 mg/ml Zymolyase. Incubate 10 min in a 37 °C waterbath and then dilute with 400 μl YPD to inhibit the reaction. Mark a line through the center of an YPD or appropriate selection plate. While holding the plate nearly vertical, place ~20 μl of Zymolyase-treated cells at the top and allow the drop to run across the surface along the line. Let the plate dry ~10 min, then, using a dissection microscope with attached micromanipulator, move a complete tetrad to the first of four clearly marked locations, separate the spores, and place one spore in each of the three remaining positions (Fig. 22.3B). Repeat until a sufficient number of tetrads have been dissected. Micromanipulators for yeast dissection typically have well-marked grids clearly denoting the four positions for depositing the spores. Each plate can accommodate ~10–20 tetrads. Incubate at 30 °C until the spores form colonies. For a description of the microdissection technique, see Treco and Winston (2008).

Note I: Handle Zymolyase-treated cells gently. Avoid vortexing or vigorous pipetting.
Note II: The concentration of Zymolyase and duration of digestion required may vary by strain, thus optimization is recommended for best results.
Note III: For statistical significance, we recommend assessing the viability of at least 25 tetrads per mutation.

22.3.3 Genotype analysis of haploid spores

To determine the effects of a specific mutation, the phenotype of the haploid spores must be correlated with genotype. In the protocols presented above, the mutated α- or β-tubulin is tightly linked to tryptophan or uracil prototrophy, respectively, and the dissection plate should be replica-plated onto SC-Trp or SC-Ura selection media to determine whether spores contain the mutated tubulin gene.

Common replica-plating devices consist of a ring that is used to spread a sterile velvet towel over a Petri dish-sized circular pedestal. Gently press the dissection plate onto the spread velvet towel. Subsequently, press the selection plate onto the same towel and mark the orientation with respect to the dissection plate. As an example, spores from a heterozygous β-tubulin mutant diploid that grow on SC-Ura plates inherited the mutant allele whereas spores that fail to grow contain the wild-type allele (Fig. 22.3B).

Note: If haploid spores containing the mutant allele are viable, they should be saved for analysis with respect to the wild-type sister spores.

22.3.4 Mating-type analysis

Interpretation of tetrad dissection results depends on the fidelity of meiotic segregation and the reliable separation of spores from complete tetrads. Tetrad dissection can be verified by following 2+/2− segregation of independent heterozygous markers and/or haploid mating type. Mating-type tester strains are prototrophic for common laboratory markers and auxotrophic for an uncommon marker. Upon conjugation with laboratory strains, the auxotrophic markers in both haploid strains are complimented and the resultant diploid can grow on minimal media.

With a single sterile velvet towel, replica-plate the colonies on a tetrad dissection plate onto two YPD plates. Using clean velvets, now replica-plate a lawn of *MAT***a** tester cells onto the first YPD plate. Repeat with a lawn of *MAT*α testers on the second YPD plate. Incubate the YPD plates for 8–24 h at 30 °C to allow conjugation, then replica-plate each onto minimal media (YNB), and grow overnight. Only newly formed diploids with complimented markers will grow on minimal media. Growth after mating with *MAT***a** tester cells indicates that the spore in the corresponding position is *MAT*α haploid. Conversely, growth after mating with *MAT*α testers denotes a *MAT***a** spore. A properly dissected, valid tetrad will generate two *MAT***a** and two *MAT*α spores. If this is not the case, the tetrad should not be considered for analysis.

Note: To prepare mating-type tester lawns, spread ~200 μl saturated culture onto YPD plates, incubate overnight at 30 °C, and store in a sealed container at 4 °C up to several months.

22.4 ASSESSING MICROTUBULE STABILITY BY DRUG SENSITIVITY

Challenging yeast cells with the microtubule destabilizing drug benomyl is a common method to assess changes in microtubule stability *in vivo*. Benomyl supersensitivity is indicative of less stable microtubules while improved resistance implies increased stability. For example, the β-tubulin C354S substitution resulted in robust benomyl resistance. The increased microtubule stability was corroborated by direct visualization of green fluorescent protein (GFP)-labeled microtubule dynamics and confirmed to be independent of any potential changes to benomyl binding and/or regulatory proteins by purifying the mutated tubulin and measuring microtubule dynamics *in vitro* (Gupta et al., 2002). Overall, the benomyl plate assay is representative of changes in microtubule stability *in vivo*.

22.4.1 Benomyl plate assay

Prepare 5-ml overnight cultures in YPD. The following day, dilute 1:50 in fresh media and grow to log phase (4–6 h). Measure optical density of the cultures at 600 nm (OD600) and equalize cell concentration by adjusting volumes with YPD. In a 96-well plate, prepare logarithmic serial dilutions of each culture in YPD. With a multichannel pipette or a "frogger," spot 3 μl of each dilution onto the assay plates. We typically use plates with 0, 4, 6, 9, 12, 15, 18, and 21 μg/ml benomyl. Incubate plates at 24 °C for 72 h. A larger dynamic readout is consistently obtained by conducting this assay at 24 °C rather than 30 °C.

Drug sensitivity is scored as the concentration that inhibits cell growth. Spots where high-density cultures were deposited often have minor but visible growth at high drug concentrations. Growth inhibition becomes more obvious where low-density cultures were spotted and individual colonies are dispersed.

Note I: Preparations of benomyl plates can vary. It is essential to include a nonmutated strain and, if possible, strains with known sensitivity or resistance as controls.
Note II: To prepare benomyl plates, autoclave the YPD and agar with a stir bar, cool to ~65 °C, and then add 10 mg/ml benomyl in DMSO dropwise while stirring.

22.5 DIRECT ANALYSIS OF MICROTUBULE DYNAMICS *IN VIVO*

Live-cell imaging of cells expressing GFP-tagged α-tubulin allows individual astral microtubules to be tracked at high temporal and spatial resolution. This approach determines how microtubule behavior and parameters of dynamic instability are altered by specific tubulin mutations.

22.5.1 Yeast strains for *in vivo* microtubule dynamics analysis

Yeast strains are typically engineered to express an exogenous copy of α-tubulin fused to GFP under the *TUB1* promoter (*pTUB1–GFP–TUB1*) because GFP–tubulin alone does not support cell viability. Two commonly used integrating plasmids, pAFS125 and pMG3, are described in Table 22.1. To integrate *GFP–TUB1*, transform cells with ~1 μg linearized plasmid and plate on appropriate SC-Ura or SC-Leu media. Identify positive transformants by visually inspecting for GFP expression. Generate clonal strains of the positive transformants by two rounds of single-colony isolation on selection media.

Note: If strains destined for *in vivo* microtubule analysis are not auxotrophic for uracil and leucine, they may be made so by disrupting the *URA3* or *LEU2* gene using "marker swap" cassettes (Cross, 1997).

22.5.2 Maintaining comparable *GFP–TUB1* expression levels

Linearized plasmids can potentially integrate as tandem repeats, introducing multiple copies of *pTUB1–GFP–TUB1*. Anecdotal evidence also suggests that increased GFP–Tub1 is correlated with increased microtubule stability. Thus, it is advised to ensure that strains used for the *in vivo* analysis of microtubule dynamics have comparable levels of GFP–Tub1.

A common and practical method of selecting strains with similar *GFP–TUB1* expression is to compare ~10 transformants after streaking for single colonies. Prepare the strains for microscopy as described in the following section. With identical excitation intensity and exposure times, compare the relative fluorescence signal from individual astral microtubules among the strains. Typically, >50% will have microtubules with similar fluorescence intensity, while a minority display noticeably increased intensity. Select strains representative of the group with the lower basal level of GFP–Tub1 fluorescence, which should be comparable to other strains created by this method. Alternately, a haploid strain containing *GFP–TUB1* can be crossed with haploid strains containing nonlethal tubulin mutations. The resultant diploids are sporulated and dissected to generate control and experimental strains, both containing an identical integration of *GFP–TUB1*. If necessary, GFP–Tub1 levels among strains can be directly compared by western blot as described below.

22.5.3 Preparation of samples for microscopy

Prepare 5-ml overnight cultures at 30 °C by inoculation from fresh plates into SC, or appropriate SC-based selection media, with 0.2 mM adenine. The following day, dilute cultures to OD600 ~0.02 in the same media and grow with shaking at room temperature 4–6 h to reach log phase. Pellet 1.5 ml for 1 min at 13,000 × *g* and resuspend in 30–50 μl of fresh media. Sediment an additional 1.5 ml if a higher cell density is required for imaging. Prepare a fresh sample from the log-phase culture after 20–30 min.

Note: Yeast strains auxotrophic for adenine (*ade*$^-$) require imaging media supplemented with adenine to avoid high background fluorescence.

Prepare microscope slides with agarose pads (Fig. 22.4). Add 1.2% agarose in SC media, boil briefly in a microwave, and pipet 40 μl onto the center of a microscope slide. Let stand for ~2 s and place another slide on top of the agarose and let cool 2 min. Remove the top slide by gently rotating ~60° and slowly sliding it away from the bottom slide. Deposit 1.5 μl of resuspended yeast sample onto the pad and apply a coverglass (22 × 22 × 1 mm). Do not apply strong pressure as this may damage the cells. Seal the coverglass edges with melted valap to prevent drying during imaging.

Note I: To prepare valap, slowly melt ingredients (Vaseline, lanolin, and paraffin 1:1:1 w/w/w) using low setting on a hot plate and mix thoroughly using a disposable heat-resistant container and stirring devise if possible. Pour into small bottles, let cool, and cap for storage.
Note II: Valap is difficult to remove from microscope optics and will degrade image quality. Do not allow valap to get onto microscope optics.
Note III: To apply, dip a wooden- or cotton-tipped applicator into melted valap and spread quickly along the edges of the microscope coverglass.
Note IV: Use caution when heating valap for extended periods. The vapor will resolidify on nearby surfaces.
Note V: Valap is only recommended for time-lapse imaging exceeding 2–3 min.

22.5.4 Microscopy and analysis of microtubule dynamics

Microtubules in live yeast are typically imaged with an automated microscope equipped with a 63× Plan Fluor 1.4 N.A. objective, a filter set optimized for GFP, and a cooled CCD camera. Typically, time-lapse series encompassing eight z-series images with a step-size increment of 0.75 μm are captured at ~8 s intervals. To minimize photobleaching, the light source should be reduced using neutral density filters and the exposure time increased, yet still allow z-series to be acquired in

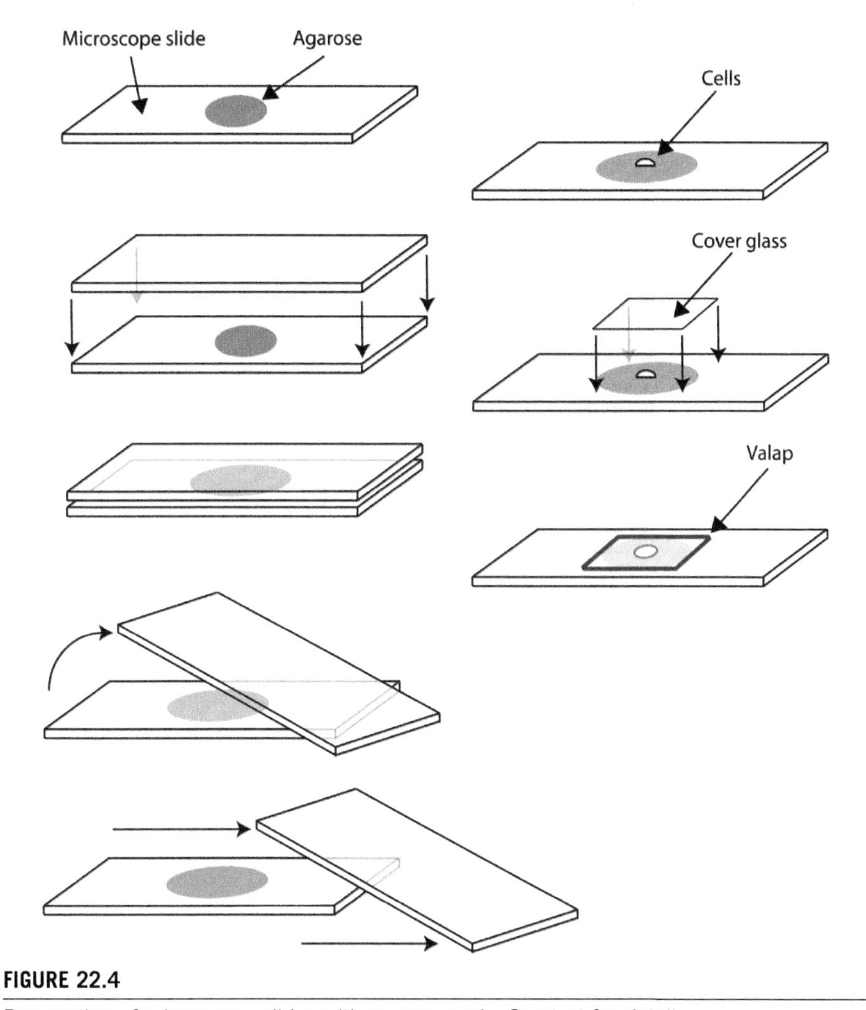

FIGURE 22.4

Preparation of microscope slides with agarose pads. See text for details.

<8 s. With these conditions, time lapses can be captured for as long as 8 min without significant photobleaching. Although precise measurement of microtubule lengths will be somewhat compromised, time lapses can be extended by reducing excitation intensity or exposure time and binning images 2×2.

Microtubule length can be measured using commercial (e.g., Slidebook/ Metamorph) or public domain (ImageJ) software. Over subsequent time points, the lengths of individual microtubules are determined from the base of the microtubule, where they anchor to the spindle pole body, to the opposite tip of the microtubule. The measurement should accommodate the three-dimensional length by utilizing the z-focal planes. Care should be taken that z-series encompass the whole cell, and only microtubules whose entire length remained within focus should be measured. The length of the microtubule throughout the time lapse is typically measured 2–3 times and the average graphed over time to construct a "lifetime history plot."

Using the lifetime history plots, microtubule polymerization and depolymerization events are typically identified as four or more consecutive time points spanning >0.5 μm with a linear regression fit of $R^2 \geq 0.85$. Attenuations, or pauses, are defined as four or more time points during which length change did not exceed ±0.2 μm. Because the criteria for specific events generally require a sustained behavior, that is, multiple consecutive time points, brief periods for some microtubules will remain unclassified. Catastrophes are scored as transitions from polymerization or paused states into depolymerization, and rescues are scored as transitions from depolymerization into paused or polymerization states. To calculate catastrophe frequency, the number of catastrophe events is divided by the total observable lifetime spent in polymerization and paused states for all analyzed microtubules. Similarly, to determine rescue frequency, the number of rescues is divided by total time spent in depolymerization.

22.6 LOCALIZATION OF MICROTUBULE-ASSOCIATED PROTEINS *IN VIVO*

Budding yeast utilizes a conserved set of microtubule-associated proteins to regulate microtubule behavior during processes such as spindle positioning and spindle elongation. Here, we outline the use of quantitative fluorescence microscopy to determine whether specific mutations alter the interactions of microtubule-associated proteins with microtubules. The technique involves tagging the endogenous copy of a microtubule regulator with yellow fluorescent protein (YFP) and monitoring cyan fluorescent protein (CFP)–Tub1-labeled microtubules in the same cell.

22.6.1 Yeast strains for localization of microtubule-associated proteins

Engineer cells to express CFP–Tub1. A useful plasmid for this purpose is pMG130, which has the CFP variant Cerulean fused to *TUB1* (Table 22.1). Digest, integrate, and screen as described for *GFP–TUB1* above.

Fuse the protein of interest to YFP to allow colocalization with CFP-labeled microtubules. The simplest method utilizes fragment-mediated homologous recombination. A fragment containing YFP and a selection marker is amplified by PCR from a plasmid template. Many plasmids useful for this are commonly available (Janke et al., 2004). The forward and reverse PCR primers are designed to contain >40 bases of homology just prior to and after the stop codon, respectively, of the targeted gene. Potential transformants should be screened for YFP fluorescence and isolated as described for *GFP–TUB1*. Successful fusion with the desired gene can be verified by PCR using primers that anneal inside the gene of interest and YFP.

Note: The functionality of fusion proteins is frequently increased by introducing a soluble flexible linker of 6–8 amino acids between the two proteins.

To increase fluorescence signal, proteins are often tagged with multiple tandem copies of YFP. A straightforward method to accomplish this uses a selectable, integrating plasmid containing tandem copies of YFP. A C-terminal fragment of the targeted gene, selected to contain >50 bp on either side of a restriction site unique to the plasmid, is cloned, without the stop codon, in-frame upstream of the YFPs. Plasmid digestion at the unique restriction site generates a linear fragment that will be targeted to the C-terminus of the endogenous locus. A major advantage of this approach is that tandem incorporation of multiple plasmid copies will simply generate additional C-terminal YFP regions, which lack the promoter and most of the ORF and, thus, are unlikely to have residual expression/activity. For an example in which the kinesin Kip3 was tagged by fusion with 3YFP (see Gupta, Carvalho, Roof, & Pellman, 2006).

The positive YFP-fusion transformants should be identified microscopically or by PCR as described above. The forward primer should anneal to a region of the tagged protein that was not cloned into the plasmid, and the reverse primer to the linker sequence between the tagged protein and the YFPs. This avoids complications from primer annealing to the repetitive sequences within the tandem YFPs. Generate clonal strains and confirm the YFP fusion before storing and/or using for analyses.

22.6.2 Maintaining equal levels of YFP-fusion proteins

Strains used for quantitative analyses must be verified by western blot to express equal amounts of the YFP-tagged protein. Prepare log-phase cultures in SC as described above.

> *Note*: We find improved consistency in quantitative imaging and western blots by preparing cultures under selective pressure for the tagged protein, for example, SC-Leu for cells containing *KIP3-3YFP-LEU2*.

Determine OD600 of the cultures, centrifuge 10 ml of cells, and resuspend in 1 ml of cold 0.1 M NaOH. Incubate 10 min on ice, centrifuge, and resuspend in ~200 μl SDS-PAGE loading buffer. The exact volume of loading buffer should reflect the OD600 ratio to maintain equal cell-to-volume ratios among the samples. Heat to 100 °C for 10 min, cool on ice, and centrifuge 10 min at 13,000 × *g*. Use equal volumes for western blotting. For low abundance fusion proteins, we find 4–12% Bis–Tris gradient gels (NuPAGE, Invitrogen) useful to concentrate and sharpen protein bands. GFP/CFP/YFP can be detected with polyclonal anti-GFP antibody (ABM). Actin (anti-β-actin, Abcam) can be used to verify equal loading.

22.6.3 Microscopy and analysis of microtubule-associated proteins

Prepare concentrated log-phase cells, mount on agarose pads, and image in the CFP and YFP channels, essentially as described for GFP–Tub1. The z-series should be acquired rapidly to minimize microtubule movements during imaging. Acquisition speeds can generally be improved by capturing the entire CFP z-series before switching filters to capture YFP, increased excitation intensity, a piezoelectric z-control, 2 × 2 binning, and increased efficiency with dedicated CFP/YFP filter sets rather

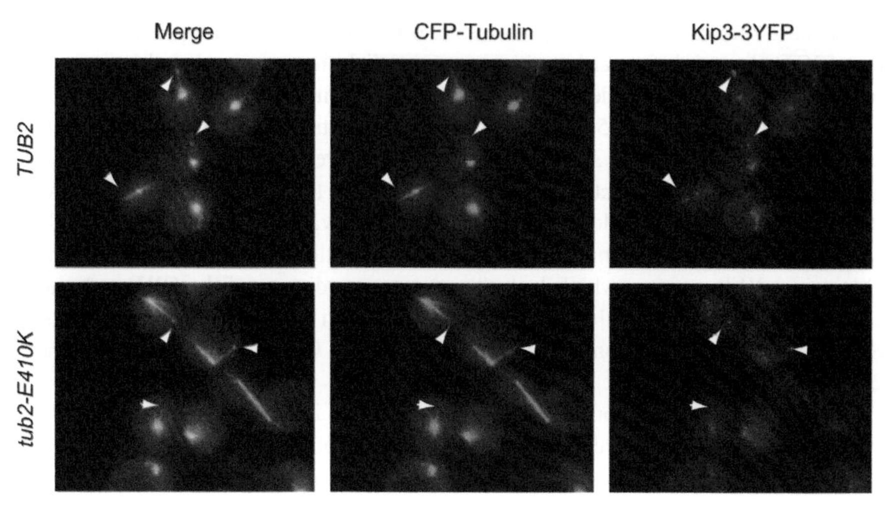

FIGURE 22.5

Microtubule plus-end localization of the kinesin Kip3 on microtubules bearing the β-tubulin E410K substitution (bottom row). Quantitative analysis determined that Kip3-3YFP at microtubule plus-ends (arrowheads) was reduced by 80% in the mutant strain relative to control cells (top row). Microtubules and Kip3 were visualized with CFP–Tub1 and Kip3-3YFP fusions, respectively (Cederquist et al., 2012). (See color plate.)

than filter wheel with shared dichroic. Care must be taken that z-series include the entire cell. Capture all images with identical settings, for example, excitation intensity, exposure times, and microscope/camera configuration.

The precise localization and dynamics of different microtubule-associated proteins can vary extensively. Thus, a detailed discussion of analyses is prohibitive. Here, we describe an example where localization of Kip3-3YFP to microtubule plus-ends was quantified (Fig. 22.5). Z-series images were deconvolved and merged to create a single summed image with Slidebook software (Intelligent Imaging Innovations). A circular region of 12 pixels was designated over the plus-ends of visible microtubules, and the average YFP intensity was calculated with the same software. Cytoplasmic background signal was determined and subtracted by calculating the YFP intensity of an identical region placed in the cytosol near each microtubule plus-end. To determine average Kip3-3YFP localization, ~100 microtubules were measured on multiple days from three wild-type or mutated tubulin clones.

Acknowledgments

We thank E. Murphy, K. Proudfoot, and R. Rizk for helpful comments on the manuscript. We regret not having space to cite many important works in this field.

References

Anders, K. R., & Botstein, D. (2001). Dominant-lethal alpha-tubulin mutants defective in microtubule depolymerization in yeast. *Molecular Biology of the Cell, 12*, 3973–3986.

Becker, D. M., & Lundblad, V. (2001). Introduction of DNA into yeast cells. *Current Protocols in Molecular Biology, 27*, 13.17.1–13.17.10.

Bode, C. J., Gupta, M. L., Suprenant, K. A., & Himes, R. H. (2003). The two alpha-tubulin isotypes in budding yeast have opposing effects on microtubule dynamics in vitro. *EMBO Reports, 4*, 94–99.

Brandt, R., Hundelt, M., & Shahani, N. (2005). Tau alteration and neuronal degeneration in tauopathies: Mechanisms and models. *Biochimica et Biophysica Acta, 1739*, 331–354.

Cederquist, G. Y., Luchniak, A., Tischfield, M. A., Peeva, M., Song, Y., Menezes, M. P., et al. (2012). An inherited TUBB2B mutation alters a kinesin-binding site and causes polymicrogyria, CFEOM and axon dysinnervation. *Human Molecular Genetics, 21*, 5484–5499.

Cross, F. R. (1997). 'Marker swap' plasmids: Convenient tools for budding yeast molecular genetics. *Yeast, 13*, 647–653.

Davis, A., Sage, C. R., Dougherty, C. A., & Farrell, K. W. (1994). Microtubule dynamics modulated by guanosine triphosphate hydrolysis activity of beta-tubulin. *Science, 264*, 839–842.

Desai, A., & Mitchison, T. J. (1997). Microtubule polymerization dynamics. *Annual Review of Cell and Developmental Biology, 13*, 83–117.

Entwistle, R. A., Rizk, R. S., Cheng, D. M., Lushington, G. H., Himes, R. H., & Gupta, M. L., Jr. (2012). Differentiating between models of epothilone binding to microtubules using tubulin mutagenesis, cytotoxicity, and molecular modeling. *ChemMedChem, 7*, 1580–1586.

Gupta, M. L., Jr., Bode, C. J., Dougherty, C. A., Marquez, R. T., & Himes, R. H. (2001). Mutagenesis of beta-tubulin cysteine residues in Saccharomyces cerevisiae: Mutation of cysteine 354 results in cold-stable microtubules. *Cell Motility and the Cytoskeleton, 49*, 67–77.

Gupta, M. L., Jr., Bode, C. J., Georg, G. I., & Himes, R. H. (2003). Understanding tubulin-Taxol interactions: Mutations that impart Taxol binding to yeast tubulin. *Proceedings of the National Academy of Sciences of the United States of America, 100*, 6394–6397.

Gupta, M. L., Jr., Bode, C. J., Thrower, D. A., Pearson, C. G., Suprenant, K. A., Bloom, K. S., et al. (2002). beta-Tubulin C354 mutations that severely decrease microtubule dynamics do not prevent nuclear migration in yeast. *Molecular Biology of the Cell, 13*, 2919–2932.

Gupta, M. L., Jr., Carvalho, P., Roof, D. M., & Pellman, D. (2006). Plus end-specific depolymerase activity of Kip3, a kinesin-8 protein, explains its role in positioning the yeast mitotic spindle. *Nature Cell Biology, 8*, 913–923.

Huang, B., & Huffaker, T. C. (2006). Dynamic microtubules are essential for efficient chromosome capture and biorientation in S. cerevisiae. *The Journal of Cell Biology, 175*, 17–23.

Jaglin, X. H., Poirier, K., Saillour, Y., Buhler, E., Tian, G., Bahi-Buisson, N., et al. (2009). Mutations in the beta-tubulin gene TUBB2B result in asymmetrical polymicrogyria. *Nature Genetics, 41*, 746–752.

Janke, C., Magiera, M. M., Rathfelder, N., Taxis, C., Reber, S., Maekawa, H., et al. (2004). A versatile toolbox for PCR-based tagging of yeast genes: New fluorescent proteins, more markers and promoter substitution cassettes. *Yeast, 21*, 947–962.

Johnson, V., Ayaz, P., Huddleston, P., & Rice, L. M. (2011). Design, overexpression, and purification of polymerization-blocked yeast alphabeta-tubulin mutants. *Biochemistry*, *50*, 8636–8644.

Jordan, M. A. (2002). Mechanism of action of antitumor drugs that interact with microtubules and tubulin. *Current Medicinal Chemistry. Anti-Cancer Agents*, *2*, 1–17.

Katz, W., Weinstein, B., & Solomon, F. (1990). Regulation of tubulin levels and microtubule assembly in Saccharomyces cerevisiae: Consequences of altered tubulin gene copy number. *Molecular and Cellular Biology*, *10*, 5286–5294.

Lei, P., Ayton, S., Finkelstein, D. I., Adlard, P. A., Masters, C. L., & Bush, A. I. (2010). Tau protein: Relevance to Parkinson's disease. *The International Journal of Biochemistry & Cell Biology*, *42*, 1775–1778.

Lewis, S. A., Tian, G., Vainberg, I. E., & Cowan, N. J. (1996). Chaperonin-mediated folding of actin and tubulin. *The Journal of Cell Biology*, *132*, 1–4.

Neff, N. F., Thomas, J. H., Grisafi, P., & Botstein, D. (1983). Isolation of the beta-tubulin gene from yeast and demonstration of its essential function in vivo. *Cell*, *33*, 211–219.

Reijo, R. A., Cooper, E. M., Beagle, G. J., & Huffaker, T. C. (1994). Systematic mutational analysis of the yeast beta-tubulin gene. *Molecular Biology of the Cell*, *5*, 29–43.

Richards, K. L., Anders, K. R., Nogales, E., Schwartz, K., Downing, K. H., & Botstein, D. (2000). Structure-function relationships in yeast tubulins. *Molecular Biology of the Cell*, *11*, 1887–1903.

Sage, C. R., Davis, A. S., Dougherty, C. A., Sullivan, K., & Farrell, K. W. (1995). beta-Tubulin mutation suppresses microtubule dynamics in vitro and slows mitosis in vivo. *Cell Motility and the Cytoskeleton*, *30*, 285–300.

Schatz, P. J., Pillus, L., Grisafi, P., Solomon, F., & Botstein, D. (1986). Two functional alpha-tubulin genes of the yeast Saccharomyces cerevisiae encode divergent proteins. *Molecular and Cellular Biology*, *6*, 3711–3721.

Sikorski, R. S., & Boeke, J. D. (1991). In vitro mutagenesis and plasmid shuffling: From cloned gene to mutant yeast. *Methods in Enzymology*, *194*, 302–318.

Straight, A. F., Marshall, W. F., Sedat, J. W., & Murray, A. W. (1997). Mitosis in living budding yeast: Anaphase A but no metaphase plate. *Science*, *277*, 574–578.

Tischfield, M. A., Baris, H. N., Wu, C., Rudolph, G., Van Maldergem, L., He, W., et al. (2010). Human TUBB3 mutations perturb microtubule dynamics, kinesin interactions, and axon guidance. *Cell*, *140*, 74–87.

Tischfield, M. A., Cederquist, G. Y., Gupta, M. L., Jr., & Engle, E. C. (2011). Phenotypic spectrum of the tubulin-related disorders and functional implications of disease-causing mutations. *Current Opinion in Genetics & Development*, *21*, 286–294.

Treco, D. A., & Lundblad, V. (2001). Preparation of yeast media. *Current Protocols in Molecular Biology*, *23*, 13.11.11–13.11.17.

Treco, D. A., & Winston, F. (2008). Growth and manipulation of yeast. *Current Protocols in Molecular Biology*, *82*, 13.12.11–13.12.12.

CHAPTER

Using MTBindingSim as a Tool for Experimental Planning and Interpretation 23

Julia T. Philip, Aranda R. Duan, Emily O. Alberico and Holly V. Goodson

Department of Chemistry and Biochemistry, Interdisciplinary Center for the Study of Biocomplexity, University of Notre Dame, Notre Dame, Indiana, USA

CHAPTER OUTLINE

Introduction 376
23.1 MTBindingSim 376
23.2 Experimental Designs in MTBindingSim 376
23.3 Binding Models in MTBindingSim 377
23.4 Using MTBindingSim Example 1: Tau–MT Binding 379
23.5 Using MTBindingSim Example 2: Does Human EB1 Bind to the MT Seam or Lattice? 381
Conclusion 383
References 383

Abstract

MTBindingSim is a program that enables users to simulate experiments in which proteins or other ligands (e.g., drugs) bind to microtubules or other polymers under various binding models. The purpose of MTBindingSim is to help researchers and students gain an intuitive understanding of binding behavior and design experiments to distinguish between different binding mechanisms. MTBindingSim is open-source, freely available software and can be found at bindingtutor.org/mtbindingsim.

This chapter first describes the capabilities of MTBindingSim, including the experimental designs and protein-binding models that it simulates, and then discusses two examples in which MTBindingSim is utilized in an experimental context. In the first, MTBindingSim is used to investigate potential explanations for unusual behavior observed in the binding of the neuronal protein Tau to microtubules, demonstrating that some potential explanations are incompatible with the experimental data. In the second example, MTBindingSim is used to design experiments to examine the question of whether the plus-end tracking protein EB1 binds preferentially to the microtubule seam.

Methods in Cell Biology, Volume 115

ISSN 0091-679X
http://dx.doi.org/10.1016/B978-0-12-407757-7.00023-2

INTRODUCTION

Microtubules (MTs) are a protein–polymer that are a fundamental part of the cell cytoskeleton and are involved in many critical cellular processes, such as cell division, maintenance of cell polarity, and cargo transport (Amos & Schlieper, 2005; Desai & Mitchison, 1997; Howard & Hyman, 2003; Lansbergen & Akhmanova, 2006). MTs interact with a vast, highly coordinated array of microtubule-associated proteins (MAPs) to carry out these functions (Lansbergen & Akhmanova, 2006). Researchers often investigate the interactions between MAPs and MTs by investigating their interactions *in vitro*. In particular, many MAP–MT interactions have been probed using binding assays (e.g., Gupta et al., 2010, 2009; Zhu et al., 2009), which are powerful experimental tools often used in studies involving MTs. However, the results of binding studies can be difficult to interpret in cases where the MAP–MT interaction is not a simple 1:1 binding interaction with a single, well-defined dissociation constant (K_D). To assist researchers in interpreting MAP–MT-binding experiments, Philip, Pence, and Goodson (2012) have written a program to model MAP–MT-binding interactions, MTBindingSim. MTBindingSim is also a useful tool to help researchers train their intuition about MAP–MT binding and to aid in the design of new experiments that would distinguish between different binding mechanisms. While the text below focuses on using MTBindingSim to study interactions between MAPs and MTs, some of the models used in MTBindingSim can be used to simulate binding to any polymer, and still others are suitable for modeling general protein–ligand interactions.

23.1 MTBindingSim

MTBindingSim simulates experimental data for various MAP–MT-binding mechanisms (described below) according to user-specified values for the total MAP and MT concentrations and the binding dissociation constant(s). MTBindingSim was written in MATLAB and can be run either in MATLAB or as an independent program that does not require MATLAB on Windows or Mac OS. MTBindingSim is open-source software and the MATLAB code and independent programs are freely available at bindingtutor.org/mtbindingsim. A user manual with detailed instructions on the operation of MTBindingSim and the mathematical calculations used in simulating the binding curves is also available. Some text of this section has been adapted from the user manual.

23.2 EXPERIMENTAL DESIGNS IN MTBindingSim

MTBindingSim can plot simulated curves from three different experimental designs. The most common experimental method for investigating MAP–MT binding is a cosedimentation assay, where MT polymer and MAPs are mixed in known quantities, incubated to allow the proteins to bind to equilibrium, and then centrifuged, which causes the MTs and any bound MAPs to separate out of solution (the supernatant) and form a pellet. The supernatant and pellet fractions are then assayed to determine the protein concentration in each fraction by SDS-PAGE and densitometry analysis. The cosedimentation method relies on the assumption (verified by appropriate controls) that the MTs will form a pellet regardless of any MAPs that may be bound and that the MAPs will not spin down in the absence of binding to MTs. While the simulated curves in MTBindingSim were designed with cosedimentation assays in mind, the data will be the same for any method of investigating MAP–MT binding, such as fluorescence anisotropy, SPR, or tryptophan fluorescence. MTBindingSim follows the standard convention of writing the concentration of MTs, [MT], to mean the concentration of polymerized tubulin dimers (the "MT polymer"). MAP–MT-binding experiments are usually performed in the presence of MT stabilizers such as taxol, so the concentration of unpolymerized tubulin dimers will be minimal and can be neglected. Experiments performed with a high concentration of unpolymerized tubulin dimers will present additional complications in interpreting binding curves and cannot (currently) be modeled with MTBindingSim.

Vary [MT]: In this assay, the concentration of A (a generic MAP) is held constant and the concentration of MT is varied. This is the "standard" binding experiment. The fraction of A bound to MT is graphed on the *y*-axis. The user can choose either the total MT concentration or the free MT concentration to be graphed on the *x*-axis. Note that when the *x*-axis is set to [MT] free, this experiment shows the familiar Langmuir isotherm. The quick method of determining K_D from the concentration of MT where 50% of A is bound *only* works when the *x*-axis plots [MT] free, not when it plots [MT] total.

Vary [A]: In this assay, the concentration of MT is held constant and the concentration of A is varied. There are three possible ways of graphing data from this kind of simulation. The first is a Scatchard plot where the *x*-axis is [A] bound and the *y*-axis is [A] bound/[A] free. In this kind of plot, binding data will be linear for a simple 1:1 interaction, and a deviation from linearity indicates cooperativity or other types of nonsimple binding. The other two plots have [A] bound on the *y*-axis and the *x*-axis set to either [A] free or [A] total.

Competition: In this assay, there are two MT-binding proteins, A and B. The concentration of A and MT is held constant and the concentration of B is varied. The *y*-axis is the fraction of A bound and the *x*-axis is the total concentration of B. In the present version of MTBindingSim, this kind of experiment can only be simulated when both A and B bind to MTs with simple 1:1 binding.

23.3 BINDING MODELS IN MTBindingSim

The binding models simulated in MTBindingSim were chosen to reflect commonly seen or hypothesized mechanisms of MAP–MT binding. While MTBindingSim was written specifically with MAP–MT binding in mind, most of its models have broader applicability. Only the "seam and lattice binding" model is MT specific as it relies on the unique structure of the MT. The "MAPs dimerize" and "pseudocooperativity" models are valid for any protein–polymer or ligand–polymer interaction. The other models are valid for any protein–protein or ligand–protein interaction. The calculations performed by MTBindingSim are based on the definitions of the dissociation constants and mass balances for each binding model. Detailed derivations of the equations used in MTBindingSim can be found in the user manual available at bindingtutor.org/mtbindingsim.

Binding ratio: Some MAPs appear to interact with MTs at ratios differing from 1 MAP:1 tubulin dimer. MTBindingSim allows for such situations by including the binding ratio, n, as a user-specified input. The binding ratio is defined as n*A:1 tubulin dimer in all models in MTBindingSim. MTBindingSim includes n in its calculations by setting the concentration of initial available binding sites to the concentration of free polymerized tubulin dimers ([MT] free) and only taking the binding ratio into account when calculating how many binding sites have been occupied by a binding event.

First-order binding (Fig. 23.1A): In this model, the MAP A binds to MTs in a 1 MAP:1 tubulin dimer fashion with a dissociation constant K_{AMT}. In this model, all A–MT interactions are identical.

Seam and lattice binding (Fig. 23.1B): In this model, A can bind to either the MT seam (α–β lateral tubulin contacts, forming 1/13 of the number of tubulin dimers) or the lattice (α–α and β–β lateral tubulin contacts, forming 12/13 of the tubulin dimers) with dissociation constants K_{AS} and K_{AL}, respectively (heterogeneity of protofilament numbers is not currently taken into account).

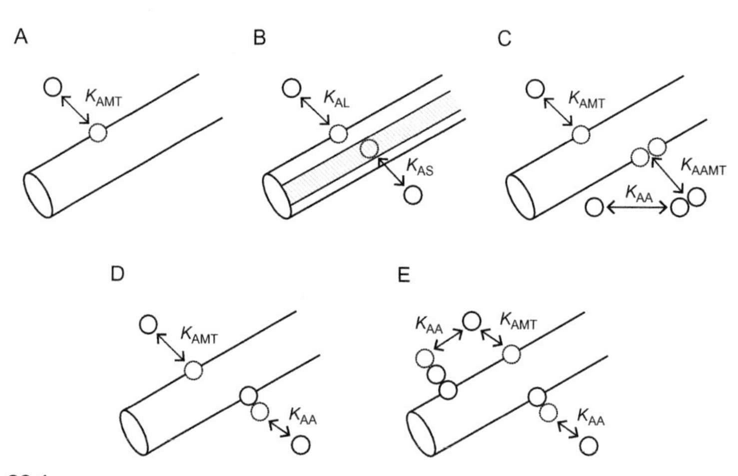

FIGURE 23.1

Cartoons illustrating binding models from MTBindingSim. (A) First-order binding. (B) Seam (1/13 of the MT polymer) and lattice (12/13 of the MT polymer) binding. (C) MAPs dimerize, where a MAP can exist either as a monomer or as a dimer, each with different affinities for the MT. (D) MAPs bind MT-bound MAPs, where the binding of a MAP to the MT creates a new MAP-binding site. (E) Two MAPs bind MT-bound MAPs, where each MAP bound to the MT surface can bind two more MAPs.

MAPs dimerize (Fig. 23.1C): In this model, A can exist either as a monomer or as a dimer, both of which can bind the MTs with dissociation constants K_{AMT} and K_{AAMT}, respectively. The dissociation constant for the A–A dimerization is K_{AA}.

Pseudocooperativity: Because the standard model of cooperative biding (i.e., that of hemoglobin) cannot be applied to polymers and models commonly used for polymers do not fit well into the mathematical framework of MTBindingSim, we have implemented a "pseudocooperativity" model. In this model, each A that binds to an MT site with a dissociation constant of K_{AMT} alters a second MT site so that it becomes MT*, which has a dissociation constant for A of K_{AMT^*}.

MAPs bind MT-bound MAPs (Fig. 23.1D): In this model, the binding of A to an MT with a dissociation constant of K_{AMT} creates a new binding site for another A to bind to the previous MT-bound A with a dissociation constant of K_{AA}. This allows a second A to bind without taking up another MT site.

Two MAPs bind MT-bound MAPs (Fig. 23.1E): This model is an extension of the "MAPs bind MT-bound MAPs" model, where the binding of A to an MT with a dissociation constant of K_{AMT} then allows two more As to bind to the MT-bound A with a dissociation constant of K_{AA}. An examination of this model and the "MAPs bind MT-bound MAPs" model shows that further A–A binding on the MT surface will follow a similar pattern, ultimately resulting in a nonconvergent infinite series, which cannot be modeled by the equations used in MTBindingSim (this would be a polymerization).

Two binding sites: In this model, there are two independent binding sites on the MT for an A with dissociation constants K_{AMT1} and K_{AMT2}.

23.4 USING MTBindingSim EXAMPLE 1: TAU–MT BINDING

Tau is a MAP that binds all along the length of MTs and stabilizes the MT structure. Despite multiple investigations into the binding of Tau to MTs, it is unclear how many tubulin dimers one Tau protein binds to or what is the Tau–MT dissociation constant (K_D) because different publications report different answers. For example, measured K_D values range from <100 nM up to >1 μM (Ackmann, Wiech, & Mandelkow, 2000; Goode, 1994; Gustke, Trinczek, Biernat, Mandelkow, & Mandelkow, 1994; Kar, Fan, Smith, Goedert, & Amos, 2003; Makrides, Massie, Feinstein, & Lew, 2004).

Ackmann et al. (2000) and more recently Duan and Goodson (2012) have shown that the amount of Tau that binds to 1 μM MTs continues to rise as the Tau concentration is increased (data from Duan and Goodson are shown in Fig. 23.2A). If a binding ratio of 1 Tau:1 tubulin dimer is assumed, the data appear to show that Tau is supersaturating the MTs (the concentrations of cosedimented Tau end up being greater than the concentrations of total MTs).

Ackmann et al. (2000) suggested that these observations could be explained by the idea that Tau binds to MT-bound Tau. Further experimental support for this idea was provided by cross-linking and AFM experiments by Makrides et al. (2003). In this proposed model, Tau binds to the MT and a second Tau protein is then able to

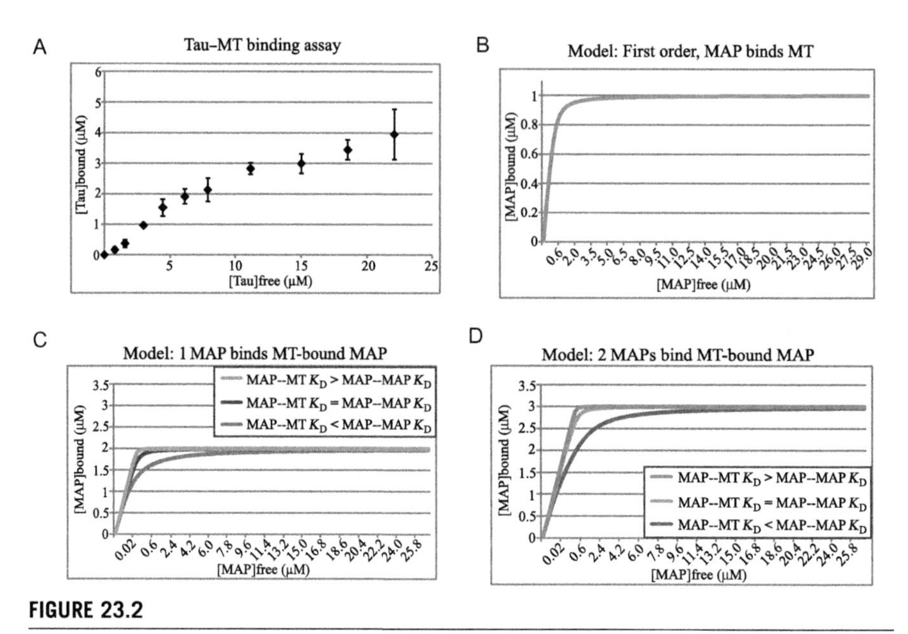

FIGURE 23.2

Tau–MT-binding assay and simulations from MTBindingSim. (A) Data from a Tau–MT cosedimentation binding assay. (B) MTBindingSim simulation of a first-order binding relationship with a K_D of 0.1 μM. (C) MTBindingSim simulation using the MAPs bind MT-bound MAPs model, with the MAP–MAP K_D less than (green line), equal to (red line), and greater than (blue line) the MAP–MT K_D. In interpreting these curves, it is important to remember that a smaller K_D value corresponds to a higher affinity. (D) MTBindingSim simulation using the two MAPs bind MT-bound MAPs model with the MAP–MAP K_D less than (orange line), equal to (teal line), and greater than (purple line) the MAP–MT K_D. (See color plate.)

Panel A: Reproduced from Duan and Goodson (2012)

bind the first, already MT-bound, Tau protein. This would result in Tau oligomers forming on the MT surface (approximately 2–3 Tau proteins binding each other on the MT surface). Both Tau–MT and Tau–Tau binding would contribute to the amount of Tau that is measured in the pellet fraction of a cosedimentation assay, which would normally be interpreted as Tau bound to MTs.

Could this type of model explain the data? Intuitively, it is appealing. MTBindingSim can be used to investigate this question in a more quantitative way. To start, it is a good idea to investigate how proteins interact according to a simple binding model. To do this, the standard "first-order binding" model is employed, in which Tau binding only to MTs is simulated. For the curve shown in Fig. 23.2B, the K_D was set to 0.1 μM based on reasonable estimates from other Tau–MT-binding experiments (data not shown). As shown in Fig. 23.2B, the amount of Tau that binds to the MTs saturates at 1 μM.

Next, the behavior of models in which Tau binds to MT-bound Tau can be examined. Figure 23.2C shows the amount of Tau that binds to 1 μM MTs using the "MAPs bind MT-bound MAPs" model. Three different scenarios are shown: the MAP–MT-binding affinity is weaker than the MAP–MAP binding, the MAP–MT- and the MAP–MAP-binding affinities are equal, and the MAP–MT-binding affinity is stronger than the MAP–MAP affinity. Figure 23.2C shows that, in any scenario in which the one MAP is allowed to bind each MT-bound MAP, all curves saturate at 2 μM bound MAP, unlike what happens with the data that are shown in Fig. 23.2A.

To test the hypothesis that the data in Fig. 23.2A can be explained by adding an additional layer of Tau–Tau interaction, the "Two MAPs bind MT-bound MAPs" model was used with the same binding affinities as those set in Fig. 23.2C (Fig. 23.2D). The results of Fig. 23.2D are similar to those shown in Fig. 23.2C in that the curves saturate; the amount of MAP bound simply saturates at 3 μM instead of 2 μM. More importantly, Fig. 23.2C and D demonstrates that, for any small number of Tau–Tau interactions at the MT surface, there will be a clear upper bound limit in the amount of Tau bound. This behavior is in contrast to the continual rise in the amount of Tau bound seen in the experimental data of Fig. 23.2A.

These observations suggest that a model based on Tau–MT interactions and finite Tau–Tau oligomerization is not able to explain the data observed in Fig. 23.2A. There must be another interaction occurring in solution between the Tau and MT proteins that has not yet been accounted for by the model. Duan and Goodson have recently used fluorescence microscopy to provide evidence that this "additional process" is formation of Tau-only filaments that are induced by the presence of MTs. The presence of these filaments confounds the MT cosedimentation assays by removing free Tau from solution (causing an apparent reduction of Tau–MT affinity) and by sedimenting independently of MTs (increasing the apparent affinity) (Duan & Goodson, 2012).

This work shows the value of MTBindingSim as a tool for interpreting binding data. Using MTBindingSim demonstrated that the observations of supersaturation in the Tau–MT experiments could not be due to Tau oligomerization at the MT surface as had previously been suggested. This conclusion opened up consideration to other possible explanations for the observed Tau–MT behavior and ultimately led to the microscopy studies that revealed the cause of the puzzling binding data. Without the aid of MTBindingSim, it would have been more difficult to confidently reject the hypothesis of Tau oligomerization on the MT surface and then move on to other possibilities.

23.5 USING MTBindingSim EXAMPLE 2: DOES HUMAN EB1 BIND TO THE MT SEAM OR LATTICE?

One class of MAPs of particular interest is the MT plus-end tracking proteins (+TIPs). One +TIP that has been of particular interest is EB1, which is considered the "core" of the +TIP network (Akhmanova & Steinmetz, 2008). There are conflicting reports in the published literature about the nature of EB1–MT-binding interactions and whether EB1 has a preference for the MT seam. The MT seam is a region where the lateral contacts between tubulin dimers are α–β, in contrast to the lattice, where the lateral contacts between tubulin dimers are α–α and β–β. The seam and lattice are thus chemically distinct. Sandblad et al. published binding data that they interpreted as evidence that EB1 binds strongly to the MT seam and weakly, if at all, to the lattice (Sandblad et al., 2006). However, more recent electron microscopy and fluorescence microscopy experiments by Maurer, Fourniol, Bohner, Moores, and Surrey (2012) have contradicted Sandblad et al.'s conclusions, providing strong evidence that EB1 binds to the entire MT and specifically avoids the seam.

How can these reports be reconciled? Using MTBindingSim simulations, it can be shown that there are other possible interpretations of the Sandblad binding data (Fig. 23.3A) (Alberico, et al., 2013). To take a biochemical approach to resolving

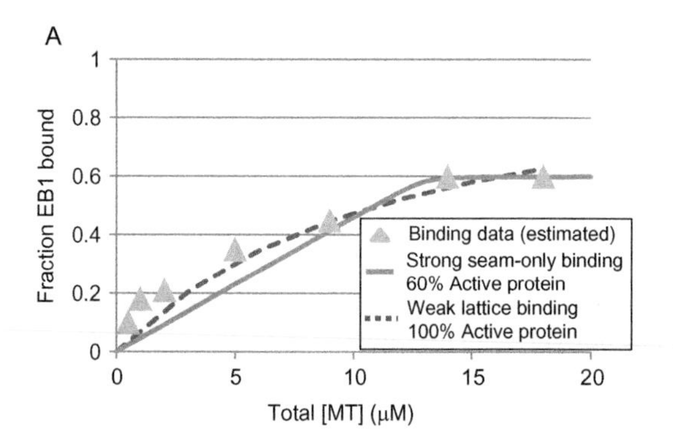

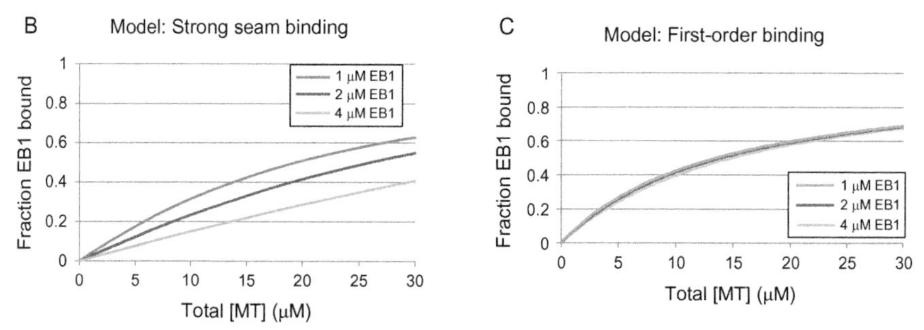

FIGURE 23.3

Does EB1 bind preferentially to the MT seam? (A) Data extracted from Sandblad et al. (2006) (green triangles), with overlaid curves corresponding to different binding models as generated by MTBindingSim. The blue curve shows a simulated curve for strong seam binding (K_D=0.001 μM) with 60% active EB1 protein, while the purple curve illustrates the data expected for weak binding (K_D=10 M) to the entire MT. These curves show that both binding models are similarly consistent with the data. (B) Curves from MTBindingSim with strong seam binding (K_D=1 μM) and weak lattice binding (K_D=1 M) at 1, 2, and 4 μM EB1. (C) Curves from MTBindingSim with first-order binding to the entire MT lattice (K_D=13 μM) at 1, 2, and 4 μM EB1. (See color plate.)

Figures are adapted from Alberico, et al., (2013)

the question of whether EB1 binds to the lattice or seam and provide additional evidence on the specific question of how human EB1 binds to MTs, (Alberico, et al., 2013) used MTBindingSim to identify a set of experimental conditions that should distinguish between a model where EB1 binds strongly to the MT seam with weak to no lattice binding, and one where EB1 binds to the entire MT with a moderate affinity. The resulting theoretical binding curves are shown in Fig. 23.3B and C (Alberico, et al., 2013). These data lead to the prediction that, if EB1 binds strongly to the seam, a moderate concentration of EB1 will saturate the available binding sites, and as the concentration of EB1 increases, the fraction of EB1 that is bound to MTs at a given MT concentration will fall. In contrast, if EB1 binds to the entire lattice with moderate or weak affinity, the overall concentration of free MT sites will change little as the concentration of EB1 moves up or down, leading to a situation where the fraction of EB1 bound to MTs stays relatively constant. Note that these graphs are plotted against total [MT] because calculating the free [MT] cannot be done without making assumptions about the EB1–MT-binding ratio. Using these simulated binding curves as a guide, (Alberico, et al., 2013) have recently provided strong biochemical evidence that human EB1 does not bind to the MT seam.

CONCLUSION

MTBindingSim is a software designed to assist researchers in designing and interpreting binding experiments as well as in training their intuition about how MAP–MT-binding interactions will behave. This program has been very helpful when investigating puzzling MAP–MT-binding behavior and in designing experiments to distinguish between different models of MAP–MT binding. For further information on how to use MTBindingSim, see the user manual at bindingtutor.org/mtbindingsim. MTBindingSim is under active development and the developers welcome suggestions from other researchers for new features or models for MTBindingSim or for changes to the user interface (see Web site for contact details). A more general version of MTBindingSim, BindingTutor, for use as an educational tool, is also currently under development and will be available at bindingtutor.org/bindingtutor when it is completed.

References

Ackmann, M., Wiech, H., & Mandelkow, E. (2000). Nonsaturable binding indicates clustering of tau on the microtubule surface in a paired helical filament-like conformation. *Journal of Biological Chemistry*, *275*(39), 30335–30343. http://dx.doi.org/10.1074/jbc.M002590200.

Akhmanova, A. S., & Steinmetz, M. O. (2008). Tracking the ends: A dynamic protein network controls the fate of microtubule tips. *Nature Reviews. Molecular Cell Biology*, *9*(4), 309–322. http://dx.doi.org/10.1038/nrm2369.

Alberico, E. O., Lyons, D. F., Murphy, R. H., Philip, J. T., Duan, A. R., Correia, J. J., & Goodson, H. V. (2013). Biochemical evidence that human EB1 does not bind preferentially to the microtubule seam. *Cytoskeleton*, *70*, 317–327. http://dx.doi.org/10.1002/cm.21108.

Amos, L. A., & Schlieper, D. (2005). Microtubules and maps. *Advances in Protein Chemistry*, *71*, 257–298. http://dx.doi.org/10.1016/S0065-3233(04)71007-4.

Desai, A., & Mitchison, T. J. (1997). Microtubule polymerization dynamics. *Annual Review of Cell and Developmental Biology*, *13*, 83–117. http://dx.doi.org/10.1146/annurev.cellbio.13.1.83.

Duan, A. R., & Goodson, H. V. (2012). Taxol-stabilized microtubules promote the formation of filaments from unmodified full-length tau in vitro. *Molecular Biology of the Cell*, *23*, 4796–4806. http://dx.doi.org/10.1091/mbc.E12-05-0374.

Goode, B. L. (1994). Identification of a novel microtubule binding and assembly domain in the developmentally regulated inter-repeat region of tau. *The Journal of Cell Biology*, *124*(5), 769–782. http://dx.doi.org/10.1083/jcb.124.5.769.

Gupta, K. K., Joyce, M. V., Slabbekoorn, A. R., Zhu, Z. C., Paulson, B. A., Boggess, B., et al. (2010). Probing interactions between CLIP-170, EB1, and microtubules. *Journal of Molecular Biology*, *395*(5), 1049–1062. http://dx.doi.org/10.1016/j.jmb.2009.11.014.

Gupta, K. K., Paulson, B. A., Folker, E. S., Charlebois, B., Hunt, A. J., & Goodson, H. V. (2009). Minimal plus-end tracking unit of the cytoplasmic linker protein CLIP-170. *Journal of Biological Chemistry*, *284*(11), 6735–6742. http://dx.doi.org/10.1074/jbc.M807675200.

Gustke, N., Trinczek, B., Biernat, J., Mandelkow, E.-M., & Mandelkow, E. (1994). Domains of tau protein and interactions with microtubules. *Biochemistry*, *33*(32), 9511–9522. http://dx.doi.org/10.1021/bi00198a017.

Howard, J., & Hyman, A. A. (2003). Dynamics and mechanics of the microtubule plus end. *Nature*, *422*(6933), 753–758. http://dx.doi.org/10.1038/nature01600.

Kar, S., Fan, J., Smith, M. J., Goedert, M., & Amos, L. A. (2003). Repeat motifs of tau bind to the insides of microtubules in the absence of taxol. *EMBO Journal*, *22*(1), 70–77. http://dx.doi.org/10.1093/emboj/cdg001.

Lansbergen, G., & Akhmanova, A. S. (2006). Microtubule plus end: A hub of cellular activities. *Traffic*, *7*(5), 499–507. http://dx.doi.org/10.1111/j.1600-0854.2006.00400.x.

Makrides, V., Massie, M. R., Feinstein, S. C., & Lew, J. (2004). Evidence for two distinct binding sites for tau on microtubules. *Proceedings of the National Academy of Sciences of the United States of America*, *101*(17), 6746–6751. http://dx.doi.org/10.1073/pnas.0400992101.

Makrides, V., Shen, T. E., Bhatia, R., Smith, B. L., Thimm, J., Lal, R., et al. (2003). Microtubule-dependent oligomerization of tau. Implications for physiological tau function and tauopathies. *The Journal of Biological Chemistry*, *278*(35), 33298–33304. http://dx.doi.org/10.1074/jbc.M305207200.

Maurer, S. P., Fourniol, F. J., Bohner, G., Moores, C. A., & Surrey, T. (2012). EBs recognize a nucleotide-dependent structural cap at growing microtubule ends. *Cell*, *149*(2), 371–382. http://dx.doi.org/10.1016/j.cell.2012.02.049.

Philip, J. T., Pence, C. H., & Goodson, H. V. (2012). MTBindingSim: Simulate protein binding to microtubules. *Bioinformatics (Oxford, England)*, *28*(3), 441–443. http://dx.doi.org/10.1093/bioinformatics/btr684.

Sandblad, L., Busch, K. E., Tittmann, P., Gross, H., Brunner, D., & Hoenger, A. (2006). The Schizosaccharomyces pombe EB1 homolog Mal3p binds and stabilizes the microtubule lattice seam. *Cell*, *127*(7), 1415–1424.

Zhu, Z., Gupta, K. K., Slabbekoorn, A., Paulson, B. A., Folker, E. S., & Goodson, H. V. (2009). Interactions between EB1 and microtubules: Dramatic effect of affinity tags and evidence for cooperative behavior. *Journal of Biological Chemistry*, *284*(47), 32651–32661. http://dx.doi.org/10.1074/jbc.M109.013466.

CHAPTER

Imaging Individual Spindle Microtubule Dynamics in Fission Yeast

24

Judite Costa[*,†], **Chuanhai Fu**[‡], **Viktoriya Syrovatkina**[*] **and Phong T. Tran**[*,†]

[*] *Cell and Developmental Biology, University of Pennsylvania, Philadelphia, Pennsylvania USA*

[†] *Cell Biology, Institut Curie, UMR 144 CNRS, Paris, France*

[‡] *Biochemistry, University of Hong Kong, Pokfulam, Hong Kong*

CHAPTER OUTLINE

Introduction **386**
24.1 Methods **387**
24.1.1 PDMS Slide Spacer 387
24.1.2 Slide Assembly 388
24.1.3 Live-cell Imaging 390
24.1.4 Data Analysis 391
Conclusion **393**
Acknowledgments **393**
References **394**

Abstract

Microtubules exhibit dynamic instability, stochastically switching between infrequent phases of growth and shrinkage. In the cell, microtubule dynamic instability is further modulated by microtubule-associated proteins and motors, which are specifically tuned to cell cycle stages. For example, mitotic microtubules are more dynamic than interphase microtubules. The different parameters of microtubule dynamics can be measured from length versus time data, which are generally obtained from time-lapse acquisition using the optical microscope. The typical maximum resolution of the optical microscope is $\sim\lambda/2$ or $\sim$300 nm. This scale represents a challenge for imaging fission yeast microtubule dynamics specifically during early mitosis, where the bipolar mitotic spindle contains many short dynamic microtubules of $\sim$1-μm scale. Here, we present a novel method to image short fission yeast mitotic

Methods in Cell Biology, Volume 115

ISSN 0091-679X
http://dx.doi.org/10.1016/B978-0-12-407757-7.00024-4

microtubules. The method uses the thermosensitive reversible kinesin-5 cut7.24^{ts} to create monopolar spindles, where asters of individual mitotic microtubules are presented for imaging and subsequent analysis.

INTRODUCTION

The fission yeast *Schizosaccharomyces pombe* has traditionally been a good genetic model organism to dissect conserved mechanisms of microtubule organization and function (Sawin & Tran, 2006). In particular, fluorescent protein-tagged tubulin can be expressed in the fission yeast cell for observing microtubule dynamics throughout the cell cycle (Snaith, Anders, Samejima, & Sawin, 2010). In particular, fission yeast interphase microtubules are organized as several discrete dynamic bundles parallel to the long axis of the cell (Drummond & Cross, 2000; Tran, Marsh, Doye, Inoue, & Chang, 2001). Each microtubule bundle in turn is composed of mainly two individual microtubules bundled at their minus ends by an overlapping region at the cell center and with their dynamic plus ends facing and interacting with the cell tips (Hoog et al., 2007). Interphase microtubules can reach lengths of ~10 μm, roughly the length scale of the cell, making imaging and subsequent analysis of microtubule dynamics relatively straightforward. In contrast, during mitosis, the spindle pole bodies organize the bipolar spindle. Each spindle pole body initially nucleates about ~25 individual microtubules (Ding, McDonald, & McIntosh, 1993), which are then cross-linked by the kinesin-5 cut7p (Hagan & Yanagida, 1992), and slide apart to elongate the spindle to a steady-state metaphase length of ~3 μm (Nabeshima et al., 1998). During the early stages of mitosis and spindle formation, individual microtubules cannot be imaged, due to the fact that they are relatively short and numerous, and the spindle poles and the spindle widths are diffraction-limited structures, that is, less than 300 nm (Ding et al., 1993). Nevertheless, given that microtubule dynamics are known to dramatically change between interphase and mitosis in animal cells (Rusan, Fagerstrom, Yvon, & Wadsworth, 2001), it would be useful to measure cell cycle differences in microtubule dynamics in fission yeast, especially in the context of mutations which affect microtubule dynamics.

Here, we present a simple method for imaging individual spindle microtubules during early mitosis. The method takes advantage of the thermosensitive mutant cut7.24^{ts} (Hagan & Yanagida, 1992). cut7p is a kinesin-5 and functions in bipolar spindle formation by cross-linking and sliding apart antiparallel microtubules emanating from opposite spindle poles (Hagan & Yanagida, 1992). As a cross-linker and slider, cut7p does not alter microtubule dynamics per se. In addition, the cut7.24^{ts} mutation has been shown to be quickly reversible between the permissive (25 °C) and nonpermissive (37 °C) temperatures, without any noticeable effects on subsequent spindle dynamics (Velve Casquillas et al., 2011). We thus reasoned that by inactivating cut7p via using the cut7.24^{ts} mutant at 37 °C, we would create monopolar spindles. These monopolar spindles could not separate their spindle poles, and their respective microtubules would appear as asters with protruding microtubules. Some of the protruding microtubules, due to their increasing radial spatial separation as they elongate, would appear as discrete individual microtubules for imaging and analysis. Thus, the cut7.24^{ts} monopolar spindle would serve as the "control" background for analyzing individual spindle microtubule dynamics. A second mutation can then be introduced into this background for comparative studies of proteins which may regulate mitotic microtubules.

We detail here the step-by-step protocol for imaging individual spindle microtubule dynamics in fission yeast.

24.1 METHODS

24.1.1 PDMS slide spacer

To prevent drying from long-term culturing and imaging at high temperature, we use a relatively thick (~1 mm) pad of media infused agarose to position the fission yeast. To have such a thick pad, a simple square well is created from the elastomer PDMS (polydimethylsiloxane). To make the PDMS spacer:

1. Pour liquid PDMS into a cup and then add curing agent for a final 9:1 ratio.
2. Stir vigorously with a plastic fork for about 3 min to homogenize the mixture. The stirring will generate small air bubbles, turning the clear mixture white.
3. Put the cup in a vacuum desiccator for 30 min to remove all air bubbles. The mixture should be again transparent at the end.
4. Pour the mixture gently onto a large Petri dish to a height of ~1 mm. Wait a few minutes to allow any new air bubbles to reach the surface and gently blow them away.
5. Put the dish into an oven at 65 °C for 2–4 h to cure and harden the mixture.
6. After the PDMS mixture is cured or hardened, place a glass specimen slide on top and using a surgical knife cut around the slide to get a slice of the same exact size (25 × 75 mm). This piece should look like a slice of transparent and flexible soft plastic.
7. Cut out a square (~10 × 10 mm) at the center of the PDMS slice. This is now ready for usage.
8. The spacer is reusable after washing with warm soap and water.

Notes

- On step 1, the ratio of PDMS to curing agent affects the stiffness of the final PDMS product. More curing agent leads to stiffer PDMS products. We find that 9:1 ratio is optimal for our work, but small changes (7:1–12:1) will not be critical for this application.
- On step 5, variation from the recommended 2–4 h curing time is not critical for our application. However, too long a curing time could lead to PDMS aging and brittleness. In this step is important to make sure that the

Petri dish is on a flat and even surface, so the slice of final PDMS is flat and even everywhere.

- The PDMS spacers are reusable and should be good for a couple years.
- When handling liquid PDMS wear gloves because it is greasy and hard to wash out; but once cured, no gloves are needed.

Materials

Oven, Fisher Scientific* Isotemp* (13-247-637G, 3.75 cu. ft./106.2L).
PDMS, Sylgard, 184 SILICONE ELASTOMER KIT (Dow Corning).
Plastic cup/plastic fork.
Vacuum desiccator (Bel-Art Products, 08-594-15B, 260 mm diameter).
145 × 20 mm Petri dishes (Thermo Scientific, No. 240401).

24.1.2 Slide assembly

Figure 24.1A shows a schematic of the final assembled slide. The PDMS slice is strongly adherent to glass, and therefore, no glue or sealing agent is needed during the assembly process:

1. Prepare melted fission yeast media infused agarose, for example, YE5S 2% agarose at 100 °C.
2. Wash the PDMS slice with 70% ethanol. (The PDMS slices are reusable, wash before and after experimental use.)
3. Dry the PDMS slice on KimWipes or paper towel.
4. Place the PDMS slice on top of a glass slide. PDMS will naturally adhere to the glass.
5. Fill up the square well completely with the melted YE5S agarose, until it forms a slight convex cap. While the agarose is still warm, immediately place a glass coverslip on top.
6. Gently, press down on the coverslip around its edges (never press down on top of the well itself). This act to flatten out the agarose surface.
7. Wait for ~1 min for agarose to solidify. Then carefully remove the glass coverslip from the well. (Excess agarose outside the well should also be removed by scrape away.)
8. Wait for ~1 min further for the opened agarose well to dry.
9. During this time, concentrate cells by centrifugation and then resuspend cell in ~20 μL.
10. Apply 1 μL of cell suspension on top of the agarose well.
11. Place a glass coverslip on top gently at a small angle to make sure cells are spread as evenly as possible on the agarose surface. The coverslip will adhere to the PDMS spacer.

Notes

- On steps 5 and 11, be careful to not create air bubbles. Bubbles in the agarose or in the cells suspension will cause unwanted drifts or movements during imaging.

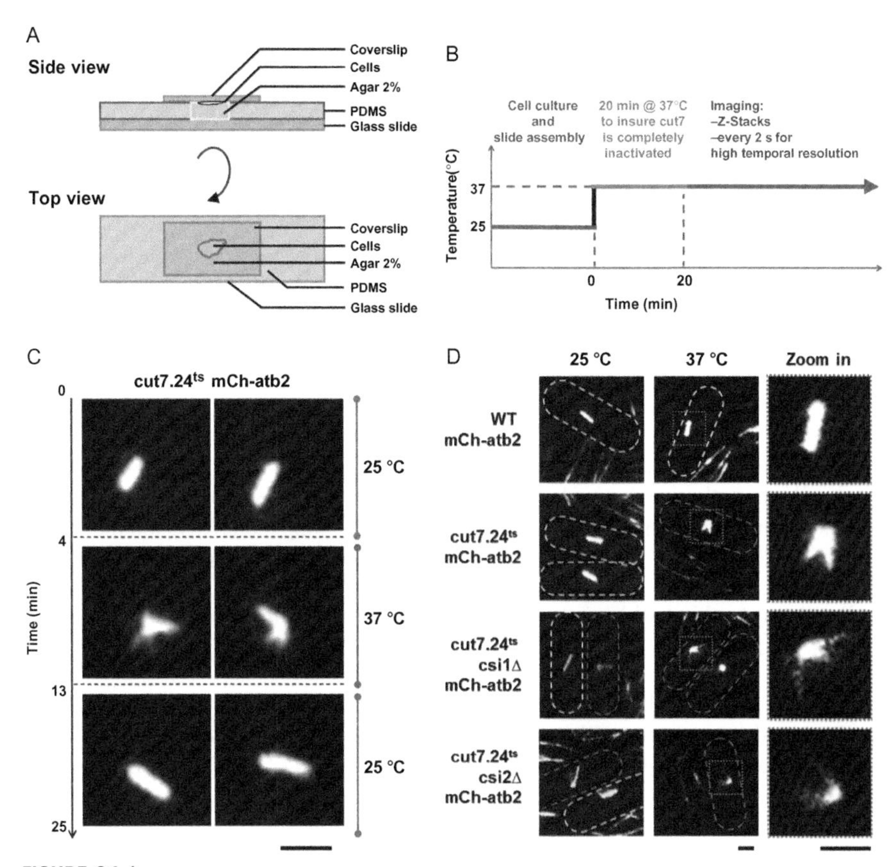

FIGURE 24.1

A method to image individual mitotic spindle microtubules in fission yeast. (A) Schematic view of the specimen slide assembly. (B) Schematic view of the temperature shift experiment. Cell culture and slide assembly are performed at ambient room temperature (25 °C) prior to shifting for 20 min to the nonpermissive temperature (37 °C) and then transferred to the microscope set at 37 °C for imaging. (C) Control cut7.24^{ts} cells expressing mCherry-atb2 (tubulin) under permissive, nonpermissive, and then permissive temperatures. Z-stacks (4 sections, 1 μm spacing) are captured at 1-min intervals. Shown are maximum projection time-lapse images. cut7.24^{ts} cells can produce reversible monopolar spindles. Bar: 2 μm. (D) Comparisons of spindle morphology between the wild-type and control cu7.24^{ts}, and the mutants cut7.24^{ts}:csi1Δ and cut7.24^{ts}:csi2Δ expressing mCherry-atb2 at permissive and nonpermissive temperatures. Z-stacks (4 sections, 0.5 μm spacing) are captured at one time point. Shown are maximum projection time-lapse images. Dashed yellow outlines show cells with normal bipolar spindles. Dashed red outlines show cells with monopolar spindles. Zoom in shows 5× magnification of the different spindles. Bar: 2 μm. (For interpretation of the references to color in this figure legend, the reader is referred to the online version of this chapter.)

- The cells used in step 9 should be previously cultured in liquid medium to mid-log phase.
- The assembled slide will hold cells in good condition for more than 5 h of imaging.

Materials

Glass slide 25 × 75 × 1.0 mm (Fisher Brand 12-550A3).
Coverslip 24 × 40–1.5 mm (Fisher Brand 12-544-C).

24.1.3 Live-cell imaging

As proof of concept, we have imaged three fission yeast strains expressing mCherry-atb2 (tubulin): cut7.24^{ts} (control), cut7.24^{ts}:csi1Δ, and cut7.24^{ts}:csi2Δ. csi1 and csi2 are novel fission yeast genes involved in bipolar spindle organization and chromosome segregation (data to be published elsewhere). Deletion of either csi1 or csi2 (csi1Δ or csi2Δ) results in transient monopolar spindles and subsequent chromosome segregation defects. We hypothesize that csi1 and/or csi2 regulates mitotic spindle microtubule dynamics because csi1Δ and csi2Δ also exhibit abnormally long metaphase spindles compared to wild-type cells. To test this hypothesis, we image the strains as follows:

1. Put the assembled slide containing cells into an incubator of 37 °C for 20 min (Fig. 24.1B).

At this time, cells which have begun to transit into mitosis, or cells which are still in prophase and metaphase, will become monopolar due to the inactivation of cut7p (Hagan & Yanagida, 1992). Cells which have already started anaphase will continue through the cell cycle (Fu et al., 2009). As incubation at 37 °C continues, more cells will enter mitosis, resulting in more monopolar spindles.

2. The microscope should have fluorescence imaging capability and, importantly, have a temperature box which can maintain a stable 37 °C throughout the imaging duration.

We typically use a Yokowaga spinning disk confocal microscope equipped with a cooled CCD- or EMCCD-camera for imaging live cells (Tran, Paoletti, & Chang, 2004). Our microscope is enclosed in a temperature box which stably tunes the temperature from ambient (22 °C) to 40 °C.

3. Quickly transfer the slide from the incubator onto the microscope which has been preset at 37 °C for live-cell imaging.

We image the cells according to the spatiotemporal requirement of the biological process being studied. For example, individual fission yeast interphase microtubules have an average life of ~3 min (Drummond & Cross, 2000; Tran et al., 2001). For this timescale, 5-s interval sampling frequency was sufficient to capture microtubule dynamics with high spatiotemporal resolution. In contrast, mitotic microtubules are expected to be more dynamic. We found empirically that individual fission yeast mitotic microtubules have an average life time of less than 1 min. This short timescale requires at least 2-s interval sampling frequency to sufficiently capture mitotic microtubule dynamics.

Notes

- For step 1, the slide can be placed directly onto the microscope stage set at 37 °C for 20 min prior to imaging. This would negate the need for an incubator.
- During the transfer of the slide from the incubator to the microscope, it is critical that the transfer time is less than 1 min. cut7.24^{ts} is a fast acting mutant (Velve Casquillas et al., 2011). Within ~1 min at the nonpermissive temperature (37 °C), cut7p is inactive; and within ~1 min at the permissive temperature (25 °C), cut7p is active again. To ensure sustained monopolar spindles, we recommend that the transfer time is completed under 30 s.
- Ultimately, the practical achievable spatiotemporal resolutions of imaging individual spindle microtubules will depend on many factors: (1) the quality and speed of the optical pathway, (2) the sensitivity of the detector of the microscope, (3) the exposure time needed to acquire a good fluorescent signal, (4) the bleaching of the fluorescent signal, etc. Compromises will be necessary to achieve optimal results.

Materials

37 °C incubator.

Optical microscope capable of fluorescent imaging and has a temperature control box.

24.1.4 Data analysis

Figure 24.1C shows two examples of the "control" cut7.24^{ts} strain expressing mCherry-atb2 (tubulin). cut7.24^{ts} is reversible, able to transform from a short bipolar spindle (represented by the "bar" of fluorescent signal) to a monopolar spindle (represented by the "aster" of microtubule protrusions) at the nonpermissive temperature, and then back to a bipolar spindle when the temperature is again permissive. This is an important control to show that the inactivation of cut7p is reversible and does not permanently alter spindle dynamics and organization, strongly implying that individual microtubule dynamics is not affected by the inactivation of cut7p. Therefore, cut7.24^{ts} is a good tool to study individual spindle microtubule dynamics.

Figure 24.1D shows comparisons of spindle morphology between the wild-type and control cut7.24^{ts}, and the mutants cut7.24^{ts}:csi1Δ and cut7.24^{ts}:csi2Δ at permissive and nonpermissive temperatures. The wild-type spindles remain bipolar at both temperatures. The control cut7.24^{ts} bipolar spindles become monopolar at the nonpermissive temperature. Both mutants cut7.24^{ts}:csi1Δ and cut7.24^{ts}:csi2Δ have transient monopolar spindle phenotype even at the permissive temperature as part of the

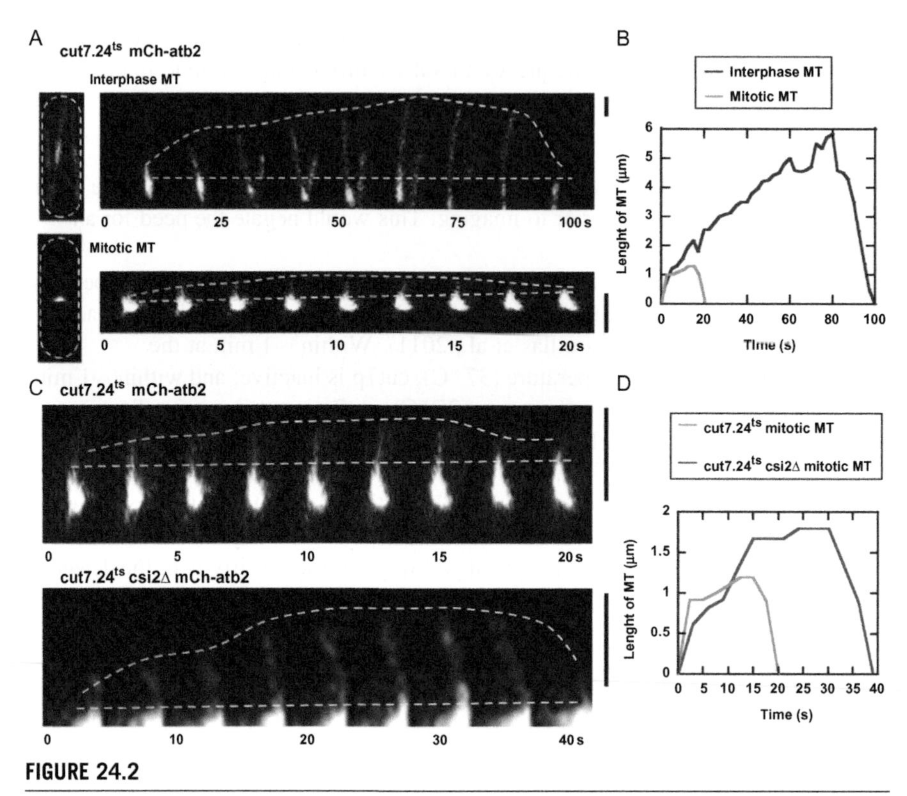

FIGURE 24.2

High-temporal resolution imaging of interphase and mitotic microtubules. (A) Control cut7.24^{ts}cells expressing mCherry-atb2 were imaged at 37 °C. Z-stacks (4 sections, 0.5 μm spacing) are captured at 2-s intervals. Shown are maximum projection time-lapse images of an interphase and mitotic microtubule. Bar: 2 μm. (B) Length versus time plot of the interphase and mitotic microtubule shown in (A). (C) Shown are maximum projection time-lapse images of a mitotic microtubule from the control cut7.24^{ts} cell, and the mutant cut7.24^{ts}:csi2Δ cell expressing mCherry-atb2 imaged at 37 °C. Bar: 2 μm. (D) Length versus time plot of the mitotic microtubules shown in (C). (For color version of this figure, the reader is referred to the online version of this chapter.)

mutation. Nevertheless, at the nonpermissive temperature, both mutants have monopolar spindles. Qualitatively, the monopolar spindles of the mutants exhibit seemingly longer microtubule protrusions than the control. This suggests that csi1p and csi2p may regulate spindle microtubule dynamics.

Figure 24.2A shows comparative length differences between an individual interphase microtubule and an individual mitotic microtubule of the control cut7.24^{ts} cell. The individual interphase microtubule, due to its long length and few adjacent confounding neighbor microtubules, is relatively easy to track. In contrast, the individual mitotic microtubule, due to its short length and numerous adjacent confounding neighbor microtubules, is relatively difficult to track. Indeed, most of the mitotic microtubules cannot be tracked with precision. However, infrequently, there are individual mitotic microtubules which grow longer and away from the dense monopolar spindle core. These longer mitotic microtubules can be tracked with precision. This does imply that individual mitotic measurements will tend to bias toward the longer microtubules. Nevertheless, useful measurements can still be made from individual mitotic microtubule dynamics as a comparison to other mutant mitotic microtubules or to interphase microtubules.

The microtubule length versus time plot provides information to derive all parameters of microtubule dynamics, including velocities of growth and shrinkage, and frequencies of catastrophe and rescue (Walker et al., 1988). Statistical significance and inference will require many measurements. Here, we choose to present one example to make a point. Figure 24.2B highlights the contrast in length and life time of an interphase and mitotic microtubule. Qualitatively, the velocities of growth and shrinkage (i.e., similar growth and shrinkage slopes) are similar between interphase and mitotic microtubules, and their time to catastrophe (i.e., inverse of catastrophe frequency) is very different. Figure 24.2C shows comparative length differences between an individual mitotic microtubule of the control cut7.24^{ts} and an individual mitotic microtubule of the mutant cut7.24^{ts}:csi2Δ. Qualitatively, the mutant mitotic microtubule appears longer. Figure 24.2D shows that the velocities of growth and shrinkage are similar between mitotic microtubules of the control and mutant cell. However, the mutant mitotic microtubule grows a longer time before catastrophe. This suggests that csi2p regulates mitotic microtubule dynamics. Specifically, csi2p positively regulates microtubule frequency of catastrophe.

CONCLUSION

We presented a simple method to image and analyze individual spindle microtubule dynamics in the fission yeast. This method overcomes the spatial limitation of having a dense microtubule structure within the diffraction-limited spindle width and spindle poles. In principle, this method could be applied generally to other yeasts and fungi having similar spindle structures, provided that they have a thermosensitive kinesin-5 as the motor controlling spindle bipolarity.

Acknowledgments

We thank Dr. Andrea Stout of the Penn CDB Imaging Core for technical support. J. C. is supported by a predoctoral fellowship from the FCT-Portugal. This work is supported by grants from the NIH and ANR.

References

Ding, R., McDonald, K. L., & McIntosh, J. R. (1993). Three-dimensional reconstruction and analysis of mitotic spindles from the yeast, Schizosaccharomyces pombe. *The Journal of Cell Biology*, *120*, 141–151.

Drummond, D. R., & Cross, R. A. (2000). Dynamics of interphase microtubules in Schizosaccharomyces pombe. *Current Biology*, *10*, 766–775.

Fu, C., Ward, J. J., Loiodice, I., Velve-Casquillas, G., Nedelec, F. J., & Tran, P. T. (2009). Phospho-regulated interaction between kinesin-6 Klp9p and microtubule bundler Ase1p promotes spindle elongation. *Developmental Cell*, *17*, 257–267.

Hagan, I., & Yanagida, M. (1992). Kinesin-related cut7 protein associates with mitotic and meiotic spindles in fission yeast. *Nature*, *356*, 74–76.

Hoog, J. L., Schwartz, C., Noon, A. T., O'Toole, E. T., Mastronarde, D. N., McIntosh, J. R., et al. (2007). Organization of interphase microtubules in fission yeast analyzed by electron tomography. *Developmental Cell*, *12*, 349–361.

Nabeshima, K., Nakagawa, T., Straight, A. F., Murray, A., Chikashige, Y., Yamashita, Y. M., et al. (1998). Dynamics of centromeres during metaphase-anaphase transition in fission yeast: Dis1 is implicated in force balance in metaphase bipolar spindle. *Molecular Biology of the Cell*, *9*, 3211–3225.

Rusan, N. M., Fagerstrom, C. J., Yvon, A. M., & Wadsworth, P. (2001). Cell cycle-dependent changes in microtubule dynamics in living cells expressing green fluorescent protein-alpha tubulin. *Molecular Biology of the Cell*, *12*, 971–980.

Sawin, K. E., & Tran, P. T. (2006). Cytoplasmic microtubule organization in fission yeast. *Yeast*, *23*, 1001–1014.

Snaith, H. A., Anders, A., Samejima, I., & Sawin, K. E. (2010). New and old reagents for fluorescent protein tagging of microtubules in fission yeast; experimental and critical evaluation. *Methods in Cell Biology*, *97*, 147–172.

Tran, P. T., Marsh, L., Doye, V., Inoue, S., & Chang, F. (2001). A mechanism for nuclear positioning in fission yeast based on microtubule pushing. *The Journal of Cell Biology*, *153*, 397–411.

Tran, P. T., Paoletti, A., & Chang, F. (2004). Imaging green fluorescent protein fusions in living fission yeast cells. *Methods*, *33*, 220–225.

Velve Casquillas, G., Fu, C., Le Berre, M., Cramer, J., Meance, S., Plecis, A., et al. (2011). Fast microfluidic temperature control for high resolution live cell imaging. *Lab on a Chip*, *11*, 484–489.

Walker, R. A., O'Brien, E. T., Pryer, N. K., Soboeiro, M. F., Voter, W. A., Erickson, H. P., et al. (1988). Dynamic instability of individual microtubules analyzed by video light microscopy: rate constants and transition frequencies. *The Journal of Cell Biology*, *107*, 1437–1448.

Index

Note: Page numbers followed by *f* indicate figures and *t* indicate tables.

A

Acetylation, 248–249, 250, 255–256
Agrobacterium tumefaciens
  acetosyringone, 237
  description, 234–235
  E. coli, 235
  sucrose and surfactant Silwet L-77, 236–237
  suspension, plant transformation, 242
  T-DNA (transfer DNA), 235
Anion exchange chromatography, 163
Arabidopsis thaliana
  A. tumefaciens (*see Agrobacterium tumefaciens*)
  building transgenic (*see* Transgenes, *A. thaliana*)
  description, 232, 234
  gamma-tubulin nucleation complexes, 233–234
  imaging
    microscope and camera/PMT settings, 241
    microtubules and MAPs, signal intensity, 241
    photobleaching and plant health, 241
    TIRF microscopy, 240–241
  interphase array organization and cell growth, 232–233, 233*f*
  kinesins, katanins, MAP65's, EB1's, XMAP215 and CLASP, 234
  plant products, 232
  sample preparation
    cell divisions, 239
    cover glass spacers, 239–240, 240*f*
    seedlings, 240
    tap water, 240
  sterile media, plant growth, 242
  types, microtubule arrays, 232–233
Axoplasm
  axons, 129–130
  biochemistry
    microtubules, 133–134
    perfusion, 134, 134*f*
    pharmacology, 135
  buffer X, squid, 130–131, 132*t*
  extrusion, 129–130
  giant axons, 127–128, 128*f*
  in situ microtubule properties, 130
  immunohistochemistry, 135–136
  KEP method, 132
  mantle, 126–127
  microtubule dynamics, 131–133
  nerve bundle, 129
  neuronal, 126
  nonprotein composition, 130, 131*t*
  polymerization, 125–126
  preparation, 126, 127*f*
  stereo dissection microscope, 129
  suitable squid, 126

B

Bacterial tubulin BtubA/B
  anion exchange chromatography, 273
  assembly
    M loop, β-tubulin, 277–278
    polymerization, sedimentation, 277, 278*f*
    requirements, 276–277
  assembly/disassembly cycle, 274
  btubA and *btubB* genes, 272–273
  BtubA and BtubB purification, 275–276, 275*f*
  cells breaking, 273
  chimera containing eukaryotic sequences, 276, 277*f*
  cryo-electron tomography, 272
  description, 280
  Escherichia coli, 271–272
  GTP and magnesium, 271–272
  protein concentration measurement, 274
  protofilaments, 272
  RMSD, 270
  size exclusion chromatography, 274
  structural alignment, 270, 271*f*
  tubulin-like proteins, 270
Benomyl plate assay, 366–367
Binding analysis, SPR
  bi-molecular interactions, 177
  Gsα and tubulin, 177–178
  tubulin immobilization, 177
Binding models, ITC
  direct and reverse titration, 287–289, 287*f*
  enthalpy and entropy, 289
  "equal noninteracting sites" model, 287–289
  one-set-of-sites, 287–289
  stathmin and tubulin, 287–289, 287*f*
  two-sets-of-sites and sequential-binding-sites models, 287–289
Bioorthogonal ligation reaction, 3
Biophysical assays, microtubule-severing enzymes *in vitro*
  binding and diffusion
    experiment, 207–208
    SpotTracker plug-in, 208, 209*t*
    squares, displacements, 208

Biophysical assays, microtubule-severing enzymes *in vitro (Continued)*
buffers, 200
depolymerization rate analysis, 206*f*, 207
epifluorescence and single molecule TIRF imaging methods, 204–205
equipment, 200
experimental chamber
buffers, 202
procedures, 202–203, 203*f*
reagents, 202
GMPCPP (*see* Guanalyl-(α,β)-methylene diphosphate (GMPCPP))
photobleaching, 208–211
polarity-marked microtubules, 201
protofilaments, 202
reagents, 200
severing frequency analysis, 207
taxol-stabilized microtubules, 200–201
two-color colocalization, 205–206, 206*f*
Buffer composition, tubulin PTMs
cell fixation
DSP-PFA method, 261–262
PFA method, 261
SDS-PAGE, α/β-tubulin separation, 262–263

C

Calibrations
description, 85, 86*f*
velocities, 85
viscosity, 85
Carboxy-terminal tail (CTT), 194
cDNA clones and vectors, 159–161
Cell line
CHO, 52, 52*f*
fluorescence intensity, 51–52
Cell morphology assay, 187
Cellular functional assays, 183, 184*f*
Centrosome-based microtubule assembly
cell plating, culture and treatment, 145
microtubule growth, 145–146
permeabilized cells, 145
Chinese hamster ovary (CHO)
proliferation, 54
and vascular endothelial cells, 51–52
CHO. *See* Chinese hamster ovary (CHO)
CLEM. *See* Correlative light-electron microscopy (CLEM)
Correlative light-electron microscopy (CLEM)
EM, 328
high-pressure freezing, 340
mitotic spindle, 328–329
protocol, 340
Coverslips and slides
chemical cleaning, 111
in vitro experiments, 111
sonication, 111
Creep analysis, MT networks, 91–92
Critical concentration, tubulin polymerization
determination, 219
spectrophotometer, 218–219
steady-state OD, 218–219
total tubulin concentration, 217
trimethylamine oxide (TMAO), 217–218, 218*f*
Cross-linking, MT networks, 79–80, 86–87
Cryo-electron microscopy (cryo-EM)
3D structure determination, 39–43
MT-binding site, 31, 32*f*
Cryo-electron tomography (cryo-ET)
B-lattice 13-pf MTs, 39
MT architectures, 37–38
CTT. *See* Carboxy-terminal tail (CTT)

D

DAPI. *See* Diamidino-phenylindole (DAPI)
DCX. *See* Doublecortin (DCX)
Detachment, MTs
apoptosis, 57–59
cell line, 51–52
centrosomes, 51
cytoplasmic microtubule network, 50
fragments, 57
GFP–MAP4 expression, 54
GFP-tubulin expression, 54
interphase, 50
MCAK, 59
measurement
cell preparation, 55
centrosome region, 56–57, 56*f*
discernable centrosome, ventral surface, 55–56
equipment, 54–55
mechanisms, 57–59, 58*f*
mitosis, 59
paclitaxel-dependent mutants, 51
rhodamine tubulin microinjection, 53
spindles, 50–51, 52–53
Detyrosination, 248–249, 250, 256–258
Diamidino-phenylindole (DAPI), 216, 217, 223, 223*f*
Dithiobis succinimidyl propionate (DSP), 249–252, 251*f*, 254*f*, 259–260, 261–262
Doublecortin (DCX)
biochemical stoichiometry, 31–32

cryo-EM (*see* Cryo-electron microscopy (cryo-EM))
cryo-ET (*see* Cryo-electron tomography (cryo-ET))
DCLK, 29–30, 29*f*
discovery, 28–29
double cortex, 28–29
ensemble gliding assay, 31
low-resolution reconstruction, 31
methods
3D structure determination, 39–43
in *E. coli*, 32–33
expression, 33
MT architectures, 37–38
purification, 33–37
in *Spodoptera frugiperda*, 32–33
in MT regulation, 29–30, 29*f*
NMR study, 30
13-protofilament (13-pf) architecture, 31
sequence analysis, 30
tubulin-bound nucleotide, sensitivity, 30
X chromosome, 28–29
Doublecortin-like kinase (DCLK), 29–30, 29*f*
DSP. *See* Dithiobis succinimidyl propionate (DSP)
3D structure determination, DCX
GTPγS, 43
kinesin-decorated 13-pf B-lattice MT, 39
M, N and H2S3 loops, 42–43
paclitaxel-stabilized tubulin assemblies, 42
pseudo-atomic model, 40–42
reconstruction, DCX–K–MTs, 40, 41*f*
sample preparation and data collection, 39
seam and image selection, 39–40
setup and automated processing, 39
single-particle image processing, 39
two-dimensional projections, 39

E

ECL. *See* Enhanced chemiluminescence (ECL)
Electron microscopy (EM)
electron crystallography, 344
microtubule type, 346
EM. *See* Electron microscopy (EM)
Enhanced chemiluminescence (ECL), 179
Epifluorescence and single molecule TIRF imaging methods
buffers, 204
equipment, 204
experimental procedures, 204–205
reagents, 204
Expression, DCX
366-amino-acid isoform, 33
baculovirus genome, 33
description, 33, 34*f*
large-scale protein expression, 33

F

Fast force switching, MT networks
description, 87
high-resolution camera, 87–88
mechanical stability, 87
tracking accuracy, 87
Fission yeast *(Schizosaccharomyces pombe)*
cut7.24^{ts} mutation, 386–387
data analysis
"control" cut7.24^{ts}, 389*f*, 391
cu7.24^{ts}:csi1Δ and cu7.24^{ts}:csi2Δ, 389*f*, 391–392
interphase *vs.* mitotic microtubules, 392–393, 392*f*
microtubule length *vs.* time plot, 392*f*, 393
velocities of growth and shrinkage, 392*f*, 393
interphase microtubules, 386
live-cell imaging, 390–391
mitosis and spindle formation, 386
PDMS slide spacer, 387–388
slide assembly, 388–390, 389*f*
Fixation and sample preparation, kinetochore-fiber ultrastructure
fixative solution osmolarity, 331
LM, 332–333
resin embedding, 333
Flexural rigidity, 14
Fluorescence
DAPI, 217
drawback, 217
multiwell plates, 227
sensitivity, optical density, 217
Fluorescence-based assays
B-lattice bond, 345–346
controlled 13-pf, controlled 14-pf microtubules, 348–349
DCX-GFP, purification, 351
electron density maps, pf numbers, 344, 345*f*
EM, 352
interprotofilament bonds, 344–345
kinesin-1, 346
materials, 347–348
microtubule architecture, 344, 345*f*
microtubules, internal 14-pf control, 349–351
13-pf microtubules, 346
protein, 346–347
single-molecule fluorescence assay, 348
TIRF, 344
tubulin and microtubule preparations, 348

Fluorescent labeling, tubulin
  assembly activity, 9–10
  bioorthogonal ligation reaction, 3
  carboxy terminus, α-tubulin, 2
  coupling reactions, 3–4
  C-terminus of α-tubulin
    BH stock preparation, 8
    biotinylation, 8
    coupling reaction, 7
    ligand preparation, 6–7
  derivatives, 3
  detection, modification in C-terminus
    biotinylated tubulin, 9
    fluorophore-labeled tubulin, 9
    tyrosine/3-fomyltyrosine, 8
  detection of
    buffers, 5
    monoclonal anti-tubulin, 5
    PBST, 5
    sample buffer, 5
    SDS-PAGE gel, 5
    streptavidin-HRP, 5
  detyrosination, 6
  fluorescent labeling
    catalyst stock, 7
    fluorophore-labeled tubulin, 8
    fluorophore stock, 7–8
  lysine-reactive probes, 2
  materials, 4–5
  polymerization activity, detection of
    DMSO, 5
    taxol, 5
  protein isolation and purification
    TTL, 6
    tubulin, 5
  recombinant human TTL, 3
  TLL expression and purification, 4
  TTL (*see* Tubulin tyrosine ligase (TTL))
  two-part procedure, 2
Flutax displacement assay
  description, 310
  docetaxel, ZMP and epothilone A, 310–312, 313*f*
  equilibrium constants, Flutax-2, 310, 310*t*
  temperature, binding constants, 310–312, 311*t*

G

GAB buffer. *See* Glycerol–EGTA (GAB) buffer
Gel filtration chromatography, 163
Gliding assay, microtubule
  experiments, 19–20
  kinesin-1, 16, 16*f*, 17
  microscope and EM-CCD camera, 20
  microtubule trajectories and persistence length, 21*f*, 22
  steps, protocol, 19–20
  stock solutions, 19
Glycerol–EGTA (GAB) buffer, 306, 308–309, 312, 316, 319
GMPCPP. *See* Guanalyl-(α,β)-methylene diphosphate (GMPCPP)
Gsα interaction, tubulin/microtubules
  binding studies, cells
    cellular consequences, 183–185, 186*f*
    immunocytochemistry, 185
    immunofluorescence techniques, 183
    immunoprecipitation, 183–185
  cellular functional assays, 183, 184*f*
  microtubule dynamics assay, cells, 185–187
GTP-bound tubulin
  biochemical studies, 142
  cell fixation and colabeling, 144
  in vitro techniques, 147–150
  imaging, centrosome-based microtubule, 145–146
  labeling operations, 142–143, 143*f*
  MAPs, 141–142
  MB11 antibody, 142
  microtubules, 141
  permeabilization, 143–144
Guanalyl-(α,β)-methylene diphosphate (GMPCPP)
  description, 200
  microtubules, 201
  polarity-marked microtubules, 201–202
  protofilament defect microtubules, 202

H

Haploid spores
  genotype analysis, 365
  microdissection, 364–365
Heterotrimeric G proteins and microtubules
  binding analysis, SPR, 177–178
  buffer compositions, 187–188
  description, 174
  Gsα, 176
  in vitro binding assays, 174–175, 175*f*
  microtubule dynamics assay, video microscopy, 181–183
  microtubule polymerization assay, 181
  peptide array membrane analysis, 178–179
  protein purification, 175–176
  pull-down assays, 177
  tubulin, 176
  tubulin–Gsα interaction, 179–181

High-performance liquid chromatography (HPLC), 312–314, 317–319, 317*f*
HPLC. *See* High-performance liquid chromatography (HPLC)
Human EB1 bind, MT seam/lattice
  MTBindingSim, 382–383, 382*f*
  proteins, 381–382

I

Image reconstruction, DCX
  cryo-EM, 31, 32*f*, 39–43
  cryo-ET, 37–38, 39
  low-resolution, 31, 32*f*
  subnanometer resolution, 37
Imaging
  individual spindle microtubule dynamics, fission yeast (*see* Fission yeast (*Schizosaccharomyces pombe*))
  microtubules and MAPs, *A. thaliana* (*see Arabidopsis thaliana*)
  MT networks
    autocorrelation, Fourier, 86–87
    confocal microscopy, 80–81, 81*f*, 82–83
    portable magnetic tweezers, 90
    time-lapsed confocal fluorescence, 81
In vitro microtubule (MT) polymerization, measurement
  buffer, Na glutamate, 223
  critical concentration, 217–219
  description, 216
  dimer and polymer size, 216
  drug effects–polymerization inhibitors, 223, 223*f*
  equipment, 224
  fluorescence (*see* Fluorescence)
  instrument, spectrophotometer, 224
  materials, 225
  MTP, 224
  polymerization curves and "turbidity coefficient"
    buffer effects, critical concentration, 221*t*, 222
    description, 219–221, 220*f*
    parameters, 219
    steady-state OD and polymer mass, 219–221
    temperature, 222
    wavelengths, OD, 222
  preparation, 225
  promoters, 222
  turbidity (*see* Turbidity, tubulin polymerization)
In vitro sample chambers, TIRF microscope
  coverslip, 113
  description, 112
  pipette, 113
In vitro techniques
  description, 147
  fluorescent-tubulin labeling, 148
  GMPCPP and GTP microtubules, 147–148
  incubation chambers, 149
  MAP-free tubulin preparation, 148
  microtubule seeds and elongation, 149–150
  recombinant KHC reconstitution, 148–149
  taxol-stabilized microtubule seed polymerization, 149
Isothermal titration calorimetry (ITC)
  binding models, 287–289
  experimental procedure, 286
  microtubules, 284–285
  MTAs, 284–285
  principles, 286
  temperature dependence, ΔH, 289–290
  thermodynamic profile, binding, 289
  tubulin and MAPS sample preparation (*see* Tubulin, and MAPS)
ITC. *See* Isothermal titration calorimetry (ITC)

K

KEP. *See* Kinetic equilibration paradigm (KEP)
KHC. *See* Kinesin heavy chain (KHC)
Kinesin
  biotinylation, 17
  density of, 16, 22
  gliding assay (*see* Gliding assay, microtubule)
Kinesin heavy chain (KHC), 148–149
Kinetic equilibration paradigm (KEP), 132
Kinetochore-fiber ultrastructure
  cell transfection and observation, 331
  CLEM, 328–329, 338–339
  EM, 328
  fixation and sample preparation, 331–333
  high-pressure freezing, 340
  imaging and sample tilting, 336–338
  LM, 328
  longitudinal sectioning, 333–336
  materials, 329–331
  mitotic cells, CLEM, 329, 330*f*
  mitotic spindle, 328
  orthogonal sectioning, 336

L

Light microscopy (LM)
  cells relocation, 332
  spindle axis, 332–333

Lissencephaly
DCX and DCLK, 28–30
point mutations, 30
symptoms, 28–29
LM. *See* Light microscopy (LM)
Longitudinal sectioning
CLEM processing, 333–335, 334*f*
EM method, 333
resin, 333–335
trimming and sectioning, resin, 335

M

Magnetic tweezers devices
calibrations methods, 85
custom-built, 77
description, 83–84, 84*f*
fast force switching, 87–88
high-force implementation, 88
magnetic field, 84–85
mechanical measurements, 83–84
MT contour length, 83–84, 83*f*
nano-Newton forces, 78
NdFeB (*see* Neodymium iron boron (NdFeB) magnets)
particle tracking, 86–87
portable magnetic tweezers, 90
ring-shaped NdFeB magnet, 89–90
sample plane and magnet range, 85
MAPs. *See* Microtubule-associated proteins (MAPs)
Mass spectrometry (MS)
description, 318
ion-filtering experiments, 319
PIS and SRM, 318
protein preparation and digestion
assembled microtubules, 319–320
compounds, 320, 321*f*
cross-linked stabilized MTs, 320
peptides, 318–319
unassembled tubulin, 320
MB11 staining, 144
MCAK. *See* Mitotic centromere-associated kinesin (MCAK)
Mechanics, MT networks
creep analysis, 91–92
oscillatory measurements, 93
Micropatterned coverslips
cell shape and size, 100–101, 100*f*
manufacturing, 100–101
Microrheology methods
active methods, 77–78
fast force switching, 87
particle tracking, 86–87
Microscopy
and microtubule dynamics analysis, 368–370
samples preparation, 368
Microtubule-associated proteins (MAPs)
DCX (*see* Doublecortin (DCX))
in vivo
budding yeast, 370
microtubule-associated proteins, 371–372
yeast strains, 370–371
YFP-fusion proteins, 371
removal, tubulin, 176
and tubulin, ITC (*see* Tubulin, and MAPS)
Microtubule protein (MTP), 224, 225
Microtubules (MTs)
active measurements, 14–15
advantage, technique, 22
architectures, DCX
B-lattice and A-lattice, 37
data collection, 38
3D reconstruction, 37
in vitro and *in vivo*, 37
sample preparation, 37–38
structure determination, 38, 38*f*
assembly
glycerol buffer, 308–309
PEDTA buffer, 309–310
transmission electron microscopy, 310
tubulin polymerization, 307–308
detachment (*see* Detachment, MTs)
dynamics assay
cell morphology assay, 187
cytoskeletons, 185
detergent-insoluble microtubules, 185–187
video microscopy, 181–183
dynamics *in vivo*
GFP–Tub1 expression levels, 367–368
microscopy, 368–370
yeast strains, 367
kinesin, density of, 22
limitations, technique, 22–23
MAPs (*see* Microtubule-associated proteins (MAPs))
mechanical properties, 14
mechanical rigidity, 14
methods
biotinylation of kinesin, 17
buffer stock solutions, 18
fluorescently labeled, polymerization of, 18–19
gliding assay (*see* Gliding assay, microtubule)
image analysis, 20
and materials, 17

oxygen scavenging solution, 18
motions measurement
coverslips and slides, 111
GMPCCP-stabilized seeds preparation, 113–115
in vitro assay, 110–111
in vitro sample chambers, TIRF microscope, 112–113
networks (*see* Structure-mechanics relationships, MT)
persistence length, theory
definition, 15
flexibility, 15
kinesin gliding assay, 16, 16*f*
single fluorophores, imaging, 16–17
tangent vectors and angles, 15, 15*f*
tip fluctuations, 16
Young's modulus, 16
polymerization assay, 181
protofilaments, 22
seeds and elongation
attachment, 150
kinesin heavy chain, 150, 151*f*
nonspecific kinesin immobilization, 149
stability, drug sensitivity
benomyl plate assay, 366–367
yeast cells, 366
stages in cellular life, 14
Microtubule-severing enzymes *in vitro*
biophysical assays, 200–211
CTT, tubulin, 194
description, 194
katanin, spastin and fidgetin, 192, 193*f*, 194
kinesin-8, kinesin-13 and kinesin-14, 192
MAPs, 192
purification, Sf9 cells, 195–200
Microtubule-stabilizing agent (MSA)
laulimalide/peloruside binding site, 304–305
ZMP (*see* Zampanolide (ZMP))
Microtubule targeting agents (MTAs), 284–285, 296–297
Mitochondria
cell seeding and spreading, 105
description, 98
individual cultured cells, 98
live-cell visualization, 103
micropatterned coverslips, 100–101
micropatterns preparation, 104–105
and microtubule labeling, 102–103
and microtubule visualization, 105–106
photomask, 99
seeding and spreading, cells, 101–102
Mitotic centromere-associated kinesin (MCAK)
depletion, 59
mitotic centromeres, 59
spindles, 59
MS. *See* Mass spectrometry (MS)
MSA. *See* Microtubule-stabilizing agent (MSA)
MTAs. *See* Microtubule targeting agents (MTAs)
MTBindingSim
binding models
first-order binding, 378, 378*f*
MAP-MT, 377–378
MAPs bind MT-bound MAPs, 379
MAPs dimerization, 379
pseudocooperativity, 379
seam and lattice binding, 378, 378*f*
tubulin dimer, 378
two binding site, 379
two MAPs bind MT-bound MAPs, 379
experimental designs
cosedimentation assay, 376–377
Langmuir isotherm, 377
MT-binding proteins, 377
human EB1 bind, MT seam/lattice, 381–383
MAP-MT interactions, 376
MATLAB, 376
Tau-MT binding, 379–381
MTP. *See* Microtubule protein (MTP)
MTs. *See* Microtubules (MTs)

N

NdFeB magnets. *See* Neodymium iron boron (NdFeB) magnets
Neodymium iron boron (NdFeB) magnets
advantages, 78
applied magnetic field, 84–85
force levels, 87
higher force applications, 88
ring-shaped, 89–90

O

Orthogonal sectioning
CLEM processing, 336, 337*f*
description, 336
ultramicrotome, 336

P

Paraformaldehyde (PFA), 249–252, 251*f*, 254*f*, 259–260, 261–262
Particle tracking, MT networks, 86–87
PDMS slide spacer. *See* Polydimethylsiloxane (PDMS) slide spacer
PEDTA buffer. *See* Phosphate–EDTA (PEDTA) buffer

Peptide array membrane analysis
ECL, 179
Gsα sequence, 178–179
TBS, 179
Permeabilization
cells, 143
MB11 antibody, 143
Persistence length, microtubules. *See* Microtubules
PFA. *See* Paraformaldehyde (PFA)
Phosphate–EDTA (PEDTA) buffer, 306, 309–310, 317
Photomask
apoptosis, 99
cell size, 99
cytoskeleton, 99
micropatterns, 99
PIS. *See* Precursor ion scanning (PIS)
Polydimethylsiloxane (PDMS) slide spacer, 387–388
Polyglutamylation, 248–249, 250–251, 258
Polyglycylation, 248–249, 251, 258–259
Portable magnetic tweezers
design and application, 90, 91*f*
measurements, 90
principle, 90
"sideways-pulling" geometry, 90
Posttranslational modifications (PTMs)
acetylation, 248–249, 250, 255–256
antibody concentration, MT
α-/β-tubulin dimers, 252
GT335, 253, 254*f*
low levels, 252
neuronal axons and dendrites, 253, 254*f*
U2OS cells, TTLL4, 252–253, 254*f*
buffer composition (*see* Buffer composition, tubulin PTMs)
cell culture and transfection, 259
cold methanol fixation, 259
description, 249
detection, antibodies, 255–259, 257*t*
detyrosination/tyrosination, 248–249, 250, 256–258
DSP-PFA method, 249–252, 251*f*, 259–260
electrophoresis and immunoblot, 260–261
fixed cells, immunocytochemistry, 260
glutamylated peptides, 249
immunoblot analysis, antibody and sample concentration, 253–255, 256*f*
MAPs, 248
mechanisms, 248–249
microscopy and image acquisition, 260
microtubules (MTs), 248
PFA fixation, 249, 259
polyglutamylation, 248–249, 250–251, 258
polyglycylation, 248–249, 251, 258–259
Precursor ion scanning (PIS), 318, 321*f*
Protein purification, Sf9 cells
buffers, 196–197
equipment, 197
procedures, 197–198, 199*f*
reagents, 197
Protein samples preparation
GFP, 120
in vivo experiments, 120–121
reaction mix, 121
sample chambers, 120
TIP, 120
PTMs. *See* Posttranslational modifications (PTMs)
Pull-down assays
6His-Gsα, 177
tubulin and Gsα, 177, 178*f*
Purification
and assembly, BtubA/B (*see* Bacterial tubulin BtubA/B)
DCX
buffers, 33, 36*t*
centrifugation, 37
cleavage, His-tagged, 36–37
cytoplasmic proteins, extraction of, 33–35
description, 34*f*, 35
His-tagged, 35
42 kDa His-DCX, 35
salt concentration, 35
Sf9 cells
concentration determination, 198
description, 198–200
protein expression, 196
protein purification, 196–198
virus amplification, 195–196

R

Resin embedding, 333
Rheology. *See* Microrheology methods
RISC. *See* RNA-induced silencing complex (RISC)
RMSD. *See* Root-mean square deviation (RMSD)
RNA-induced silencing complex (RISC), 67–68
Root-mean square deviation (RMSD), 270

S

Samples preparation, MT networks
entangled MT networks, 79–80
tubulin proteins, 78–79
Selective reaction monitoring (SRM), 318
Single-molecule fluorescence assay, 348

Spindles poles
MCAK, 59
meiotic spindles, 50–51
microtubule detachment, 52–53
nucleation rate, 50–51
SPR. *See* Surface plasmon resonance (SPR)
SRM. *See* Selective reaction monitoring (SRM)
Stathmin–tubulin interaction, ITC
formation, T2S complex (K_{T2S}), 296–297
nonequal interacting sites, 295–296, 296*f*
one-set-of-sites binding model, 295
temperature, 292*f*, 295–296
thermodynamic parameters, 296–297
vinblastine, 296–297
Stereo dissection microscope, 129
Structure determination, MT networks
average filament length, 82–83
confocal microscopy images, 80–81, 81*f*
dynamic network formation, 81
network mesh size, 81–82
steady-state structures, 80
Structure-mechanics relationships, MT
active microrheology methods, 77–78
custom-built magnetic tweezers devices, 77
cytoskeleton, 76–77
magnetic tweezers (*see* Magnetic tweezers devices)
mechanics (*see* Mechanics, MT networks)
optical trapping methods, 77–78
samples preparation, 78–80
Surface plasmon resonance (SPR)
binding analysis, 177–178
Gsα and tubulin, 177–178
tubulin immobilization, 177

T

Tau-MT binding assay
"first-order binding" model, 380
MTBindingSim, 380*f*, 381
oligomerization, 381
oligomers, 379–380
Tau-Tau interaction, 381
tubulin dimers, 379
Tau–tubulin interaction, ITC, 294*f*, 297–298
TBCD
anion exchange chromatography, 166
chromatography, hydroxylapatite, 166–167
expression, 165–166
HeLa cell lysates preparation, 166
size exclusion chromatography, 167
TBCE
purification
anion exchange chromatography, 164–165
chromatography, hydroxylapatite, 165
insect cell lysates preparation, 164
size exclusion chromatography, 165
recombinant baculovirus construction, 163
Tetrad viability
haploid spores, 364–365
mating-type analysis, 365–366
spores, 363
and spore viability analysis, 363, 364*f*
sporulation, 364
TUB1 and TUB2, 363–364
TIRFM. *See* Total internal reflection fluorescence microscope (TIRFM)
TIRF microscopy. *See* Total internal reflection fluorescence (TIRF) microscopy
Total internal reflection fluorescence (TIRF), 204–205, 207, 344
Total internal reflection fluorescence microscope (TIRFM)
gliding assay calculation, 21*f*, 22
single fluorophores, 16–17, 20
Total internal reflection fluorescence (TIRF) microscopy
chemically functionalizing sample chamber, 117–118
EB1 comets, HeLa cell, 121–122, 122*f*
MT system, 115–116
75 nM GFP-EB3, 121–122, 122*f*
oxygen scavenger system, 116
protein samples preparation, 120–121
required materials, 115
seeds and blocking attachment, 118–119
stable heating, sample, 121–122
Transgenes, *A. thaliana*
Agrobacterium tumefaciens, 234–235
construction, 235–236
recovery
antibiotics, 237
hydrogen peroxide/ethanol solution, 237
selection process, 238
T2 seed collection, 237, 238*f*
screening T2 seed lines, 235–236
transformation
acetosyringone, 237
axillary floral stems, 236–237
"floral dipping" protocol, 236
seed collection, 236–237
Silwet concentration, 236–237
TTL. *See* Tubulin tyrosine ligase (TTL)

Tubulin. *See also* Yeast tubulin
brains, 176
expression regulation, micro-RNAs
cells and tissues, 65
measuring activity, 67–68
miR-200 family, 64–65
next-generation sequencing, 66–67
PCR arrays, 66
RISC immunoprecipitation, 67–68
RNA extraction and qRT-PCR, 65–66
transcription and processing, 64
transfections, premiRNAs, 67
tubulin-binding drugs, 65
β-tubulin isotypes and miR-100, 68
tumor suppressor, 64
fluorescent labeling (*see* Fluorescent labeling, tubulin)
and MAPS
buffer conditions, 293
description, 284–285, 285*f*
determination, protein concentrations, 291–292
equilibration, 290
FtsZ assembly, 294–295
reaction volume and duration, 293–294, 294*f*
removal, 176
stathmin–tubulin interaction, 295–297
tau–tubulin interaction, 294*f*, 297–298
temperature, 292–293, 292*f*
mutations
directed mutagenesis, 360
gene replacement, yeast β-tubulin, 360, 361*f*
modified tubulin DNA transformation, 360–362
ORF, 360
plasmids, genetic manipulation, 359*t*, 360
"plasmid shuffle", 360
prototrophy gene, 360
verification, yeast, 362–363
yeast β-tubulin, 360, 361*f*
polymerization (*see In vitro* microtubule (MT) polymerization, measurement)
Tubulin–Gsα interaction
GTPase assay protocols, 179, 180*f*
nucleotide state, 179
single-turnover tubulin GTPase assay, 180–181
steady-state tubulin GTPase assay, 179–180
α/β-Tubulin heterodimer
actin, 157–159
anion exchange chromatography, 163
bacterial cell growth and induction, 161
bacterial lysates preparation, 162
CCT-dependent folding assay, 159
cDNA clones and vectors, 159–161
chaperone-dependent reaction, 159
chromatography, 162–163
cytosolic chaperonin, 157
description, 156–157
E. coli cells, 157
gel filtration chromatography, 163
in vitro folding reactions, 167, 168*f*
microtubule destruction, 159, 160*f*
structural data, 169, 169*t*
TBCA, TBCB and TBCC, 161
tubulin folding and heterodimer assembly pathway, 158*f*, 159
β-Tubulin isotypes
and miR-100
Akt/mTOR pathway, 68, 71*f*
MCF7 breast cancer cells, 68
paclitaxel treatment, 68, 69*f*
and miR-200c
breast cancers, 68–70
MDA-MB-231 breast cancer cells, 68–70, 73*f*
paclitaxel, 68–70
qRT-PCR, 68–70, 71*f*
ZEB1 and relative changes, 68–70, 72*f*
Tubulin tyrosine ligase (TTL)
flexibility, 3
and GST, 3, 7
Sephadex G-75, 6
Turbidity, tubulin polymerization
analysis, 227
cuvettes, 225–226
description, 216
drawbacks, 216–217
multiwell plates, 226–227
Tyrosination, 248–249, 250, 256–258

V

Virus amplification
description, 196
equipment, 195
reagents, 195
recombinant Baculovirus stocks, 195–196
Sf9 transfection, 195

Y

Yeast tubulin
brain, 357
budding yeast, 357
microtubule-associated proteins *in vivo*, 370–372
microtubule-dependent processes, 357–358
microtubule dynamics *in vivo*, 367–370
microtubule polymer, 356
microtubule stability, drug sensitivity, 366–367
mutagenesis, 357

mutant proteins, 357
reagents and equipment, 358–360
tetrad viability, 363–366
tubulin mutations, 360–363
tubulin polypeptides, 356–357

Z

Zampanolide (ZMP)
binding kinetics
Flutax-2 method, 316
HPLC analysis, 317–318
description, 305
Flutax displacement assay, 310–312
GAB buffer, 306
laulimalide/peloruside competition assay, 312–314
laulimalide/peloruside site, 304–305
MS (*see* Mass spectrometry (MS))
MT assembly (*see* Microtubules (MTs), assembly)
PEDTA buffer, 306
P-glycoprotein (P-gp), 304
software, data analysis, 307
stoichiometry, 314–316, 315*f*
tubulin and stabilized MTs, 305–306
ultracentrifugation, 307
visualization, covalent binding
adherent cell lines, 322
cells, suspension, 322–323, 323*f*
experiments, characterization, 324
Flutax, 320–322
ZMP. *See* Zampanolide (ZMP)

VOLUMES IN SERIES

Founding Series Editor
DAVID M. PRESCOTT

Volume 1 (1964)
Methods in Cell Physiology
Edited by David M. Prescott

Volume 2 (1966)
Methods in Cell Physiology
Edited by David M. Prescott

Volume 3 (1968)
Methods in Cell Physiology
Edited by David M. Prescott

Volume 4 (1970)
Methods in Cell Physiology
Edited by David M. Prescott

Volume 5 (1972)
Methods in Cell Physiology
Edited by David M. Prescott

Volume 6 (1973)
Methods in Cell Physiology
Edited by David M. Prescott

Volume 7 (1973)
Methods in Cell Biology
Edited by David M. Prescott

Volume 8 (1974)
Methods in Cell Biology
Edited by David M. Prescott

Volume 9 (1975)
Methods in Cell Biology
Edited by David M. Prescott

Volume 10 (1975)
Methods in Cell Biology
Edited by David M. Prescott

Volume 11 (1975)
Yeast Cells
Edited by David M. Prescott

Volume 12 (1975)
Yeast Cells
Edited by David M. Prescott

Volume 13 (1976)
Methods in Cell Biology
Edited by David M. Prescott

Volume 14 (1976)
Methods in Cell Biology
Edited by David M. Prescott

Volume 15 (1977)
Methods in Cell Biology
Edited by David M. Prescott

Volume 16 (1977)
Chromatin and Chromosomal Protein Research I
Edited by Gary Stein, Janet Stein, and Lewis J. Kleinsmith

Volume 17 (1978)
Chromatin and Chromosomal Protein Research II
Edited by Gary Stein, Janet Stein, and Lewis J. Kleinsmith

Volume 18 (1978)
Chromatin and Chromosomal Protein Research III
Edited by Gary Stein, Janet Stein, and Lewis J. Kleinsmith

Volume 19 (1978)
Chromatin and Chromosomal Protein Research IV
Edited by Gary Stein, Janet Stein, and Lewis J. Kleinsmith

Volume 20 (1978)
Methods in Cell Biology
Edited by David M. Prescott

Advisory Board Chairman
KEITH R. PORTER

Volume 21A (1980)
Normal Human Tissue and Cell Culture, Part A: Respiratory, Cardiovascular, and Integumentary Systems
Edited by Curtis C. Harris, Benjamin F. Trump, and Gary D. Stoner

Volume 21B (1980)
Normal Human Tissue and Cell Culture, Part B: Endocrine, Urogenital, and Gastrointestinal Systems
Edited by Curtis C. Harris, Benjamin F. Trump, and Gray D. Stoner

Volume 22 (1981)
Three-Dimensional Ultrastructure in Biology
Edited by James N. Turner

Volume 23 (1981)
Basic Mechanisms of Cellular Secretion
Edited by Arthur R. Hand and Constance Oliver

Volume 24 (1982)
The Cytoskeleton, Part A: Cytoskeletal Proteins, Isolation and Characterization
Edited by Leslie Wilson

Volume 25 (1982)
The Cytoskeleton, Part B: Biological Systems and *In Vitro* Models
Edited by Leslie Wilson

Volume 26 (1982)
Prenatal Diagnosis: Cell Biological Approaches
Edited by Samuel A. Latt and Gretchen J. Darlington

Series Editor
LESLIE WILSON

Volume 27 (1986)
Echinoderm Gametes and Embryos
Edited by Thomas E. Schroeder

Volume 28 (1987)
***Dictyostelium discoideum:* Molecular Approaches to Cell Biology**
Edited by James A. Spudich

Volume 29 (1989)
Fluorescence Microscopy of Living Cells in Culture, Part A: Fluorescent Analogs, Labeling Cells, and Basic Microscopy
Edited by Yu-Li Wang and D. Lansing Taylor

Volume 30 (1989)
Fluorescence Microscopy of Living Cells in Culture, Part B: Quantitative Fluorescence Microscopy—Imaging and Spectroscopy
Edited by D. Lansing Taylor and Yu-Li Wang

Volume 31 (1989)
Vesicular Transport, Part A
Edited by Alan M. Tartakoff

Volume 32 (1989)
Vesicular Transport, Part B
Edited by Alan M. Tartakoff

Volume 33 (1990)
Flow Cytometry
Edited by Zbigniew Darzynkiewicz and Harry A. Crissman

Volume 34 (1991)
Vectorial Transport of Proteins into and across Membranes
Edited by Alan M. Tartakoff

Selected from Volumes 31, 32, and 34 (1991)
Laboratory Methods for Vesicular and Vectorial Transport
Edited by Alan M. Tartakoff

Volume 35 (1991)
Functional Organization of the Nucleus: A Laboratory Guide
Edited by Barbara A. Hamkalo and Sarah C. R. Elgin

Volume 36 (1991)
***Xenopus laevis:* Practical Uses in Cell and Molecular Biology**
Edited by Brian K. Kay and H. Benjamin Peng

Series Editors

LESLIE WILSON AND PAUL MATSUDAIRA

Volume 37 (1993)
Antibodies in Cell Biology
Edited by David J. Asai

Volume 38 (1993)
Cell Biological Applications of Confocal Microscopy
Edited by Brian Matsumoto

Volume 39 (1993)
Motility Assays for Motor Proteins
Edited by Jonathan M. Scholey

Volume 40 (1994)
A Practical Guide to the Study of Calcium in Living Cells
Edited by Richard Nuccitelli

Volume 41 (1994)
Flow Cytometry, Second Edition, Part A
Edited by Zbigniew Darzynkiewicz, J. Paul Robinson, and Harry A. Crissman

Volume 42 (1994)
Flow Cytometry, Second Edition, Part B
Edited by Zbigniew Darzynkiewicz, J. Paul Robinson, and Harry A. Crissman

Volume 43 (1994)
Protein Expression in Animal Cells
Edited by Michael G. Roth

Volume 44 (1994)
***Drosophila melanogaster:* Practical Uses in Cell and Molecular Biology**
Edited by Lawrence S. B. Goldstein and Eric A. Fyrberg

Volume 45 (1994)
Microbes as Tools for Cell Biology
Edited by David G. Russell

Volume 46 (1995)
Cell Death
Edited by Lawrence M. Schwartz and Barbara A. Osborne

Volume 47 (1995)
Cilia and Flagella
Edited by William Dentler and George Witman

Volume 48 (1995)
***Caenorhabditis elegans:* Modern Biological Analysis of an Organism**
Edited by Henry F. Epstein and Diane C. Shakes

Volume 49 (1995)
Methods in Plant Cell Biology, Part A
Edited by David W. Galbraith, Hans J. Bohnert, and Don P. Bourque

Volume 50 (1995)
Methods in Plant Cell Biology, Part B
Edited by David W. Galbraith, Don P. Bourque, and Hans J. Bohnert

Volume 51 (1996)
Methods in Avian Embryology
Edited by Marianne Bronner-Fraser

Volume 52 (1997)
Methods in Muscle Biology
Edited by Charles P. Emerson, Jr. and H. Lee Sweeney

Volume 53 (1997)
Nuclear Structure and Function
Edited by Miguel Berrios

Volume 54 (1997)
Cumulative Index

Volume 55 (1997)
Laser Tweezers in Cell Biology
Edited by Michael P. Sheetz

Volume 56 (1998)
Video Microscopy
Edited by Greenfield Sluder and David E. Wolf

Volume 57 (1998)
Animal Cell Culture Methods
Edited by Jennie P. Mather and David Barnes

Volume 58 (1998)
Green Fluorescent Protein
Edited by Kevin F. Sullivan and Steve A. Kay

Volume 59 (1998)
The Zebrafish: Biology
Edited by H. William Detrich III, Monte Westerfield, and Leonard I. Zon

Volume 60 (1998)
The Zebrafish: Genetics and Genomics
Edited by H. William Detrich III, Monte Westerfield, and Leonard I. Zon

Volume 61 (1998)
Mitosis and Meiosis
Edited by Conly L. Rieder

Volume 62 (1999)
Tetrahymena thermophila
Edited by David J. Asai and James D. Forney

Volume 63 (2000)
Cytometry, Third Edition, Part A
Edited by Zbigniew Darzynkiewicz, J. Paul Robinson, and Harry Crissman

Volume 64 (2000)
Cytometry, Third Edition, Part B
Edited by Zbigniew Darzynkiewicz, J. Paul Robinson, and Harry Crissman

Volume 65 (2001)
Mitochondria
Edited by Liza A. Pon and Eric A. Schon

Volume 66 (2001)
Apoptosis
Edited by Lawrence M. Schwartz and Jonathan D. Ashwell

Volume 67 (2001)
Centrosomes and Spindle Pole Bodies
Edited by Robert E. Palazzo and Trisha N. Davis

Volume 68 (2002)
Atomic Force Microscopy in Cell Biology
Edited by Bhanu P. Jena and J. K. Heinrich Hörber

Volume 69 (2002)
Methods in Cell–Matrix Adhesion
Edited by Josephine C. Adams

Volume 70 (2002)
Cell Biological Applications of Confocal Microscopy
Edited by Brian Matsumoto

Volume 71 (2003)
Neurons: Methods and Applications for Cell Biologist
Edited by Peter J. Hollenbeck and James R. Bamburg

Volume 72 (2003)
Digital Microscopy: A Second Edition of Video Microscopy
Edited by Greenfield Sluder and David E. Wolf

Volume 73 (2003)
Cumulative Index

Volume 74 (2004)
Development of Sea Urchins, Ascidians, and Other Invertebrate Deuterostomes: Experimental Approaches
Edited by Charles A. Ettensohn, Gary M. Wessel, and Gregory A. Wray

Volume 75 (2004)
Cytometry, 4th Edition: New Developments
Edited by Zbigniew Darzynkiewicz, Mario Roederer, and Hans Tanke

Volume 76 (2004)
The Zebrafish: Cellular and Developmental Biology
Edited by H. William Detrich, III, Monte Westerfield, and Leonard I. Zon

Volume 77 (2004)
The Zebrafish: Genetics, Genomics, and Informatics
Edited by William H. Detrich, III, Monte Westerfield, and Leonard I. Zon

Volume 78 (2004)
Intermediate Filament Cytoskeleton
Edited by M. Bishr Omary and Pierre A. Coulombe

Volume 79 (2007)
Cellular Electron Microscopy
Edited by J. Richard McIntosh

Volume 80 (2007)
Mitochondria, 2nd Edition
Edited by Liza A. Pon and Eric A. Schon

Volume 81 (2007)
Digital Microscopy, 3rd Edition
Edited by Greenfield Sluder and David E. Wolf

Volume 82 (2007)
Laser Manipulation of Cells and Tissues
Edited by Michael W. Berns and Karl Otto Greulich

Volume 83 (2007)
Cell Mechanics
Edited by Yu-Li Wang and Dennis E. Discher

Volume 84 (2007)
Biophysical Tools for Biologists, Volume One: In Vitro Techniques
Edited by John J. Correia and H. William Detrich, III

Volume 85 (2008)
Fluorescent Proteins
Edited by Kevin F. Sullivan

Volume 86 (2008)
Stem Cell Culture
Edited by Dr. Jennie P. Mather

Volume 87 (2008)
Avian Embryology, 2nd Edition
Edited by Dr. Marianne Bronner-Fraser

Volume 88 (2008)
Introduction to Electron Microscopy for Biologists
Edited by Prof. Terence D. Allen

Volume 89 (2008)
Biophysical Tools for Biologists, Volume Two: In Vivo Techniques
Edited by Dr. John J. Correia and Dr. H. William Detrich, III

Volume 90 (2008)
Methods in Nano Cell Biology
Edited by Bhanu P. Jena

Volume 91 (2009)
Cilia: Structure and Motility
Edited by Stephen M. King and Gregory J. Pazour

Volume 92 (2009)
Cilia: Motors and Regulation
Edited by Stephen M. King and Gregory J. Pazour

Volume 93 (2009)
Cilia: Model Organisms and Intraflagellar Transport
Edited by Stephen M. King and Gregory J. Pazour

Volume 94 (2009)
Primary Cilia
Edited by Roger D. Sloboda

Volume 95 (2010)
Microtubules, in vitro
Edited by Leslie Wilson and John J. Correia

Volume 96 (2010)
Electron Microscopy of Model Systems
Edited by Thomas Müeller-Reichert

Volume 97 (2010)
Microtubules: In Vivo
Edited by Lynne Cassimeris and Phong Tran

Volume 98 (2010)
Nuclear Mechanics & Genome Regulation
Edited by G.V. Shivashankar

Volume 99 (2010)
Calcium in Living Cells
Edited by Michael Whitaker

Volume 100 (2010)
The Zebrafish: Cellular and Developmental Biology, Part A
Edited by: H. William Detrich III, Monte Westerfield and Leonard I. Zon

Volume 101 (2011)
The Zebrafish: Cellular and Developmental Biology, Part B
Edited by: H. William Detrich III, Monte Westerfield and Leonard I. Zon

Volume 102 (2011)
Recent Advances in Cytometry, Part A: Instrumentation, Methods
Edited by Zbigniew Darzynkiewicz, Elena Holden, Alberto Orfao, William Telford and Donald Wlodkowic

Volume 103 (2011)
Recent Advances in Cytometry, Part B: Advances in Applications
Edited by Zbigniew Darzynkiewicz, Elena Holden, Alberto Orfao, Alberto Orfao and Donald Wlodkowic

Volume 104 (2011)
The Zebrafish: Genetics, Genomics and Informatics 3rd Edition
Edited by H. William Detrich III, Monte Westerfield, and Leonard I. Zon

Volume 105 (2011)
The Zebrafish: Disease Models and Chemical Screens 3rd Edition
Edited by H. William Detrich III, Monte Westerfield, and Leonard I. Zon

Volume 106 (2011)
Caenorhabditis elegans: Molecular Genetics and Development 2nd Edition
Edited by Joel H. Rothman and Andrew Singson

Volume 107 (2011)
Caenorhabditis elegans: Cell Biology and Physiology 2nd Edition
Edited by Joel H. Rothman and Andrew Singson

Volume 108 (2012)
Lipids
Edited by Gilbert Di Paolo and Markus R Wenk

Volume 109 (2012)
Tetrahymena thermophila
Edited by Kathleen Collins

Volume 110 (2012)
Methods in Cell Biology
Edited by Anand R. Asthagiri and Adam P. Arkin

Volume 111 (2012)
Methods in Cell Biology
Edited by Thomas Müler Reichart and Paul Verkade

Volume 112 (2012)
Laboratory Methods in Cell Biology
Edited by P. Michael Conn

Volume 113 (2013)
Laboratory Methods in Cell Biology
Edited by P. Michael Conn

Volume 114 (2013)
Digital Microscopy, 4th Edition
Edited by Greenfield Sluder and David E. Wolf

Color Plate

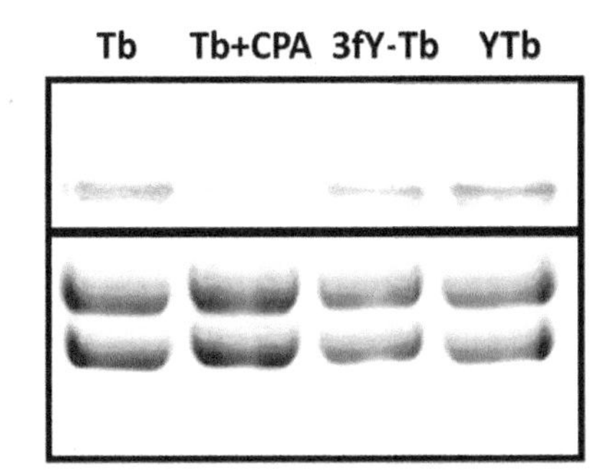

PLATE 1 (Fig. 1.1 on page 9 of this volume).

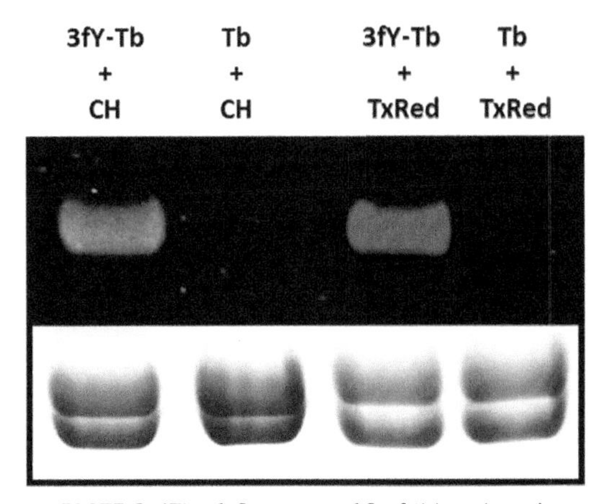

PLATE 2 (Fig. 1.2 on page 10 of this volume).

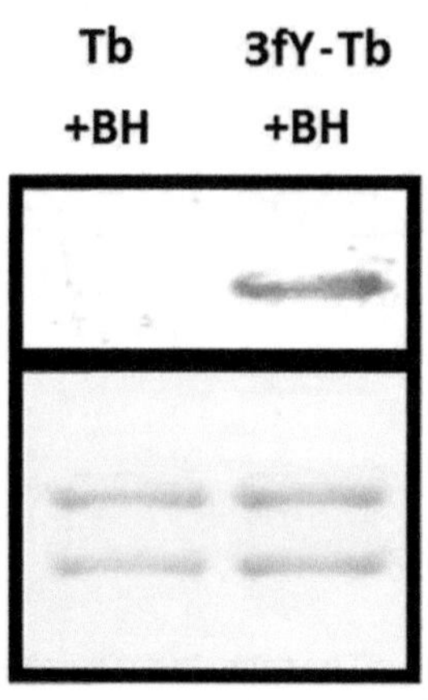

PLATE 3 (Fig. 1.3 on page 10 of this volume).

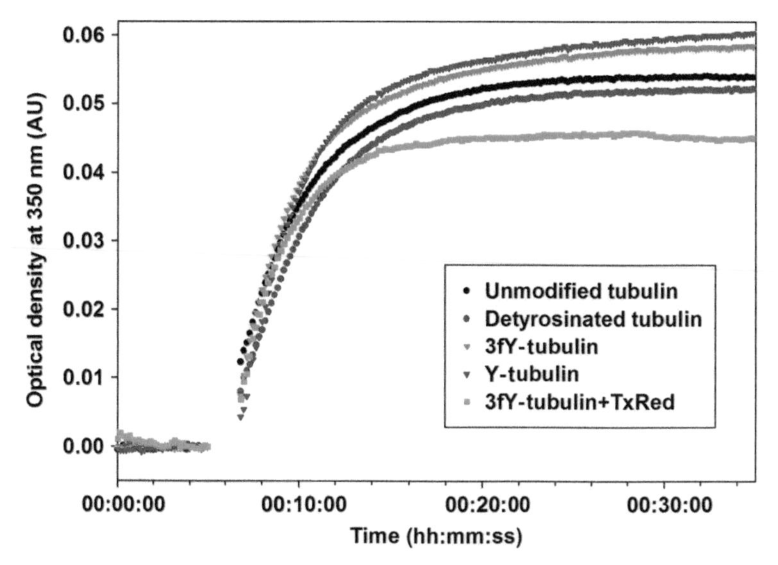

PLATE 4 (Fig. 1.4 on page 11 of this volume).

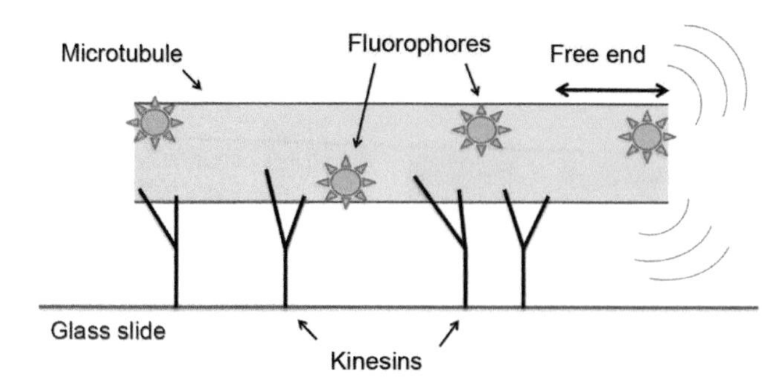

PLATE 5 (Fig. 2.2 on page 16 of this volume).

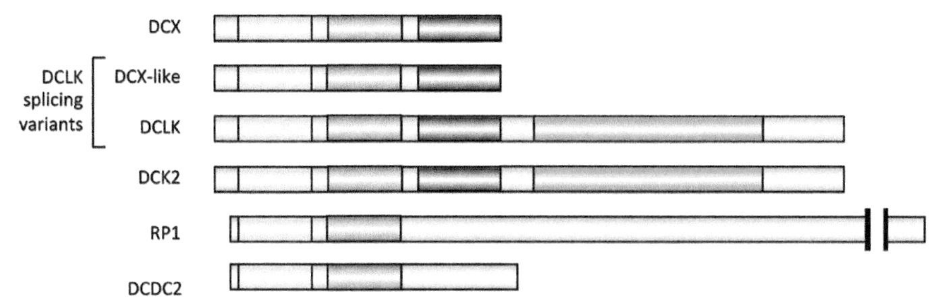

PLATE 6 (Fig. 3.1 on page 29 of this volume).

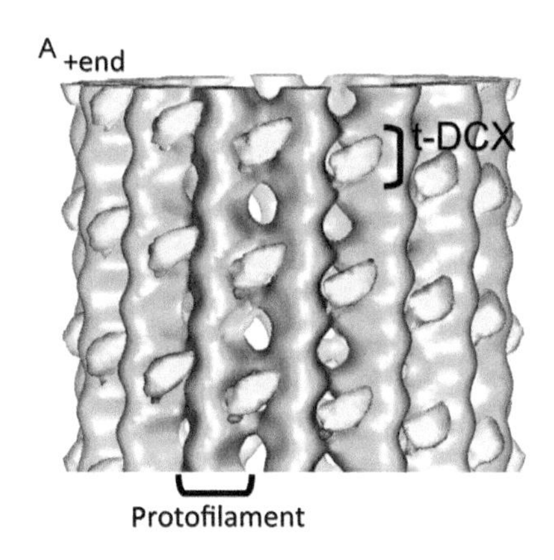

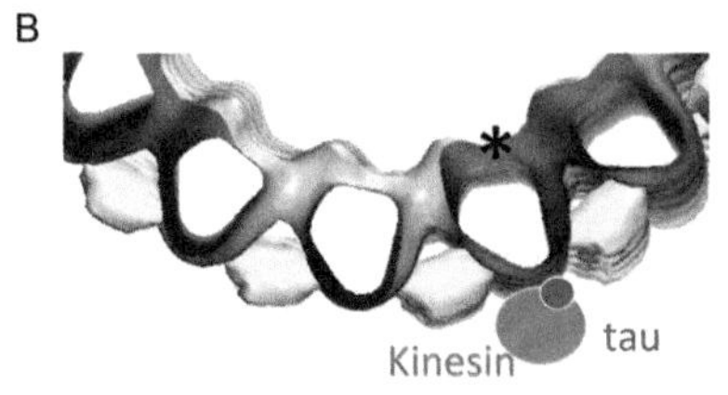

PLATE 7 (Fig. 3.2 on page 32 of this volume).

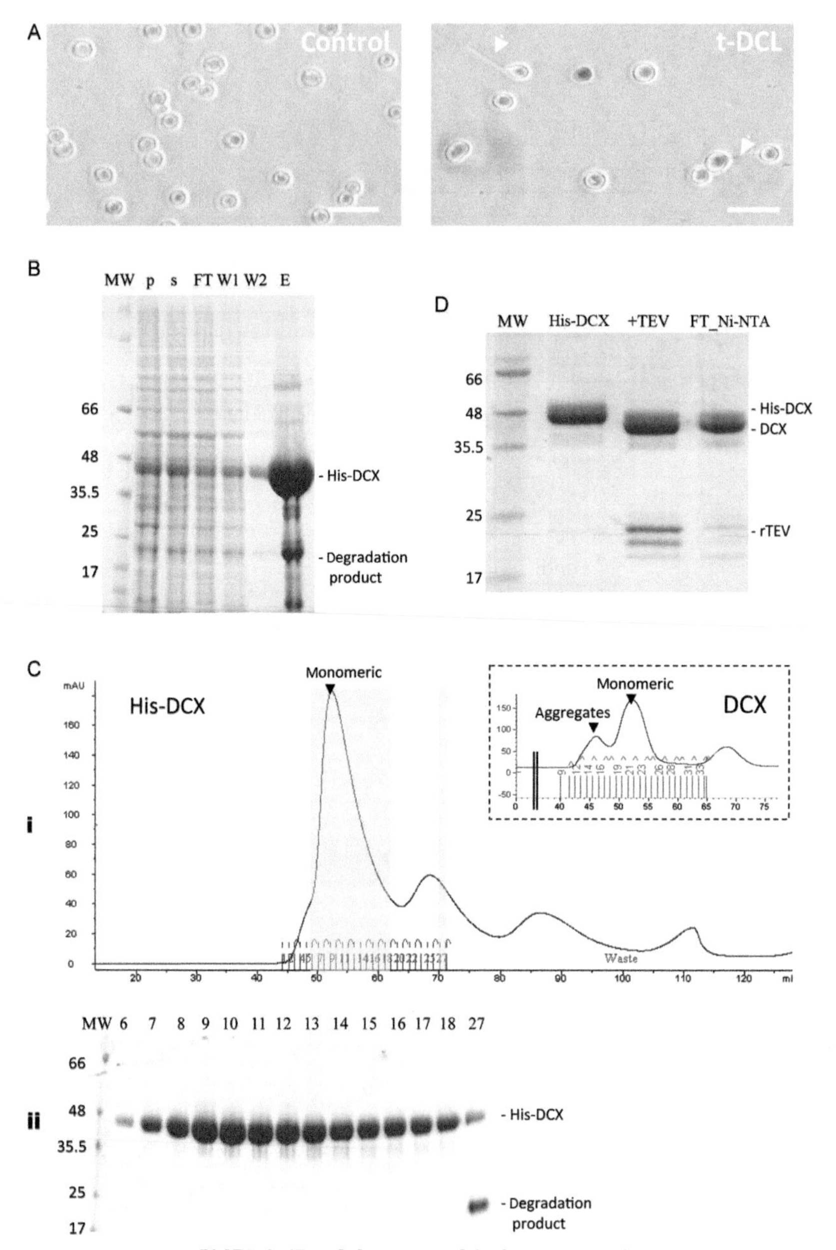

PLATE 8 (Fig. 3.3 on page 34 of this volume).

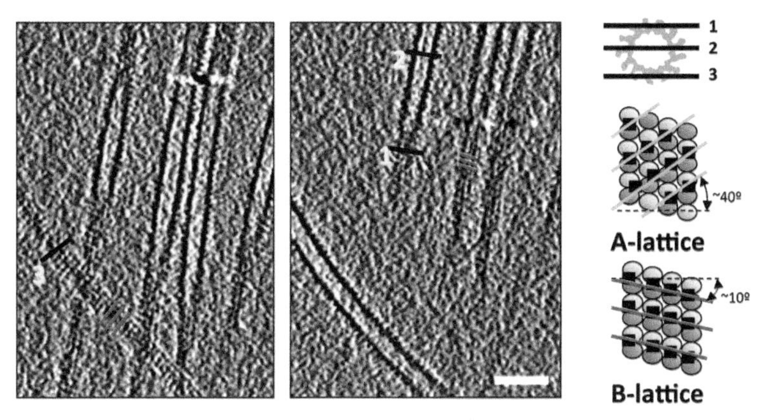

PLATE 9 (Fig. 3.4 on page 38 of this volume).

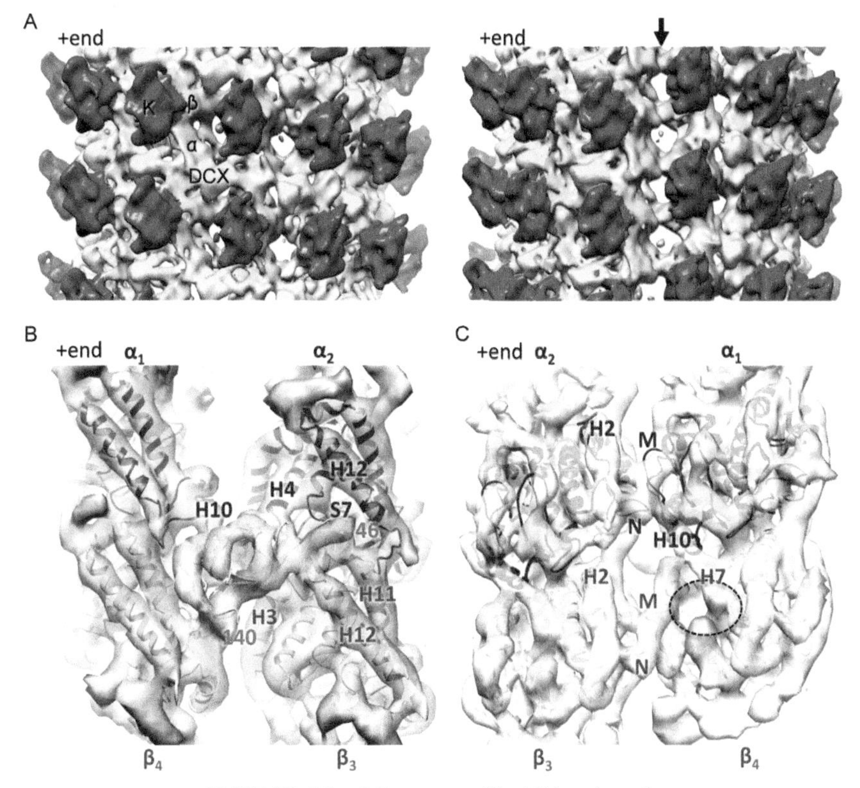

PLATE 10 (Fig. 3.5 on page 41 of this volume).

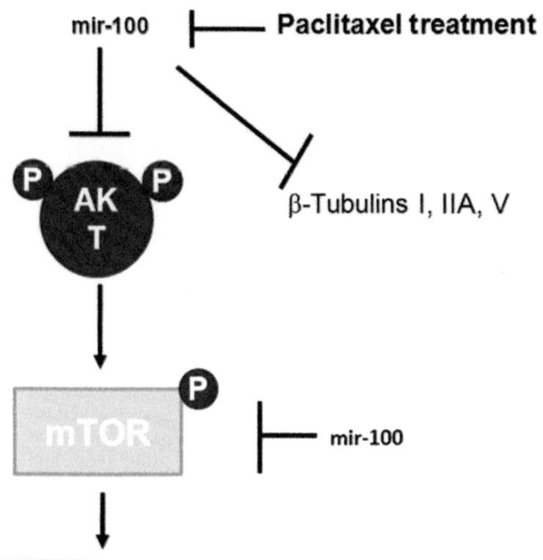

PLATE 11 (Fig. 5.2 on page 71 of this volume).

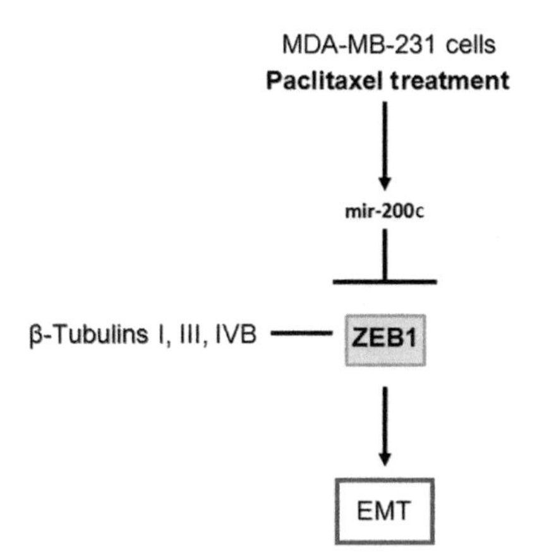

PLATE 12 (Fig. 5.5 on page 73 of this volume).

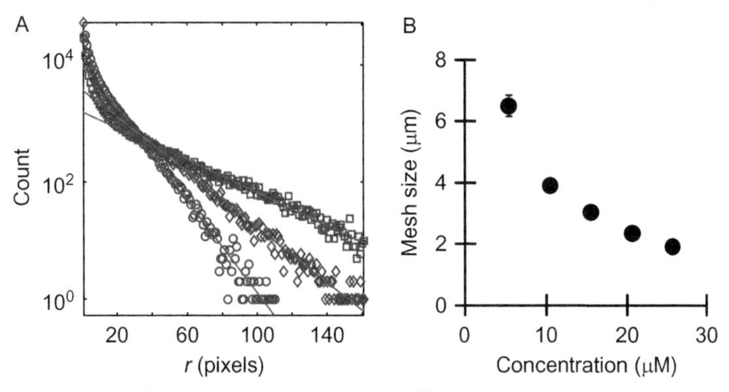

PLATE 13 (Fig. 6.2 on page 82 of this volume).

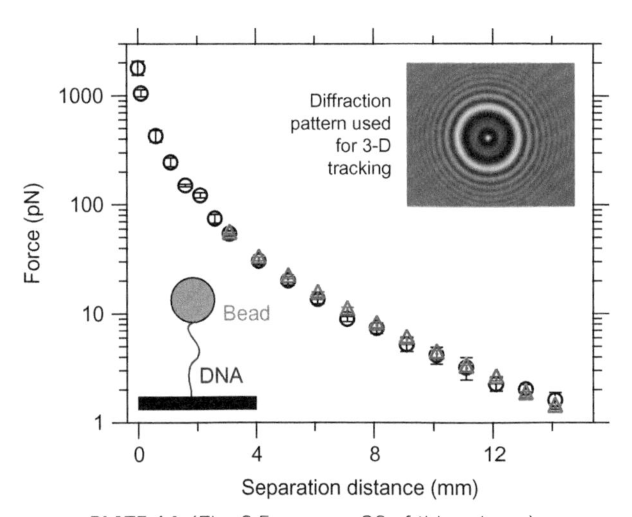

PLATE 14 (Fig. 6.5 on page 86 of this volume).

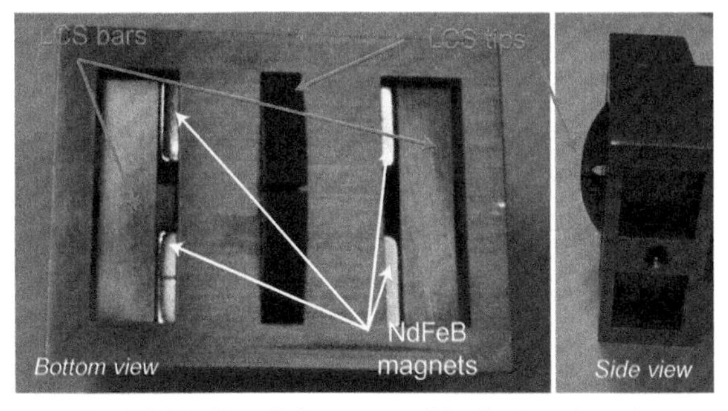

PLATE 15 (Fig. 6.6 on page 88 of this volume).

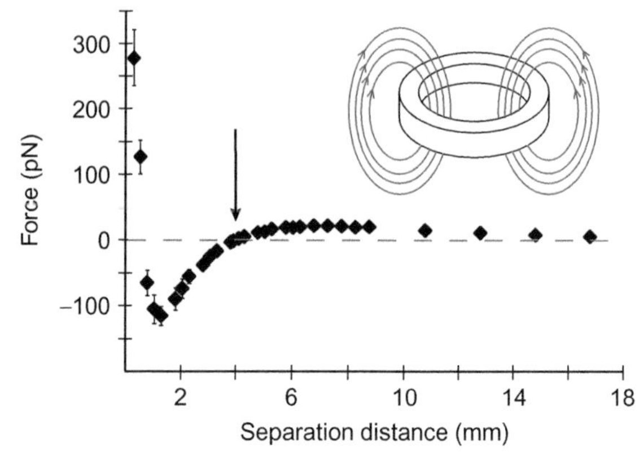

PLATE 16 (Fig. 6.7 on page 89 of this volume).

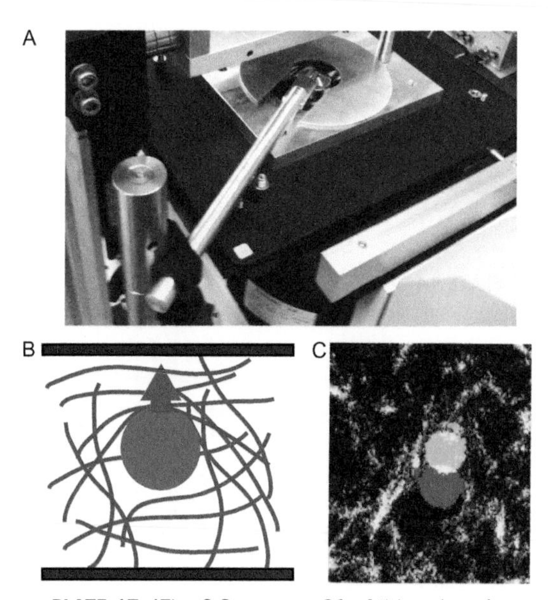

PLATE 17 (Fig. 6.8 on page 91 of this volume).

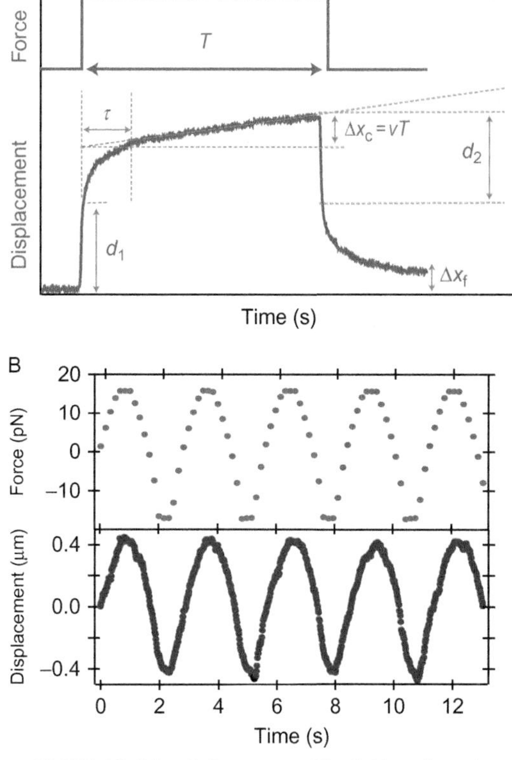

PLATE 18 (Fig. 6.9 on page 92 of this volume).

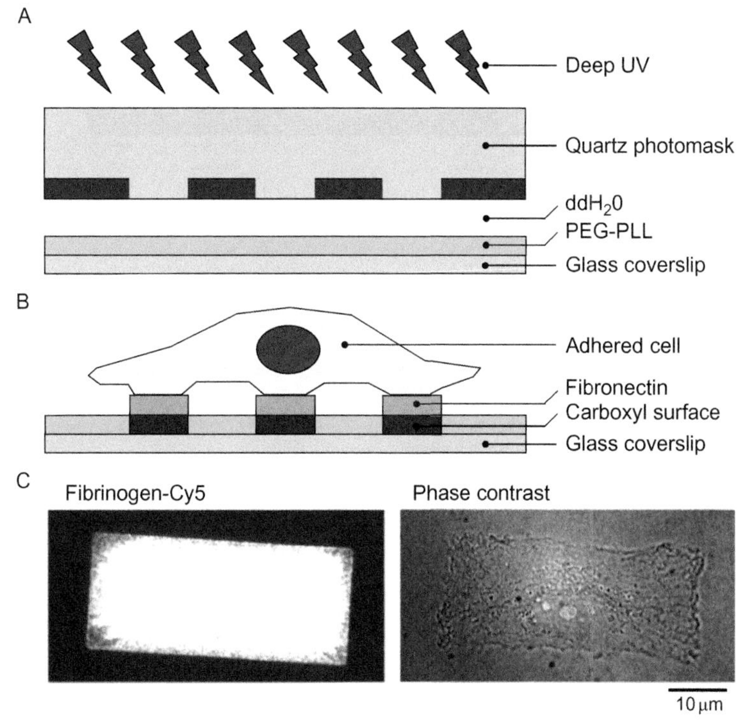

PLATE 19 (Fig. 7.1 on page 100 of this volume).

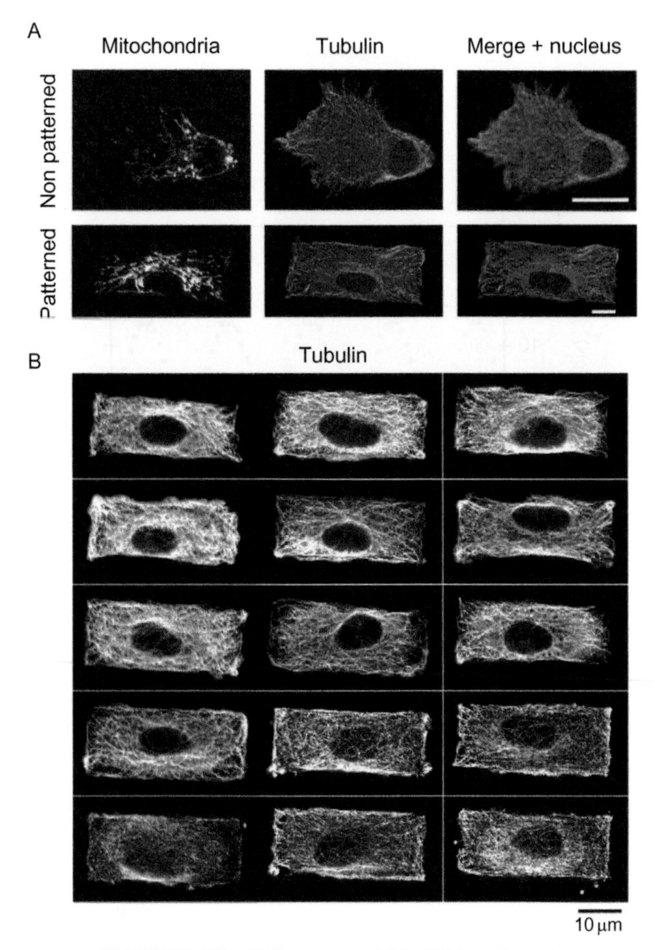

PLATE 20 (Fig. 7.2 on page 104 of this volume).

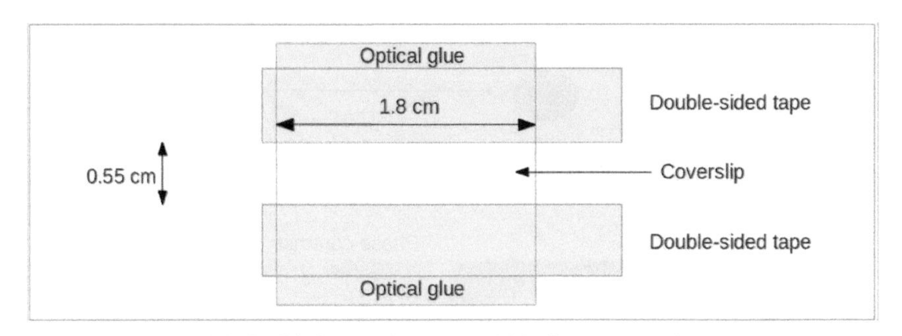

PLATE 21 (Fig. 8.2 on page 113 of this volume).

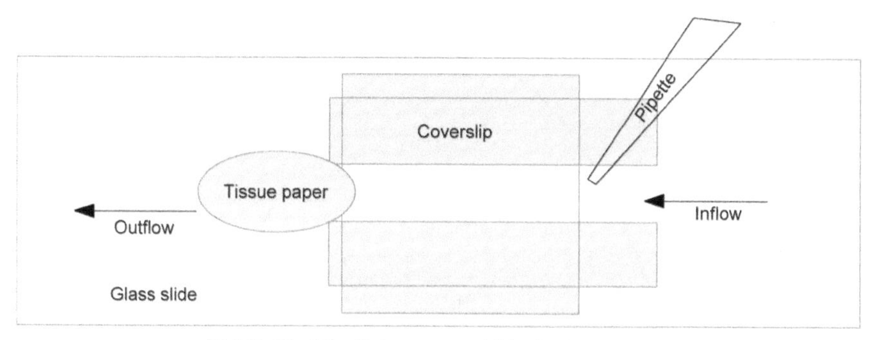

PLATE 22 (Fig. 8.4 on page 118 of this volume).

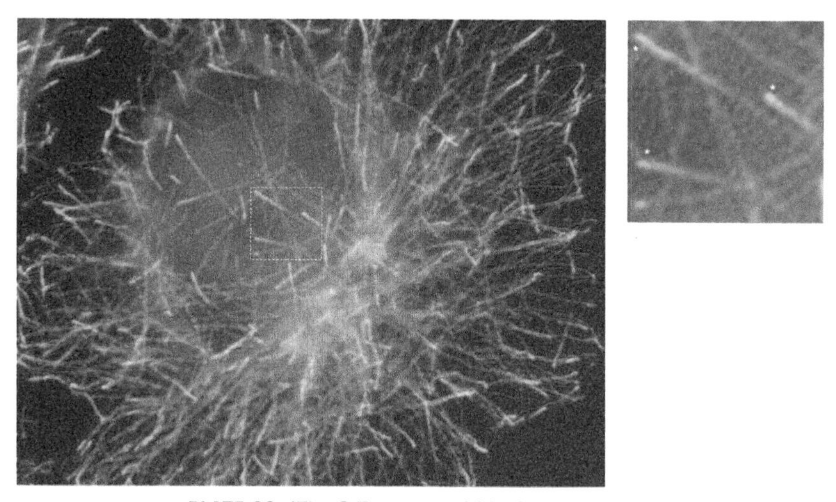

PLATE 23 (Fig. 8.7 on page 122 of this volume).

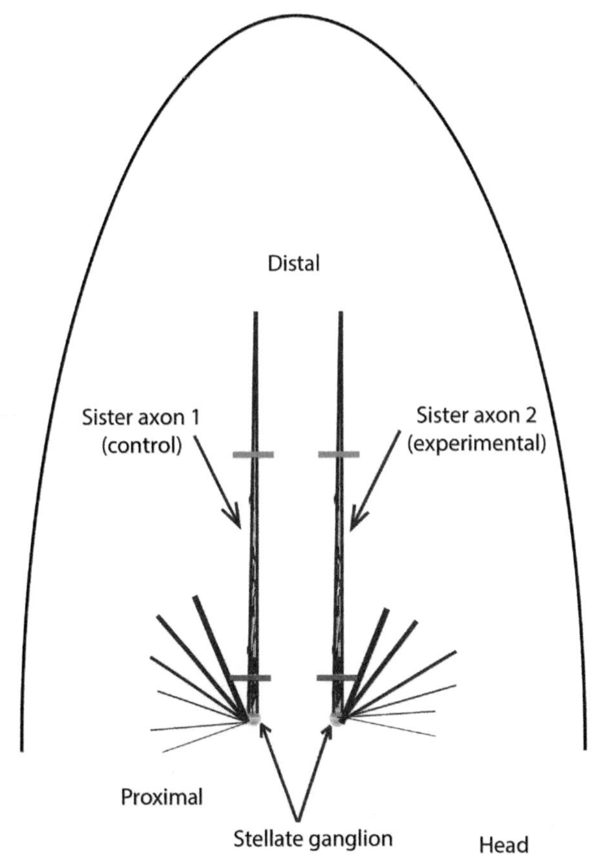

PLATE 24 (Fig. 9.2 on page 128 of this volume).

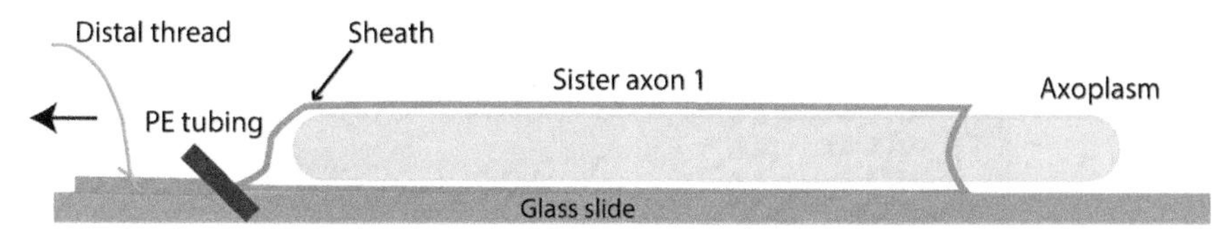

PLATE 25 (Fig. 9.3 on page 129 of this volume).

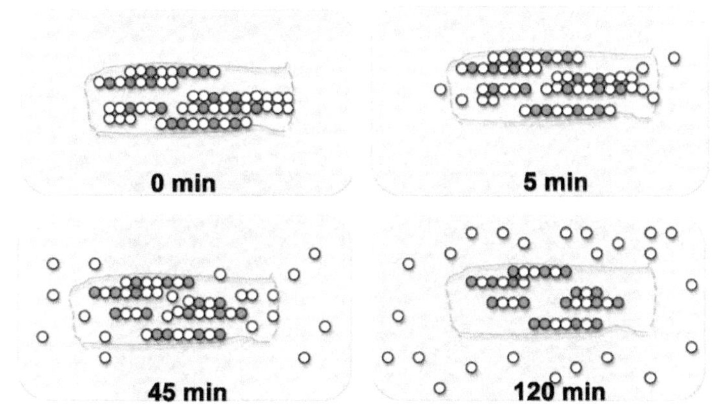

PLATE 26 (Fig. 9.4 on page 133 of this volume).

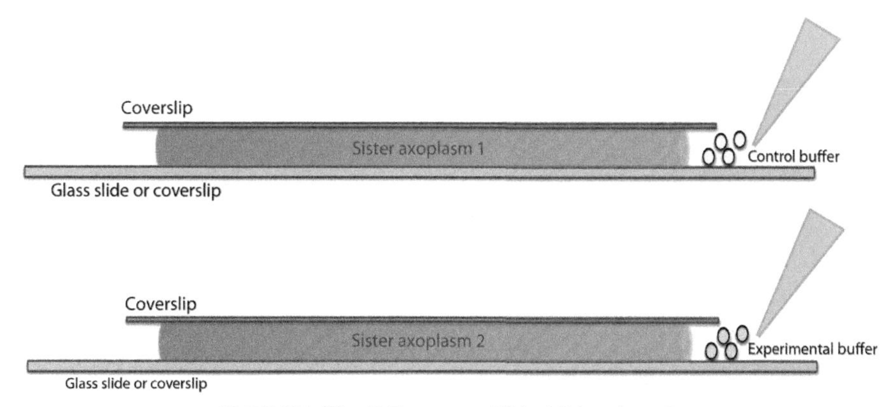

PLATE 27 (Fig. 9.5 on page 134 of this volume).

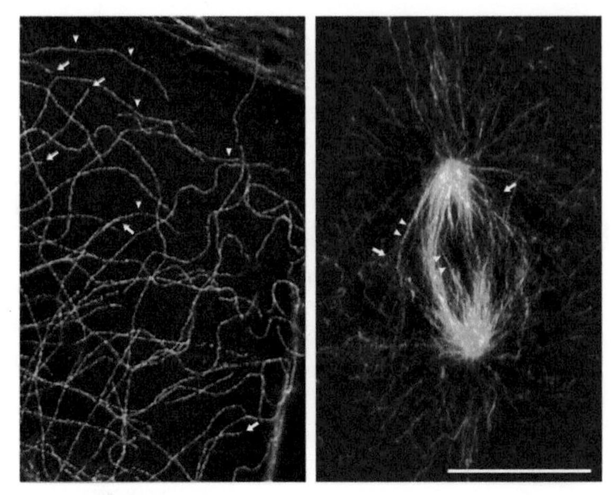

PLATE 28 (Fig. 10.1 on page 143 of this volume).

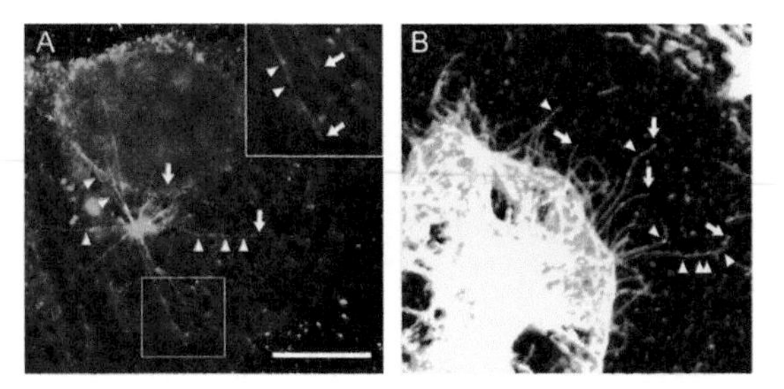

PLATE 29 (Fig. 10.2 on page 146 of this volume).

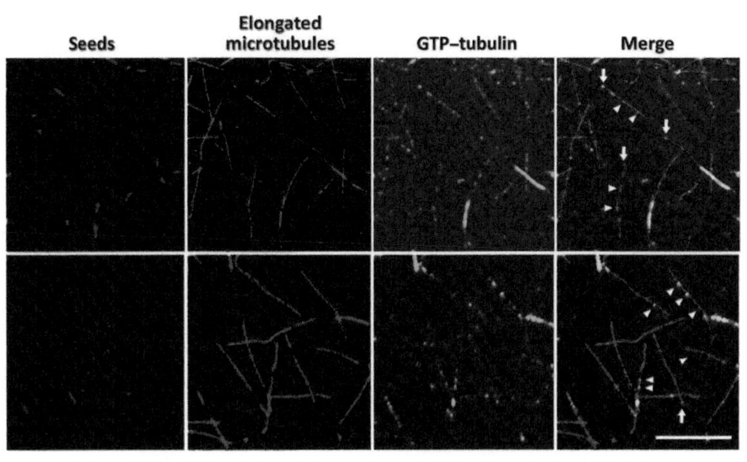

PLATE 30 (Fig. 10.3 on page 151 of this volume).

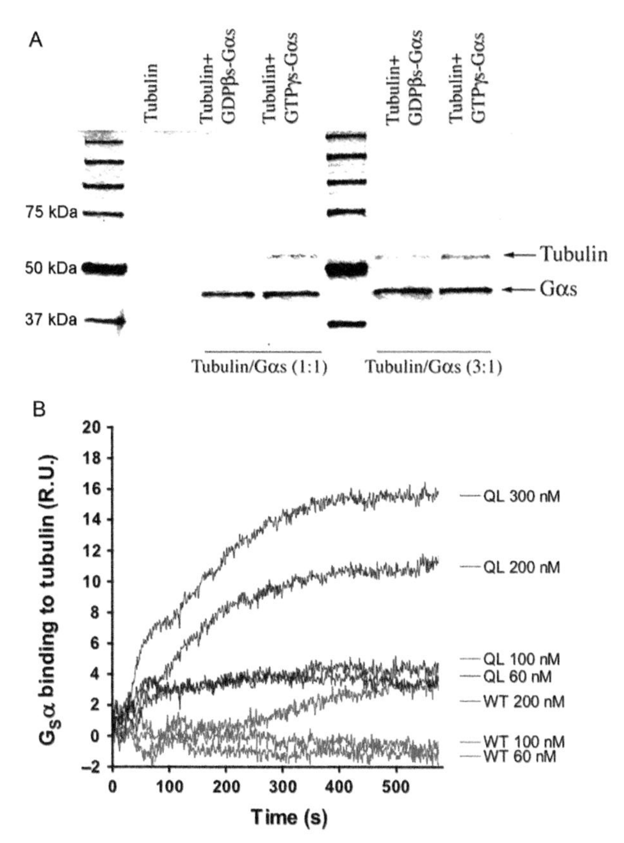

PLATE 31 (Fig. 12.2 on page 178 of this volume).

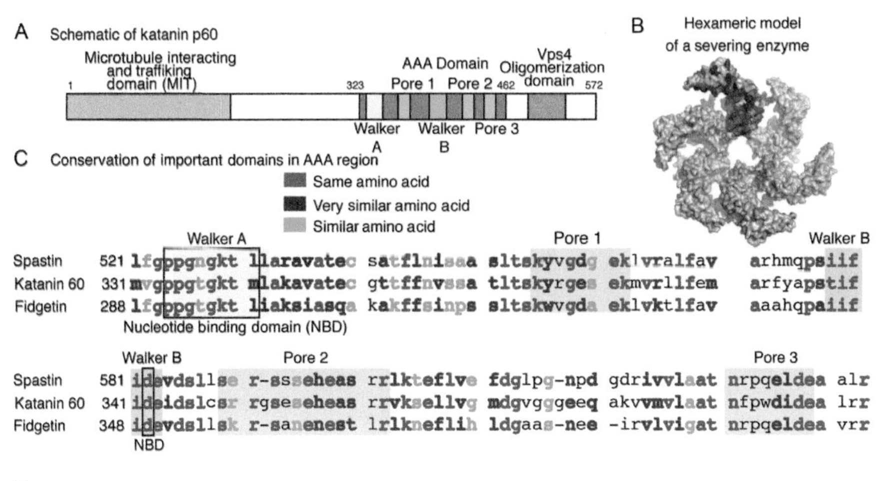

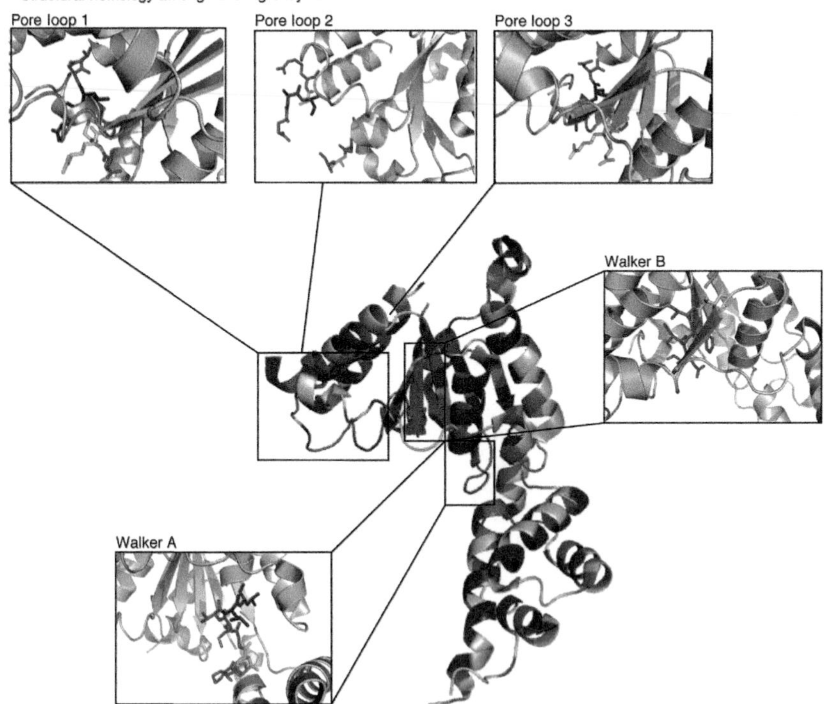

PLATE 32 (Fig. 13.1 on page 193 of this volume).

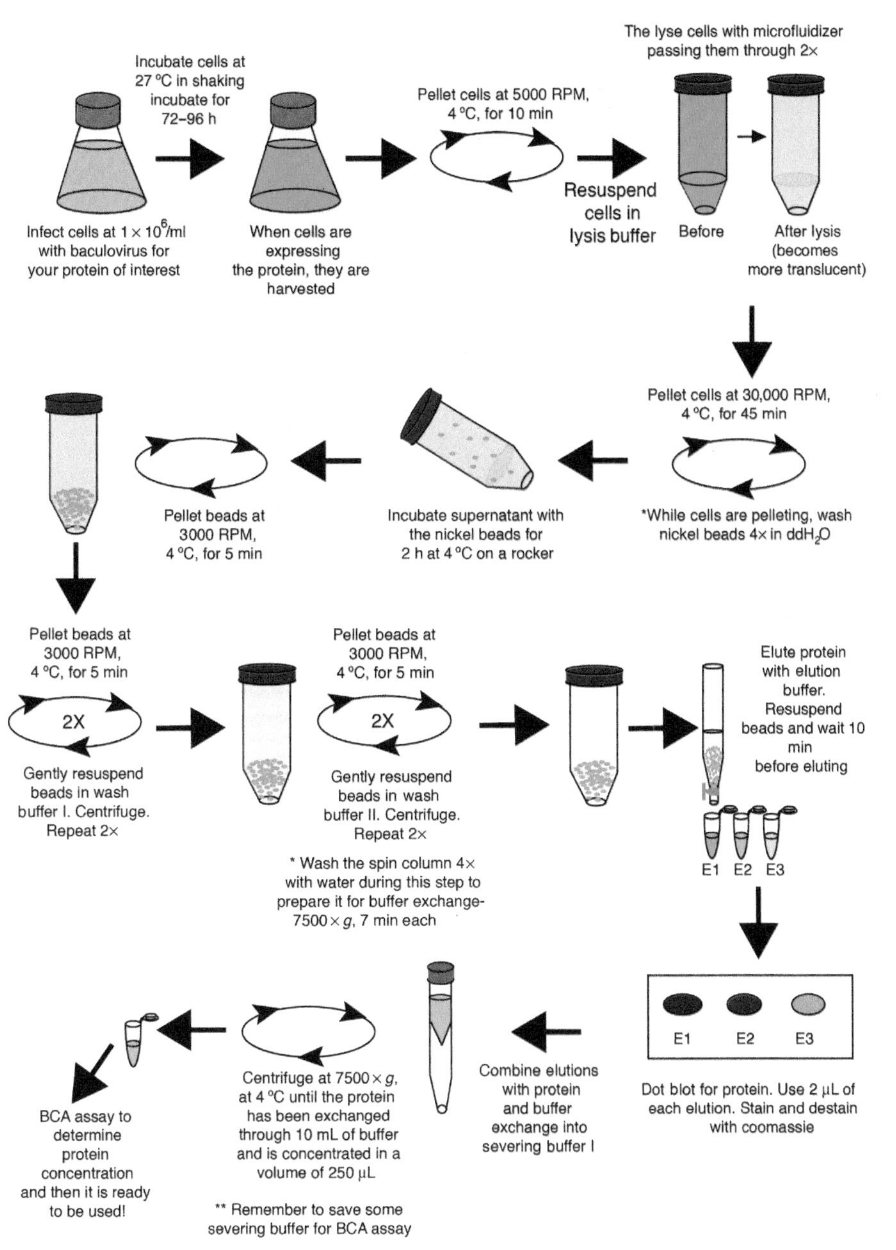

PLATE 33 (Fig. 13.2 on page 199 of this volume).

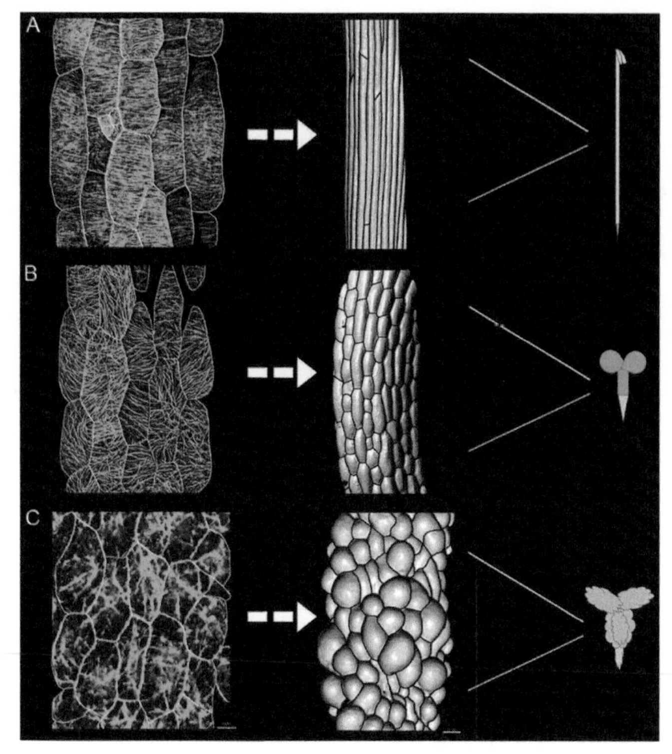

PLATE 34 (Fig. 15.1 on page 233 of this volume).

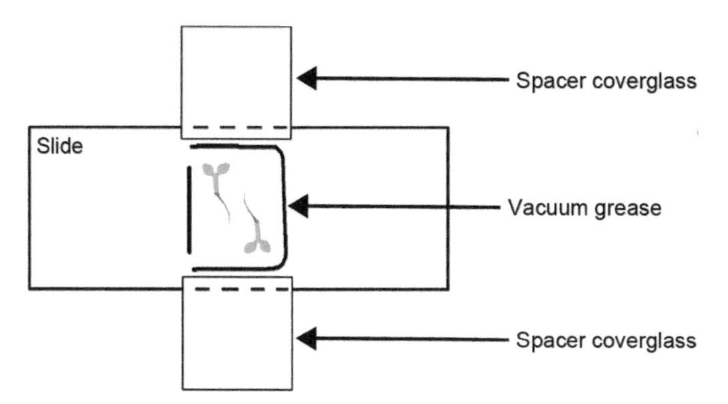

PLATE 35 (Fig. 15.3 on page 240 of this volume).

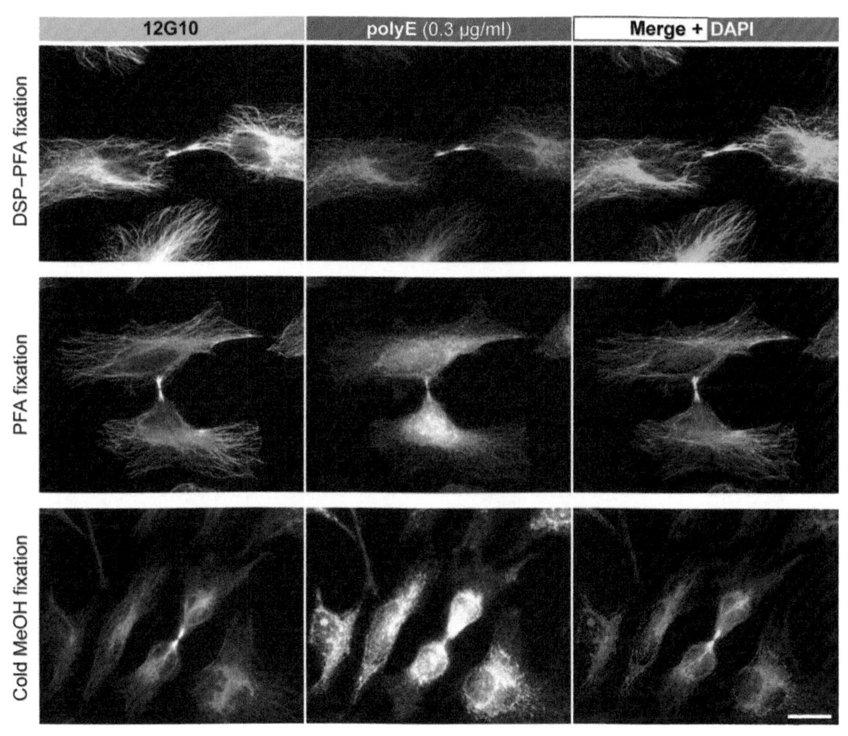

PLATE 36 (Fig. 16.1 on page 251 of this volume).

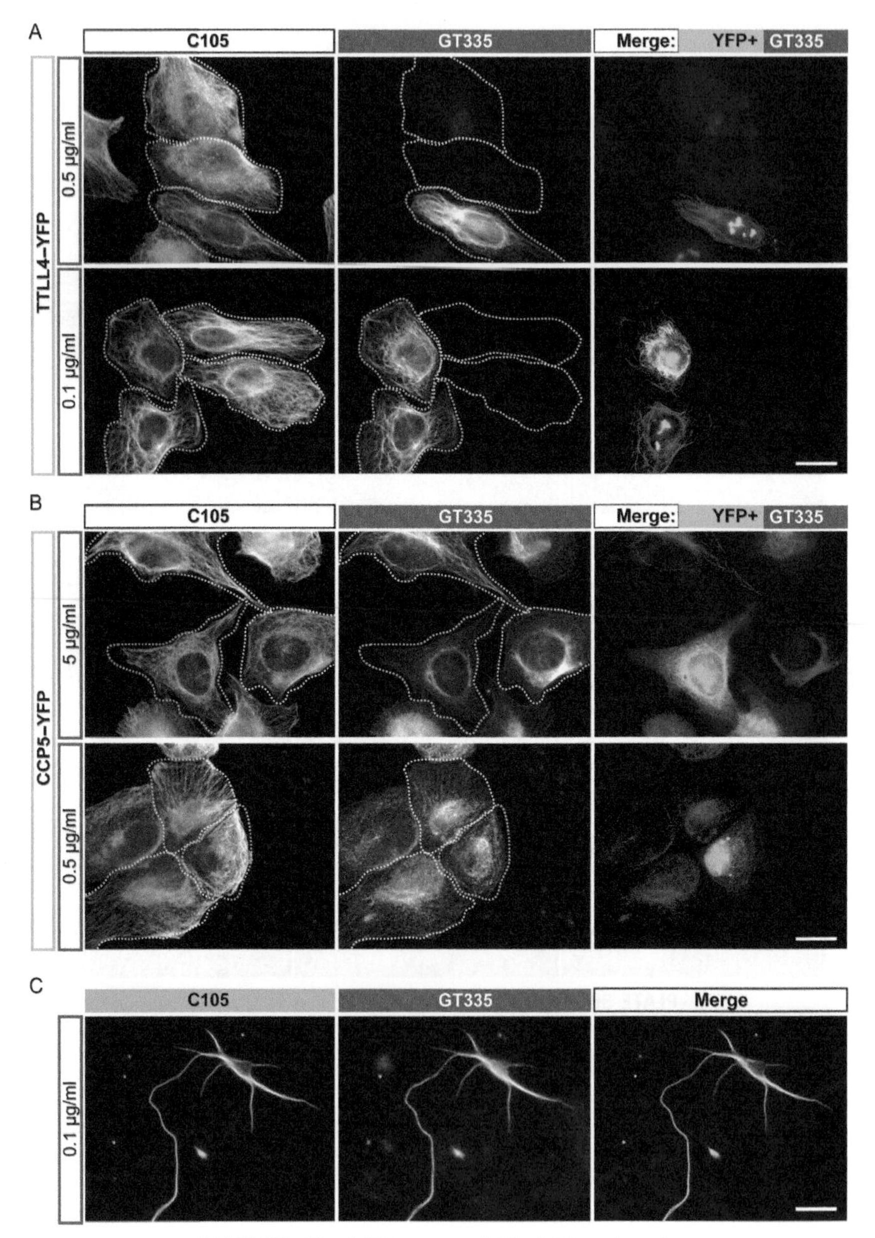

PLATE 37 (Fig. 16.2 on page 254 of this volume).

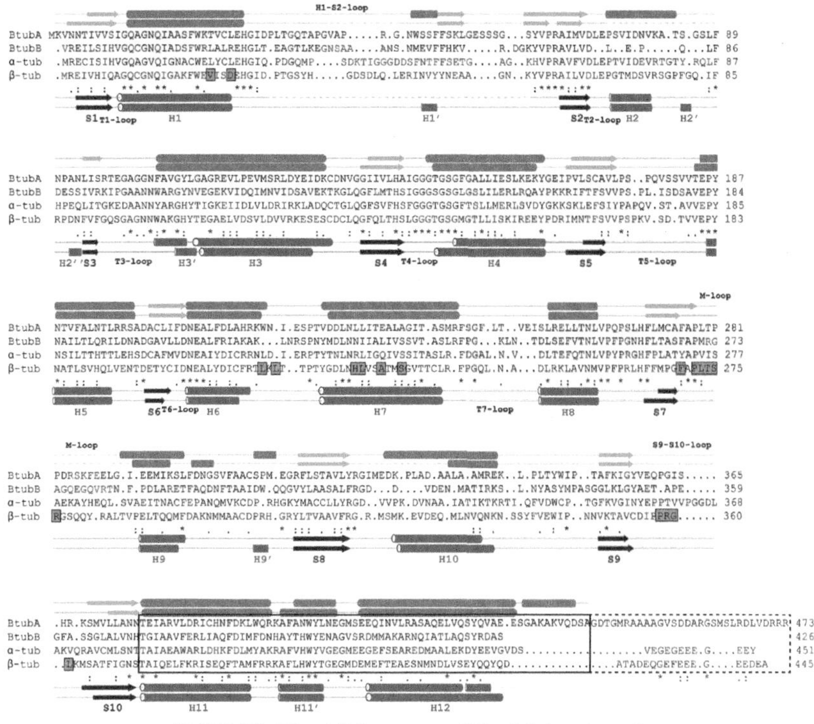

PLATE 38 (Fig. 17.1 on page 271 of this volume).

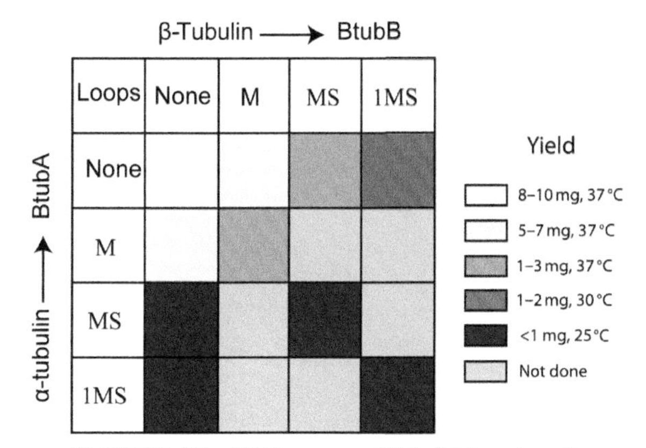

PLATE 39 (Fig. 17.3 on page 277 of this volume).

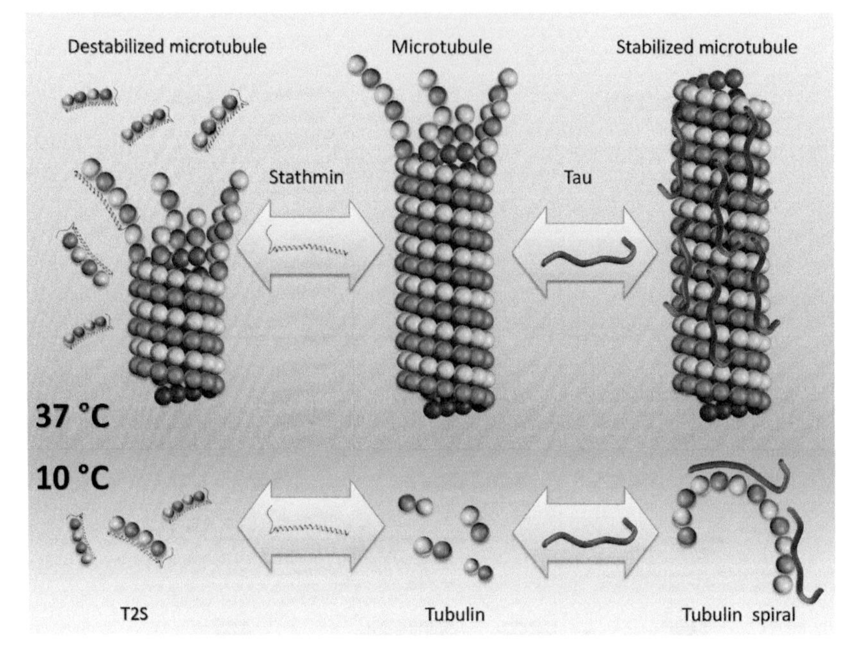

PLATE 40 (Fig. 18.1 on page 285 of this volume).

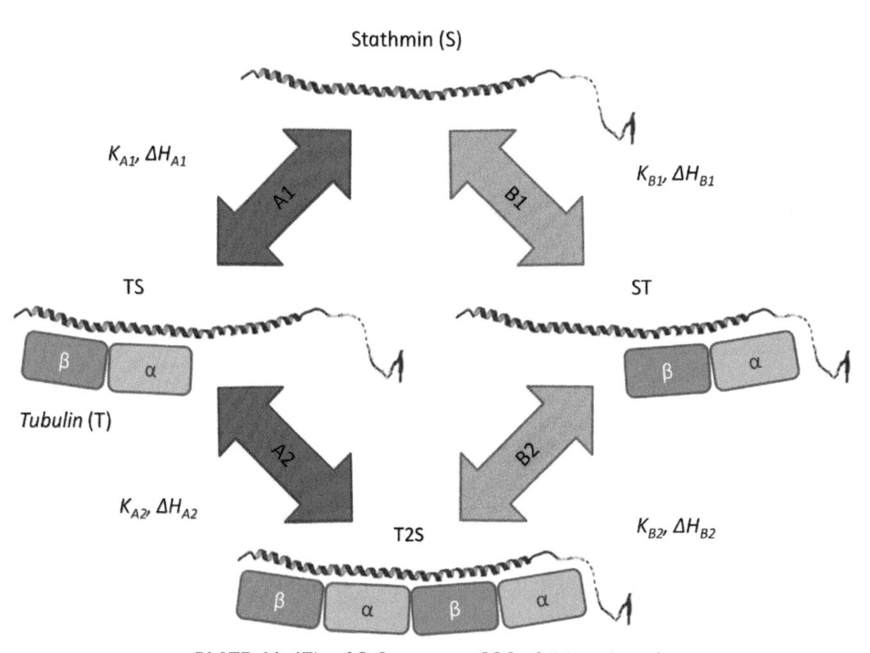

PLATE 41 (Fig. 18.6 on page 296 of this volume).

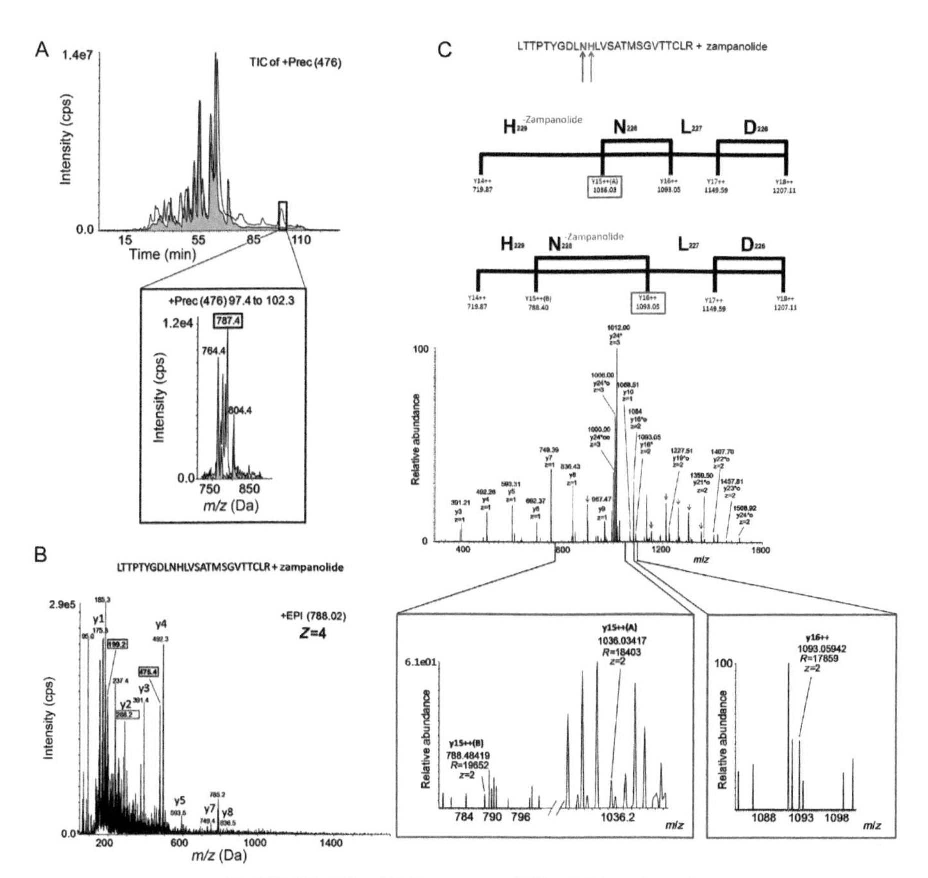

PLATE 42 (Fig. 19.5 on page 321 of this volume).

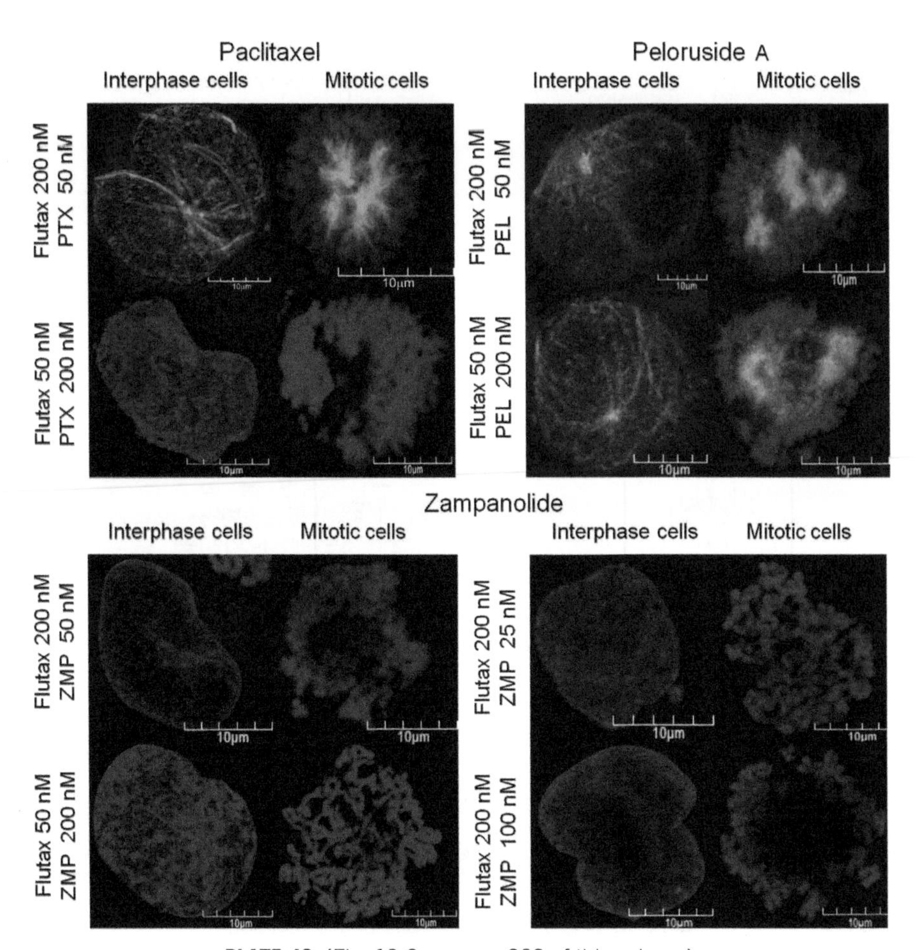

PLATE 43 (Fig. 19.6 on page 323 of this volume).

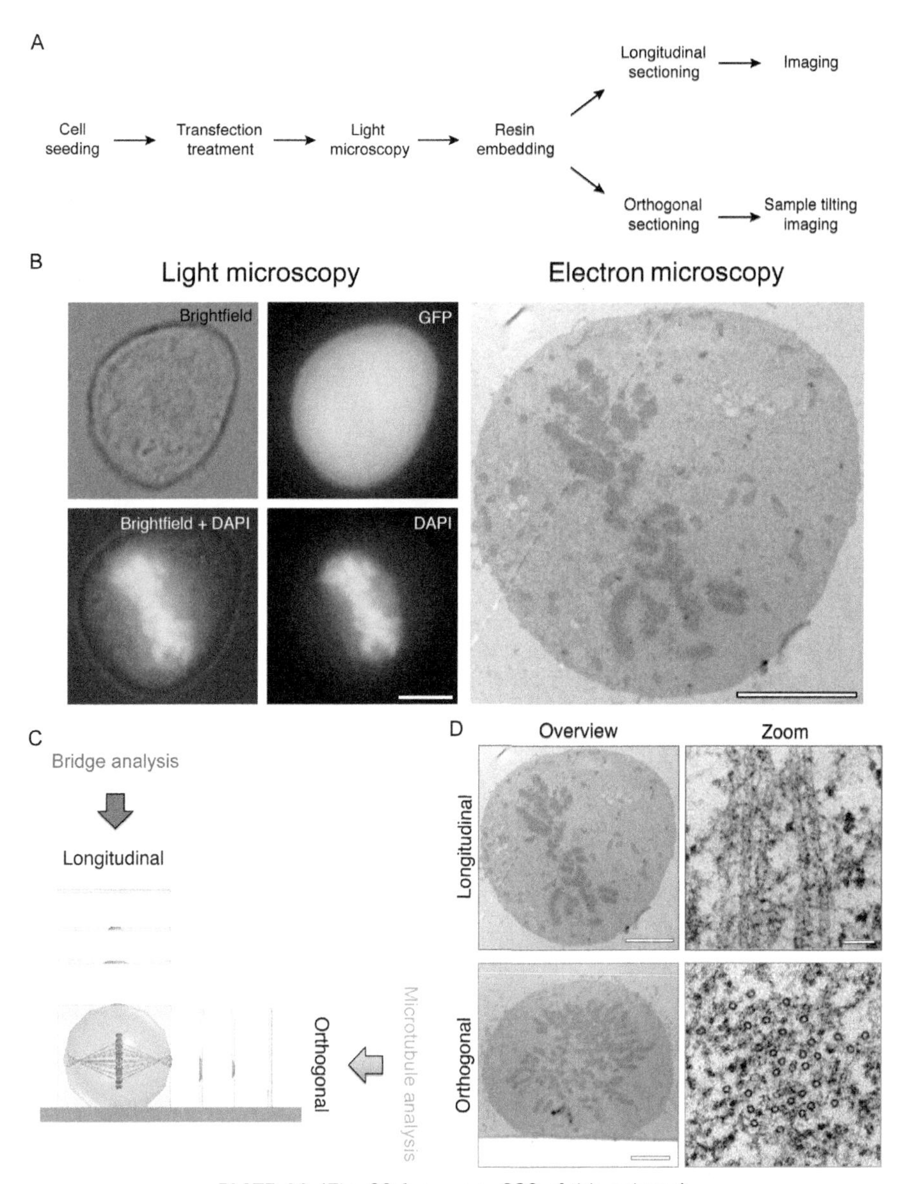

PLATE 44 (Fig. 20.1 on page 330 of this volume).

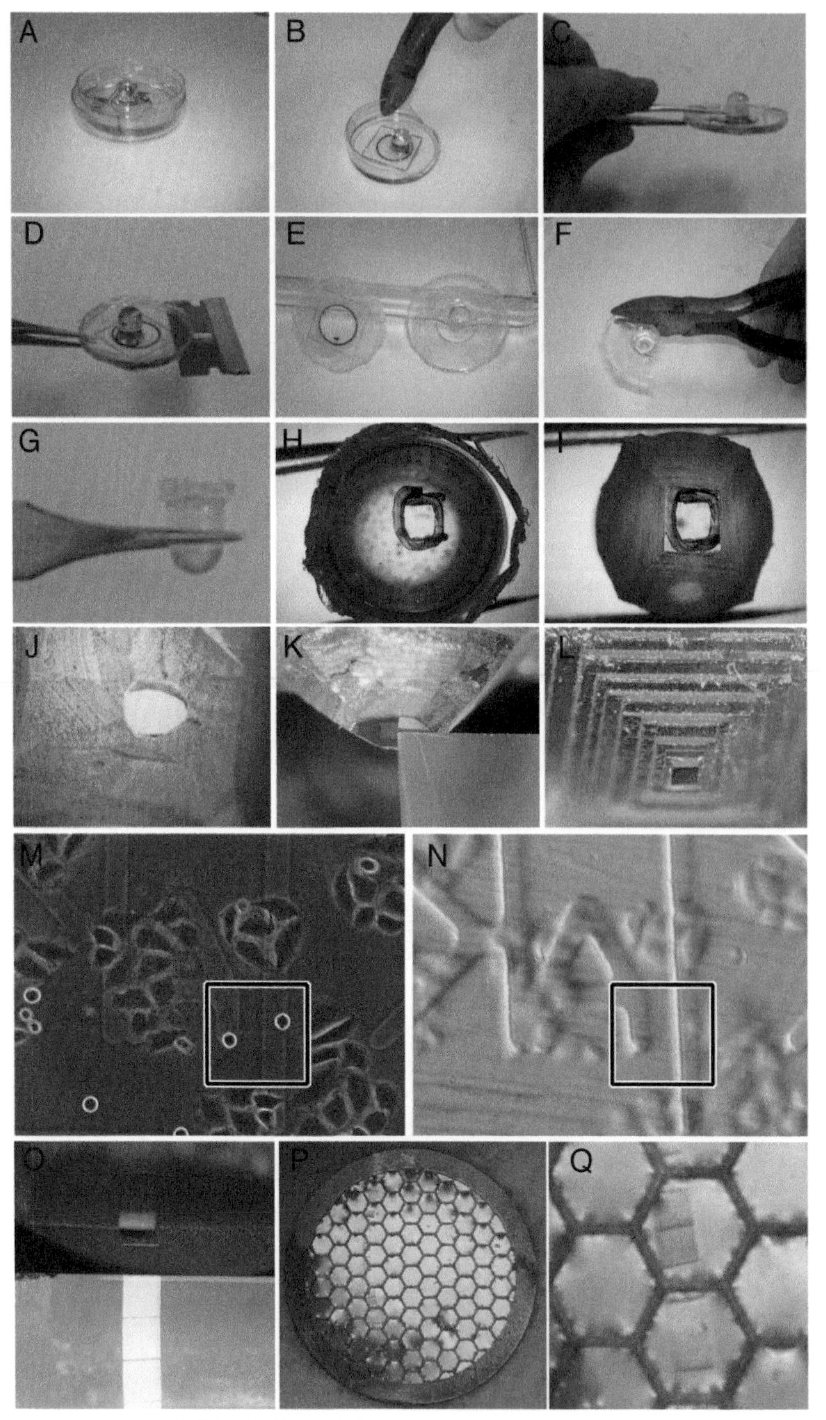

PLATE 45 (Fig. 20.3 on page 334 of this volume).

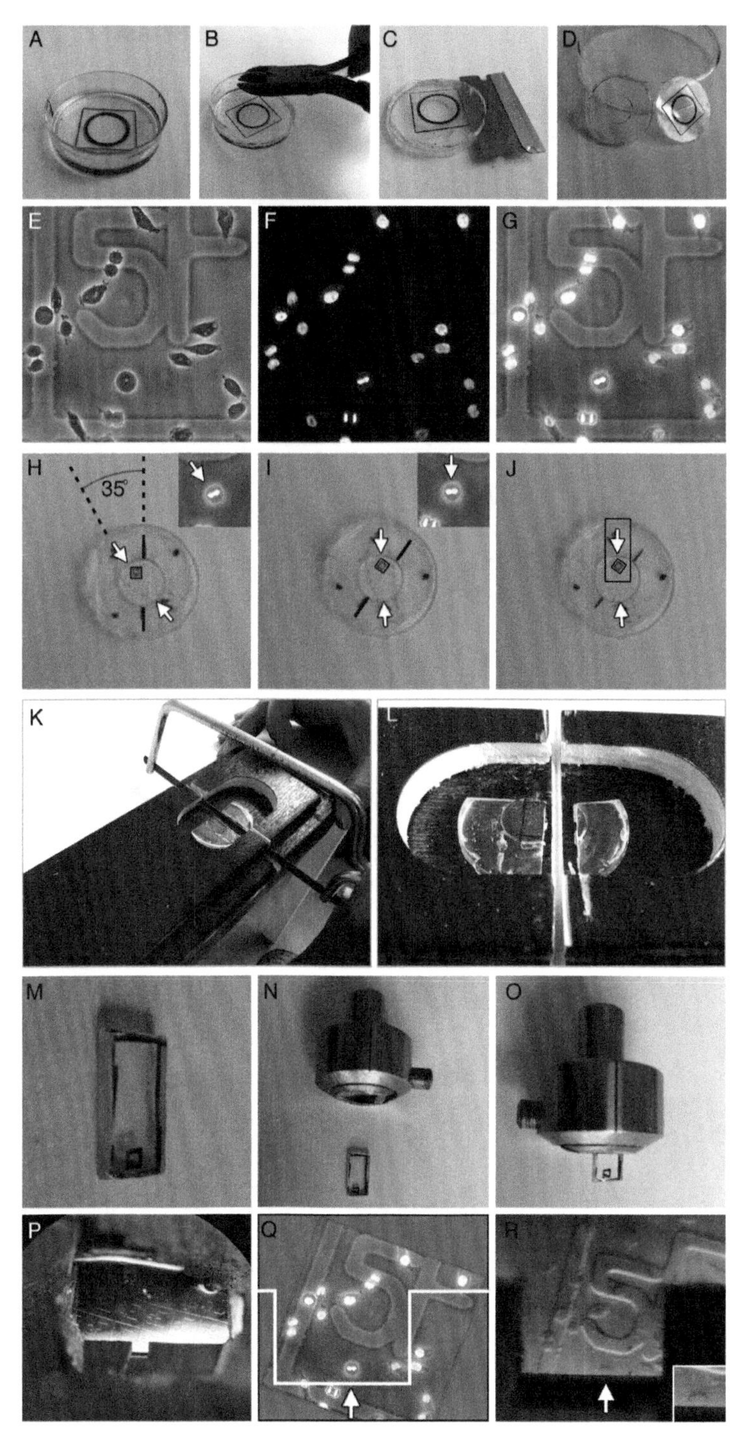

PLATE 46 (Fig. 20.4 on page 337 of this volume).

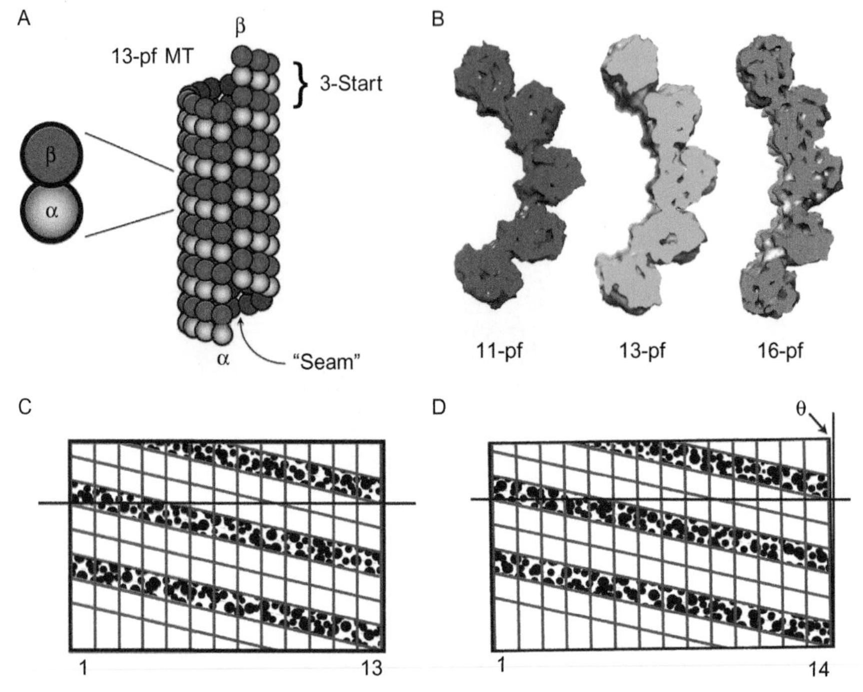

PLATE 47 (Fig. 21.1 on page 345 of this volume).

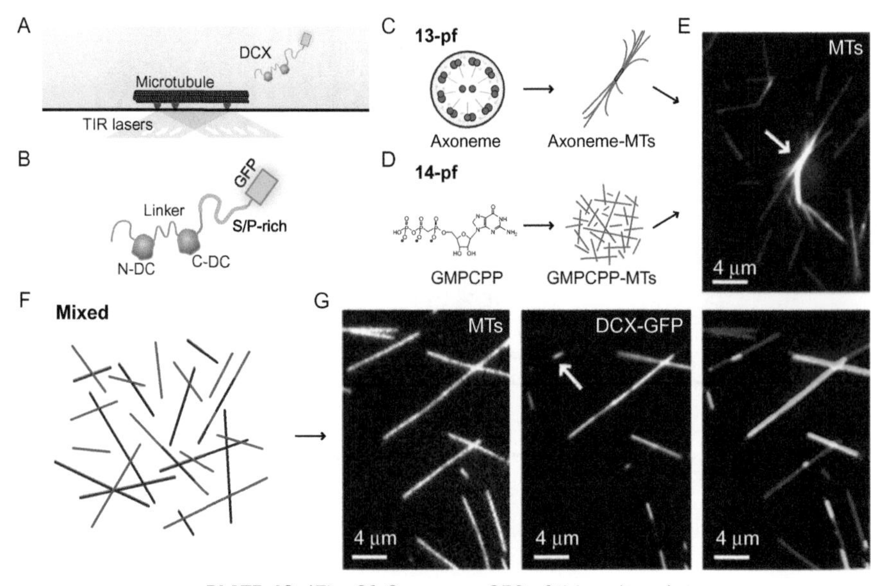

PLATE 48 (Fig. 21.2 on page 350 of this volume).

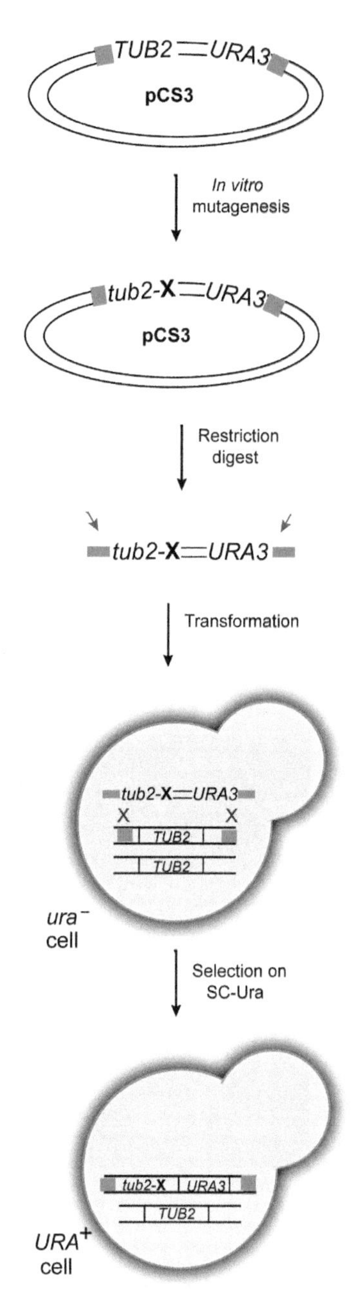

PLATE 49 (Fig. 22.1 on page 361 of this volume).

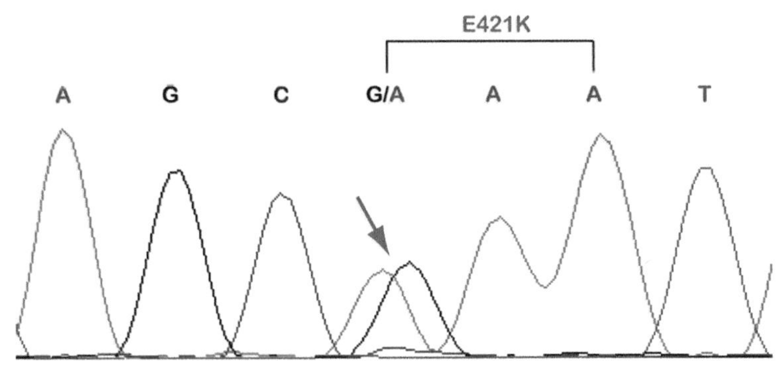

PLATE 50 (Fig. 22.2 on page 363 of this volume).

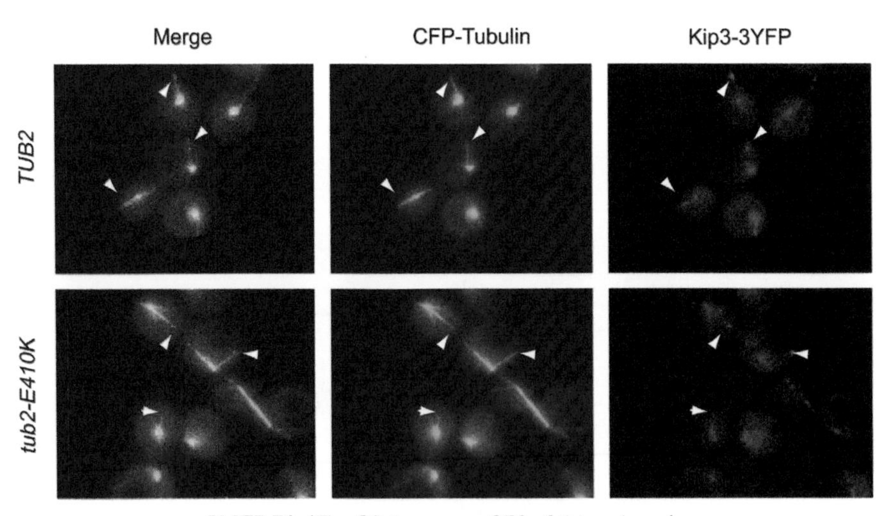

PLATE 51 (Fig. 22.5 on page 372 of this volume).

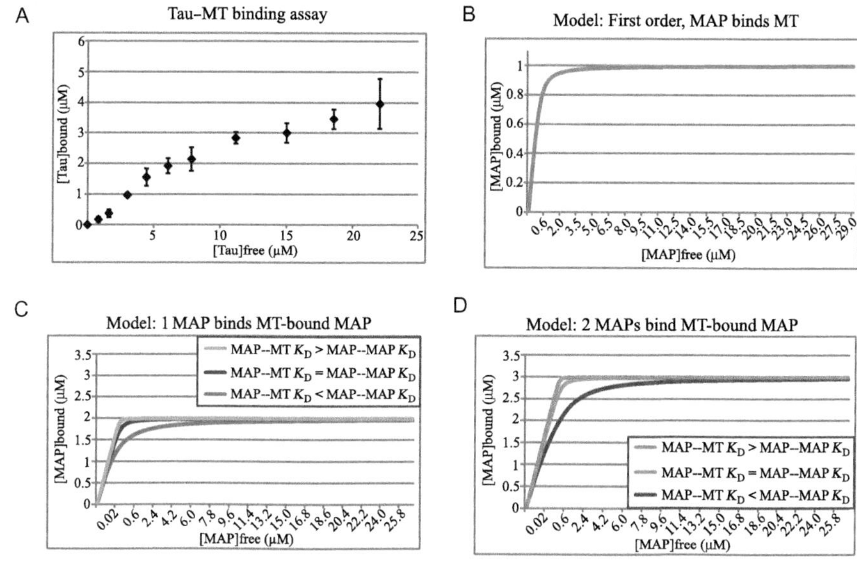

PLATE 52 (Fig. 23.2 on page 380 of this volume).

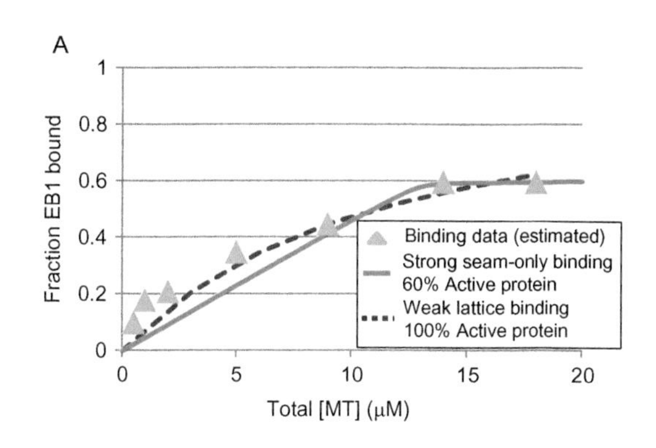

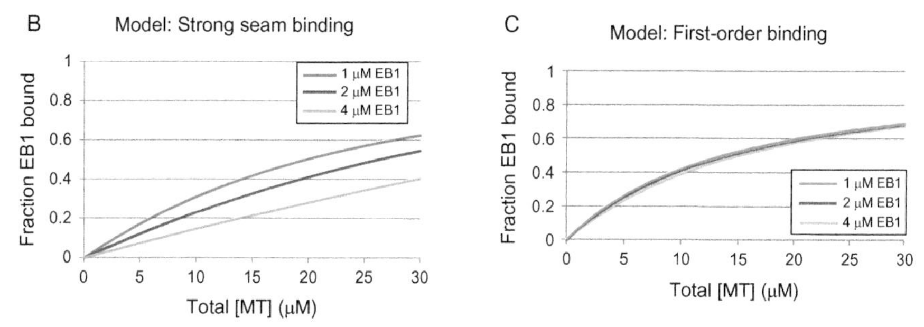

PLATE 53 (Fig. 23.3 on page 382 of this volume).

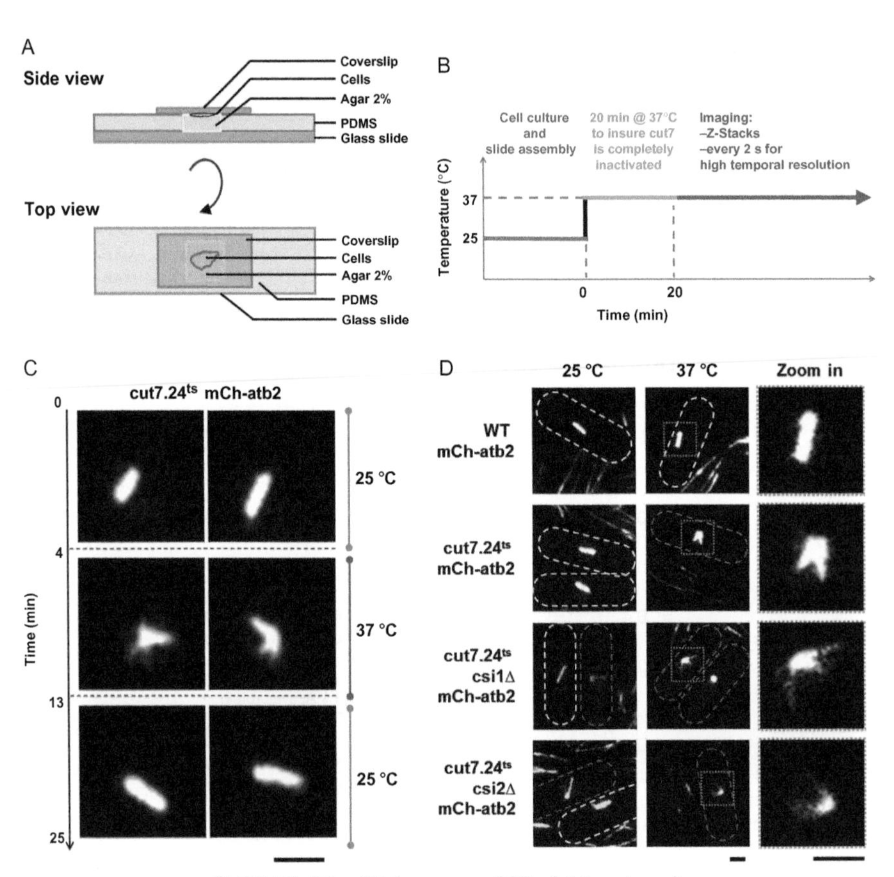

PLATE 54 (Fig. 24.1 on page 389 of this volume).

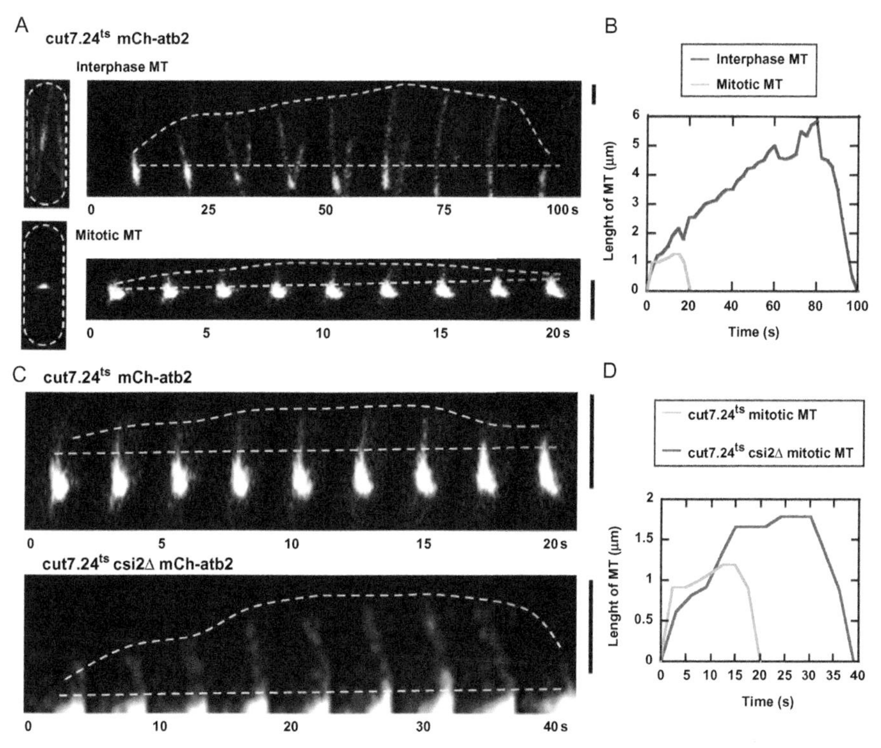

PLATE 55 (Fig. 24.2 on page 392 of this volume).

Printed and bound by CPI Group (UK) Ltd, Croydon, CR0 4YY
08/05/2025
01864959-0002